Anelastic and Dielectric
Effects in Polymeric Solids

Anelastic and Dielectric Effects in Polymeric Solids

BY

N. G. McCRUM
Department of Engineering Science
Oxford University, England

B. E. READ
Division of Materials Metrology
National Physical Laboratory, Teddington, England

AND

G. WILLIAMS
Department of Chemistry
University College of Swansea, Swansea, Wales

DOVER PUBLICATIONS, INC.
New York

Published in Canada by General Publishing Company, Ltd., 30 Lesmill
Road, Don Mills, Toronto, Ontario.
Published in the United Kingdom by Constable and Company, Ltd., 3 The
Lanchesters, 162–164 Fulham Palace Road, London W6 9ER.

This Dover edition, first published in 1991, is an unabridged, slightly
corrected republication of the work first published by John Wiley & Sons,
London, 1967.

Manufactured in the United States of America
Dover Publications, Inc., 31 East 2nd Street, Mineola, N.Y. 11501

Library of Congress Cataloging-in-Publication Data

McCrum, N. G.
    Anelastic and dielectric effects in polymeric solids / by N. G. McCrum,
B. E. Read, and G. Williams. — Dover ed.
        p.   cm.
    Reprint. Originally published: London : Wiley, 1967.
    Includes bibliographical references and indexes.
    ISBN 0-486-66752-9
    1. Polymers.   2. Dielectric relaxation.   3. Stress relaxation.   I. Read,
(Bryan Eric)   II. Williams, G. (Graham), Ph.D.   III. Title.
QD381.M3   1991
620.1'9204297—dc20   91-9088
                                                              CIP

Preface

The purpose of this book is to present a systematic account of theoretical and experimental studies of polymeric solids using mechanical and dielectric relaxation. It will be of direct interest to chemists, physicists and electrical, mechanical and chemical engineers engaged in research. We hope also that polymer technologists concerned with development and applications will find elements of the book relevant to their work. It is true that the experiments described in this book rarely duplicate any 'in-use' condition. Nevertheless a strong thread connects our subject with such complicated but important phenomena as electrical insulation, toughness, fatigue, hardness and friction. These qualities of polymeric solids are due basically to the molecular movements within the solid which are the essential object of those who study mechanical and dielectric relaxation.

The book is so arranged that the reader interested in a general understanding of mechanical and dielectric relaxation will not have to read beyond Chapter 7. The early chapters describe the chemical and physical structure and properties of polymeric solids, the phenomenological and molecular theories of relaxation, and experimental methods for the study of anelastic and dielectric relaxation. The treatment is not intended to be exhaustive but to give an introductory account. The later chapters present a comprehensive account of mechanical and dielectric relaxation in all the important classes of polymeric solid.

The discipline is young, fast moving and concerned with a class of material chosen for study because it is of technological interest, not because it is simple or structurally well understood. Indeed it is only in the last few years that guidance has been obtained to any great extent from optical and electron microscopy and x-ray diffraction. The molecular interpretation of relaxation has therefore tended to be qualitative. Fortunately there is a firm framework in the linear theory of mechanical and dielectric relaxation. We hope that we have maintained a proper balance between the firmly established parts of the subject and the more speculative.

The reader may find some of our chosen symbols contrary to his custom.

v

For instance we have not used ε for strain since it is used for dielectric constant, nor have we used τ for shear stress since it is used for relaxation time. Students of dielectric and mechanical relaxation frequently use the same symbols for different quantities so that we have had to offend convention in some instances.

We have used Zener's nomenclature for the limiting modulii and compliances. For example J_U and J_R for the unrelaxed and relaxed compliances. This terminology we prefer to the more commonly used J_∞ and J_0. J_∞ and J_0 are not of course the compliances measured at infinite and zero frequencies: they are the compliance values approached asymptotically by $J'(\omega)$ at frequencies above and below the relaxation region. For uniformity of treatment we have used ε_U and ε_R for the limiting dielectric constants.

We have used normalized distribution functions for two reasons. First, by doing so we have been able to discuss both mechanical and dielectric relaxation with exactly analogous equations. Second, normalized distribution functions are of considerable convenience for discussion of the temperature dependence of relaxations.

Many of our colleagues assisted by reading chapters of the book. In particular we would like to thank for their comments, Professor G. Allen, Professor H. Fröhlich, Dr. J. Heijboer, Dr. K. H. Illers, Dr. G. Link, Dr. D. W. McCall, Dr. E. L. Morris, Dr. B. K. P. Scaife, Professor W. H. Stockmayer, Dr. L. R. G. Treloar and Dr. I. M. Ward.

Grateful acknowledgment is made to the following publishers and societies for permission to reproduce illustrations.

Book or Journal	Publisher	Figure
Polymer (London)	Butterworth, London	3.1, 7.4, 10.7, 10.37, 14.8, 14.9, 14.10, 14.11, 14.12, 14.20
J. Res. Nat. Bur. Stds. (Washington D.C.)		7.6, 9.9, 10.30, 10.31, 11.38, 11.39, 11.40, 11.41, 11.43, 11.44, 11.45, 11.46, 11.47, 12.5, 12.10, 12.11
Gen. Radio Exptr.	General Radio Co. West Concord, Mass.	7.9
Soc. Plastics Engrs. Trans.	Society of Plastics Engineers, Stamford, Conn.	10.34
J. Appl. Phys.	American Institute of Physics, New York	2.12, 2.13, 2.17, 4.8, 4.12, 7.13
Proc Inst. Elec. Engrs.	Institution of Electrical Engineers, London	10.27
J. Inst. Elec. Engrs.	Institution of Electrical Engineers, London	7.14, 7.15

Book or Journal	Publisher	Figure
J. Colloid. Sci.	Academic Press, New York	6.7, 8.2, 9.5, 10.23, 11.32, 12.6, 13.5, 13.7, 13.16
Brit. J. Appl. Phys.	Institute of Physics and the Physical Society, London	8.4
Physics of Non Crystalline Solids	North Holland, Amsterdam	8.5, 8.12, 8.26, 8.43
Polymer Sci. U.S.S.R.	Pergamon Press, Oxford	8.9, 8.10
Rheology of Elastomers	Pergamon Press, New York	9.25
Makromol. Chem.	Huthig and Wepf, Basel	2.8, 2.14, 8.15, 8.16, 8.18, 8.27, 8.28, 8.44, 8.45, 10.12, 11.34, 11.36, 12.9, 12.12, 12.13, 12.14
Kolloidz	Steinkopff, Darmstadt	1.7, 1.8, 1.9, 1.10, 1.11, 2.9, 6.26, 8.19, 8.20, 8.21, 8.22, 8.34, 8.35, 8.36, 8.39, 9.1, 9.2, 9.3, 9.10, 9.11, 9.13, 9.14, 9.15b, 9.18, 9.23, 9.24, 9.25, 9.26, 10.2, 10.4, 10.11, 10.13, 10.14, 10.15, 11.1, 11.2, 11.3, 11.5, 11.6, 11.7, 11.8, 11.11, 11.12, 11.14, 11.15, 11.16, 11.18, 11.19, 11.23, 11.28, 11.29, 11.30, 12.4, 13.4, 13.6, 13.9, 13.10, 13.11, 13.15
Rheol. Acta	Steinkopff, Darmstadt	10.33
Rept. Progr. Polymer Phys. (Japan)	Gakujutu Bunken Fukyu-Kai Oh'okayama, Meguroka, Tokyo	8.41, 9.30, 9.31, 9.35, 9.36, 11.9, 11.13, 14.4
J. Phys. Chem.	American Chemical Society, Washington D.C.	9.6, 9.7, 9.8, 10.29
J. Am. Chem. Soc.	American Chemical Society, Washington D.C.	11.10
Polymer Preprints	American Chemical Society, Washington D.C.	13.13, 13.14
Ergeb. Exakt. Naturw.	Springer, Heidelberg	9.12
Simp. Intern. Chim. Macromol.	Consiglio Nazionale delle Rocerche, Rome	9.15a
Mem. Fac. Eng. Kyushu Univ.	Faculty of Engineering Kyushu University, Fukuoka	9.20, 9.21, 9.22, 12.1, 13.2, 14.5

Book or Journal	Publisher	Figure
Pure Appl. Chem.	International Union of Pure and Applied Chemistry, and Butterworth, London	11.24
Advan. Polymer Sci.	Springer, Heidelberg	12.7
Trans. Faraday Soc.	The Faraday Society, London	3.7, 5.4, 5.5, 10.10, 13.1, 14.17, 14.18, 14.21, 14.24, 14.26
Progress in Dielectrics	Heywood, distributed by Iliffe, London	13.1
J. Appl. Chem.	Society of Chemical Industry (London)	14.13, 14.16
Rheology, Vols. II and III	Academic Press, New York	Table 4.1, 5.2
Elec. Res. Assoc. Private Rept.	British Electrical and Allied Industries Research Association, Leatherhead	12.2
J. Chem. Phys.	American Institute of Physics, New York	3.3, 3.4, 12.8
Sov. Phys. Tech. Phys.	American Institute of Physics, New York	8.6, 8.17
Polythene	Iliffe, London	2.7
Proc. Roy. Soc.	Royal Society, London	2.1, 3.2
The Physics of Rubber Elasticity	Clarendon Press, Oxford	3.5, 5.1, 5.6
Die Physik de Hochpolymeren	Springer, Heidelberg	1.6
Proc. 2nd Congr. Rheology	Butterworth, London	6.2a, 6.3
J. Acoust. Soc. Am.	American Institute of Physics, New York	6.4, 6.5
Rev. Sci. Instr.	American Institute of Physics, New York	6.6
Angew. Chem.	Verlag Chemie, Berlin	10.1
Symp. Markromol.	Verlag Chemie, Berlin	10.6

and also to Mr. W. Reddish, (13.1); Dr. J. Heijboer, (8.42); Professor R. M. Fuoss, (11.10); Dr. A. J. Kovacs, (9.6, 9.7, 9.8); Professor M. Takayanagi, (9.20, 9.21, 9.22, 12.1, 13.2, 14.5); Professor J. D. Ferry, (10.29); Dr. R. H. Boyd, (12.8); Dr. A. H. Willbourn, (14.17); Dr. S. Roberts, (7.13).

N. G. McCrum
B. E. Read
G. Williams

Contents

Contents   xiii

1

Introduction

If a weight is suspended from a polymeric filament the strain will not be constant but will increase slowly. The effect is due to a molecular rearrangement in the solid induced by and proportional to the stress. On release of the stress, the molecules slowly recover their former spatial arrangement and the strain simultaneously returns to zero. The processes by which the molecules rearrange are thermally activated and so proceed at a rate which increases with temperature. The mechanical parameters defined to describe such a solid under stress must therefore depend on both time and temperature. Many polymers contain polar groups, and for these the dielectric 'constants' will also depend on time and temperature.

We first review briefly the parameters* used to specify the mechanical and dielectric properties of an isotropic medium in the *absence* of relaxation effects. Consider the mechanical properties in a shear experiment. At sufficiently small stresses a linear relationship exists between the shear stress σ and the shear strain γ,

$$\sigma = G\gamma \quad \text{or} \quad \gamma = J\sigma : G = J^{-1} \tag{1.1}$$

where G and J are material constants known, respectively, as the shear modulus and shear compliance. In the absence of relaxation there is no variation of γ with elapsed time after the application of the stress. G and J are therefore independent of time. If, instead of shear conditions, we consider simple extension, then the appropriate modulus and compliance are E (Young's modulus) and its reciprocal D. Similarly, for pure volume deformation we consider the bulk modulus K or its reciprocal B (compressibility). Since simple extension is a combination of both shear

---

* For a complete development of the parameters from first principles the reader is referred to standard texts, e.g. Timoshenko and Goodier (1951), Fröhlich (1958).

(constant volume) and volume deformation, a relationship exists, for example, between G, E and K (see, for example, Jaeger, 1964)

$$E = \frac{3G}{1 + G/3K}$$

For convenience, the theoretical treatment of relaxation in this chapter and in Chapter 4 is restricted to shear deformation but the argument will apply equally to longitudinal and bulk deformations. Non-linear behaviour (i.e. modulus dependent on stress amplitude) will not be considered in this book.

In order to describe the dielectric properties of a material (again in the absence of relaxation) consider a thin parallel plate condenser of area A and plate separation $\alpha(A \gg \alpha^2)$ held in a vacuum. If the plates are given equal and opposite charges of magnitude q_0 a potential is established across the condenser

$$V_0 = \frac{q_0}{C_0}$$

C_0 is the capacitance of the condenser in vacuo. If a dielectric is inserted between the plates and the charge q_0 maintained the voltage is reduced to V. The ratio of the voltages defines the dielectric constant of the dielectric, ε,

$$\varepsilon = \frac{V_0}{V} = \frac{C}{C_0} \tag{1.2}$$

in which C is the capacitance of the condenser with the dielectric inserted.

The voltage V across the capacitance with the dielectric inserted may be written

$$V = \frac{q_0}{\varepsilon C_0}$$

$$= \frac{q_1}{C_0}$$

in which we have replaced q_0/ε by q_1. q_0 is known as the true charge since it is the charge actually placed on the condenser plate. q_1 is known as the *free* charge since it is the portion of the true charge which contributes to the voltage. The difference between the true and free charges is the *bound* charge $(q_0 - q_1)$. This charge is 'bound' by an adjacent charge of equal magnitude but opposite sign which lies in the surface of the dielectric.

The surface density of the bound charge is equal to the polarization of the dielectric P. Hence, if we assume that the condenser is of large area so that edge effects may be ignored,

$$P = \frac{1}{A}(q_0 - q_1)$$

$$\therefore \quad P = \sigma_0 - \sigma_1 \tag{1.3}$$

where σ_0 and σ_1 are the surface densities of true and free charges. These two charge densities define the electric displacement D and the electric field strength E,

$$D = 4\pi\sigma_0 \tag{1.4}$$

$$E = 4\pi\sigma_1 \tag{1.5}$$

Hence,

$$D = E + 4\pi P \tag{1.6}$$

It also follows by algebraic manipulation that,

$$D = \varepsilon E \tag{1.7}$$

In the foregoing we have assumed that the polarization P does not depend on time so that D, E and ε are also independent of time. Physically this implies that the sole contribution to P arises from the displacement of molecular electrons with respect to their nuclei, a process which takes place almost instantaneously on application of the field.

The constants defined in the preceding paragraphs are insufficient to describe the properties of real polymeric solids (other than perhaps at temperatures close to $0°\text{K}$). The reason for this is that polymer molecules when in the solid state are extremely mobile. Consequently, the application of a mechanical stress leads to a slow movement of matter (creep): the application of an electric field leads to a slow increase in polarization (time dependent dielectric constant). The parameters which have been adopted were developed to rationalize the results of two types of experiment. In the first, the so-called *step-function* experiment, the response of the material to a constant stress, strain or electric field is determined as a function of time. In the second experiment the response to a sinusoidal stress, strain or electric field is observed as a function of frequency. This experiment is often known as the *dynamic* experiment.

1.1   STEP-FUNCTION   EXPERIMENTS

The results of step-function experiments, somewhat idealized, are shown in Figures 1.1 and 1.2. In a mechanical creep experiment a stress σ_0 is applied to the specimen rapidly at zero time and then held constant until some time later, t_1, when it is removed (experimental arrangements are described in Chapter 6). The strain is commonly found by experiment to increase with time in the manner shown in Figure 1.1(a). After removal of the stress the strain returns slowly with time to zero. Now there are circumstances (high strain, high temperature) in which the strain will not return ultimately to zero and the specimen is said to exhibit permanent set.

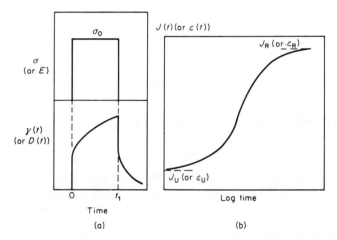

Figure 1.1. Results of a typical creep experiment. (a) Constant stress σ_0 applied to specimen leads to strain $\gamma(t)$; (b) $J(t) = \gamma(t)/\sigma_0$ plotted against $\log t$. When the creep experiment is terminated at time t_1, Figure 1.1(a), the specimen recovers slowly. These graphs also describe the dielectric quantities, E (electric field strength) and D (electric displacement) in a step function experiment.

However, permanent set is usually negligible for creep strains below $\gamma = 10^{-3}$ and at temperatures below the melting point (for crystalline polymers) or the glass transition (for amorphous polymers). The latter statement needs qualification, e.g. (a) crystalline polymers at temperatures close to the melting point must be annealed, (b) amorphous polymers at temperatures close to the glass transition must be annealed and of high molecular weight or crosslinked. Subject to these qualifications we may generalize by saying that for polymeric solids undergoing creep a time-dependent compliance $J(t)$ may be defined

$$J(t) = \frac{\gamma(t)}{\sigma_0} \tag{1.8}$$

where $\gamma(t)$ is the time-dependent strain and that for small strains ($\gamma(t) < 10^{-3}$), $J(t)$ is found to be independent of the magnitude of σ_0. $J(t)$ when plotted against $\log t$ has a sigmoidal shape, Figure 1.1(b).

If instead of stress the strain γ_0 is held constant then the stress $\sigma(t)$ is time dependent, Figure 1.2(a). This is the stress relaxation experiment. The time-dependent modulus $G(t)$, defined by

$$G(t) = \frac{\sigma(t)}{\gamma_0} \tag{1.9}$$

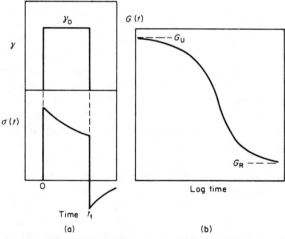

Figure 1.2. Results of a typical stress relaxation experiment. (a) Constant strain γ_0 applied to specimen leads to stress $\sigma(t)$. If the strain is returned from γ_0 to zero at time t_1 the stress recovers to zero as shown. (b) $G(t) = \sigma(t)/\gamma_0$ plotted against $\log t$.

is found to have a sigmoidal form when plotted against $\log t$, Figure 1.2(b).

For very short times $J(t)$ tends asymptotically to a constant value known as the unrelaxed compliance, J_U. At short times also $G(t)$ tends to the unrelaxed modulus G_U. It is found from experiment that $J_U = G_U^{-1}$. At very long times both $J(t)$ and $G(t)$ tend to limiting values, the relaxed compliance J_R and the relaxed modulus G_R. It is found also from experiment that $J_R = G_R^{-1}$. This is not surprising since for times effectively outside the relaxation region the solid is Hookean and therefore Equation (1.1) holds.

However, within the relaxation region $J(t) \neq G(t)^{-1}$. Nor are the times at the points of inflection in the creep and stress relaxation experiments the same. This experimental fact is grist for the theoretician's mill (see Chapter 4).

Both $J(t)$ and $G(t)$ depend very strongly on temperature. Increasing temperature moves the point of inflection to shorter times. The inflection point at one temperature in a creep experiment may occur at one year, in which case only the upward swing of the $J(t)$ curve would be observed in experiments lasting for 10^3 hours. At a higher temperature the inflection may occur at a time of one hour so that if measurements were made from 10^{-3} to 10^3 hours a fair portion of the sigmoidal shape would be resolved (as, for instance, Figure 1.6). At a higher temperature still the inflection may occur at 1 second, in which case only a portion of the $J(t)$ curve above the inflection would be observed if the first strain measurement is made at 10^{-3} hours. It is also possible that at even higher temperatures the point of inflection may occur at say 10^{-6} seconds, so that in a creep experiment the first measurement at say 10 seconds after the application of the stress would determine J_R, i.e. the whole relaxation would have occurred between the application of the stress and the first measurement.

For polymers containing dipolar groups dielectric relaxation effects are observed owing to the time required for the orientation of the permanent dipoles subsequent to the application of the electric field. If, for example, we consider a polar dielectric between the plates of a plane condenser, and we keep the condenser plates at a constant potential difference between $t = 0$ and $t = t_1$, then the electric field E is prescribed as in Figure 1.1(a). The resulting curve for the displacement $D(t)$ is shown also in Figure 1.1(a). The time-dependent dielectric constant defined

$$\varepsilon(t) = \frac{D(t)}{E_0} \qquad (1.10)$$

when plotted against $\log t$ is found to have the sigmoidal form shown in Figure 1.1(b). The inflection point always coincides more or less with inflections in the $J(t)$ and $G(t)$ curves. (The times of the inflections in the $J(t)$ and $G(t)$ curves, although not equal, are comparable.) At short times

$\varepsilon(t)$ tends to ε_U, the unrelaxed dielectric constant, and at long times to ε_R, the relaxed dielectric constant (see Figure 1.1b).

In the constant electric field experiment the time dependence of D is due to the time dependence of P (Equation 1.6). After the field E_0 is switched on polarization develops in the following steps. First the electronic polarization already mentioned. Second paraelectric polarization due to the reorientation of dipoles. The latter yields the sigmoid curve for $\varepsilon(t)$ versus $\log t$ shown in Figure 1.1(b). The inflection point in Figure 1.1(b) may occur at very short times or very long times so that only a portion of the sigmoid curve may in fact be observed. It is quite possible, as in the creep experiments already mentioned, for the inflection to occur at say 10^{-6} seconds, so that if the first measurement of $\varepsilon(t)$ is made at $t = 10$ seconds after the application of E_0 no time dependence is observed, i.e. at times of the order of seconds $\varepsilon(t) = \varepsilon_R$.

When the inflection points in either $J(t)$ versus $\log t$ or $\varepsilon(t)$ versus $\log t$ curves occur at short time (e.g. 10^{-6} seconds) the step-function experiment is of little value. The so called dynamic experiment under these circumstances is of great use.

1.2 DYNAMIC EXPERIMENTS

If a sinusoidal stress is applied to a solid exhibiting relaxation, then the strain response generally lags behind the applied force by some phase angle δ_G. This phase lag results from the time necessary for molecular rearrangements, and is analogous to the time lags observed in the step-function experiments. The subscript G indicates that δ_G is the phase angle determined in a mechanical shear experiment. If the stress is expressed in complex form,

$$\sigma^* = \sigma_0 \exp(i\omega t) \tag{1.11}$$

where σ_0 is the stress amplitude and ω the angular frequency, then the strain is given by

$$\gamma^* = \gamma_0 \exp i(\omega t - \delta_G) \tag{1.12}$$

where γ_0 is the strain amplitude.

The stress–strain relationship may be written,

$$\gamma^* = J^*\sigma^* = (J' - iJ'')\sigma^* \tag{1.13}$$

J' and J'' are the real and imaginary components respectively of the complex shear compliance J^*, and are given by,

$$J' = |J| \cos \delta_G; \qquad J'' = |J| \sin \delta_G \qquad (1.14)$$

where $|J| = (J'^2 + J''^2)^{1/2} = \gamma_0/\sigma_0$.
    The loss tangent is

$$\tan \delta_G = J''/J' \qquad (1.15)$$

The complex modulus G^* is given,

$$G^* = 1/J^* = (J' + iJ'')/|J|^2 \qquad (1.16)$$

Writing $G^* = G' + iG''$ it follows that

$$G' = J'/|J|^2; \qquad G'' = J''/|J|^2 \qquad (1.17)$$

An illustration of the commonly found behaviour of $J'(\omega)$, $J''(\omega)$ and $\tan \delta_G$ with frequency is shown in Figure 1.3. $J''(\omega)$ passes through a maximum at an angular frequency Ω_J. At this frequency $J'(\omega)$ passes through an inflection. For $\omega \ll \Omega_J$, $J'(\omega)$ tends to J_R and for $\omega \gg \Omega_J$, $J'(\omega)$ tends to J_U. The analogous G' and G'' versus log ω curves are also shown in Figure 1.3. Tan δ_G passes through its maximum at a frequency Ω_δ, $\Omega_G > \Omega_\delta > \Omega_J$.

It is to be noted that whereas $J(t)$ and $G(t)$ curves, Figures 1.1 and 1.2, come from entirely different experiments, $J^*(\omega)$ and $G^*(\omega)$ come from the same experiment.

In dielectric experiments an alternating electric field produces an alternating electric polarization which, in the case of polar solids, will lag behind the applied field by some phase angle δ_ε. By analogy with Equations (1.11) and (1.12) the complex field (E^*) and displacement (D^*) are given, respectively, by

$$E^* = E_0 \exp(i\omega t) \qquad (1.18)$$

$$D^* = D_0 \exp(i\omega t - \delta_\varepsilon) \qquad (1.19)$$

where E_0 and D_0 are the field and displacement amplitudes respectively.
    Furthermore, we have,

$$D^* = \varepsilon^* E^* = (\varepsilon' - i\varepsilon'')E^* \qquad (1.20)$$

and

$$\varepsilon' = |\varepsilon| \cos \delta_\varepsilon; \qquad \varepsilon'' = |\varepsilon| \sin \delta_\varepsilon; \qquad \frac{\varepsilon''}{\varepsilon'} = \tan \delta_\varepsilon \qquad (1.21)$$

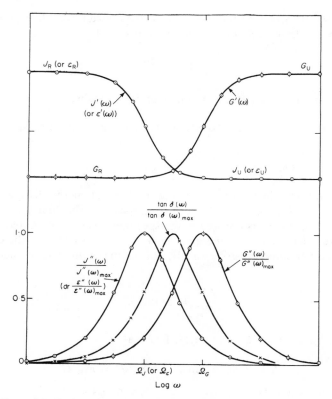

Figure 1.3. Typical dependence of relaxation parameters J', J'', G', G'', and tan δ on angular frequency ω. The results of typical dynamic dielectric experiments are also represented.

ε' and ε'' are the real and imaginary components of the complex dielectric constant ε^*. ε'' is usually called the dielectric loss factor and tan δ_ε is termed the loss tangent.

In a relaxation region ε' is found to decrease with frequency from a value of ε_R to ε_U (Figure 1.3). This decrease, which is called the dispersion of the dielectric constant, corresponds to the decrease of J' with frequency in a mechanical experiment. The difference $(\varepsilon_R - \varepsilon_U)$ is known as the magnitude of the relaxation and is a measure of the orientation polarization. In the relaxation region ε'' passes through a maximum at a frequency Ω_ε.

The significance of J'', G'' and tan δ_G is well illustrated by calculating the energy absorbed by the specimen in the quarter cycle from $t = 0$ to $t = \frac{1}{4}(2\pi/\omega)$. In a mechanical experiment this is W

$$W = \int_0^{\pi/2\omega} \sigma\left(\frac{d\gamma}{dt}\right) dt$$

Making the substitution $\sigma = \sigma_0 \sin \omega t$ and $\gamma = \gamma_0 \sin(\omega t - \delta_G)$,

$$W = \omega\sigma_0\gamma_0 \int_0^{\pi/2\omega} [\cos \omega t \sin \omega t \cos \delta_G + \sin^2 \omega t \sin \delta_G] \, dt$$

$$= \sigma_0\gamma_0 \left[\frac{\cos \delta_G}{2} + \frac{\pi \sin \delta_G}{4}\right]$$

The second term in the bracket is the mechanical energy dissipated per quarter cycle (for $\delta_G = 0$ this term disappears). The first term is the maximum elastic energy stored in the specimen. When integrated over a complete cycle the stored energy is zero. If we write ΔW for the energy dissipated per complete cycle and W_{st} for the maximum stored energy we have,

$$\frac{\Delta W}{W_{st}} = 2\pi \tan \delta_G \qquad (1.22a)$$

It also follows from Equations 1.14, 1.17 that

$$\Delta W = \pi\sigma_0\gamma_0 \sin \delta_G = \pi\sigma_0^2 J'' = \pi\gamma_0^2 G''$$

The quantity $\Delta W/W_{st}$ is known as the specific loss.

A much used measure of mechanical loss is the logarithmic decrement, Λ. If, for example, a body is set into torsional oscillation and then the free decay of amplitude observed (see Chapter 6), Λ is defined by

$$\Lambda = \ln \frac{A_n}{A_{n+1}} \qquad (1.23)$$

where A_n and A_{n+1} are successive amplitudes of oscillation. For low values of Λ (i.e. $A_n/A_{n+1} \sim 1$) we have from Equation 1.23

$$\Lambda \doteqdot \frac{1}{2} \frac{A_n^2 - A_{n+1}^2}{A_n^2}$$

The square of the amplitude is proportional to the stored energy so that

$$\Lambda \doteqdot \frac{1}{2} \frac{\Delta W}{W_{st}}$$

From Equation 1.22a we have therefore

$$\Lambda = \pi \tan \delta_G \qquad (1.22b)$$

The specific loss is also related in a simple manner to two other measurable quantities, the 'Q' of the system, and the attenuation coefficient. (Kolsky, 1953; Nowick, 1953.) If a specimen is driven into oscillation in the region of a resonant frequency by an oscillating force of constant amplitude and variable frequency then the specific loss is given by,

$$\frac{\Delta W}{W_{st}} = \frac{2\pi}{\sqrt{3}} \frac{\Delta\omega}{\omega_0} = 2\pi Q^{-1} \qquad (1.22c)$$

in which ω_0 is the resonant frequency and $\Delta\omega$ is the half-width of the resonant peak, that is, the difference between the frequencies at which the amplitude of oscillation has fallen to half the resonant value. The attenuation coefficient, α, of high-frequency sound waves of wavelength λ is related to the specific loss

$$\frac{\Delta W}{W_{st}} = 2\alpha\lambda \qquad (1.22d)$$

In the dielectric experiment the energy dissipation per second (Fröhlich, 1958) is given by

$$L = \frac{\omega}{8\pi^2} \int_0^{2\pi/\omega} E\left(\frac{dD}{dt}\right) dt \qquad (1.24)$$

from which it follows that

$$L = \frac{\omega}{8\pi} E_0 D_0 \sin \delta_\varepsilon = \frac{\omega}{8\pi} E_0^2 \varepsilon'' \qquad (1.25)$$

In the mechanical and dielectric experiments the energy dissipation is thus proportional to J'' (or G'') and ε'' respectively. It is likewise proportional to the sine of the phase angle, which in dielectrics is referred to as the power factor. In the absence of relaxation effects the phase lag is zero and no energy is dissipated. In this case J'', G'' and ε'' are zero and $J' = |J|$, $G' = |G|$, $\varepsilon' = |\varepsilon|$.

1.3 SINGLE RELAXATION TIME MODEL

Since the basic event in relaxation is the movement of a molecular segment, it is to be expected *a priori* that the several time and frequency parameters should be interrelated. For instance, that in dynamic and step-function experiments at one temperature the observed Ω_J, Figure 1.3, and time of inflection in the $J(t)$ curve, Figure 1.1b, should be related. This is found by experiment to be so but the relationships are usually fairly complex. There are, however, a few relaxations, both mechanical and dielectric, which conform closely to a very simple model known as the single relaxation time model. In the same sense that the properties of real gases are understood in terms of deviations from the ideal gas model so the properties of real relaxations are understood in terms of deviations from the ideal relaxation: that is, a relaxation with a single relaxation time. The single relaxation time model is thus of major significance to the theory of mechanical and dielectric relaxation.

The following insight into the physical basis of the single relaxation time model in mechanical relaxation is due to Nowick (1953). A parallel argument can be made for dielectric relaxation (Fröhlich, 1949).

Let the time-dependent portion of the strain in a creep experiment $(J(t) - J_U)\sigma$ (see Figure 1.1b), be proportional to an internal parameter p.

$$(J(t) - J_U)\sigma = \lambda p \tag{1.26}$$

λ is a constant. The parameter p is a measure of the degree to which the molecular arrangement departs from the equilibrium arrangement for the solid under zero stress. Let the equilibrium value of p, denoted $\bar{p}$, be proportional to the stress,

$$\bar{p} = \kappa\sigma \tag{1.27}$$

in which κ is a constant. If, at a particular time t and stress σ the value of p is not equal to $\bar{p}$ then assume the rate of change of p to be

$$\dot{p} = -\frac{(p - \bar{p})}{\tau} \tag{1.28}$$

τ is a constant, called here the relaxation time. For a creep experiment in which a stress σ_0 is imposed on the specimen at $t = 0$ we have, by integration of Equation (1.28),

$$p = \bar{p}(1 - e^{-t/\tau}) \tag{1.29}$$

Multiplying both sides by λ we have, from Equation (1.26),

$$J(t) = J_U + (J_R - J_U)(1 - e^{-t/\tau}) \tag{1.30}$$

in which we have used $\lambda\bar{p} = (J_R - J_U)\sigma$. A plot of $J(t)$ as a function of $\log(t/\tau)$ is given in Figure 1.4 for a model with $J_U = 1{\cdot}00 \times 10^{-10}\ \mathrm{cm^2/dyn}$ and $J_R = 1{\cdot}01 \times 10^{-10}\ \mathrm{cm^2/dyn}$. Note that $J(t)$ exhibits an inflection

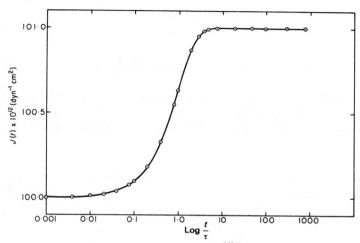

Figure 1.4. Dependence of $J(t)$ on $\log\left(\dfrac{t}{\tau}\right)$ for a single relaxation time model with $J_U = 1{\cdot}00 \times 10^{-10}$ and $J_R = 1{\cdot}01 \times 10^{-10}$ cm²/dyn (Equation 1.20).

$[\mathrm{d}^2 J(t)/\mathrm{d}(\log t)^2 = 0]$ at $t = \tau$. For a solid obeying this model, clearly the observed inflection time would be a measure of τ.

For an imposed oscillating stress

$$\sigma = \sigma_0\, e^{i\omega t} \tag{1.31}$$

the time dependence of the equilibrium value of p is,

$$\bar{p} = \kappa\sigma_0\, e^{i\omega t} \tag{1.32}$$

The solution to Equation (1.28) is clearly

$$p^* = \frac{\bar{p}}{1 + i\omega\tau} \tag{1.33}$$

Multiplying through by λ,

$$(J^*(\omega) - J_U) = \frac{J_R - J_U}{1 + i\omega\tau} \tag{1.34}$$

Equating real and imaginary parts of Equation (1.34),

$$J'(\omega) = J_U + \frac{(J_R - J_U)}{1 + \omega^2\tau^2} \tag{1.35}$$

$$J''(\omega) = \frac{(J_R - J_U)\omega\tau}{1 + \omega^2\tau^2} \tag{1.36}$$

$$\tan \delta_G = \frac{J''(\omega)}{J'(\omega)} = \frac{(J_R - J_U)\omega\tau}{J_R + J_U\omega^2\tau^2} \tag{1.37}$$

Equations (1.35) and (1.36) are plotted in Figure 1.5 for a model system with $J_U = 1.00 \times 10^{-10}$ cm^2/dyn and $J_R = 1.01 \times 10^{-10}$ cm^2/dyn. Note that for $\omega = 1/\tau$, J'' goes through a maximum and J' an inflection.

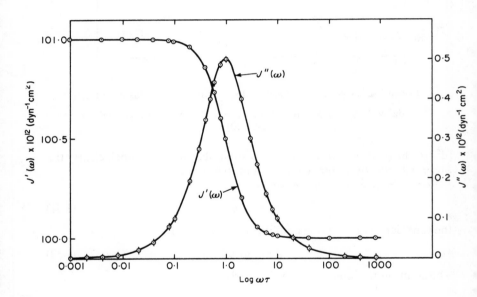

Figure 1.5. Dependence of $J'(\omega)$ and $J''(\omega)$ on $\log \omega\tau$ for a single relaxation time model with $J_U = 1.00 \times 10^{-10}$ and $J_R = 1.01 \times 10^{-10}$ cm^2/dyn.

The connection between step-function and dynamic experiments for the single relaxation time model is illustrated by Equations (1.30) and (1.35) and (1.36). Both experiments are governed by the physical constants J_U, J_R and τ. The two experiments, therefore, yield equivalent information. In practice, however, the step-function and dynamic experiments usually cover different time scales. For instance, at a given temperature if $\tau = 10^{-3}$

seconds then in order to observe relaxation in the dynamic experiment, experiments must be performed in the region of $\omega = 1/\tau = 10^3 \text{ sec}^{-1}$, which is an easily accessible frequency range. In the step-function experiments measurements must be made for values of t of the order of $t = \tau = 10^{-3}$ seconds. It is extremely difficult to perform step-function experiments at such short times. Conversely, if $\tau = 10^3$ seconds the step-function experiments are much easier to perform than dynamic experiments. These experimental techniques are described in Chapter 6.

For some relaxations in non-polymeric materials the simple single relaxation time theory is found to reproduce faithfully the observed behaviour. That is, adjustment of the parameters J_U, J_R and τ leads to exact agreement between theory and experiment. Single relaxation times are observed in molecular liquids, and one pleasing explanation is that the mechanism occurs within one molecule (Davies and Lamb, 1957). This has never been observed in polymers. In Figure 1.6 the measured $J(t)$ for the α relaxation in polyisobutylene is compared with the single relaxation time theory. The observed curve is broader than the theoretical curve. It is invariably found in polymers that the observed curves although similar in form to the single relaxation time curves, are never exactly duplicated by them.

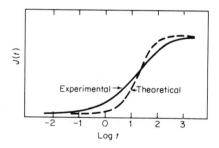

Figure 1.6. Compliance of polyisobutylene showing the α relaxation. Full line is experimental data. Dashed line shows an attempted fit using the single relaxation time model (Equation 1.30) (After Schwarzl and Staverman, 1956).

The reader may wonder why the single relaxation time model, since it cannot be used for real polymeric solids, should be of such significance. There are two reasons. First, the model is capable of generalization to obtain fit between theory and experiment. This is done by using several or, in the limit, a distribution of relaxation times (see Chapter 4). Second, both mechanical and dielectric relaxations are observed by experiment to be thermally activated. It follows from this, as we will now show, that the hypothesis implicit in Equation (1.28) for the rate of change of p towards equilibrium is correct in the sense that any other hypothesis would be unreasonable.

Consider for simplicity N identical molecular segments which have two potential wells of equal energy available to them. Let the wells be a distance a apart. The flux of segments from well 1 to well 2, ϕ_{12}, is (Zener, 1950)*

$$\phi_{12} = N_1 v \, e^{\frac{S}{k}} e^{\frac{-H}{kT}} \tag{1.38a}$$

and the flux from well 2 to 1, ϕ_{21}, is

$$\phi_{21} = N_2 v \, e^{\frac{S}{k}} e^{\frac{-H}{kT}} \tag{1.38b}$$

N_1 and N_2 are the number of molecular segments in wells 1 and 2, v is the frequency of vibration of the segments and S and H the change in entropy and enthalpy when one segment is brought to the top of the enthalpy barrier between the wells. The probability per second for the transition of a molecular segment from well 1 to 2, Γ_{12} is,

$$\Gamma_{12} = \frac{\phi_{12}}{N_1} \tag{1.39a}$$

and from well 2 to 1

$$\Gamma_{21} = \frac{\phi_{21}}{N_2} \tag{1.39b}$$

Clearly if $N_1 \neq N_2$ the fluxes will be such as to cause the populations to equalize. Thus the stable state is for $N_1 = N_2$ and both fluxes are

$$\phi = \frac{N}{2} v_0 \exp -\frac{H}{kT} \tag{1.40}$$

and both transition probabilities are

$$\Gamma = v_0 \exp -\frac{H}{kT} \tag{1.41}$$

in which ($v \exp S/k$) is replaced by v_0.

* Above the glass-transition temperature the jump processes are probably activated by a fluctuation in free volume around the segment rather than by fluctuation in thermal energy (Cohen and Turnbull, 1959). The argument in this section presents the process to be thermally activated and must be reformulated for temperatures above the glass transition.

If a force F acts parallel to the line between wells 1 and 2 the jump enthalpy at the barrier is increased or decreased by $\frac{1}{2}aF$ depending on whether the segments move against or along the direction of F. The transition probabilities are now

$$\Gamma_{12} = v_0 \exp -\left(\frac{H - aF/2}{kT}\right) \tag{1.42a}$$

$$\Gamma_{21} = v_0 \exp -\left(\frac{H + aF/2}{kT}\right) \tag{1.42b}$$

Hence

$$\Gamma_{12} = \Gamma_{21} \exp \frac{aF}{kT}$$

which for values of $aF \ll kT$ becomes approximately

$$\Gamma_{12} = \Gamma_{21}\left(1 + \frac{aF}{kT}\right) \tag{1.43}$$

Clearly ϕ_{12} will increase with respect to ϕ_{21} as soon as the force F is established, and there will be a net flux of segments out of well 1 into well 2. But as N_2 thereby increases with respect to N_1, the disparity between ϕ_{12} and ϕ_{21} will decrease since the fluxes are equal to the product of the transition probability and the well population (Equation 1.38). The rate of change of $(N_2 - N_1)$ can be easily shown (Fröhlich, 1958) to be given by

$$\frac{d}{dt}(N_2 - N_1) = -\frac{[(N_2 - N_1) - \overline{(N_2 - N_1)}]}{(2\Gamma)^{-1}} \tag{1.44}$$

$$\overline{(N_2 - N_1)} = \frac{aFN}{2kT} \tag{1.45}$$

The solution to Equation (1.44) is

$$(N_2 - N_1) = \overline{(N_2 - N_1)}(1 - e^{(-t/2\Gamma)}) \tag{1.46}$$

Clearly $\overline{(N_2 - N_1)}$ represents the equilibrium difference in populations between wells 1 and 2 due to the force F. The equilibrium is attained exponentially with time, the time constant being $(2\Gamma)^{-1}$.

If $(N_2 - N_1)$ is identified with p (Equation 1.26), the molecular rearrangement induced by the force F will lead to a mechanical relaxation controlled by a single relaxation time. In this case the relaxation time defined in Equation (1.28) is

$$\tau = (2\Gamma)^{-1} \qquad (1.47)$$

By a parallel argument, if F is an electrical force and if the polarization is proportional to $(N_2 - N_1)$ then the molecular rearrangement leads to dielectric relaxation controlled by a single relaxation time. This model is far too simple to find application in the relaxation of polymeric solids but it serves as a conceptional illustration. A similar model with two wells was used by Snoek (1941) to give a complete atomic description of the relaxation of interstitial impurity atoms in body-centred cubic metals.

One crucial point in linear relaxation theory is that the applied force F is taken to be sufficiently small that the work done, aF, in the jump of one molecular segment is much less than kT. That is, the applied force merely biases the flux of segments from one well to another, the flux itself being caused by thermal fluctuations.*

1.4 TEMPERATURE DEPENDENCE

So far for both mechanical and dielectric relaxation we have discussed the variables time or frequency, it being assumed that the temperature is constant. Relaxation can be studied, however, by performing measurements at constant time or frequency by varying the temperature. The relationship between the time or frequency dependence of relaxation parameters and temperature dependence is easily understood for the single relaxation time model of the last section.

The temperature dependence of τ is, from Equations (1.47) and (1.41),

$$\tau = \tau_0 \exp \frac{H}{kT} \qquad (1.48)$$

in which τ_0 is written for $(2\nu_0)^{-1}$. Consequently, with increasing temperature the dispersion regions move to higher frequencies and shorter times. This is illustrated in Figure 1.7, which shows the variation of tan δ with frequency for polycyclohexyl methacrylate (Heijboer, 1960b). The measurements were made at temperatures between $-90°$C and $+18°$C. Note that the peak moves to higher frequencies as the temperature is increased; at $-80°$C the peak occurs at 1 c/s and at $-30°$C at ca. 800 c/s. The temperature dependence of the relaxation time is of great use in studying mechanical relaxation for the following reason.

* See footnote on page 16.

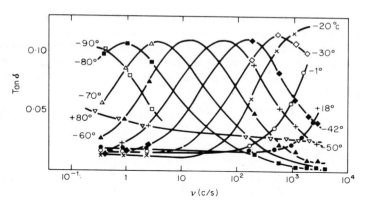

Figure 1.7. Dependence of tan δ (mechanical) on frequency for polycyclohexyl methacrylate at different temperatures (After Heijboer, 1956).

For values of $(J_R - J_U)$, small compared to J_U, we rewrite Equation (1.37).

$$\tan \delta_G \doteqdot \frac{J_R - J_U}{J_U} \frac{\omega\tau}{1 + \omega^2\tau^2} \tag{1.49}$$

The fraction $(J_R - J_U)/J_U$ is known as the relaxation strength (Zener, 1948).

In order to determine the relaxation strength and relaxation time by experiment it is necessary to measure tan δ_G as a function of ω over a sufficiently wide frequency range to delineate the relaxation. For the single relaxation time model the peak half-width is 1·14 decades (Equation 1.37). In order to include the whole relaxation it would be necessary to measure over at least four or five decades of frequency. Furthermore this span of frequency is insufficient to span the relaxation region of the relaxations observed in polymers which are considerably broader than the predictions of the single relaxation time model. Now, dynamic mechanical experiments are extremely difficult to make over a wide frequency range. For instance, in Figure 1.7 note the lack of data between 10 and 100 c/s. The reason for this is that there are three main technical methods which operate in the regions 1, 10^3 and 10^6 c/s (see Chapter 6). Therefore, in order to make measurements of over six decades it is necessary to use three quite different and involved techniques, thus introducing systematic error.

Happily there is a way out of this dilemma which was first recognized by Snoek (1941). Equation (1.49) contains τ as a variable in addition to ω. It is therefore possible to measure tan δ as a function of $\omega\tau$ by keeping ω constant and varying τ by changing the temperature. If tan $\delta_{G,\,max}$ be the maximum value of tan δ_G, we have for a single relaxation time (from Equation 1.49)

$$\frac{\tan \delta_G}{\tan \delta_{G,\,max}} = \frac{2\omega\tau}{1+\omega^2\tau^2} = \text{sech} \ln \omega\tau \qquad (1.50)$$

Therefore, using Equation (1.48),

$$\frac{\tan \delta_G}{\tan \delta_{G,\,max}} = \text{sech} \left[\frac{H}{kT} + \ln \omega\tau_0 \right] \qquad (1.51)$$

The function $[H/kT + \ln \omega\tau_0]$ can be varied continuously over a wide range by changing T with no variation in ω. The data are normally presented by plotting tan δ against T, although $(1/T)$ would be better theoretically.

The temperature variation experiment at constant frequency is not so important in dielectric relaxation since frequency variation is more easily achieved (Chapter 7).

1.5 MECHANICAL AND DIELECTRIC RELAXATION IN POLYMERS CONTAINING THE CYCLOHEXYL GROUP

As an example of mechanical and dielectric relaxation in polymers we describe some of the experiments of Heijboer (1960b) on polymers containing the cyclohexyl unit (1).

$$R-CH \underset{CH_2-CH_2}{\overset{CH_2-CH_2}{<}} CH_2$$

(1)

Karpovich (1954) and Lamb and Sherwood (1955) had observed a mechanical relaxation in liquid methyl cyclohexane $(1; R = CH_3)$ and attributed it to a mechanism in which the molecule flipped from one isomer to another. In cyclohexane $(1; R = H)$ both chair and boat isomers occur but the chair isomer predominates since it has the lower energy. Six of the hydrogen atoms of the cyclohexane molecule lie close to the general plane of the ring and are said to be in the equatorial position. Of the other six H atoms, three are above the plane of the ring and three below; these are said to be in the axial position (Beckett, Pitzer and Spitzer, 1947)*. For each cyclohexane ring *two chair forms* are possible, and in changing from one chair form to the other equatorial atoms become axial atoms and vice versa. At any given temperature there is thus a constant interchange back and forth of hydrogen atoms from the axial to the equatorial positions.

The temperature dependence of tan δ for polycyclohexyl methacrylate **(2)** is shown in Figure 1.8 (Heijboer, 1956).

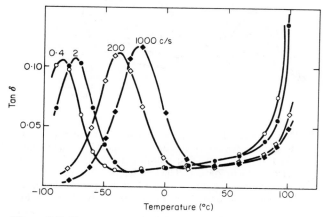

(2)

Figure 1.8. Temperature dependence of tan δ (mechanical) for poly-cyclohexyl methacrylate at several measuring frequencies. The low-temperature peak is due to the flip mechanism of the cyclohexyl group. The rise in tan δ below 100°c is due to the glass–rubber relaxation (After Heijboer, 1956).

* The nomenclature used here differs from that of Beckett and coworkers (see Klyne, 1954).

The measurements were made at different frequencies which are marked on the graph. The low-temperature peak is caused by the axial–equatorial mechanism of the cyclohexyl group. The activation energy (H, Equation 1.48) of the relaxation is 11·5 kcal/mole. From this value of the activation energy, Heijboer (1956) showed the identity of the low-temperature loss mechanism in polycyclohexyl methacrylate with the loss mechanism in liquid methyl cyclohexane and other low molecular weight cyclohexyl derivatives studied by Karpovich (1954) and Lamb and Sherwood (1955).

The temperature and frequency of the cyclohexyl loss peak is largely independent of the structure of the polymer containing the group. This is illustrated in Figure 1.9 (Heijboer, 1960b) which shows the cyclohexyl peak for copolymers of cyclohexyl methacrylate and methyl methacrylate.

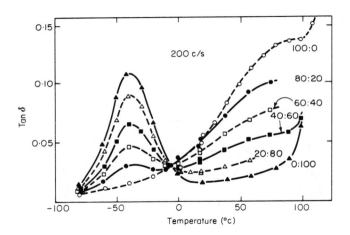

Figure 1.9. Dependence of tan δ (mechanical) on temperature for copolymers of methyl methacrylate and cyclohexyl methacrylate. Contents marked in per cent, first number methyl methacrylate. Frequency of measurement 200 c/s. The low-temperature peak is due to the flip of the cyclohexyl group, and the high-temperature peak is due to the relaxation of the oxycarbonyl group (After Heijboer, 1956).

This copolymer is a mixture of the methyl methacrylate unit (**3**) and the cyclohexyl methacrylate unit (**4**). The way in which the magnitude but not the temperature of the peak is influenced by the cyclohexyl content is striking. The peak occurs (lowered by 4°C) if the cyclohexyl group is present in monomeric form, Heijboer (1956).

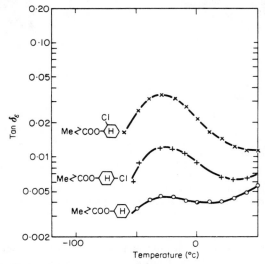

(3)   (4)

Dielectric experiments on chlorine-substituted derivatives of poly-cyclohexyl methacrylate confirm the interpretation of the mechanical experiments in all details. The variation of dielectric loss tangent with temperature at 300 c/s is shown in Figure 1.10 for polycyclohexyl methacrylate, poly-4-chlorocyclohexyl methacrylate and poly-2-chlorocyclohexyl methacrylate (Heijboer, 1960b). The activation energy of the dielectric peak is 11·5 kcal/mole, which is the same as the activation energy of the internal friction peak in both polycyclohexyl methacrylate and the chlorine derivatives.

Figure 1.10. Dependence of tan δ_ε on temperature for polycyclohexyl methacrylate, poly-4-chlorohexyl methacrylate and poly-2-chlorohexyl methacrylate. Frequency of measurement 300 c/s (After Heijboer, 1960b).

The loss mechanism for the 4-chlorocyclohexyl derivative is illustrated in Figure 1.11 (Heijboer, 1960b). Only half the substituents are shown so as to simplify the model. The oxycarbonyl group is attached to the molecular backbone and is shown in the same position for the two alternative chair conformations of the 4-chlorocyclohexyl group. In addition to the movement of mass caused by the flip from one chair conformation to the other, there is also a change of orientation of the dipole at the carbon–chlorine bond. Thus the process leads to both mechanical and dielectric loss.

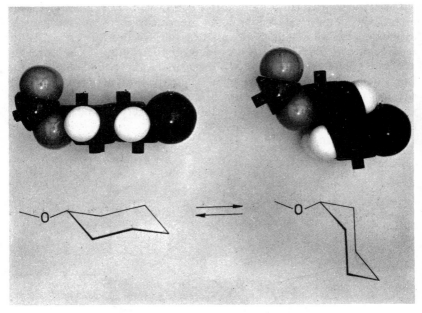

Figure 1.11. Molecular model of the two alternative 'chair' forms of the trans-4-chlorocyclohexyloxycarbonyl group. Only half the substituents are indicated.

The difference in the magnitudes of the dielectric loss peaks between the 2- and 4-substituted derivatives, Figure 1.10, is considered by Heijboer (1960b) to be caused directly by the position of the chlorine atom. Note from the molecular models in Figure 1.11 that after the flip the change in orientation of the chlorine atom in the 4-position is not large, whereas if it were in the 2-position it would be much larger. To elaborate, although in both 4- and 2-positions the chlorine atoms move from axial to equatorial positions as the group flips, the orientation of the plane of the cyclohexyl group simultaneously flips, since the oxycarbonyl group and the polymer backbone remain fixed in position.

The copolymer of cyclohexyl methacrylate and methyl methacrylate exhibits two internal friction peaks, Figure 1.9. As the methyl methacrylate content is increased, the low-temperature peak decreases and the high-temperature peak increases in magnitude. This evidence (and further evidence described in Chapter 8) leads to the hypothesis that the high-temperature peak is due to the relaxation of the methyl methacrylate portion of the copolymer and, in particular, to the relaxation of the oxycarbonyl side-chain (3).

The presence of more than one relaxation is typical of homopolymers as well as copolymers. Under this circumstance J_U and J_R take on new meaning. Each of the several relaxations of a polymer will be described by a J_U and a J_R. For example, at a given temperature in a creep experiment, as one relaxation relaxes to completion and $J(t) \rightarrow J_R$ another will, perhaps, be initiated. The J_R of the first relaxation thus becomes the J_U of the second. The relaxations may indeed be partially superposed as are the relaxations in Figure 1.9.

The reader will notice that this assignment of the low-temperature relaxation of polycyclohexyl methacrylate to the relaxation of the cyclohexyl unit is not supported by a quantitative molecular theory. There is, of course, little doubt that the assignment of Heijboer is the correct one. But there is no molecular theory which enables us to predict the mechanical relaxation frequency in terms of the frequency at which the molecule flips from one chair position to the other. Nor is there a theoretical prediction of the relaxation magnitude.

2

The Chemical and Physical Structure of Polymeric Solids

Probably the most powerful stimulus for the study of mechanical or dielectric relaxation is the possibility of determining a molecular mechanism. There are two ways of doing this. The first is quantitative and involves measuring relaxation parameters (relaxation strength, relaxation time) and then predicting the values from first principles or in terms of other experimental results. A good example of this is the relaxation in body-centred cubic metals due to the movement of interstitial impurity atoms (Snoek, 1941). For this relaxation the exact relationship between the mechanical relaxation time and the rate of atomic movement is well understood. The second method of determining a molecular mechanism is qualitative and therefore less satisfying. This involves measuring relaxation parameters and their dependence on structure (e.g. chemical structure of the polymer repeat unit, copolymer content, density) and then by hypothesis postulating a mechanism. For relaxations in polymeric solids the qualitative method alone has been used, although there exist crude molecular theories of the dynamics of the glass–rubber relaxation (see Chapter 5). The lack of a quantitative relaxation theory is due to the difficulty in setting up a profitable model for a polymeric solid since, compared to metals and molecular crystals, relatively little is known of the ordering of the molecules in the solids which have been studied. Single crystals of a number of polymers have been prepared in recent years but very few measurements have yet been made on them since their size is very small, usually in the region of 10^{-3} cm. Even the arrangement of structural groups within the molecules is not always known with absolute certainty, nor is much known of the barriers restricting rotation about chain linkages which is the basic event in both mechanical and dielectric relaxation. We therefore review in this chapter the present understanding of the chemical structure of high polymers (Section 2.1), the barriers to internal rotation (Section 2.2) and the arrangement of the molecules in the solid state (Section 2.3).

2.1 CHEMICAL STRUCTURE

It is useful to subdivide polymers into two classes: *condensation* and *addition* polymers. This classification, which was suggested by Carothers (1929), is based on two different mechanisms of polymerization. The synthesis of condensation polymers proceeds by the stepwise reactions of functional groups, at least two of which are contained by the monomers and growing polymer chains. Each reaction proceeds with the elimination of a small molecule, usually water. Addition polymers are formed by the addition of unsaturated monomers to the growing chain without the elimination of *water* or other small molecules. Hence, condensation polymers comprise structural units which lack certain elements present in the monomers from which they are formed, whereas addition polymers are composed of structural units which contain the same number of atoms as the corresponding monomer.

2.1a Condensation Polymers

The use of bifunctional monomers in condensation polymerizations leads to *linear* condensation polymers, typical examples of which are the polyesters and polyamides. Polyesters are formed from intermolecular condensation of hydroxy acids (HO—X—$COOH$) or from dihydroxy compounds (**1**) and dibasic acids (**2**).

$$n\left[HO-X-OH\right] + n\left[HO-\underset{\underset{O}{\|}}{C}-Y-\underset{\underset{O}{\|}}{C}-OH\right] \xrightarrow[-nH_2O]{} HO\left[-X-O-\underset{\underset{O}{\|}}{C}-Y-\underset{\underset{O}{\|}}{C}-O-\right]_n H$$

$\qquad$ **(1)** $\qquad\qquad\qquad$ **(2)**

Likewise polyamides are prepared either from amino acids (H_2N—Y—$COOH$) or from equimolar mixtures of a diamine (H_2N—X—NH_2) with a diabasic acid. In the latter case the resulting structure may be written as (**3**).

$$H\left[-\underset{\underset{H}{|}}{N}-X-\underset{\underset{H}{|}}{N}-\underset{\underset{O}{\|}}{C}-Y-\underset{\underset{O}{\|}}{C}-\right]_n OH$$

$$\textbf{(3)}$$

A well-known polyester, polyethylene terephthalate (**4**) (Terylene

$$HO\left[-CH_2CH_2OOC-\bigcirc\!\!\!-COO-\right]_n H$$

$$\textbf{(4)}$$

Dacron), may be prepared from ethylene glycol and terephthalic acid.

Nylon 66 (5), prepared from hexamethylenediamine and adipic acid,

$$H[-NH(CH_2)_6NHCO(CH_2)_4CO-]_nOH$$

(5)

is a typical polyamide. Examples of polyesters and polyamides which have been investigated mechanically and dielectrically are presented in Chapters 12 and 13.

In the above formulae the structural units are shown in square brackets and the number of structural units per polymer molecule is denoted by n, the degree of polymerization. The molecular weight of the polymer is clearly nM_0, where M_0 is the molecular weight of the structural unit.

In general, linear condensation polymers are partially crystalline since sections of their chain molecules are able to pack closely together in parallel ordered arrays. This is due mainly to two factors. Firstly, the structural units usually have no possibility for existing in different isomeric forms so that the chains are perfectly regular and, secondly, polar linkages occur within the main chain. A diagram of the arrangement of chain segments in the crystal of Nylon 66 is shown in Figure 2.1 for illustration (Bunn and Garner, 1947). Polyamides generally have higher melting points than the corresponding polyesters on account of the strong ability of the amide links to form intermolecular hydrogen bonds (shown as dotted lines in Figure 2.1). When the molecular weights of condensation polymers exceed about 10,000 they may frequently be drawn into tough fibres in which the crystallites are oriented.

Non-linear condensation polymers are formed from monomers with a functionality exceeding two. These polymers are usually highly branched or crosslinked into a three-dimensional network, and are thus hard infusible and insoluble resins.

2.1b Addition Polymers

The synthesis of addition polymers from unsaturated monomers may proceed by either a free-radical or an ionic mechanism. The free-radical processes are characteristic of chain reactions and involve at least three processes, namely (a) *chain initiation* by which free radicals are introduced into the system either by the action of heat or ultraviolet light upon the monomer or by the decomposition of an added catalyst such as an organic peroxide, (b) *chain propagation* which consists of successive additions of unsaturated monomer to a growing free radical with electronic rearrangements at each step to regenerate a new radical and (c) *chain termination* in which the growing chain radicals are destroyed. The propagation

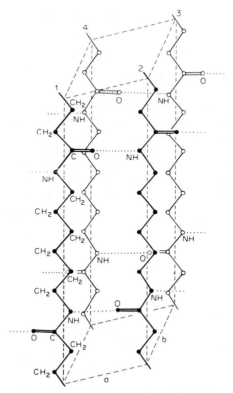

Figure 2.1. Packing of nylon 66 in crystal (After Bunn and Garner, 1947).

reaction in the polymerization of vinyl monomers may be represented generally as follows,

$$\overset{X}{\underset{|}{RCH_2}}\overset{X}{\underset{|}{-CH\cdot}} + CH_2 =\overset{X}{\underset{|}{CH}} \longrightarrow RCH_2-\overset{X}{\underset{|}{CH}}-CH_2-\overset{X}{\underset{|}{CH\cdot}}$$

Typical examples of addition polymers are polystyrene ($X = C_6H_5$), polyethylene ($X = H$), and polyvinyl chloride ($X = Cl$). Disubstituted monomers may also be polymerized and yield polymers such as polymethyl methacrylate (PMMA) (See Chapter 1, p. 23). Chain termination may occur by recombination of growing radicals,

$$RCH_2\overset{X}{\underset{|}{CH}}\cdot + \cdot\overset{X}{\underset{|}{CH}}-CH_2R' \longrightarrow RCH_2\overset{X}{\underset{|}{CH}}-\overset{X}{\underset{|}{CH}}CH_2R'$$

or by disproportionation through transfer of a hydrogen atom,

$$\underset{\substack{| \\ X}}{RCH_2CH \cdot} + \cdot \underset{\substack{| \\ X}}{CH}-CH_2R' \longrightarrow RCH_2CH_2 + \underset{\substack{| \\ X}}{CH} = \underset{\substack{| \\ }}{CHR'}$$

Generally, recombination appears to be the dominant terminating mechanism. Disproportionation is appreciable, however, in the case of PMMA as evidenced by the observation that a large number of chains are terminated by double bonds (Bevington, Melville and Taylor, 1954). Certain unsaturated monomers may also polymerize by an *ionic* addition mechanism which may be either cationic or anionic depending on the nature of the monomer and the catalyst employed. In cationic polymerization the chain propagation mechanism involves, essentially the addition of the double bond of the monomer to a carbonium ion situated at the end of the growing chain. Chain termination by recombination of growing chains is thus excluded. Monomers which are able

$$\underset{\substack{| \\ Y}}{RCH_2-\overset{\substack{X \\ |}}{C}{}^+} + \underset{\substack{| \\ Y}}{CH_2}=\overset{\substack{X \\ |}}{C} \longrightarrow RCH_2-\underset{\substack{| \\ Y}}{\overset{\substack{X \\ |}}{C}}-CH_2-\underset{\substack{| \\ Y}}{\overset{\substack{X \\ |}}{C}}{}^+$$

to donate electrons, such as isobutylene ($X = Y = CH_3$) and styrene ($X = C_6H_5$, $Y = H$), can be polymerized by this method, and acidic catalysts are required. The propagation step in an *anionic* polymerization involves a *negative* charge at the extremity of the growing chain. In this case X and/or Y must be an electron withdrawing group that will stabilize a negative charge by resonance or inductive charge distribution (e.g. —COOR, —C_6H_5, —CN, —CH = CH_2), and basic catalysts are employed. The subject of anionic polymerization has been reviewed by Mulvaney, Overberger and Schiller (1961).

A few anionic polymerizations do not exhibit any termination reaction and, in the absence of impurities, the polymerization will continue if more monomer is added. The best-known example of these so-called 'living polymers' is polystyrene prepared from styrene monomer using a sodium naphthalene complex as catalyst. This type of polystyrene is particularly interesting since it has a very narrow molecular-weight distribution (Szwarc, 1960).

The oxide polymers (see Chapter 14), which contain both carbon and oxygen atoms in the main chain, are usually prepared by ionic addition polymerizations of certain aldehydes or cyclic ethers. These polymers are of interest from the dielectric point of view owing to the dipolar C—O bonds *within* the chain backbone. For example, formaldehyde

$(CH_2 = O)$ can be polymerized by both acidic and basic catalysts (Schweitzer and coworkers, 1959) to give polyformaldehyde, known more often as polyoxymethylene, $[-CH_2-O-]_n$. This polymerization apparently involves the stepwise addition of formaldehyde to a growing chain bearing a positive or negative charge, the carbonyl double bond of the monomer being opened at each step to regenerate the charge. The growing chains are thought to be terminated by transfer with water molecules to yield hydroxy end-groups. The thermal stability of polyoxymethylene is increased by esterifying these end-groups, and small amounts of antioxidant are frequently added. The thermal stability of polyacetaldehyde, $[-CH(CH_3)-O-]_n$, is similarly enhanced.

The best-known example of a polymer formed from the (ring-opening) addition polymerization of a cyclic ether is polyethylene oxide $[-CH_2CH_2-O-]_n$. The classical anionic polymerization of ethylene oxide (6) proceeds slowly and yields quite low molecular weights (< 10,000)

$$CH_2-CH_2$$
$$\diagdown \; O \diagup$$

(6)

(Gee and coworkers, 1959). However, molecular weights exceeding a million have been obtained using a strontium carbonate catalyst (Hill and coworkers, 1958). Polytrimethylene oxide $[-CH_2CH_2CH_2-O-]_n$ and polytetramethylene oxide $[-CH_2CH_2CH_2CH_2-O-]_n$ are synthesized from oxetane (7) and tetrahydrofuran (8) respectively. The

$$CH_2-CH_2 \qquad CH_2-CH_2$$
$$CH_2-O \qquad CH_2 \quad CH_2$$
$$\diagdown \; O \diagup$$

(7)   (8)

mechanism of the polymerization of cyclic ethers has been reviewed by Eastham (1960), and Furukawa and Saegusa (1963) have reviewed the polymerization studies of both aldehydes and cyclic ethers. As might be expected from their symmetrical and polar repeat units the oxide polymers of general formula $[-(CH_2)_p-O-]_n$ are usually highly crystalline.

A very important class of addition polymerizations, which in some cases may be described as coordinated anionic polymerizations, take place in the presence of multicomponent (Ziegler) catalysts. A typical such catalyst system (Natta, 1959) consists of a metal alkyl in conjunction with a crystalline compound such as titanium tetrachloride. At low temperatures the polymerization probably proceeds by a stepwise ionic addition to the metal alkyl and is thought to occur on the surface of the crystalline compound (i.e. a heterogeneous mechanism). The presence of the ordered crystalline support is of great importance since it leads to *stereoregular* polymers (see below) which are often highly crystalline. Natta and collaborators have been largely responsible for the development of the Ziegler catalyst system and in a number of publications have postulated polymerization mechanisms and described the determination of the polymer structure.

The properties of addition polymers are dependent to a large extent upon the structural order or symmetry exhibited by the macromolecules. Generally, dissymmetry may arise from the possibility of the existence of isomers (positional, structural, geometric, steric, etc.), and also from the possibility of the occurrence of side-branches and occasional crosslinks within the structure.

2.1c   Positional Isomers

Successive additions of vinyl monomers ($CH_2{=}CHX$) may lead to a polymer chain bearing the substituents either on alternate carbon atoms (known as the head-to-tail structure),

$$-CH_2-\underset{X}{CH}-CH_2-\underset{X}{CH}-$$

or on adjacent carbon atoms which yields the head-to-head tail-to-tail structure,

$$-CH_2-\underset{X}{CH}-\underset{X}{CH}-CH_2-CH_2-\underset{X}{CH}-\underset{X}{CH}-$$

The majority of vinyl polymers have predominantly the head-to-tail configuration, and are thus positionally ordered (Marvel, 1943). Important exceptions are the haloacrylate polymers. However, a small percentage of head-to-head linkage in a predominantly head-to-tail polymer is often observed. For example, Flory and Leutner (1948) observed the head-to-head content of polyvinyl alcohol to vary between 1·23 and 1·95 % depending on the method of polymerization. The disorder introduced by occasional head-to-head linkages is unlikely to have much effect on the polymer properties. In the case of crystalline polymers it may have the effect of reducing the degree of crystallinity by hindering the parallel packing of chain segments.

2.1d   Steric Isomers

When the structural units of olefin polymers contain unlike substituents $(-CH_2-CXY-)_n$ and the substituents are arranged along the chain in the head-to-tail configuration, the molecules are subject to another form of dissymmetry. The substituted carbon atoms are, in this case, asymmetric and both d and l configurations of the structural units are possible. When the steric configurations of the structural units are together ordered according to some rule the polymer chains are said to be *stereo-regular*. The three main types of steric arrangements are illustrated in Figure 2.2 which shows a three-dimensional model of a vinyl polymer. The black spheres represent carbon atoms, the white ones substituent atoms,

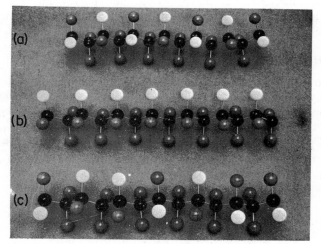

Figure 2.2.   Three steric isomers of a vinyl polymer. From top to bottom (a) syndiotactic; (b) isotactic; (c) atactic.

and the others hydrogen atoms. The carbon chain is arranged in the planar zig-zag form. In model (a) successive substituent atoms are placed on opposite sides of the vertical plane containing the carbon atoms. In this configuration the molecule is termed 'syndiotactic'. In (b) the substituents all occur on the same side of the plane of the carbon atoms, this arrangement being called 'isotactic'. The substituent atoms in model (c) occur randomly on either side of the reference plane, and this is known as the 'atactic' configuration. The recognition and description of these steric isomers is a recent development (Natta and Corradini, 1955, 1956). Natta and Danusso (1959) have fully discussed the nomenclature relating to sterically ordered structures. It should be emphasized that although the conformation of the polymer chain is readily varied by rotation about main-chain bonds the steric order remains unchanged. Also, it should be noted that if the substituents are identical, as, for example, in polyethylene or polytetrafluoroethylene ($[-CF_2-]_n$) the molecules are not subject to this form of dissymmetry.

Variations in stereoregularity have a large effect on the properties of polymeric solids. For example, polystyrene is an amorphous solid when prepared by the conventional free-radical method, but is a partially crystalline solid when prepared in the isotactic form using stereospecific (Ziegler) catalysts. This is also true of many other polymers including polymethyl methacrylate, polypropylene and polybutene. Apparently, the polymer chain may only be fitted into a regular crystalline lattice when the substituent side-groups are placed in a regular manner.

2.1e Structural and Geometric Isomers

Certain polymers are subject to other structural irregularities owing to the possibility of the existence of structural and geometric isomers. Of particular importance within this class are the polymers derived from diene monomers such as butadiene $(CH_2\!=\!CH\!-\!CH\!=\!CH_2)$, isoprene $(CH_2\!=\!C(CH_3)\!-\!CH\!=\!CH_2)$ and chloroprene $(CH_2\!=\!C(Cl)\!-\!CH\!=\!CH_2)$. Considering polybutadiene, all or some of the following structural units may be present in the polymer.

| 1,2-Structure | *cis*-1,4-Structure | *trans*-1,4-Structure |

The 1,2-structure may be further subdivided into its steric isomers due to the asymmetric carbon atom. In the case of synthetic polybutadiene, infrared analysis has been used to determine the proportions of the various structural units present, using model olefins for calibration purposes (Richardson and Sacher, 1953). Synthetic diene polymers vary considerably in structure depending on the method of polymerization (i.e. whether free-radical or ionic and, in the case of an ionic mechanism, which catalyst is employed) (Foster and Binder, 1957). The geometric configuration of the structural units can have a large influence on the polymer properties. For example, natural rubber (*cis*-1,4-polyisoprene) is an amorphous rubber at room temperature whereas gutta percha (*trans*-1,4-polyisoprene) is partially crystalline since the more extended *trans* units may readily pack together in ordered arrays.

2.1f Chain Branching and Crosslinking

So far we have considered the symmetry of polymer chains arising from structural features of the chain units and their steric relationships. Larger-scale effects such as chain branching and crosslinking may also influence the structural regularity. Chain branching in vinyl polymers can arise during the polymerization reaction due to the transfer of a hydrogen atom from a monomer or from a polymer molecule to a growing chain radical. The chain transfer with polymer, for example, may result in the formation of a free radical *within* the polymer chain which then yields a branched structure by addition of monomer.

$$
R\cdot + H\overset{\overset{\textstyle CH_2}{|}}{\underset{|}{C}}\!-\!X \longrightarrow RH + \cdot\overset{\overset{\textstyle CH_2}{|}}{\underset{|}{C}}\!-\!X \underset{+\,Monomer}{\longrightarrow} \overset{\overset{\textstyle CH_2}{|}}{\underset{|}{C}}\!-\!X
$$

This reaction is a further example of a termination mechanism for the chain radical R·. Vinyl polymers generally contain only about one branch for every 10^4 monomer units in the chain and are therefore essentially linear. Polyvinyl acetate, however, has an unusually high rate of chain transfer and may be fairly highly branched (Flory, 1953). It is unlikely that a few long side-branches would significantly affect the relaxation properties of solid polymers, except perhaps in the irreversible flow region at high temperatures, which involves the motion of complete molecules.

The presence of side-branches in polyethylene is known to have a significant effect on its physical properties. When prepared by the free-radical polymerization of ethylene at high pressures and at temperatures above 200°C, the resulting 'low-density' or 'high-pressure' polyethylene is highly branched. The branches include a small number, perhaps one per molecule, of long branches produced by intermolecular chain transfer as discussed above. However, most of the branches are very short and appear to be mainly ethyl ($-CH_2CH_3$) and butyl ($-(CH_2)_3CH_3$) side-groups. The butyl groups are believed to arise from intramolecular chain transfer as follows (Roedel, 1953),

$$
\begin{array}{ccc}
\overset{\textstyle CH_2}{\diagup\;\diagdown} & CH_3 & CH_3 \\[2pt]
\overset{|}{CH_2}\quad \overset{|}{CH_2} & (\overset{|}{CH}_2)_3 & (\overset{|}{CH}_2)_3 \\[2pt]
-\overset{|}{C}H_2\quad \overset{|}{C}H_2 \underset{\cdot}{\;} \xrightarrow{\quad} -\overset{|}{C}H\cdot \xrightarrow{+\,Monomer} -\overset{|}{C}H\!-\!CH_2\!-\!CH_2-
\end{array}
$$

Willbourn (1959) has postulated an extension of this mechanism to account for the ethyl branches. The proposed extension occurs by an additional intramolecular transfer reaction also involving a six-membered ring. The short side-branches, of which there may be about 30 for every 1000 chain atoms, reduce both the density and the melting point of the polymer (Bunn, 1957). Sperati and coworkers (1953) have given quantitative relationships between density, branching and degree of crystallinity in polyethylene. The number of methyl and ethyl groups in polyethylene, due to the side-branches, is best measured by infrared absorption at 7.25μ and 11.2μ respectively (Bryant and Voter, 1953).

Polyethylene prepared using a Ziegler catalyst (Ziegler and coworkers, 1955) or a catalyst of the Phillips type (Smith, 1956) differs from the conventional high-pressure polyethylene in having considerably fewer short and long chain branches. For example, using a Phillips catalyst, the resulting polymer, (e.g. Marlex 50), contains less than 1.5 methyl groups for every 1000 carbon atoms. On account of the low number of side-branches, polyethylenes prepared in these ways have higher densities, are harder and have higher melting points than the high-pressure polyethylenes.

When diazomethane (CH_2N_2) is decomposed by strong Lewis acids linear polymethylene $[-CH_2-]_n$ is produced (Buckley and Ray, 1952). Bawn and coworkers (1959) have reported on the mechanism of this polymerization using boron trifluoride as catalyst. Polymethylene differs significantly from branched polyethylene in its mechanical properties (see Section 10.1).

The formation of crosslinked networks in polymers may also have some effect on polymer properties. Diene monomers readily form crosslinked polymers due to the residual double bond in the structural unit (either 1,4 or 1,2), and about one crosslink in four polymer molecules is sufficient to yield an infinite network (Flory, 1953). Divinyl monomers such as divinylbenzene and ethylene glycol dimethacrylate polymerize to very highly crosslinked structures. The curing of natural and synthetic polymers, whereby linear molecules are crosslinked together by heating with additives such as sulphur or peroxides or by treating with ionizing radiations, is well known. This curing process effects the polymer properties at temperatures above the melting point (if crystalline) or the glass transition of the polymer. Specifically, amorphous polymers are transformed from viscous liquids or rubber-like materials which show irreverisible behavior (e.g. permanent set) to highly elastic and reversible rubbers.

Polyethylene undergoes crosslinking under the action of ionizing radiation. Many other polymers undergo crosslinking by irradiation, although chain scission predominates in some cases. The effects of radiation on polymers are discussed in a textbook by Bovey (1958).

2.1g Molecular Weight and Molecular-Weight Distribution

Since polymerization reactions usually yield polymers having a very broad distribution of chain lengths, a polymer is ordinarily characterized by some average molecular weight. The two most commonly used averages, the number average ($\overline{M}_n$) and the weight average ($\overline{M}_w$), may be determined from osmotic pressure and light-scattering measurements, respectively, on dilute solutions of the polymer (Flory, 1953). The number average molecular weights of condensation polymers are commonly determined by a chemical analysis of the number of functional end-groups. The most convenient method of molecular-weight determination, however, is from solution viscosity measurements, providing that the viscosity–molecular-weight relationship is known. An empirical measure of the width of the molecular-weight distribution is provided by the ratio $\overline{M}_w/\overline{M}_n$ which should be unity for a monodisperse polymer (Flory, 1953).

The average molecular weight can have a significant effect on the properties of a polymer. In the case of amorphous polymers, and crystalline polymers above their melting points, many physical properties show changes at a critical (weight average) number of chain atoms somewhat less than 1000. For example, the bulk viscosity–molecular-weight relationship exhibits an abrupt change (Fox, Gratch and Loshaek, 1956). The irreversible viscoelastic properties are also molecular-weight dependent (Ferry, 1961). The density and glass-transition temperature of amorphous polystyrene have been observed to increase rapidly with increasing molecular weight toward asymptotic limits which are reached at a molecular weight of about 15,000 (Fox and Flory, 1954). In the case of crystalline (solid) polymers (above a critical molecular weight) the density is generally found to depend inversely on molecular weight providing the rate of cooling during crystallization is constant. For a given molecular weight the density is higher the slower the rate of cooling from the melt. Since density has a marked effect on the relaxation properties of crystalline polymers, these properties are consequently affected by changes in molecular weight.

2.1h Copolymers

The simultaneous polymerization of two or more monomers yields copolymers in which the different monomeric units are present within the same chain. The molar composition of the copolymer usually differs from that

of the initial monomer composition (Mayo and Walling, 1950) and usually
the distribution of units within the chain is random,

$$-M_1-M_1-M_2-M_1-M_2-M_2-M_1-M_1-M_1-M_2-$$

Occasionally, however, a copolymer is formed in which the different units
alternate along the chain,

$$-M_1-M_2-M_1-M_2-M_1-M_2$$

Indirect methods can also be used to synthesize polymers containing
uninterrupted sequences of like units. For example *graft* copolymers may
be prepared in which the backbone chain consists of M_1 units and carries
branches of M_2 units,

$$-M_1-M_1-M_1-\underset{\underset{M_2-M_2-M_2}{|}}{M_1}-M_1-M_1-$$

Also *block* copolymers may be prepared which consist of long sequences
of like units within the same chain,

$$-M_1-M_1-M_1-M_1-M_1-M_2-M_2-M_2-M_2-M_2-$$

Smets and Hart (1960) have reviewed the methods by which graft and block
copolymers can be synthesized.

The properties of random copolymers generally differ from those of the
homopolymers synthesized from the individual monomers. For example,
in the case of amorphous copolymers the glass-transition temperature is
dependent on the composition of the copolymer, an effect that has been
treated quantitatively by Gordon and Taylor (1952) and others (see
Section 2.3a). Also, the presence of non-crystalline units in an otherwise
crystalline polymer depresses the melting point of the crystalline polymer
(Flory, 1953). The incorporation of non-crystalline units (e.g. vinyl
acetate) into the polyethylene chain has provided useful indications of the
nature of the mechanical relaxation properties of polyethylene (see
Section 10.1). The physical properties of block and graft copolymers,
however, are generally more similar to those of a mechanical mixture of
both homopolymers.

2.2 ROTATIONAL ISOMERISM

The structure of a polymer when written in the usual fashion is ambiguous
to the extent that hindered rotation is possible around most of the single
bonds. By this mechanism, the polymer chain may assume many conformations in space. This is true of molecules in both liquid and solid states.

Rotational isomerism has been studied in some detail in small molecules using various techniques including infrared, Raman and microwave spectroscopy, electron diffraction, dipole moment measurements, ultrasonic attenuation and high-resolution nuclear magnetic resonance (Mizushima, 1954; Wilson, 1959). There are two aspects to the problem of conformational analysis; firstly, the study of the various rotational isomers and their energy differences and, secondly, the rate at which the different rotational isomers interconvert. The differences in energy levels between the various rotational isomers govern the fractions of each isomer. The interconversion rate depends on the barrier heights.

At present, even in simple molecules, there is no satisfactory theory of internal rotation which meets the most elementary requirement, that of predicting barriers (Wilson, 1959). There is, however, always the possibility of inferring the conformation and barriers to hindered rotation in polymers from experiments on simple analogues. It is, for instance, clear that rotation never occurs around double bonds. In addition, a barrier will be high if the bond possesses partial double-bond character due to resonance with an adjacent double bond (Phillips, 1958). If resonance is absent, then steric repulsion between substituent atoms appears to be the most important factor in determining the most stable conformation (Mizushima, 1952).

The variation of the potential energy in ethane due to the rotation of one methyl group with respect to the other is shown in Figure 2.3(a) and is of the form

$$U = \tfrac{1}{2} U_0 (1 - \cos 3\phi) \tag{2.1}$$

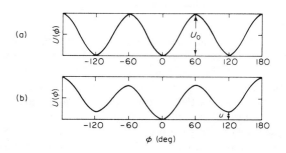

Figure 2.3. Potential energy for rotation. (a) Around the C—C bond in ethane; (b) Around the central C—C bond (Figure 2.5) in n-butane.

ϕ is the angle of rotation measured from the lowest energy state, which is illustrated in Figure 2.4(a). The potential energy is a maximum in the eclipsed state, Figure 2.4(b). The best value for the barrier U_0 lies in the range 2·7 to 3 kcal/mole (Wilson, 1959).

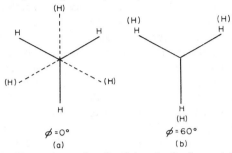

$\phi = 0°$ $\phi = 60°$
(a)   (b)

Figure 2.4. Views down the C—C bond of ethane. (a) The lowest energy state in which each C—H bond bisects the angle formed by the two C—H bonds of the other C atom; (b) The highest energy (eclipsed) state.

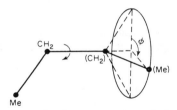

Figure 2.5. Illustrating the interconversion of the conformational isomers of n-butane.

The way in which the conformation of n-butane can be varied is illustrated in Figure 2.5. Consider three of the carbon atoms (in the groups Me, CH_2, (CH_2)) to lie in the plane of the paper. The other methyl group (Me) can take positions on the circle (Figure 2.5) by rotations around the central C—C bond. Four of the conformational isomers are illustrated in Figure 2.6. The energy of the *trans* conformation (a) is exceeded by the *gauche* (b), eclipsed (c) and fully eclipsed (d) conformations by 0·8, 2·9 and 3·6 kcal/mole (Barton and Cookson, 1956). The potential governing the conformation of the n-butane is shown in Figure 2.3(b). The angle ϕ is measured from the *trans* conformation (Figure 2.6a). The other two minima at $\phi = \pm 120°$ are for the two *gauche* conformations (Figure 2.6b).

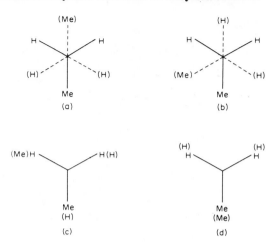

Figure 2.6. Four conformations of the n-butane molecule looking along the central C—C bond, Figure 2.5. (a) *trans*, (b) *gauche*, (c) eclipsed, (d) fully eclipsed.

The conformation of the polymer molecule in the crystal is a good indication of the most stable conformation. Polyethylene crystallizes in the *trans* conformation Figure 2.7 (Bunn, 1939). Isotactic polymers with bulky substituents usually crystallize in a helix (Natta and Corradini, 1955). The probable reason for this is that in the *trans* form the substituents in the isotactic polymers interfere sterically with each other. The relationships between the conformation of a polymer and the potential for rotation around its bonds have been exhaustively developed by Volkenstein and coworkers (for references, see Volkenstein, 1958, 1963). The distribution of rotational isomers and the difference in their energy levels can be measured by the infrared method (Ptitsyn, 1959).

It is possible to draw conclusions about the rate of hindered rotation in polymers by studying simple analogues. Phillips (1955) has shown that there are high barriers restricting rotation about the C—N bond in amides of the type (9). For instance, in *N,N*-dimethylformamide the rate of

$$R_1 \diagdown \qquad \diagup O$$
$$N—C$$
$$R_2 \diagup \qquad \diagdown R_3$$

(9)

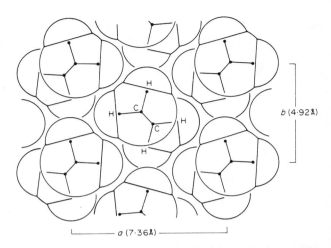

Figure 2.7. Structure of polyethylene crystal. The atoms have their correct external radii (After Bunn, 1960).

rotation around the N—C bond is less than 38 sec^{-1} at room temperature. This low frequency is caused by the C—N bond possessing partial double-bond character due to resonance with the neighbouring double bond. There is every reason to believe that the rotation about C—N bonds in proteins and polyamides is equally hindered. On the other hand, the rotational frequency in low molecular weight hydrocarbons can be very high. For instance, Young and Petrauskas (1956) observed that the relaxation frequency of 2-methylbutane in the liquid state was of the order of 10^7 sec^{-1} at 200°K. This relaxation is due to rotation around the central C—C bond.

If rotation, hindered or free, does not occur then a polymer molecule cannot move. Kauzmann and Eyring (1940) demonstrated in a very convincing fashion that linear hydrocarbon polymers in the liquid state flow by a mechanism which involves the cooperative movement of segments of the order of 20 to 25 carbon atoms. The movement involves rotation around main-chain C—C bonds. In both glassy and crystalline polymers the molecules can also undergo hindered rotation, which provides the mechanism for the transport of mass or the rotation of dipoles in the solid state. Many of the unique properties of polymeric substances, glassy and crystalline as well as liquid, are due to the flexibility of the long chain molecules conferred by internal rotation around C—C bonds.

2.3 THE ARRANGEMENT OF THE MOLECULES IN POLYMERIC SOLIDS

2.3a Amorphous Polymers

The majority of high polymers form transparent glasses when cooled from the liquid or rubber-like state. It is usually assumed that in the amorphous state (glass or rubber) the molecules are tangled up in a completely random manner. There is no unambiguous evidence for this hypothesis (the random-chain hypothesis) which has of late been attacked by Kargin and his school (Kargin, Kitaigorodskii and Slonimskii, 1957). According to these authors, amorphous polymers are partially ordered systems and are made up of globules or bundles of polymer chains. Within the bundle the polymer chains are ordered but the bundles are tangled.

Whether the molecule, or bundle is the basic element in the structure of the amorphous solid is uncertain, there being insufficient evidence to reject either hypothesis. Happily, the major transition of an amorphous polymer, the glass transition, can be explained qualitatively by both hypotheses (see Treloar, 1949; Volkenstein, 1959a).

The variation of the specific volume with temperature for amorphous polystyrene is shown in Figure 2.8, curve a (Natta, Danusso and Moraglio, 1958). The $V-T$ curve exhibits an abrupt change in slope at 90°C, which is termed the glass-transition temperature, T_g. The observed T_g depends

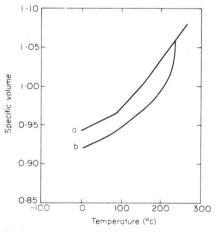

Figure 2.8. Volume–temperature curves for polystyrene. Curve a atactic (amorphous); Curve b isotactic (partially crystalline). (After Natta, Danusso and Moraglio, 1958).

to some extent on the rate of cooling: the lower the rate of cooling the lower T_g (Kovacs, 1958). Figure 2.8 is typical of all amorphous polymers and an explanation can be stated as follows.

The polymer liquid ($T > T_g$) is considered to be quasi-crystal. That is, although the motion of the chain segments in the liquid is chaotic, at any given instant the segments are grouped in small quasi-crystals with holes in between. The difference in density between crystal and liquid is due mainly to the holes (Frenkel, 1946). The molecules, or molecular segments, vary their positions by discrete jumps so that a grouping of molecules at one instant moves rapidly into another grouping a short time later. In the liquid the equilibrium volume contracts with temperature due mainly to a decrease in the density of holes. For each temperature there is an equilibrium hole density, i.e. an equilibrium volume. Above the melting point the hole density can change very rapidly so that it is not possible to change the temperature fast enough to obtain a non-equilibrium volume. But, if the liquid can be supercooled (cooled below the melting point without crystallization) then it is possible to cool to temperatures at which the internal reordering takes place so slowly that the volume cannot contract at a rate sufficient to maintain equilibrium. This leads to a knee in the V–T curve at a temperature identified as T_g, the glass-transition temperature. Above T_g the hole density increases with increasing temperature and decreases with decreasing temperature. Below T_g the hole density does not change when the temperature is changed. The hole density appropriate to a temperature close to T_g is 'frozen-in'. This explanation clearly accounts for the observed dependence of T_g on the rate of cooling. The recorded T_g is usually the value obtained at the slowest rates, roughly $1°c$/hour. For a partially crystalline polymer T_g is usually one-half to two-thirds of the melting point (Kauzmann, 1948; Boyer, 1952; Beaman, 1953).

This theory of the glass transition, the so-called kinetic theory, explains the abrupt variation in other physical properties at T_g. For instance, the specific heat of polyisobutylene (**10**) is compared with the coefficient of

$$\left\{ \begin{array}{c} \ \ \ \ \overset{CH_3}{|} \\ \overset{H}{\underset{H}{-C}}-\overset{|}{C}- \\ \ \ \ \ \overset{|}{CH_3} \end{array} \right\}_n$$

(10)

thermal expansion in Figure 2.9(b, c) (Ferry and Parks, 1936). Both exhibit discontinuities which place T_g at $\sim -70°C$. Above T_g the specific heat consists of the vibrational (Debye) specific heat plus the conformational component, due to the variation with temperature of the potential

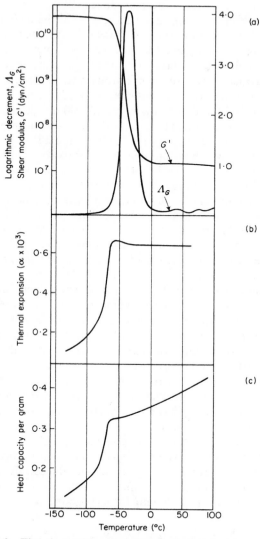

Figure 2.9. The glass-rubber transition in polyisobutylene. (a) Dependence of logarithmic decrement and shear modulus at frequencies ~1 c/s on temperature (After Schmieder and Wolf, 1953); (b) Dependence of thermal expansion on temperature (After Ferry and Parks, 1936); (c) Dependence of heat capacity on temperature (After Ferry and Parks, 1936).

energy of interaction of the molecules (Jones, 1956). Below T_g the conformational component disappears (i.e. there is no change in hole density) leading to an abrupt drop in specific heat at T_g. As illustrated in Figure 2.9(a), T_g is always accompanied by a large mechanical relaxation, with $G_U/G_R \sim 10^3$. Below the relaxation temperature the modulus is that of a glassy polymer, $\sim 10^{10}$ dyn/cm^2; above the relaxation temperature the modulus is that of a rubbery polymer, $\sim 10^7$ dyn/cm^2.

According to Gibbs and DiMarzio (1956, 1958) the glass transition is a manifestation of a thermodynamic second-order transition predicted by them at a temperature denoted by T_2. In Gibbs and DiMarzio's theory T_2 is the temperature at which the number of conformations W available to an assembly of polymer molecules becomes unity (or very small) so that the conformational entropy $S_c = k \ln W$ essentially vanishes. Both W and S_c depend on the fraction of 'holes' in the polymer and, more particularly, on the energy difference u (see Figure 2.3b) between the stable rotational conformations of the main-chain bonds. From statistical mechanical calculations, Gibbs and DiMarzio have shown that T_2 is determined largely by the energy difference u, which they term the 'flex' energy. They have also argued that at temperatures a few degrees above T_2 the number of available conformations will be small, and widely separated in phase space, so that conformational rearrangements will occur very slowly. On account of the resulting relaxation effects, the observed T_g would be several degrees higher than T_2. Hence T_2 is essentially unattainable. However, if the theory of Gibbs and DiMarzio is accepted then T_g, like T_2, should be related to the energy difference u which is a measure of the *equilibrium* chain flexibility. This concept of the glass transition differs fundamentally from that in the kinetic theory. It should be added that the Gibbs and DiMarzio theory has not been conclusively verified and it has been criticized by Volkenstein (1959).

A problem of some importance in connection with the assignment of mechanical loss peaks in polymers concerns the relationship between the glass-transition temperature of a copolymer and the copolymer composition. For several two-component copolymers the glass temperature T_g is given by an equation of the following form (Wood, 1958),

$$(1 - C_2)(T_g - T_{g,1}) + KC_2(T_g - T_{g,2}) = 0 \qquad (2.2)$$

where C_2 is the weight fraction of component 2, $T_{g,1}$ and $T_{g,2}$ are the glass-transition temperatures of components 1 and 2, respectively, and K is a constant ($K > 0$). An equation equivalent to (2.2) was first derived by Gordon and Taylor (1952) assuming ideal volume additivity of the different repeat units. Equation (2.2) is thus often termed the Gordon–Taylor equation. As shown by Wood (1958), Equation (2.2) is also equivalent to the relationships derived by Fox (1956) and by Mandelkern, Martin and Quinn (1957) on the basis of 'free-volume' (Section 5.3c) considerations. DiMarzio and Gibbs (1959) have derived a similar expression based on their statistical thermodynamic theory of glass formation. Each of these derivations differs in the significance to be attached to the constant K, but for present purposes it is sufficient to regard K as an arbitrary *positive* parameter. Now since $K > 0$ and $1 > C_2 > 0$ it follows from (2.2) that T_g must be intermediate between $T_{g,1}$ and $T_{g,2}$, as found for many copolymer systems. However for a few random copolymers, notably acrylonitrile–methyl methacrylate (Beevers and White, 1960), styrene–methyl methacrylate (Beevers, 1962) and vinylidene chloride–methyl acrylate (Illers, 1963) this condition is not fulfilled and either minima or maxima have been observed in plots of T_g against C_2. It must be concluded that Equation (2.2) is not universally valid. Deviations from this equation have been discussed by Illers (1963) in terms of a theory proposed by Kanig (1963).

2.3b   Crystalline Polymers

There are some polymers which crystallize when cooled from the melt. Operationally, a crystalline polymer is distinguished from an amorphous polymer by (a) the presence of sharp x-ray lines superposed on an amorphous halo; amorphous polymers yield the halo alone, (b) the presence of polycrystalline aggregates known as spherulites (Figure 2.10) which are usually large enough to be seen in the optical microscope, and (c) the occurrence of a first-order transition, the melting point. The melting point of crystalline (mainly isotactic) polystyrene is shown in the volume–temperature curve in Figure 2.8 at approximately 220°C. Note that the partially crystalline polystyrene also exhibits a glass transition at 90°C, as does the amorphous (atactic) polystyrene.

The sharp x-ray lines lead to the determination of the unit cell, a grouping of atoms which by translation in three dimensions defines the crystal structure. Determination of the unit cell permits calculation of the perfect crystal density (Bunn, 1946). There are, however, several properties of the crystal which are not obtained easily from x-ray diffraction.

First, x-rays 'observe' the time-average lattice and are relatively insensitive to movement of the molecules. Second, the presence of defects in the lattice, such as side-branches or vacancies, is not easily observed in x-ray diffraction. Third, x-ray theory is not able to give an unambiguous

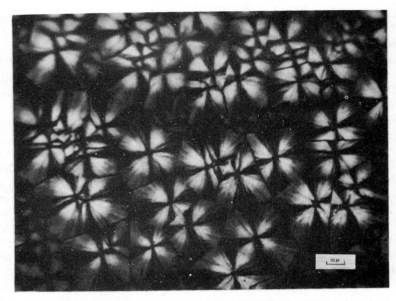

Figure 2.10. Optical micrograph of spherulites in polyethylene viewed
under crossed polaroids (After Palmer, 1965).

interpretation of the amorphous halo nor the fact that the 'sharp' crystalline lines are broader than in crystals of low molecular weight. Consequently in structural studies x-ray diffraction has been supplemented to a
considerable extent by other techniques, particularly electron microscopy.

The x-ray evidence alone leads to two conflicting models for the
structure of highly crystalline polymers. One of the models assumes
that the solid consists of *two* phases, crystalline and amorphous (Gerngross, Hermann and Abitz, 1930). This model, known variously as the
'fringed-micelle' model or 'two-phase' model (Stuart, 1959), is illustrated
in Figure 2.11. The crystals are small and for this reason the crystalline
x-ray reflections are not so narrow as they might be otherwise. The
amorphous regions in between the crystals diffract x-rays into the amorphous halo. That is, the x-ray photograph contains reflections from two
physically separate regions. The natural extension of this interpretation
is to determine the volume ratios of the crystalline and amorphous portions
of the solid. The simplest way of doing this is to determine the specific

Figure 2.11. Schematic representation of the two-phase or fringed-micelle model of a partially crystalline polymer.

volume v, and assume that it is related to the specific volumes of the crystalline and amorphous regions v_c, v_a and the crystalline fraction x,

$$v = xv_c + (1 - x)v_a \qquad (2.3)$$

The volume fraction of crystalline polymer is sometimes known as the 'crystallinity' and expressed as a percentage. Many other methods of measuring the crystallinity have been described in the literature, including x-ray diffraction (Matthews, Peiser and Richards, 1949), infrared absorption (Miller and Willis, 1956), calorimetry (Dole and coworkers, 1952) and nuclear magnetic resonance (Wilson and Pake, 1953). The two-phase model was undisputed following its inception for almost three decades but has of late been under attack (for review, see Geil, 1963).

Stuart (1959) has rejected the two-phase model for polymers of high crystallinity such as high-density polyethylene. According to this author the polymeric solid consists entirely of a crystalline phase containing structural distortions and grain boundaries which cause the bulk density to be less than the density determined from the unit cell. The distortions are due to irregularities in the chain such as branching, head-to-head sequences or atactic disorder, as well as those caused by imperfect crystallization. The latter type of distortion can be largely eliminated by annealing. The various distortions can be described by the paracrystal model introduced by Hosemann and coworkers (Hosemann, 1950; Hosemann, Bonart and Schoknecht, 1956; Hosemann and Bonart, 1957). These authors simulated the diffraction effects of various disorders with two-dimensional Bragg–Lipson analogues. Supporters of the paracrystalline model reject the two-phase model as an explanation of the x-ray diffraction patterns of highly crystalline polymers which they consider to be amply interpreted by the paracrystalline model.

The attack on the fringed-micelle model has been based largely on evidence obtained with the electron microscope. Polyethylene when crystallized from dilute solution forms single crystals (Till, 1957; Keller, 1957; Fischer, 1957). An electron micrograph of polyethylene single crystals crystallized from a dilute solution in ethylene tetrachloride is shown in Figure 2.12 (Reneker and Geil, 1960). The single crystals are observed as lamellae with dimensions of the order of 10×20 microns laterally and of thickness ca. 100 Å. Electron diffraction experiments have shown that the molecules are oriented approximately normal to the plane of the lamellae. Since polyethylene molecules are usually of the order of a micron in length it is clear that the molecules must be folded (Keller, 1957). A photograph of a Stuart–Briegleb model is shown in Figure 2.13 illustrating the manner in which polyethylene molecules may fold to form a single crystal similar to those shown in Figure 2.12 (Reneker and Geil, 1960). The molecules are folded in (110) planes and the surface of the crystal is a (112) plane.

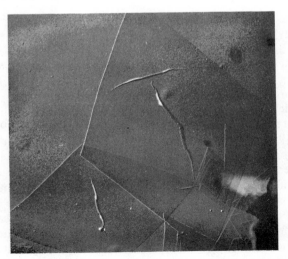

Figure 2.12. Electron micrograph of a single crystal of linear poly-ethylene crystallized from dilute solution (After Reneker and Geil, 1960).

Figure 2.13. Stuart–Briegleb model showing fold planes of a polyethylene crystal (After Reneker and Geil, 1960).

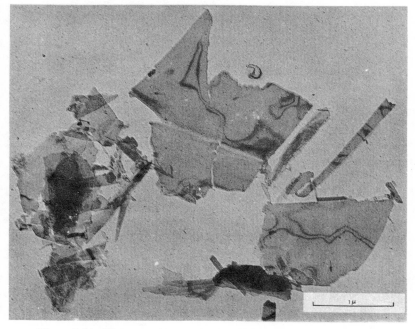

Figure 2.14. Electron micrograph of lamellar debris obtained by the etching of bulk linear polyethylene (After Palmer and Cobbald, 1964).

The single-crystal experiments have led to the belief that in many bulk crystalline polymers the basic structural block is a lamella not very different to those grown in dilute solution, i.e. with folded chains. The clearest evidence is obtained from nitric acid etching experiments (Palmer and Cobbald, 1964; Keller and Sawada, 1964) and from electron micrographs of surface replicas. A good example of lamellar debris from the nitric acid etch experiment is shown in Figure 2.14 (Palmer and Cobbald, 1964). In the surface replica technique the surface studied may either be a

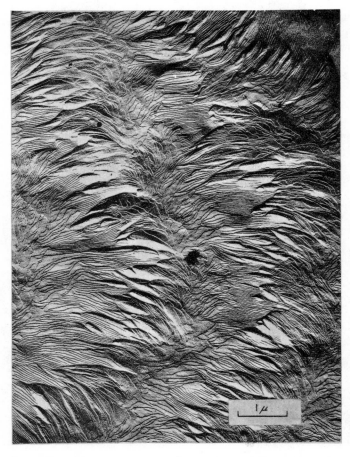

Figure 2.15. Electron micrograph of the surface of Ziegler polyethylene showing lamellae (After Fischer, 1966).

fracture (interior) surface or an exterior surface. Good examples of lamellae observed on the surface of polyethylene are shown in Figure 2.15 (Fischer, 1966). The lamella thicknesses observed by electron microscopy have been frequently correlated with a characteristic spacing observed in the small-angle diffraction of x-rays.

Spherulites, despite the name, are polyhedra which fit together tightly as do grains in a metal. Spherulites are usually studied by optical microscopy (for review, see Keller, 1958) but have also been studied by electron microscopy (Keller, 1959; Keith and Padden, 1959; Price, 1959), light scattering (Stein, 1964) and x-ray diffraction (Keller, 1955). Although many types of spherulites are known, the following model (based on Matsuoka, 1962, and Keith and Padden, 1964a, b) of the genesis of a spherulite accounts for their principal properties.

When a polymer melt is cooled crystallization is initiated at nucleii (usually heterogeneous) at different points in the specimen. The crystallization proceeds by the growth of the spherulites, each spherulite having a nucleus at its centre. The spherulite is composed of densely packed, branched fibrils oriented along the radii. The spherulite expands radially at a constant rate. This stage of crystallization is known as primary crystallization and is illustrated in Figure 2.16, which shows spherulites

Figure 2.16. Optical micrograph showing at two spherulites of polyethylene oxide growing into the melt at 10°C supercooling (After Price, 1959).

of polyethylene oxide growing into the melt (Price, 1959). Radial growth ceases when the spherulite impinges on its neighbours. Residual melt is trapped between the fibrils. Keith and Padden's (1964b) model of fibrils (lamellae) in polyethylene is shown in Figure 2.17.

Primary crystallization is followed by the crystallization of some of the interfibrillar melt, a process known as secondary crystallization. If the polymer melt contains impurities or atactic molecules then the secondary crystallization will be retarded or completely inhibited. The more the secondary crystallization proceeds to completion the more the solid density approaches the density of the crystal.

The foregoing description of the genesis of a spherulite is based on the two-phase model. However, according to Stuart (1959) and Geil (1960) this model cannot be applied to a highly crystalline polymer. For a highly crystalline polymer, such as linear polyethylene, the one-phase model applies, so that the spherulite of linear polyethylene is a disordered single crystal (Geil, 1960). If this is true, how are we to interpret those of the properties of linear polyethylene which, on the two-phase model, are ascribed to the amorphous regions?

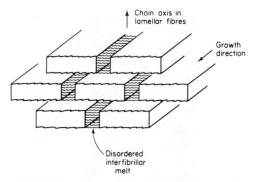

Figure 2.17. Schematic representation of distribution of residual melt after completion of primary crystallization in polyethylene. Small amounts of residual melt probably occur between the broad faces of the fibrils (After Keith and Padden, 1964).

This interesting question is best considered by examining the V–T curves of linear polyethylene which have been determined from $-150°$ to $100°$C for the crystal and the partially crystalline solids Figure 2.18. The crystal specific volume was obtained by x-ray diffraction (Cole and Holmes, 1960)

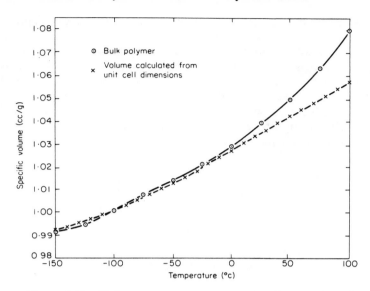

Figure 2.18. Volume–temperature curves for linear polyethylene (After Cole and Holmes, 1960).

and the specific volume of the partially crystalline solid by dilatometry (Quinn and Mandelkern, 1958). The V–T curve for the crystalline solid exhibits two knees, one at ca. $-120°$C and the other ca. $-20°$C. The V–T curve for the crystal runs close to the solid curve at low temperatures but diverges above $-20°$C. Thus both solid and crystal exhibit knees at $-120°$C, and the solid a second knee at $-20°$C. According to the two-phase model, this clearly implies that the glass transition of amorphous polyethylene occurs at $-20°$C (Cole and Holmes, 1960). This result is confirmed by measurements of the temperature dependence of the diffraction angle of the main amorphous halo, which, according to the two-phase model, is a measurement of the interchain separation in the amorphous regions (Ohlberg and Fenstermaker, 1958). Now, according to the one-phase model, the divergence in the V–T curves above $-20°$C, Figure 2.18, cannot by definition be due to a glass transition. Whether or not thermal expansion of a paracrystal would fully account for it is not known. At present there is no doubt that those properties of highly crystalline polymers attributed to the amorphous regions have not been rationalized by the one-phase model.

The two-phase model has an undeniable advantage in that it is applicable to polymers with crystallinities from 0 up to 100%. The one-phase model, if true at all, is only true for crystallinities close to 100%. If the one-phase model is accepted for these highly crystalline polymers it is necessary to specify at what lower crystallinity the model must be changed to the two-phase model. According to Keith (1963) the description of a spherulite in a highly crystalline polymer as a disordered single crystal belies the relationship between spherulites in these polymers and spherulites in less crystalline polymers.

3

Theories of the Limiting Moduli
and the Static Dielectric Constant

The purpose of this chapter is to present a description of the existing calculations of the limiting moduli and the static dielectric constant (ε_R) of polymers. There will be no discussion here of relaxation time or any other rate parameter.

Each mechanical relaxation in shear is characterized by two limiting moduli, G_U and G_R, and an ideal theory would yield both, or their ratio (G_R/G_U). An example of the latter type of theory is the theory for the relaxation of grain boundaries in metals due to Zener (1941) which, for aluminium, yields the values $G_R/G_U = 0.636$ (Kê, 1947). There is at present no satisfactory theory of this type applicable to polymers. There are, however, several calculations of moduli for effectively immobile molecules in an ordered crystalline array (Sections 3.1 and 3.2) and for mobile molecules in a disordered amorphous (rubber-like) array (Section 3.3).

Section 3.1 contains an account of tensile modulus calculations of polymer fibres. The calculations are based on the assumption that the strain is due entirely to distortions within the molecule. This theory is in fair agreement with modulus measurements on fibres in which a large proportion of the molecules lie parallel to the fibre axis. The moduli are found to be very high, of the order of the modulus of diamond ($E = 1.0 \times 10^{13}$ dyn/cm^2, $\langle 111 \rangle$ direction, Kelly 1966). This calculation tells us little about the modulus of polymeric solids in which the crystals are randomly arranged; in, for instance, solids formed by moulding. For these solids the major distortion under hydrostatic stress will be caused by changes in the intermolecular spacing since the van der Waals forces between molecules are far weaker than the covalent forces between atoms within a molecule. The theories of Müller and Brandt, which take intermolecular distortion into account, are described in Section 3.2. These theories give the bulk modulus of a polymer crystal and are in reasonable agreement with experiment. There is no comparable calculation of the compressibility of a glassy amorphous polymer. Section 3.3 contains an account of the theory of rubber elasticity. When a crosslinked polymer is heated above the glass transition, or if crystalline, the melting point, it displays the remarkable mechanical property known variously as rubber elasticity, high elasticity or entropy elasticity. The most obvious property is that the rubber, whilst being completely form stable, has an extremely low modulus of the order of 10^7 to 10^8 dyn/cm^2. If the polymer is stretched it extends without breaking to as much as 10 times its original length. On release of the stress the original shape is completely restored. The phenomenon of rubber elasticity is displayed by chemically crosslinked high polymers and also by high molecular weight polymers which are effectively crosslinked by molecular entanglements. It has been established beyond doubt that rubber elasticity is due to the vigorous micro-Brownian movements of the molecules between the crosslinks.

Section 3.4 describes theories of the dipolar contribution to the dielectric constant. By means of these theories the mean square dipole moment of a polymer molecule can be determined experimentally from measured values of ε_R and ε_U. A comparison of the experimental dipole moment with values estimated theoretically for model chains (Section 3.5) can yield valuable information about the conformations of the polymer molecules.

3.1 THEORY OF THE MODULUS OF A FIBRE

Treloar (1960a) calculated the tensile modulus along the c axis of the crystal on the assumption that the total deformation is due to intramolecular dilation. The problem is thereby reduced to the calculation of the effective force constant of a single molecule along its axis. The required information comprises the unit cell dimensions, bond lengths and angles and the force constants for bond-length and bond-angle deformation.

3.1a Polyethylene

Let the bond lengths be l and the angle between successive bonds α, as shown in Figure 3.1. Forces F are applied at the two ends of n segments along the axis of the molecule. Bunn's structure for the polyethylene crystal is used, Figure 2.7. The cross-sectional area of each chain $A = 18 \cdot 25$ Å^2. The C—C bond length $l = 1 \cdot 53$ Å and valence angle $\alpha = 112°$.

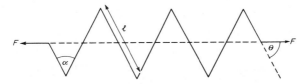

Figure 3.1. Model for the calculation of the tensile modulus of a polymer chain. The forces F are directed along the axis of the chain (After Treloar, 1960).

Let the bonds make an angle θ with the direction of F, $\theta = \frac{1}{2}(180° - 112°) = 34°$. If the total length of the molecule be $L = nl \cos \theta$ where n is the number of carbon atoms, then the total deformation ΔL under a force F is

$$\Delta L = n(\cos \theta \, \Delta l - l \sin \theta \, \Delta\theta) \tag{3.1}$$

The deformation of each bond, Δl, is produced by the force $F \cos \theta$ and is given by

$$\Delta l = \frac{F}{k_l}\cos \theta \tag{3.2}$$

where k_l is the force constant governing the separation of two carbon atoms. The deformation $\Delta\theta = -\Delta\alpha/2$, is produced by a torque of $\frac{1}{2}Fl \sin \theta$ acting around each of the bond angles and is given by

$$\Delta\theta = -\frac{1}{4}\frac{Fl}{k_\alpha}\sin \theta \tag{3.3}$$

where k_α is the valence-angle deformation constant. The tensile modulus is defined

$$E = \frac{(F/A)}{(\Delta L/L)} \tag{3.4}$$

From Equations (3.1), (3.2) and (3.3),

$$\frac{\Delta L}{F} = n\left[\frac{\cos^2 \theta}{k_l} + \frac{l^2 \sin^2 \theta}{4k_\alpha}\right] \tag{3.5}$$

This equation is simplified if the valence-angle force constant k_α is replaced by another force constant k_p. When the valence angle opens by $\Delta\alpha$, one of the carbon atoms moves a distance $l \Delta\alpha$ with respect to the other. The energy involved in the distortion, Δw, may be written in terms of $l\Delta\,\alpha$,

$$\Delta w = \tfrac{1}{2}k_\alpha(\Delta\alpha)^2 = \tfrac{1}{2}k_p(l\,\Delta\alpha)^2$$

$$\therefore\quad k_\alpha = k_p l^2 \tag{3.6}$$

Substituting k_p in Equation (3.5),

$$\frac{\Delta L}{F} = n\left[\frac{\cos^2\theta}{k_l} + \frac{\sin^2\theta}{4k_p}\right] \tag{3.7}$$

Therefore, the modulus is given by

$$E = \frac{l\cos\theta}{A}\left[\frac{\cos^2\theta}{k_l} + \frac{\sin^2\theta}{4k_p}\right]^{-1} \tag{3.8}$$

For values of $k_l = 4{\cdot}36 \times 10^5\,\mathrm{dyn/cm}$ and $k_p = 0{\cdot}35 \times 10^5\,\mathrm{dyn/cm}$ (Rasmussen, 1948) we have

$$E = 1{\cdot}82 \times 10^{12}\,\mathrm{dyn/cm^2}$$

Shimanouchi, Asahina and Enomoto (1962) obtain the theoretical value

$$E = 3{\cdot}4 \times 10^{12}\,\mathrm{dyn/cm^2}$$

using force constants determined from low-frequency Raman shifts of normal hydrocarbons (Mizushima and Shimanouchi, 1949).

3.1b   Polyhexamethylene adipamide (Nylon 66)

The calculation is based on the unit cell of Bunn and Garner (1947), Figure 2.1. The chain is planar and again the force F is directed parallel to the chain along the axis. If all bonds make the same angle with the axis then from Equation (3.7) we write immediately

$$\frac{\Delta L}{F} = \cos^2\theta \sum \frac{n_i}{k_{li}} + \sin^2\theta \sum \frac{n_j}{k_{pj}} \tag{3.9}$$

where n_i is the number of bonds of type i and force constant k_{li} and n_j is the number of valence angles of type j of deformation constant k_{pj}. The number and types of bond and their force constants are given in Table 3.1, (see also, Figure 2.1). This leads to a value of Young's modulus

$$E = 1{\cdot}97 \times 10^{12}\,\mathrm{dyn/cm^2}$$

if θ is given the approximate value 35°.

Table 3.1. Types of bond and valence angle in Nylon 66 and their force constants, (see Figure 2.1).

i	Bond	n_i	l_i (Å)	k_{li} (dyn/cm)	j	Angle	n_j	k_{pj} (dyn/cm)
1	C—C	10	1·53	$4·36 \times 10^5$	1	C—C—C	8	$0·35 \times 10^5$
2	OC—NH	2	1.40	$7·8 \times 10^5$	2	C—C—N	2	$0·36 \times 10^5$
3	H_2C—NH	2	1.47	$5·74 \times 10^5$	3	C—N—C	2	$0·68 \times 10^5$
					4	C—CO—N	2	$0·38 \times 10^5$

Treloar has also calculated Young's modulus for the crystals of poly-ethylene terephthalate, $E = 1·22 \times 10^{12}$ dyn/cm^2 (1960b), and cellulose, $E = 5·65 \times 10^{11}$ dyn/cm^2 (1960c). The pioneer calculations for cellulose by Meyer and Lotmar (1936) gave values between $7·7 \times 10^{11}$ and $1·21 \times 10^{12}$ dyn/cm^2. Lyons (1959) has criticized the values of Meyer and Lotmar and gives the value $1·80 \times 10^{12}$ dyn/cm^2. Because Treloar has removed several errors which occurred in the earlier calculations, his values are the more reliable.

It is difficult to perform experiments which can be compared with the theoretical crystal moduli. The best procedure has been to prepare fibers in which it is hoped the molecules have crystallized parallel to the fibre axis. The modulus is then determined by a sonic method (Meyer and Lotmar, 1936) or by measurement of the change in lattice spacing with stress (Dulmage and Contois, 1958). In both cases it is difficult to know whether the stress is entirely absorbed by bond- and valence-angle deformation.

The agreement between experiment and theory is reasonable. For instance, Sakurada, Itoho and Nukushina (1961) give for polyethylene the value $E = 2·6 \times 10^{12}$ dyn/cm^2, which lies between the theoretical estimates of Treloar and Shimanouchi and others. Asahina and Enomoto (1962) calculate the value $E = 4·9 \times 10^{11}$ dyn/cm^2 for the helical molecule of isotactic polypropylene. The calculation assumes that the deformation is entirely due to internal rotation. This result is in good agreement with the measured value of $E = 4·1 \times 10^{11}$ dyn/cm^2 (Sakurada and others, 1961). The measurements of E for cellulose are complicated by water absorption. For dry cellulose Mann and Roldan-Gonzalez (1962) obtain $E = 7·1 \pm 0·7 \times 10^{11}$ dyn/cm^2 by the x-ray method, in reasonable agree-ment with Treloar's theoretical value.

3.2 COMPRESSIBILITY THEORIES

3.2a Müller's Theory

Müller's calculation is for the crystals of the n-paraffins. The calculation should hold also for the polyethylene crystal. The calculation is based on the determination of the unit cell of $C_{29}H_{60}$ (Müller, 1928). From the first law of thermodynamics the compressibility of a body of volume V at $T = 0°K$ is given by

$$\beta = \left[V\left(\frac{\partial^2 \Phi}{\partial V^2}\right) \right]^{-1} \tag{3.10}$$

in which Φ is the lattice energy. The compressibility is therefore known when the dependence of the lattice energy on volume is known. The lattice energy Φ is

$$\Phi = V_A + V_R \tag{3.11}$$

in which V_A is the attractive van der Waals energy and V_R the energy due to the repulsive forces.

The van der Waals forces in a lattice containing two types of molecule 1 and 2 yield a potential V_A (London, 1937)

$$V_A = \frac{3}{2}mc^2 \left[\alpha_1 \chi_1 \sum \frac{1}{R_1^{\,6}} + \frac{2\chi_1\chi_2}{\chi_1/\alpha_1 + \chi_2/\alpha_2} \sum \frac{1}{R_{12}^{\,6}} + \alpha_2\chi_2 \sum \frac{1}{R_2^{\,6}} \right] \tag{3.12}$$

where m is the mass of the electron, c the velocity of light, α_1 and α_2 are the polarizabilities and χ_1 and χ_2 the diamagnetic susceptibilities of the molecules. The molecules 1 and 2 are presumed to be spherically symmetric and small. The distances R are between the centres of the molecules. A polymer, or a long chain paraffin, has obviously to be subdivided into a number of equivalent small molecules or force centres. The question arises whether the method of subdivision critically affects the calculation. Müller (1936) answers this question by subdividing the long chain paraffin in three different ways and calculating the potential for each. The modes of subdivision are:

(a) The force centres are placed at the nuclei of the hydrogen and carbon atoms.

(b) The force centres are placed in the middle of each bond.

(c) The force centres are placed at the centre of each (CH_2) group.

The appropriate polarizabilities and susceptibilities in Equation (3.12) for each mode of subdivision are (a) hydrogen and carbon atoms, (b)

C—H and C—C bonds and (c) the CH_2 group. The summations over R^{-6} in Equation (3.12) are performed using the structure for the n-paraffin crystal (Müller, 1928). The geometric parameters used by Müller are shown in Figure 3.2. The values used were

$a = 7\cdot426\,\text{Å}$
$b = 4\cdot956\,\text{Å}$
$s = 2\cdot51\,\text{Å}$ (double the distance between two consecutive CH_2 groups)
$w = 0\cdot886\,\text{Å}$ (distance between carbon rows in the molecule)
$\delta = 1\cdot10\,\text{Å}$ (distance between the hydrogen and carbon molecules)
$\phi = 30°$ (setting angle)

The potentials V_A calculated for the three modes of subdivision are given in Table 3.2.

Table 3.2. Dispersion potential of a CH_2 group in an n-paraffin lattice
(Müller, 1936)

Subdivision (a)	$-2\cdot58 \times 10^{-13}$ erg
Subdivision (b)	$-1\cdot77 \times 10^{-13}$ erg
Subdivision (c)	$-1\cdot57 \times 10^{-13}$ erg

The sublimation energy of the crystal is approximately equal to $-1\cdot62 \times 10^{-13}$ erg per CH_2 group. Consequently, subdivisions (a), (b) and (c) all give results which are of the correct order of magnitude, i.e. they agree approximately with the sublimation energy. Therefore the subdivision may be made in the manner which most simplifies computation. For an n-paraffin the most convenient subdivision is to take the CH_2 group as the force centre.

The repulsive potential. V_R, is taken to be similar to Slater's helium potential

$$V_R = M \exp -\varepsilon r \qquad (3.13)$$

where r is the distance between nuclei, and M and ε are constants. The justification for this is that the outer electronic configuration of a CH_2 group resembles that of a noble gas. The repulsive potential is found to be $V_R \sim 0\cdot2 \times 10^{-13}$ erg. Strictly speaking it is the sum of the attractive potential and V_R which should be compared to the sublimation energy. However, the addition of V_R to the much larger potential energies shown in Table 3.2 is a correction which does not affect the order of magnitude argument of the last paragraph.

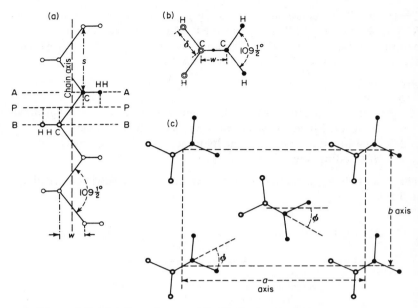

Figure 3.2. The dimensions used by Müller in calculating the van der Waals potential of a paraffin lattice (After Müller, 1936).

Müller simplifies the calculation of the compressibility by taking the crystal to be isotropic in the plane normal to the chain axis. The crystal is considered to be incompressible along the chain axis. A linear compressibility is defined by

$$\text{linear compressibility} = q/l \times (\text{second derivative of potential})^{-1} \quad (3.14)$$

where q is the cross-section of one CH_2 group measured in a plane parallel to the chain axes and l the distance between successive planes containing the axes of the chains and being parallel to the a or the b axes of the crystal. By averaging in the a and b directions of the unit cell a mean value is obtained $q/l = 2 \cdot 7$ Å. Two values of the linear compressibility are calculated for extreme values of ε (Equation 3.13). Using Equations (3.12), (3.13) and (3.14) with extreme values of ε (5Å^{-1} and $2 \cdot 7 \text{Å}^{-1}$) Müller (1941) calculates linear compressibilities of $3 \cdot 8 \times 10^{-12}$ and $10 \cdot 8 \times 10^{-12}$ cm²/ dyn. Müller tested these values by measuring the change in the lattice parameters Δa and Δb caused by a hydrostatic pressure p. The measured values of

$$\frac{1}{p}\frac{\Delta a}{a} \quad \text{and} \quad \frac{1}{p}\frac{\Delta b}{b}$$

for three paraffins are given in Table 3.3. These measurements agree quite well with the theory.

Table 3.3. Linear compressibilities at room temperature, determined by experiment (Müller, 1941). All values in units of 10^{-12} cm^2/dyn

Substance	$\dfrac{1}{p}\dfrac{\Delta a}{a}$	$\dfrac{1}{p}\dfrac{\Delta b}{b}$
n-$C_{23}H_{48}$	10	11
Commercial wax	9	8
n-$C_{29}H_{60}$	2·5	3

3.2b   Brandt's Theory

Brandt (1957) calculates the compressibility of crystals of linear polymers taking into account the dilation along the chain. The calculation is made for $T = 0°K$, i.e. lattice motion is ignored. The polymer chains are assumed to be linear strings of van der Waals' force centres, F, evenly spaced out along the chain axis. The strings are packed side by side with all the force centres in planes at right angles to the chain axes.

The model is illustrated in Figure 3.3. Take cylindrical coordinates $x\phi z$ such that the z axis coincides with the chain axis. Let the force centres be specified by the numbers kl where $\sqrt{k}x_0$ is the equilibrium radius of the kth ring of force centres from the reference centre, F_{00}, and (lz_0) is the height of the lth plane above the plane in which F_{00} lies, Figure 3.3. The equilibrium distance between the force centres F_{00} and F_{kl} is then

$$r_{kl}(\text{equilibrium}) = [kx_0{}^2 + (lz_0)^2]^{1/2} \tag{3.15}$$

Under pressure both x_0 and z_0 are diminished. However, since the dilation in z is considered much smaller than that in x the distance of the force centres under pressure is written

$$r_{kl} = [kx^2 + (lz_0)^2]^{1/2}$$

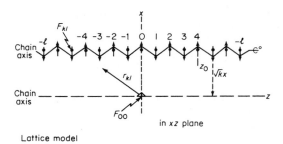

in xz plane

Lattice model

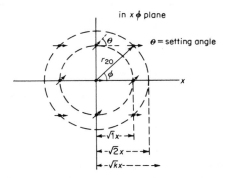

Figure 3.3. Model for chain interaction (After Brandt, 1957).

The value of z_0 is taken from crystallographic data. The value of x_0 is obtained from the value of the density ρ_0 extrapolated to $T = 0°$K. In this way allowance is made for defects in the crystal structure. If the mass of the repeat unit be M_R, the number of force centres per repeat unit q and the volume per repeat unit defined $V_0 = x_0^2 z_0/\gamma$ where γ is a constant which depends on the unit cell, then

$$x_0 = \left[\frac{\gamma M_R}{\rho_0 q z_0} \right]^{1/2}$$

The potential of the repeat unit F_{00} in the field of all the other repeat units is given

$$\Phi_{00} = \tfrac{1}{2} \sum_{k=1}^{\infty} \sum_{l=-\infty}^{\infty} n_k \phi_{0kl} \tag{3.16}$$

where ϕ_{0kl} is the interaction between force centres F_{00} and F_{kl}; n_k is the number of chains in the kth shell. Thus the interactions are assumed to be additive. The dispersion and repulsive forces are approximated by the Lennard-Jones equation. Dipole interactions are taken into account by including the energy of interaction of dipole μ_{kl} with the dipole $\mu_{00\,eff}$

$$\mu_{00\,eff} = \mu_{00} + \mu_{00\,ind} \tag{3.17}$$

where μ_{00} is the paraelectric dipole of F_{00} and $\mu_{00\,ind}$ the dipole induced in F_{00} by the surrounding lattice dipoles. The complete interaction energy of F_{00} with F_{kl} is written

$$\phi_{0kl} = 4\varepsilon_{0kl}\left[\left(\frac{\sigma_{0kl}}{r_{kl}}\right)^{12} - \left(\frac{\sigma_{0kl}}{r_{kl}}\right)^{6}\right]f_{0kl}$$

$$- [\mu_{00\,eff} \cdot \mu_{kl} - 3(\mu_{00\,eff} \cdot \mathbf{r}_{kl})(\mu_{kl} \cdot \mathbf{r}_{kl})r_{kl}^{-2}]r_{kl}^{-3} \tag{3.18}$$

ε_{0kl} and σ_{0kl} are the Lennard-Jones force constants. The factor f_{0kl} accounts for the asymmetry of the force field of the pair F_{00} and F_{kl}. The calculations are made only for linear polymers in which $q = 2$, i.e. polymers such as polyethylene. Thus the force centre is a CH_2 group and the repeat unit a (CH_2CH_2) group. Therefore, the lattice energy per repeat unit, $\Phi_R = \Phi_{00} + \Phi_{01}$. In addition to Φ_R there is also the intramolecular, covalent interaction, which is written

$$\Phi(z) = -E_0 + \tfrac{1}{2}k(z)(z - z_0)^2 \tag{3.19}$$

Note that changes in z due to pressure, affect $\Phi(z)$ but are not considered to materially affect r_{kl}. Thus the external forces acting in the z direction are resisted by the covalent forces, and the external forces in the x direction by the van der Waals forces. The absence of crossterms is considered justified by the large anisotropy of the polymer lattice.

The calculation of Φ_R is greatly simplified on two counts. Firstly, the dispersive interaction falls off sufficiently fast compared to the interchain spacing to permit the sum over all but nearest neighbours to be replaced by an integration. Secondly, the mutual orientation of dipoles causes cancellation so that the effective dipolar interaction between chains decays very rapidly with distance. Also the induced dipole turns out to be practically zero ($< 10^{-2}$ debye).

When $\Phi_R(x)$ and $k(z_0)$ are known it is possible to calculate the compressibility. Consider the element of volume $x^2 z/\gamma$ containing one repeat unit. The constant γ depends on the unit cell; $\gamma = 1$ for a tetragonal lattice, $\gamma = 2\sqrt{3}$ for a hexagonal lattice. From the first law of thermodynamics at $T = 0°\text{K}$, the pressure p is given by

$$p = -\frac{d\Phi_R}{dx}\frac{\partial x}{\partial V} = -\frac{d\Phi(z)}{dz}\frac{\partial z}{\partial V} \qquad (3.20)$$

$$\therefore \quad p = -\frac{d\Phi_R}{dx}\frac{\gamma}{2x_0 z_0} = -\frac{\gamma k(z_0)(z - z_0)}{x_0^2} \qquad (3.21)$$

Furthermore, for small values of $(V - V_0)$,

$$\frac{V - V_0}{V_0} = \frac{2(x - x_0)}{x_0} + \frac{(z - z_0)}{z_0} \qquad (3.22)$$

From Equation (3.21) it follows that for $\Delta V = V - V_0$

$$\frac{\Delta V}{V_0} = \frac{2(x - x_0)}{x_0} + \frac{x_0}{2z_0^2}\frac{d\Phi_R/dx}{k(z_0)} \qquad (3.23)$$

The compressibility is given by

$$\beta = -\frac{1}{V_0}\left(\frac{dV}{dp}\right) \qquad (3.24)$$

$$= -\frac{2}{x_0}\frac{dx}{dp} - \frac{1}{z_0}\frac{dz}{dp} \qquad (3.25)$$

and from Equations (3.21),

$$\beta = \frac{4z}{\gamma(d^2\Phi_R/dx^2)} + \frac{x^2}{\gamma z_0 k(z_0)} \qquad (3.26)$$

Brandt (1957) extrapolates compression data on polyethylene to $T = 0°\text{K}$ and compares the theory with experiment, Figure 3.4. The agreement is excellent up to 40,000 kg/cm^2, especially since no adjustable parameter is used.

3.3 THEORY OF THE RUBBER ELASTIC MODULUS

The purpose of this section is to consider the reversible deformation mechanism of rubber-like polymers, and to review those theories which relate the rubbery modulus to the polymer structure. For a more complete account of these problems the reader is referred to the works of Treloar (1958), Flory (1953) and Ciferri (1961).

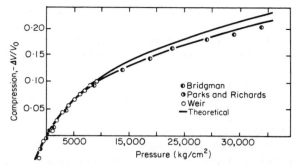

Figure 3.4. Theoretical and experimental values for the compression of polyethylene as a function of pressure. Upper curve—chains compressible along their axes. Lower curve—chains incompressible along their axes (After Brandt, 1957).

3.3a Thermodynamic Analysis

At equilibrium, the tensile force f required to maintain an elongated elastic material at a length L and volume V is given by the derivative of Helmholtz free energy A with respect to length (Treloar, 1958)

$$f = \left(\frac{\partial A}{\partial L}\right)_{T,V} = \left(\frac{\partial U}{\partial L}\right)_{T,V} - T\left(\frac{\partial S}{\partial L}\right)_{T,V} \tag{3.27}$$

$$= \left(\frac{\partial U}{\partial L}\right)_{T,V} + T\left(\frac{\partial f}{\partial T}\right)_{V,L} \tag{3.28}$$

where U and S are the internal energy and the entropy respectively. A knowledge of the thermodynamic derivatives $(\partial U/\partial L)_{T,V}$ and $(\partial S/\partial L)_{T,V}$ is useful for interpreting the elastic deformation mechanism on a molecular basis. From Equation (3.28) it is seen that these derivatives could be obtained experimentally from measurements of $(\partial f/\partial T)_{V,L}$, i.e. from the variation of f with temperature for a material held at constant (extended) length and constant *volume*. Such an experiment is unfortunately difficult since it requires the application of a varying external pressure to maintain a constant volume with changing temperature. The simplest experiment is performed at constant (atmospheric) pressure. For this reason the following approximation, proposed independently by Elliott and Lippmann (1945) and by Gee (1946), has often been employed in the evaluation of $(\partial f/\partial T)_{V,L}$,

$$\left(\frac{\partial f}{\partial T}\right)_{V,L} \cong \left(\frac{\partial f}{\partial T}\right)_{P,\lambda} \tag{3.29}$$

where λ, the extension ratio, is defined as the ratio of the extended to the unextended length. The determination of $(\partial f/\partial T)_{P,\lambda}$ from measurements of $(\partial f/\partial T)_{P,T}$ is straightforward if the linear thermal expansion coefficient of the *unstrained* polymer is known. Equations (3.28) and (3.29) give

$$\left(\frac{\partial U}{\partial L}\right)_{T,V} \cong f - T\left(\frac{\partial f}{\partial T}\right)_{P,\lambda} \tag{3.30}$$

so that an approximate determination of $(\partial U/\partial L)_{T,V}$ and $(\partial S/\partial L)_{T,V}$ is possible with the aid of (3.28) and (3.30).

Early experiments on vulcanized natural rubber (Anthony, Caston and Guth, 1942; Gee, 1946) showed that for extensions up to at least 100%, f was essentially proportional to $T(f = aT)$ at constant P and λ. It was concluded from (3.30) that $(\partial U/\partial L)_{T,V}$ was very small and that the tension was due almost entirely to the entropy derivative $(\partial S/\partial L)_{T,V}$. Rubbers for which $(\partial U/\partial L)_{T,V}$ is strictly zero, so that $f = -T(\partial S/\partial L)_{T,V}$, are termed 'ideal rubbers'. This definition is analogous to the definition of an ideal gas for which the pressure is given by $P = T(\partial S/\partial V)_T$.

It should be added that when natural rubber is extended under normal conditions of constant atmospheric pressure, a small increase in volume occurs owing to the fact that the tensile stress has a hydrostatic component. This effect is largely responsible for a reversal in the usual thermoelastic behaviour (f increasing with T at constant length) at extensions below about 6% (Treloar, 1958). Both internal energy and entropy changes result from the volume increase, but these changes almost exactly cancel each other out. Hence the volume change has a negligible effect on the free energy and therefore on the stress. The stress is thus determined primarily by the entropy decrease as represented by $-T(\partial S/\partial L)_{T,V}$. For shear deformations the problem associated with the volume increase does not arise since, by definition, a shear deformation is one at constant volume.

From a molecular point of view, the entropy decrease associated with the extension of a rubber may be interpreted briefly as follows. A long flexible polymer molecule is capable of assuming many spatial conformations. Statistically, the number of conformations available to the polymer chain w is a function of its end-to-end distance r and, in particular, w decreases as r increases. The conformational entropy of the chain $s_c = k \ln w$ must therefore decrease as r increases. Extending this argument to the network of polymer chains which constitute the rubber, it follows that the entropy of the network must decrease when the polymer is extended since the extension produces, on average, an increase in r

values. It is assumed in this discussion that forms of entropy other than the conformational entropy are unaffected by the deformation. On account of the thermal energy the distribution of chain conformations is, of course, continuously fluctuating about some average distribution. The retractive force is therefore of kinetic origin, and is determined by the tendency of the chains to assume their most probable coiled up conformations. The basic mechanism by which the conformational rearrangements occur involves rotations around chemical bonds within the chains. If these rotations are free, or if the stable rotational states have the same potential energy, then all conformations will have the same internal energy and $(\partial U/\partial L)_{T,V}$ will vanish. By way of contrast we should mention that for crystalline and glassy solids the modulus is determined predominantly by the large internal energy changes resulting from the deformation. This arises from the strong interatomic cohesive forces which resist the deformation. The interchain forces in a rubber are relatively weak, the molecular arrangement and mobility resembling more closely that of a liquid than of a solid. At constant volume, these forces may reasonably be assumed to be the same for the extended and unextended states.

Flory and coworkers (Flory, Ciferri and Hoeve, 1960; Flory, 1961) have argued that the approximation represented by Equation (3.29) is inadequate. From network theory (Equation 3.60 below) they have proposed the following alternative equation for converting force–temperature data at constant pressure and length to conditions of constant volume and length,

$$[\partial \ln (f/T)/\partial T]_{V,L} = [\partial \ln (f/T)/\partial T]_{P,L} + \alpha(\lambda^3 - 1)^{-1} \qquad (3.31)$$

where α is the bulk thermal expansion coefficient. An important fact emerging from thermoelastic data analysed with the aid of Equation (3.31) is that the energy contribution to f, $f_e = (\partial U/\partial L)_{T,V}$, is generally not negligible for rubbery polymers. For crosslinked natural rubber the ratio

$$\frac{f_e}{f} = -T[\partial \ln (f/T)/\partial T]_{V,L} \qquad (3.32)$$

was found to have a value of about 0.2 (Ciferri, 1961), so that the internal energy contribution to the total retractive force is about 20%. For crosslinked polydimethylsiloxane, polyisobutylene, and polyethylene above its melting point, the observed f_e/f values were approximately 0.16, -0.03 and -0.5 respectively (Ciferri, 1961). Allen, Bianchi and Price (1963) have recently succeeded in making direct measurements of $(\partial f/\partial T)_{V,L}$ for crosslinked natural rubber by the application of varying hydrostatic pressures. Their f_e/f values (≈ 0.2) are in good agreement with those deduced by Ciferri, and their data were shown to be consistent with the use of Equation (3.31). Physically, the above results suggest that the internal energy of a polymer molecule is a function of its conformation. This result is not, of course, unexpected on the grounds that the stable states of rotation around main-chain bonds (such as the *trans* and *gauche*) do not in general have the same potential energy, owing to steric repulsions between neighbouring chain substituents (see Figure 2.3b).

Molecular theories of rubber elasticity generally involve two stages. Firstly, the statistical properties of a single chain are considered in order to obtain an expression for its free energy as a function of the end-to-end separation. Secondly, the free energy of deformation is calculated for the polymer network as a function of its extension. The stress–strain relationships follow from the network free-energy expression.

3.3b   Statistical Properties of a Single Polymer Molecule

The average dimensions and the probability distribution function for a polymer chain depend on the number of links in the chain, the length of each link and steric factors such as the fixed valence angles, energy differences between the stable rotational states around each bond, and the interdependence between rotations around neighbouring bonds. A consideration of all these factors is very complicated and most theories of chain conformation usually consider some approximate model in place of the real molecular chain. Many of the earliest theories considered an idealized random chain in which each link can freely assume any random direction with respect to its neighbouring links. If one end of this model chain is placed at the origin of a rectangular coordinate system (Figure 3.5) then, in the absence of external forces, the probability that the other end is located at the point x, y, z in the volume element $dx\,dy\,dz$ is given by (Kuhn, 1934, 1936; Guth and Mark, 1934)

$$P(x, y, z)\,dx\,dy\,dz = \frac{b^3}{\pi^{3/2}}\exp\left[-b^2(x^2 + y^2 + z^2)\right]dx\,dy\,dz \quad (3.33)$$

where the parameter $b^2 = 3/2nl^2$, n being the number of links in the chain and l the length of each link. Equation (3.33) may be written in the form

$$P(r)\,dv = \frac{b^3}{\pi^{3/2}}\exp(-b^2r^2)\,dv \quad (3.34)$$

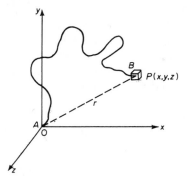

Figure 3.5. Illustration of a statistically kinked chain and the probability that one chain end should fall in the volume element $dV = dx\,dy\,dz$ a distance r from the other end, which is at the origin (After Treloar, 1958).

where r is the chain end-to-end distance ($r^2 = x^2 + y^2 + z^2$) and dv the small volume element. The Gaussian probability distributions, represented by (3.33) or (3.34), are valid only if n is very large and $r \ll nl$. The number of conformations available to the chain (w), subject to the above restrictions, is proportional to $P(r)\,dv$. Since the conformational entropy $s_c = k \ln w$ it follows from (3.34) that

$$s_c = \text{const.} - 3kr^2/2nl^2 \tag{3.35}$$

where the constant includes the size of the volume element dv.

For our freely jointed model chain, all conformations must have the same internal energy. It follows that a force f_c, given by

$$f_c = -T\frac{ds_c}{dr} = \frac{3kTr}{nl^2} \tag{3.36}$$

acts along r tending to pull the chain ends together. The chain therefore acts as a Hooke's law (stress proportional to strain) spring of zero unstretched length.

The mean square end-to-end distance of the random Gaussian chain is given by (Treloar, 1958)

$$\overline{r^2} = \frac{b^3}{\pi^{3/2}} \int_0^\infty r^2 \exp\left(-b^2 r^2\right) 4\pi r^2\,dr \tag{3.37}$$

giving

$$\overline{r^2} = 3/(2b^2) = nl^2 \tag{3.38}$$

The random chain is, of course, an extremely idealized model of an actual polymer molecule. More realistic models should include factors such as the fixed valence angles and hindrances to internal rotation. Several workers have performed calculations of $\overline{r^2}$ for model chains taking account of such factors. These calculations are based on the geometric addition of the chain bond vectors. For a chain containing n bonds each of length l then, denoting the ith bond by the vector $\mathbf{l}_i$, $\overline{r^2}$ is given by

$$\overline{r^2} = \overline{(\mathbf{l}_1 + \mathbf{l}_2 + \ldots + \mathbf{l}_n) \cdot (\mathbf{l}_1 + \mathbf{l}_2 + \ldots + \mathbf{l}_n)}$$

$$= \sum_{i=1}^{n} \sum_{j=1}^{n} \overline{\mathbf{l}_i \cdot \mathbf{l}_j} = nl^2 + 2 \sum_{i=2}^{n} \sum_{j=1}^{i-1} \overline{\mathbf{l}_i \cdot \mathbf{l}_j} \tag{3.39}$$

As illustrated in Figure (2.3), hindrances to rotation around a main-chain bond are described by a plot of the potential $u(\phi)$ against ϕ. The rotational angle ϕ is generally taken as zero in the *trans* conformation. The average value of $\cos \phi$ is written

$$\eta = \overline{\cos \phi} = \frac{\int_0^\pi \cos \phi \exp\left[-u(\phi)/kT\right] \mathrm{d}\phi}{\int_0^\pi \exp\left[-u(\phi)/kT\right] \mathrm{d}\phi} \tag{3.40}$$

For polymers in which the main-chain atoms contain symmetric side-substituents (e.g. polyethylene) we may reasonably assume that $u(\phi)$ is a symmetric function of ϕ about $\phi = 0$ so that $\overline{\sin \phi} = 0$. Assuming that rotations around neighbouring chain bonds are independent, and denoting the valence angle by α, then it can be shown on the basis of Equation (3.39) (Benoit, 1947; Taylor, 1948) that $\overline{r^2}$ is given by (for large n)

$$\overline{r^2} = nl^2 \frac{1 - \cos \alpha}{1 + \cos \alpha} \frac{1 + \eta}{1 - \eta} \tag{3.41}$$

If the chain valence angles are tetrahedral, as in polyethylene and other polymers containing only C—C bonds in the backbone, then $\cos \alpha = -\frac{1}{3}$ and Equation (3.41) becomes

$$\overline{r^2} = 2nl^2 \frac{1 + \eta}{1 - \eta} \tag{3.42}$$

For chains such as polyethylene in which the stable rotational states correspond to the *trans* $(\phi = 0)$ and two *gauche* $(\phi = \pm 120°)$ conformations, as exemplified by Figure 2.3(b) the parameter η can be related to the energy difference u between the *trans* and *gauche* states (assuming that these states have the same entropy). It is readily shown (Volkenstein, 1963) that in this case

$$\eta = \frac{1 - \exp(-u/kT)}{1 + 2\exp(-u/kT)} \tag{3.43}$$

giving, from (3.42),

$$\overline{r^2} = \tfrac{2}{3}nl^2[1 + 2\exp(u/kT)] \tag{3.44}$$

and

$$\frac{\mathrm{d}\ln\overline{r^2}}{\mathrm{d}T} = -\frac{u}{kT^2}\frac{2\exp(u/kT)}{1 + 2\exp(u/kT)} \tag{3.45}$$

Hence if $u = 0$, or for free rotation, $\overline{r^2}$ for the unperturbed chain will be independent of temperature. If u is positive (i.e. the *trans* state is of lower energy than the *gauche*) then $\overline{r^2}$ should decrease with increasing temperature.

Volkenstein (1963) has described in detail matrix methods for the calculation of $\overline{r^2}$ for various types of local steric interaction. In particular, he has considered the effects of tacticity in polymer chains containing asymmetrically substituted atoms ($\sin\phi \neq 0$) and also the effect of correlations between rotations around neighbouring bonds. The most recent, and general, calculations of this kind are those of Flory and co-workers (see, for example, Flory and Jernigan, 1965; Mark and Flory, 1965). An important effect considered in Flory's calculations is that successive *gauche* conformations of opposite direction (i.e. $-120°$, $+120°$) are often disallowed for steric reasons.

A comparison of Equations (3.38) and (3.41) shows that (for large n) $\overline{r^2}$ is proportional to n whether or not fixed valence angles and hindrances to rotation are introduced into the model. In fact this proportionality holds, irrespective of the nature of the local steric effects considered, providing that the chain is of sufficient length.* A related fact is that the statistical behaviour of any real molecular chain can be approximated by that of a freely jointed chain (the so-called 'equivalent random chain') if the real chain is sufficiently long (Kuhn, 1936, 1939; Treloar, 1958). Specifically, it has been argued (Flory, Hoeve and Ciferri, 1959) that the Gaussian distribution function, derived initially for the random chain, is also valid

---

* Deviations from the proportionality of $\overline{r^2}$ to n arise from 'excluded volume' effects due to van der Waals' interactions between non-bonded groups separated by many chain atoms. In bulk rubber-like polymers the excluded volume effect is thought to vanish, the chains assuming conformations unperturbed by these 'long-range' interactions (Flory, 1953; Kurata and Stockmayer, 1963).

for the more realistic chain models if $r \ll nl$ and n is large. The parameter b^2 in Equation (3.33) must, however, be defined generally by

$$b^2 = 3/(2\overline{r^2}) \tag{3.46}$$

rather than $b^2 = 3/(2nl^2)$. Using methods of statistical mechanics developed by Volkenstein and Ptitsyn (1955), Flory, Hoeve and Ciferri (1959) have derived and expression for the conformational free energy of a polymer chain (A_c) without assuming that this is determined entirely by the entropy term. Their result may be written (Ciferri, 1961),

$$A_c = \frac{3kTr^2}{2\overline{r^2}} + \text{const.} \tag{3.47}$$

giving

$$f_c = 3kTr/\overline{r^2} \tag{3.48}$$

for the retractive force acting on a chain. Equation (3.48) is a more general form of Equation (3.36).

3.3c Stress–Strain Relationships for a Rubber-like Polymer

Theories of the modulus of rubbery polymers are based on a model in which the chain molecules are arranged at random, and linked together (either by chemical means or by irradiation) at relatively few points into a three-dimensional network. The theoretical modulus of such a network (at small strains) is obtained from the stress–strain relationship which, in turn, is calculated from the change in elastic free energy A_{el} with deformation (Equation 3.28). Since the network is formed by connecting the single chains through the crosslinks or junction points, the calculation of A_{el} involves the summation of the conformational free energy A_c (Equation 3.47) of all chains in the network.

The more elementary versions of the kinetic theory of rubber elasticity (Treloar, 1958) assume that the free energy A_{el} is determined entirely by conformational entropy changes resulting from the deformation of the network. It is also assumed that each network is sufficiently long such that the Gaussian distribution is valid both in the deformed and the undeformed state. The latter assumption seems reasonable providing that the strain is not too high and the degree of crosslinking not excessive. The entropy of the undeformed network is calculated by summing the entropy contributions from all network chains (i.e. chains between crosslinks). In this calculation it is usually assumed that the mean square vector length of a network chain is the same as that of a free chain (containing the same number of links) unrestricted by the crosslinks. In calculating the entropy of the deformed network it is assumed that the components of the end-to-end vector of each network chain are deformed in the same ratio as the bulk sample. This type of deformation is termed an 'affine deformation'. Based on these assumptions, and a few others, the elementary theory yields the following equation for the elastic free energy,

$$A_{el} = -T\Delta S = \tfrac{1}{2}N_V kT(\lambda_x^2 + \lambda_y^2 + \lambda_z^2 - 3) \tag{3.49}$$

where N_V is the number of network chains in the volume V, ΔS is the network entropy change and λ_x, λ_y and λ_z are the extension ratios in three mutually perpendicular directions. For a unidirectional extension or compression, in which the ratio of the strained to the unstrained length (L/L_i) is denoted by λ, then, at constant volume $\lambda_x = \lambda$ and $\lambda_y = \lambda_z = \lambda_x^{-1/2}$. Hence

$$A_{el} = \frac{1}{2}N_V kT\left(\lambda^2 + \frac{2}{\lambda} - 3\right) \tag{3.50}$$

Differentiating Equation (3.50) with respect to L according to Equation (3.28) gives, for the tension f,

$$f = \left(\frac{\partial A_{el}}{\partial L}\right)_{T,V} = \frac{N_V kT}{L_i}\left(\lambda - \frac{1}{\lambda^2}\right) \tag{3.51}$$

The tension f_i per unit initial (or unstrained) cross-sectional area is obtained by dividing (3.51) by V/L_i,

$$f_i = NkT\left(\lambda - \frac{1}{\lambda^2}\right) \tag{3.52}$$

where N is the number of network chains per unit volume. The true tensile stress t, per unit *strained* cross-sectional area, is obtained by dividing (3.51) by V/L,

$$t = NkT\left(\lambda^2 - \frac{1}{\lambda}\right) \tag{3.53}$$

In terms of the tensile strain, $e = (L - L_i)/L_i = \lambda - 1$, Equation (3.53) becomes

$$t = NkT[3e + e^3(1+e)^{-1}] \tag{3.54}$$

which predicts that the stress is proportional to strain only at low strains where $e^3(1 + e)^{-1}$ is small compared with $3e$. In this case Young's modulus is given by

$$E = t/e \simeq 3NkT \tag{3.55}$$

In the case of simple shear it can be shown (Treloar, 1958) that the shear stress σ is related to the shear strain γ by

$$\sigma = NkT\gamma \tag{3.56}$$

Thus the stress–strain relationship in shear is predicted to be linear, the shear modulus G being given by

$$G = NkT \simeq E/3 \tag{3.57}$$

The elementary theory thus predicts that both G and E should be proportional to the number of crosslinks per unit volume and to absolute temperature. The moduli should be independent of the chemical structure of the polymer repeat unit if the real molecular chain is sufficiently long and flexible to justify the theoretical model. Experimental observations on rubber-like polymers have confirmed in general, the form of the above equations for relatively small strains (Figure 3.6). Deviations from Equation (3.52) occur for vulcanized natural rubber at extension ratios above about 1·5 (Figure 3.6a). These deviations have been treated extensively by Treloar (1958) and Ciferri (1961) and are beyond the scope of this book.

Most refinements or criticisms of the elementary theory suggest that the factor NkT in the above equations should be multiplied by an additional small numerical factor. Flory (1944, 1953) has considered the effect of 'network defects', such as closed loops or chain entanglements and also chain ends, i.e. chains which are linked to the network at one end only, and which do not therefore contribute to the retractive force. His treatment of the chain end effect resulted in the following expression for G,

$$G = NkT(1 - 2M_c/M) \tag{3.58}$$

where M is the molecular weight of the molecules prior to crosslinking and M_c the average molecular weight between crosslinks. James and Guth (1943, 1947) have criticized the assumption in the elementary theory that the mean square end-to-end distance for the network chains (in the undeformed state) is the same as that for a corresponding assembly of

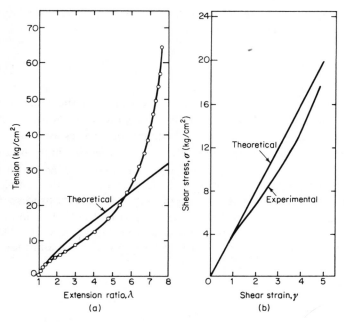

Figure 3.6. (a) Typical experimental plot ($\bigcirc$) of the tension versus extension ratio for a natural vulcanized rubber. The theoretical curve is drawn according to Equation (3.52) with $NkT = 4 \cdot 0 \, \text{kg/cm}^2$; (b) Comparison of an experimental stress–strain curve for natural vulcanized rubber in simple shear with the linear theoretical form predicted by Equation (3.56) with $NkT = 4 \cdot 0 \, \text{kg/cm}^2$ (After Treloar, 1958).

free chains. Considering the effect of the successive introduction of cross-links on the mean positions of already existing junction points they have proposed the following equation,

$$G \simeq \tfrac{1}{2} NkT \tag{3.59}$$

which differs by the factor $\tfrac{1}{2}$ from Equation (3.57).

In the more recent calculations of Flory, Hoeve and Ciferri (1959) the assumption was also avoided that the mean square end-to-end distance of a network chain is the same as that for a corresponding free chain in the unstrained state. By using Equation (3.47) for the conformational free energy of a single network chain, they also considered both intramolecular energy and entropy changes during the deformation. Their calculations yield, in place of Equation (3.51),

$$f = \frac{N_V kT}{L_i} \frac{\overline{r_i^2}}{\overline{r^2}} \left(\lambda - \frac{1}{\lambda^2} \right) \tag{3.60}$$

where $\overline{r_i^2}$ is the mean square end-to-end distance in the undeformed isotropic state for all the network chains having the same contour length, and $\overline{r^2}$ is the mean square end-to-end distance for a corresponding set of free chains. The significance of the factor $\overline{r_i^2}/\overline{r^2}$ is discussed in the original paper and also by Ciferri (1961). We note here that $\overline{r_i^2}$ depends on the volume of the polymer and on such factors as whether the sample is swollen or oriented during or subsequent to the crosslinking process. On the other hand, $\overline{r^2}$ is characteristic of the polymer molecule, depending on bond lengths, valence angles and hindrances to rotation. Equation (3.60) has also been proposed and discussed by Tobolsky, Carlson and Indictor (1961). Equations (3.32) and (3.60) yield

$$\frac{f_e}{f} = -T[\partial \ln (f/T)/\partial T]_{V,L}$$

$$= T \frac{\mathrm{d} \ln \overline{r^2}}{\mathrm{d}T} \tag{3.61}$$

With the aid of (3.45) and (3.61) Flory, Hoeve and Ciferri (1959) have estimated from tension–temperature measurements on polyethylene (at temperatures above its melting point) that the energy difference u between the *trans* and *gauche* conformations is about 540 cal/mole. Equation (3.61) has also been employed in conformational studies on polyisobutylene (Hoeve, 1960), polydimethylsiloxane (Flory, Crescenzi and Mark, 1964), polyoxymethylene (Flory and Mark, 1964) and polyethylene oxide (Mark and Flory, 1965).

To summarize, it is clear from the thermodynamic studies that the predominant effect on deforming a rubber at constant volume is the decrease in entropy. The elementary molecular theory, based entirely on calculations of conformational entropy changes, shows that the modulus of a rubber-like polymer is determined essentially by its degree of crosslinking, and temperature, as represented by NkT in Equations (3.55) and (3.57). Recent experiments have shown that small internal energy changes also occur during a deformation at constant volume. More refined molecular theories, considering both intramolecular energy and entropy changes on deformation, suggest that the factor $\overline{r_i^2}/\overline{r^2}$ should precede NkT as in Equation (3.60). With the aid of these theories, measurements of the modulus and its temperature dependence have been interpreted in terms of the conformational properties of the polymer molecules. The theories so far have ignored energies of interaction *between* neighbouring chains, assuming that at constant volume these do not vary with deformation. Intermolecular interactions could, however, influence the elastic behaviour of a bulk polymer owing to the possibility of correlations between the conformations adopted by neighbouring chains. Such correlations would imply some degree of local order, as envisaged in the proposed 'bundle-like' molecular arrangement discussed in Chapter 2. If the 'bundle' concept is accepted then, according to Volkenstein (1963), a revision of the molecular theories would be required in which the mode of deformation of the bundle, rather than the individual chain molecule, is considered. Finally, it should be emphasized that the above theories concern the modulus of a rubber under conditions of equilibrium. Attempted extensions of these theories to account for the time or frequency dependence of the modulus in non equilibrium relaxation experiments are reviewed in Section 5.3a.

3.3d   Application of the Theory of Rubber Elasticity to Partially Crystalline Polymers

The modulus theories presented so far in this chapter are based on two types of model. The first assumes effectively immobile molecules (Sections 3.1 and 3.2) and the second completely mobile molecules (Section 3.3). Both types of model have one thing in common and this is that they assume the solid to be homogeneous. But crystalline polymers are not homogeneous, being composed of both crystalline and amorphous regions (Chapter 2). Since the geometry of the boundaries between the crystalline and amorphous regions is not known, there is at present considerable difficulty in setting up an appropriate model.

   Calculations have been made of the modulus to be expected from a rubber elastic network structure crosslinked by crystals (Bueche, 1956; Nielsen and Stockton, 1963; Jackson, Flory, Chaing and Richardson, 1963). The model is designed to calculate the modulus of a crystalline polymer at temperatures above the glass-transition temperature of the amorphous regions. The assumption is that the modulus of the crystals greatly exceeds that of the amorphous regions. The calculated values are found to be greater than the observed values, usually by an order of magnitude (Nielsen and Stockton, 1963; Jackson and others, 1963).

Nielsen and Stockton (1963) consider their theory to be lacking because (1) at high crystallinities the crystals impinge on one another, (2) at high crystallinities the chains in the amorphous regions are so short that they could not obey Gaussian statistics, (3) at low crystallinities the effect of chain entanglements is neglected.

Jackson and others (1963), consider their theory to be lacking because (1) the crystals probably do not act as crosslinks in the same sense that the chemical bonds in rubbers act as crosslinks, (2) the space occupied by the crystals cannot be simultaneously occupied by the amorphous molecules which are therefore constrained in a manner which is not found in amorphous rubbers.

3.4 CALCULATION OF THE STATIC DIELECTRIC CONSTANT

3.4a Introduction

The application of a steady electric field to a system of dipolar molecules in the liquid or gaseous phase results in a net orientation of the dipoles in line with the applied field. The static dielectric constant ε_R (which here comprises contributions from *all* orientation processes) is related to the actual dipole moment of a molecule. For the case of a gaseous system Debye (1929) derived the relation

$$\frac{\varepsilon_R - 1}{\varepsilon_R + 2} = \frac{4}{3}\pi N \left[\alpha_e + \frac{\mu_0{}^2}{3kT} \right] \qquad (3.62)$$

μ_0 is the dipole moment of a molecule. N is the number of molecules per unit volume, and α_e is the 'deformation' polarizability which arises due to the elastic displacement of the molecules. α_e is given by the Mosotti–Clausius relation

$$\frac{\varepsilon_U - 1}{\varepsilon_U + 2} = \frac{4}{3}\pi N \alpha_e \qquad (3.63)$$

ε_U is the dielectric constant measured at frequencies so high that the dipole orientation contributions have vanished. The Debye equation (3.62) has been used successfully for very dilute dipolar gases, and has also been applied to very dilute solutions of dipolar molecules in non-polar solvents (Smith, 1955). It cannot be used to evaluate dipole moments for dense gases or pure liquids.

Onsager (1936) gave the first derivation of the static dielectric constant for a condensed phase of dipolar molecules. The derivation introduced the important factor known as the Reaction Field. This arises due to the polarization of the medium surrounding a molecule by the molecule itself, and this additional polarization 'reacts' back on the molecule. The Onsager theory applies only to systems of rigid molecules where there is no orientation correlation between the individual molecules. Kirkwood (1939) attempted to extend the Onsager theory by taking into account short-range orientation correlations of the molecule. Unfortunately this theory omits the deformation polarizability of the molecules in the statistical mechanical treatment, and this term is empirically added to the final equation. Fröhlich (1949) described a generalization of the Onsager theory which included the orientation correlation between molecules, and simultaneously included the deformation polarizability in the correct manner. The final relation includes the correlation quantity 'g' which will be described in detail below. For no orientation correlation between molecules, $g = 1$, and for this case the Fröhlich equation reduces to the Onsager equation, as it should do. The Fröhlich equation accounts very satisfactorily for the dielectric constant of simple associated liquids (e.g. water and the alcohols).

The static dielectric constant of condensed phases of rigid dipolar molecules is thus understood in terms of the Onsager and Fröhlich theories. The Onsager theory does not apply to systems made up of polymer molecules. The dipole moment of a polymer molecule at a given instant in time will be given by the vector sum of the component moments (suitably defined) of the chain. At ordinary temperatures a polymer molecule in a liquid polymer phase will be in continual Brownian motion, and the moment of the chain will vary continually in time. Also, there will be a strong correlation in orientations between the component moments of the chain. The theory of the static dielectric constant of a liquid polymer medium must take into account these factors. Also, a real liquid polymer medium will be made up of an intermingled network of polymer molecules which makes it difficult to take a single polymer chain as the basic dipole unit in the theory of the static dielectric constant. It is therefore appropriate to take the basic dipole unit to be a chemical repeat unit or a basic group moment of the chain, and then use the Fröhlich theory of the static dielectric constant for the polymer system. The chemical repeat unit of moment **p**, say, may have this moment rigidly attached to the main chain as in polyvinyl chloride

$$\left[\begin{matrix} CH_2 - CH \\ | \\ Cl \end{matrix} \right]_n$$

or it may be situated in a side-group as in polymethyl methacrylate. The theory to be outlined below is general to both classes of polymers, but does not specifically take into account the stereoregularity of the polymer chain. In other words, all basic dipole units are assumed to be equivalent, which requires that the orientation correlations between a unit and its neighbours along the chain is on average, the same for all units in the chain. The theory does not apply to dipolar copolymers where each comonomer is dipolar.

A brief derivation of the Onsager equation will be given in Section 3.4b. The Kirkwood and Fröhlich relations will be discussed in Section 3.4c, and the application of the Fröhlich theory to polymeric systems will be given in Sections 3.4d, 3.4e and 3.4f.

3.4b The Onsager Theory

Consider an infinite homogeneous dielectric medium of static dielectric constant ε_R. Let an empty spherical cavity of radius 'a' be introduced into the medium. If a uniform electric field E is applied to this system, the potential at all positions must satisfy Laplace's equation

$$\nabla^2 \psi = 0 \qquad (3.64)$$

If r is the distance from the centre of the sphere, and θ is the angle between $\mathbf{r}$ and the field direction, it follows that

$$\psi = -Er\cos\theta - \frac{M}{r^2}\cos\theta \qquad (r > a) \qquad (3.65a)$$

$$\psi = -Gr\cos\theta \qquad (r < a) \qquad (3.65b)$$

where,

$$M = \frac{\varepsilon_R - 1}{2\varepsilon_R + 1}Ea^3$$

and

$$G = \frac{3\varepsilon_R}{2\varepsilon_R + 1}E$$

G is known as the 'cavity field'.

If the spherical cavity has a rigid point dipole of moment m at its centre, then the potential in this system in the *absence* of an applied electric field is given by

$$\psi = \frac{m\cos\theta}{r^2} - Rr\cos\theta \qquad (r < a) \qquad (3.66a)$$

$$\psi = \frac{m^*\cos\theta}{\varepsilon_R r^2} \qquad (r > a) \qquad (3.66b)$$

where

$$m^* = \frac{3\varepsilon_R}{2\varepsilon_R + 1} m$$

$$R = \frac{2(\varepsilon_R - 1)}{(2\varepsilon_R + 1)} \frac{m}{a^3}$$

R is the field acting upon a dipole due to electric displacements induced by its own presence in the system. R is known as the 'reaction field.'

Consider now the case of a dielectric medium of dielectric constant ε_R which contains a spherical cavity of radius 'a' which has a point dipole at its centre. If an electric field E is applied to the system, the total field acting on the sphere is F,

$$\mathbf{F} = \mathbf{G} + \mathbf{R} \tag{3.67}$$

The total moment of the cavity in the presence of the field is $\mathbf{m}$

$$\mathbf{m} = \mu_0 \mathbf{u} + \alpha_e \mathbf{F} \tag{3.68}$$

μ_0 is the permanent dipole moment in the cavity, $\mathbf{u}$ is the unit vector in the direction of the permanent dipole moment. α_e is the deformation polarizability of the cavity. The cavity is identified with one molecule having a permanent moment μ_0, and a deformation polarizability α_e. Assuming that a given volume of the dielectric equals the sum of the volume of the molecules it contains, it follows from Equation (3.63) that α_e is the average polarizability per molecule, and

$$\alpha_e = \frac{\varepsilon_U - 1}{\varepsilon_U + 2} a^3 \tag{3.69}$$

It is emphasized that ε_U is the dielectric constant measured at frequencies so high that the permanent dipole contribution to the dielectric constant has vanished.

Combination of Equations (3.67), (3.68) and (3.69) with the aid of the definitions of G and R gives

$$\mathbf{m} = \mu \mathbf{u} + \varepsilon_R \frac{(\varepsilon_U + 2)}{(2\varepsilon_R + \varepsilon_U)} \alpha_e \mathbf{E} \tag{3.70}$$

where

$$\mu = \frac{(\varepsilon_U + 2)}{3} \frac{(2\varepsilon_R + 1)}{(2\varepsilon_R + \varepsilon_U)} \mu_0$$

μ is known as the 'liquid moment' of the molecule. It is the effective moment of the molecule in the liquid state and differs from μ_0 which is the effective moment of one molecule surrounded by vacuum. **R** and **m** are parallel, and **G** and **E** are parallel. The orienting force couple **M** acting on a molecule is

$$\mathbf{M} = \mathbf{F} \times \mathbf{m} = \mathbf{G} \times \mathbf{m} = \mathbf{G} \times (\mu\mathbf{u})$$

$$= \frac{3\varepsilon_R}{(2\varepsilon_R + 1)}\mu\mathbf{E} \times \mathbf{u} \qquad (3.71a)$$

Dropping the vector notation, the magnitude of **M** is

$$M = \frac{3\varepsilon_R}{2\varepsilon_R + 1}\mu E \sin \theta \qquad (3.71b)$$

The work of orientation w is given by $w = + \int M \, d\theta$, hence from Equation (3.71b),

$$w = -\frac{3\varepsilon_R}{2\varepsilon_R + 1}\mu E \cos \theta \qquad (3.72)$$

The average orientation of the molecules in the field is obtained using the Boltzmann relation

$$\overline{\cos \theta} = \frac{\int \cos \theta e^{-w/kT} \sin \theta \, d\theta \, d\phi}{\int e^{-w/kT} \sin \theta \, d\theta \, d\phi}$$

$$= L\left[\frac{3\varepsilon_R}{2\varepsilon_R + 1} \frac{\mu E}{kT}\right] \qquad (3.73)$$

$L(x)$ is the Langevin function of x

$$L(x) = \coth x - \frac{1}{x} \qquad (3.74)$$

For low field intensities, Equation (3.73) becomes

$$\overline{\cos \theta} = \frac{3\varepsilon_R}{(2\varepsilon_R + 1)} \cdot \frac{\mu E}{3kT} \qquad (3.75)$$

The average value of **m** in line with the field follows from Equation (3.70) with Equation (3.75)

$$\overline{\mathbf{m}} = \frac{\mu^2}{3kT} \frac{3\varepsilon_R}{(2\varepsilon_R + 1)}\mathbf{E} + \frac{\varepsilon_R(\varepsilon_U + 2)}{(2\varepsilon_R + \varepsilon_U)}\alpha_e\mathbf{E} \qquad (3.76)$$

The polarization per unit volume $\mathbf{P}$ is given by $\mathbf{P} = N\overline{\mathbf{m}}$, where N is the number of molecules per unit volume. Also

$$\mathbf{P} = \frac{(\varepsilon_R - 1)\mathbf{E}}{4\pi} \tag{3.77}$$

Hence from Equations (3.76), (3.77) and substituting for α_e using Equation (3.63)

$$4\pi N \frac{\mu^2}{3kT} = \frac{(2\varepsilon_R + 1)^2(\varepsilon_R - \varepsilon_U)}{3\varepsilon_R(2\varepsilon_R + \varepsilon_U)} \tag{3.78}$$

Converting μ into μ_0 according to its definition in Equation (3.70), the final Onsager relation is obtained

$$\mu_0^2 = \frac{3kT}{4\pi N} \frac{2\varepsilon_R + \varepsilon_U}{3\varepsilon_R} \left(\frac{3}{\varepsilon_U + 2}\right)^2 (\varepsilon_R - \varepsilon_U) \tag{3.79}$$

Equation (3.79) has been used very successfully in simple liquids. Values of μ_0 calculated using Equation (3.79) are usually in good agreement with those values obtained from measurements on very dilute gases using Equation (3.62). Examples of the application of the Onsager equation are given by Smyth (1955) and Smith (1955). As is seen from the Onsager model the molecules are not considered to have correlations in orientations. Thus Equation (3.79) applies only to systems of rigid non-associating molecules, and as a result cannot be applied to a condensed phase of polymer molecules.

3.4c The Kirkwood–Fröhlich Theories

Kirkwood (1939) developed a theory of the static dielectric constant which takes into account the short-range interactions between molecules in the liquid state. The following relation was obtained,

$$\frac{(\varepsilon_R - 1)(2\varepsilon_R + 1)}{9\varepsilon_R} = \frac{4}{3}\pi N \left[\alpha_e + \frac{g\mu^2}{3kT}\right] \tag{3.80}$$

μ is the moment of the molecule in the liquid state, and was related to the vacuum moment μ_0 by the Onsager relation defined in Equation (3.70). g is the Kirkwood correlation function and takes into account the short-range orientation correlations between a reference molecule and its nearest neighbours. If the reference molecule is surrounded by z equivalent nearest neighbours

$$g = 1 + z \overline{\cos \gamma} \tag{3.81}$$

$\cos \gamma$ is the average of the cosine of the angle γ made between the reference molecule and one of its z nearest neighbours. The deformation polarization term in Equation (3.80) was not included in Kirkwood's statistical mechanical theory, but was merely added in an empirical manner. As a result, for $g = 1$, that is, no orientation correlation, Equation (3.80) does not reduce (as it should do) to the Onsager Equation (3.79).

Fröhlich (1949) developed a very general dielectric theory which included the short-range correlations considered by Kirkwood, and also included the deformation polarization at the correct stage of the theory. Fröhlich considers an infinite homogeneous dielectric specimen and selects a macroscopic spherical region of volume V within the specimen. This region is regarded as a continuum of dielectric constant ε_U which contains elementary dipole units which follow the laws of statistical mechanics. The advantages of this approach are that the deformation polarization is included at the initial stage of the theory, and the cavity and reaction fields are treated in a macroscopic manner, which is more realistic than the model considered by Onsager (1936). This macroscopic sphere contains N_0 elementary dipole units, so the number of dipole units per unit volume $N = N_0/V$. Each elementary dipole unit contributes an equal amount to the average polarization of the sphere when an external field is applied. The general equation obtained by Fröhlich using this model was

$$(\varepsilon_R - \varepsilon_U) = \frac{3\varepsilon_R}{(2\varepsilon_R + \varepsilon_U)} 4\pi N \frac{\overline{\mathbf{m} \cdot \mathbf{m}^*}}{3kT} \tag{3.82}$$

$\mathbf{m}$ is the moment of a reference dipole unit fixed in a particular configuration. $\mathbf{m}^*$ is the average moment of the macroscopic sphere when the reference unit is fixed in the configuration which leads to $\mathbf{m}$. The average of $\mathbf{m} \cdot \mathbf{m}^*$ is taken over all configurations of the reference dipole unit.

In order to use the general relation (3.82) a molecular model must be specified. Fröhlich identified the dipole unit with a single spherical molecule. The molecule in a vacuum corresponds to a sphere of dielectric constant ε_U having a dipole moment μ at its centre. Hence, in Equation (3.82), $\mathbf{m} = \mu$ and $\mathbf{m}^* = \mu^*$, where μ^* is the vector sum of μ and the average of the sum of all the moments in the macroscopic sphere when the reference molecule is kept in the fixed configuration given by μ. Consider for example a reference molecule i fixed in a configuration having moment μ_i. Thus the product $\mu_i \cdot \mu^*_i$ is given by

$$\mu_i \cdot \mu^*_i = \mu^2 \left[1 + \sum_{\substack{j \\ j \neq i}}^{N_0} \overline{\cos \gamma_{ij}} \right] \tag{3.83}$$

μ is the magnitude of the dipole moment and $\overline{\cos \gamma_{ij}}$ is the average of the cosine of the angle γ_{ij} made between the reference molecule i and a molecule j in the sphere. Since all molecules contribute equally to the average polarization, the product $\mu_{i'} . \mu_{i'}^*$ for a different reference molecule i' will be identical with $\mu_i . \mu_i^*$. In the liquid all directions of a molecule are equally probable hence $\overline{\mu . \mu^*} = \mu . \mu^*$, and Equation (3.82) becomes with the aid of Equation (3.83)

$$(\varepsilon_R - \varepsilon_U) = \frac{3\varepsilon_R}{(2\varepsilon_R + \varepsilon_U)} \frac{4\pi N}{3kT} g\mu^2 \qquad (3.84)$$

Here the orientation correlation function g is given by

$$g = 1 + \sum_{\substack{j \\ j \neq i}}^{N_0} \overline{\cos \gamma_{ij}} \qquad (3.85)$$

The Fröhlich definition of μ differs from the definition of μ in the Onsager theory (see Fröhlich, 1949). Fröhlich gives the following relation between μ and the vacuum moment μ_0.

$$\mu = \mu_0 \frac{(\varepsilon_U + 2)}{3}: \qquad (3.86)$$

μ_0 is the external moment of the molecule when surrounded by vacuum. Combination of Equations (3.84) and (3.86) yields the Fröhlich equation

$$(\varepsilon_R - \varepsilon_U) = \frac{3\varepsilon_R}{(2\varepsilon_R + \varepsilon_U)} \cdot \frac{4\pi N}{3kT} \left(\frac{\varepsilon_U + 2}{3} \right)^2 g\mu_0^2 \qquad (3.87)$$

In the absence of orientation correlations between molecules, $g = 1$, and Equation (3.87) reduces to the Onsager equation (3.79). If each of z correlating dipole units adjacent to the reference dipole unit makes the same contribution to $\overline{\cos \gamma}$ then Equation (3.85) becomes Equation (3.81).

Fröhlich's consideration of dipole orientation correlations together with his realistic treatment of the deformation polarization represents an advance on the theories of Onsager and Kirkwood. It is noted, however, that for the special case $\varepsilon_R \gg \varepsilon_u$, the numerical values of $g\mu_0^2$ obtained from the Kirkwood equation (3.80) and from the Fröhlich equation (3.87) will not differ significantly.

Oster and Kirkwood (1943) calculated g for water by neglecting orientation correlations beyond the first shell of water molecules surrounding a

reference molecule. Assuming free rotation around unbendable hydrogen
bonds together with values for the coordination number z taken from
x-ray scattering data on liquid water, they calculated ε_R in the range 0
to 83°C, using the calculated g values and Equation (3.80). Fairly good
agreement was obtained between the calculated and experimentally
observed ε_R values. At 0°C, ε_R (calc) = 84·2, and ε_R (observed) = 88·0,
at 83°C, ε_R (calc) = 67·5 and ε_R (observed) = 59·9. Pople (1951) improved
the calculation of g by taking longer-range orientation correlations into
account, and also allowed the hydrogen bonds to bend. He found that
the ε_R values calculated using Equation (3.80) were lower than the experimental values, but the calculated temperature dependence of ε_R agreed
with the experiment value. Pople attributed the low ε_R values obtained
from the theory to an underestimation of the quantity μ/μ_0. Fröhlich
(1949) applied Equation (3.87) to liquid water, taking into account
association in the first shell only, and obtained a reasonable value for
ε_R at 23°C. Oster and Kirkwood (1943) also applied Equation (3.80)
to liquid alcohols with reasonable success.

Thus the Kirkwood theory, corrected by Fröhlich gives a satisfactory
description of the static dielectric constant of systems of small molecules
which possess intermolecular dipole orientation correlations. The
Fröhlich theory should not be applied to liquid polymer systems without
an appraisal of the structure of polymeric chains in relation to the specific
model required for the relation (3.82).

3.4d Application of the Fröhlich Theory to Polymer Systems

Consider an isolated polymer chain in a medium of dielectric constant
ε_R. Assuming that the deformation polarization of the elementary units
along the chain is the same as the deformation polarization of the medium,
the chain may be regarded as a succession of dipole units each having a
magnitude p. The chain will undergo Brownian motion and the total
moment of the chain $\mathbf{P}$ will fluctuate. At a given instant the moment $\mathbf{P}$
will be given by the vector sum of the n component moments of the chain
at that instant

$$\mathbf{P} = \sum_{1}^{n} \mathbf{p}_k \qquad (3.88)$$

The mean square moment $\overline{P^2}$ is given by

$$\overline{P^2} = \overline{\sum_{i=1}^{n} \mathbf{p}_i \sum_{j=1}^{n} \mathbf{p}_j}$$

$$= \sum_{i=1}^{n} \sum_{j=1}^{n} \overline{\mathbf{p}_i \cdot \mathbf{p}_j}$$

$$= p^2 \left[n + \sum_{\substack{i \\ i \neq j}}^{n} \sum_{j}^{n} \overline{\cos \gamma_{ij}} \right] = g_m p^2 \qquad (3.89)$$

Here $\overline{\cos \gamma_{ij}}$ is the average of the cosine of the angle γ_{ij} made between a dipole unit i and another dipole unit along the chain j. The form of $\overline{\cos \gamma_{ij}}$ depends upon such factors as valence-angle restriction, the tacticity of the chain and the steric hindrance to internal rotation of the chain. These factors have been considered by Volkenstein (1963).

In applying the Fröhlich theory of the static dielectric constant to polymer systems, the requirements of this general theory must be fulfilled. It is recalled from Section 3.4c that each basic unit in the macroscopic sphere must make the same average contribution to the polarization of the sphere. Therefore if the system is composed of monodisperse polymer molecules, where each molecule is distinct and separate from the other molecules, i.e. no interpenetration of molecules occurs, then the basic unit for application to the Fröhlich theory is a whole molecule. If, however, the polymer molecules interpenetrate, and form a three-dimensional 'network' of chains, the identity of a whole molecule is lost and it seems more appropriate to take a suitably chosen small repeat unit of the chain to be the basic unit in the Fröhlich theory.

Thus, on the one hand, the basic unit has a mean square moment $\overline{P^2}$, and correlations between dipole units are expressed in terms of Equation (3.89). On the other hand, the basic unit is the repeat unit of moment $\mathbf{p}$ which correlates with its neighbours along the chain and its neighbours from other chains. In Section 3.4e the basic unit is identified with the whole molecule, and in 3.4f the basic unit is identified with the repeat unit of the chain.

3.4e The Static Dielectric Constant of a System of Distinct Monodisperse Polymer Molecules

Consider an assembly of monodisperse polymer molecules where the molecules are 'distinct', i.e. there is no interpenetration of the spheres containing single molecules. The basic dipole unit in the Fröhlich theory is taken to be a single polymer molecule. Consider a macroscopic spherical region in the system, of volume V, containing N_0 molecules. At a particular instant in time a reference molecule i, say, will have a moment $\mathbf{P}_i$, which is identified with $\mathbf{m}_i$ in the Fröhlich relation, Equation

(3.82). The quantity $\mathbf{m}^*{}_i$ is given by

$$\mathbf{m}^*{}_i = \mathbf{P}_i + \sum \mathbf{P}_j \qquad (3.90)$$

Here the summation extends over all molecules in the macroscopic sphere except i, and the average is taken when molecule i is kept in its fixed configuration.

In the absence of correlations between polymer molecules,

$$\mathbf{m}_i \cdot \mathbf{m}^*{}_i = \mathbf{P}_i \cdot \mathbf{P}_i \qquad (3.91)$$

Averaging over all possible values of $\mathbf{P}_i$ gives

$$\overline{\mathbf{m}_i \cdot \mathbf{m}^*{}_i} = \overline{P_i{}^2} \qquad (3.92)$$

The same product is obtained for each molecule in the macroscopic sphere, so the subscript i is dropped and combination of Equations (3.82) and (3.92) gives

$$(\varepsilon_R - \varepsilon_U) = \frac{3\varepsilon_R}{(2\varepsilon_R + \varepsilon_U)} \frac{4\pi N}{3kT} \overline{P^2} \qquad (3.93)$$

ε_R is again the static dielectric constant of the liquid assembly, and comprises contributions from *all* orientation processes in the polymer. Now $\overline{P^2}$ of the molecule is defined according to Equation (3.89), and Equation (3.93) becomes

$$(\varepsilon_R - \varepsilon_U) = \frac{3\varepsilon_R}{(2\varepsilon_R + \varepsilon_U)} \cdot \frac{4\pi N}{3kT} g_m p^2 \qquad (3.94)$$

p is the moment of a repeat unit of the chain.

The repeat unit is assumed to be a spherical continuum having the dipole $\mathbf{p}$ at its centre. The moment of the repeat unit when surrounded by vacuum is $\mathbf{p}_0$ and is given by (Fröhlich, 1949)

$$\mathbf{p}_0 = \frac{3}{(\varepsilon_U + 2)} \mathbf{p} \qquad (3.95)$$

For n repeat units in the molecule, the number of repeat units per unit volume, $N_r = nN$ and Equations (3.94) and (3.95) give

$$(\varepsilon_R - \varepsilon_U) = \frac{3\varepsilon_R}{(2\varepsilon_R + \varepsilon_U)} \frac{4\pi N_r}{3kT} \left(\frac{\varepsilon_U + 2}{3}\right)^2 \left(\frac{g_m}{n}\right) p_0{}^2 \qquad (3.96)$$

(g_m/n) is the effective Kirkwood–Fröhlich g quantity per repeat unit of the chain, which will be abbreviated as g_r. From Equation (3.89)

$$g_r = \left(\frac{g_m}{n}\right) = 1 + \frac{1}{n} \sum_{\substack{i \\ i \neq j}}^{n} \sum_{j}^{n} \overline{\cos \gamma_{ij}} \tag{3.97}$$

It is seen from Equation (3.97) that g_r values different from unity arise due to the orientation correlations of the dipole units along the polymer chains.

For real polymer chains, the orientation correlations between units decreases rapidly with increasing separation of the units along the chain. If the polymer is of high molecular weight, end-group contributions may be ignored. If the n units of the chain contribute equally to $(g_r - 1)$, we may write

$$\sum_{\substack{i \\ i \neq j}}^{n} \sum_{j}^{n} \overline{\cos \gamma_{ij}} \simeq n \sum_{j=2}^{n} \overline{\cos \gamma_{1j}} \tag{3.98}$$

γ_{1j} is the angle between a unit of the chain (1) and a different unit j along the chain. In this special case g_r becomes g'_r

$$g'_r = 1 + \sum_{j=2}^{n} \overline{\cos \gamma_{1j}} \tag{3.99}$$

Equations (3.96) and (3.99) were obtained from the Fröhlich general theory for the special case of a polymer liquid with no correlation of orientations *between* polymer molecules, and are very similar in form to the Fröhlich relations, Equations (3.85) and (3.87), which are used for associated liquids of low molecular weight.

It would be unrealistic to apply Equation (3.96) and (3.99) to real polymer liquids or rubbers since the entanglement of chains is not allowed in the model. Therefore, Equations (3.96) and (3.99) only apply to mono-disperse polymer molecules in dilute solution where the condition of distinct and separate polymer molecules is fulfilled. It is noted that Equations (3.96) and (3.97) are general to isotactic, syndiotactic and atactic polymer molecules, since the basic unit in the theory is taken as the whole molecule. Tacticity considerations apply when the actual value of g_r is computed separately using a molecular model of the chain. Volkenstein (1963) has given a comprehensive account of the evaluation of $\overline{P^2}$ for model chains of different chemical structure and tacticity.

It is possible to derive Equations (3.96) and (3.99) by taking the basic dipole unit in the Fröhlich theory to be the repeat unit of the chain. In order to satisfy the Fröhlich theory each basic unit must make the same contribution to the average polarization of the macroscopic sphere in the presence of an applied field. The repeat unit will only satisfy this condition if the molecular weight is large so that repeat units near the end of the chain can be ignored, and also the orientation correlations between two repeat units along the chain should decrease rapidly with increasing separation of the units along the chain. Each repeat unit is assumed to contribute equally to the average polarization, so tacticity effects are excluded in the following derivation.

For this model, the moment $\mathbf{p}_i$ of a particular repeat unit is identified with $\mathbf{m}_i$ in the Fröhlich theory. $\mathbf{m}^*_i = \mathbf{p}^*_i$ is given by

$$\mathbf{p}^*_i = \mathbf{p}_i + \sum_{\substack{j \\ j \neq i}} \mathbf{p}_j \tag{3.100}$$

Here the summation is taken over all units of the macroscopic sphere excluding the reference unit i. The average is taken over all configurations of the j units keeping the reference unit i in the fixed configuration.

Since the model used for the derivation of Equations (3.96) and (3.99) excludes intermolecular orientation correlations, it follows that the summation in Equation (3.100) is taken only over units within the molecule which contains the reference unit.

Hence,

$$\mathbf{m}_i \cdot \mathbf{m}^*_i = \mathbf{p}_i \cdot \mathbf{p}^*_i = p^2 \left[1 + \sum_{\substack{j \\ j \neq i}}^n \overline{\cos \gamma_{ij}} \right] \tag{3.101}$$

γ_{ij} is the angle made between the reference unit i and the jth unit along the chain. Since in a liquid all dipolar directions are equivalent, and also since the only variable is the direction of the reference unit, it follows that

$$\overline{\mathbf{m}_i \cdot \mathbf{m}^*_i} = \mathbf{m}_i \cdot \overline{\mathbf{m}^*_i} \tag{3.102}$$

All units are equivalent, thus the subscript i may be dropped and Equations (3.95), (3.101) and (3.102) together with Equations (3.82) give

$$(\varepsilon_R - \varepsilon_U) = \frac{3\varepsilon_R}{(2\varepsilon_R + \varepsilon_U)} \frac{4\pi N_r}{3kT} \left(\frac{\varepsilon_U + 2}{3} \right)^2 g'_r p_0^2 \tag{3.103}$$

$$g'_r = 1 + \sum_{j=2}^n \overline{\cos \gamma_{1j}} \tag{3.104}$$

Equations (3.103) and (3.104) are equivalent to Equations (3.96) and (3.99) respectively. It is emphasized that Equation (3.99) is a special case of the more general relation, Equation (3.97).

3.4f   The Static Dielectric Constant of a System of Entangled Polymer Molecules

A real polymer liquid or rubber will consist of an entangled network of polymer chains. Under these circumstances the identity of a single polymer molecule is lost, and it seems more appropriate in the application of the Fröhlich theory to these systems to choose a small unit of the chain as the basic dipole unit in the theory. Each basic unit must contribute equally to the average polarization of a macroscopic sphere of the dielectric in an applied field.

We shall choose the basic dipole unit to be a chemical repeat unit (or group within the repeat unit) of the chain. For a wholly isotactic chain (... *ddddd* ... or ... *llll* ...) each chemical repeat unit will make the same contribution to the polarization. For the syndiotactic chain (... *dldldl* ...) this would not be true. For the atactic chain (... *dlldddl* ...) the random nature of the stereoregularity means that each repeat unit will not have the same g_r factor and therefore will not conform to the basic requirement of the Fröhlich theory.

The model is chosen to exclude tacticity considerations. The polymer chain is assumed to have n basic dipole units each of which make the same contribution to the polarization. This requirement is satisfied if $\overline{\cos \gamma_{1j}} \ll 1$ for $|1 - j| \gg 1$, and if the polymer is of high molecular weight so that groups near the end of the chain do not contribute significantly.

Consider a macroscopic sphere of volume V containing N_0 basic dipole units. The moment of a reference unit i is $\mathbf{p}_i$ at a given instant in time. $\mathbf{p}^*_i$ is given by the sum of $\mathbf{p}_i$ and the average of the sum of the $(N_0 - 1)$ basic units in the macroscopic sphere when the reference unit i is kept in the configuration corresponding to $\mathbf{p}_i$.

The reference unit i is part of a molecule 1 say, so we write $\mathbf{p}_i = {}^{\mathrm{I}}\mathbf{p}_i$ and $\mathbf{p}^*_i = {}^{\mathrm{I}}\mathbf{p}^*_i$.

$$^{\mathrm{I}}\mathbf{p}_i \cdot {}^{\mathrm{I}}\mathbf{p}^*_i = {}^{\mathrm{I}}\mathbf{p}_i \left[\overline{{}^{\mathrm{I}}\mathbf{p}_i + \sum_{\substack{j \\ j \neq i}}^{n} {}^{\mathrm{I}}\mathbf{p}_j + \sum_j {}^{\mathrm{II}}\mathbf{p}_j} \right] \tag{3.105}$$

Here $\sum_j {}^{\mathrm{I}}\mathbf{p}_j$ refers to basic units within the molecule I to which i belongs and $\sum_j {}^{\mathrm{II}}\mathbf{p}_j$ refers to units from all other molecules in the macroscopic sphere. Since all basic units are assumed to have the same dipole moment magnitude, $|\mathbf{p}| = p$, Equation (3.105) becomes

$$\mathbf{^I p}_i \cdot \mathbf{^I p^*}_i = p^2 \left[1 + \sum_{\substack{j \\ j \neq i}}^{n} \mathbf{^I \overline{\cos \gamma_{ij}}} + \sum_{j} \mathbf{^{II} \overline{\cos \gamma_{ij}}} \right] \tag{3.106}$$

Here $^I\overline{\cos \gamma_{ij}}$ is the average of the cosine of the angle γ_{ij} made between the reference unit i and a unit j within the same polymer chain. $^{II}\overline{\cos \gamma_{ij}}$ is the average of the cosine of the angle γ_{ij} made between the reference unit i and a unit j which does not belong to the polymer chain which contains the reference unit i.

Since in a liquid all directions of the unit i are equally probable, and we are only concerned with the direction of $^I\mathbf{p}_i$, it follows that $\overline{^I\mathbf{p}_i \cdot {}^I\mathbf{p^*}_i} = {}^I\mathbf{p}_i \cdot {}^I\mathbf{p^*}_i$. Also each unit makes the same contribution so the subscript i can be dropped. Insertion of Equation (3.106) into (3.82) where $\mathbf{m} \cdot \mathbf{m^*} = \mathbf{p}_i \cdot \mathbf{p^*}_i$, gives, together with Equation (3.95),

$$(\varepsilon_R - \varepsilon_U) = \frac{3\varepsilon_R}{(2\varepsilon_R + \varepsilon_U)} \frac{4\pi N_r}{3kT} \left(\frac{\varepsilon_U + 2}{3} \right)^2 g_r p_0^2 \tag{3.107}$$

where now

$$g_r = 1 + \sum_{j=2}^{n} {}^I\overline{\cos \gamma_{1j}} + \sum_{j} {}^{II}\overline{\cos \gamma_{1j}} \tag{3.108}$$

For the special case where intermolecular correlations of orientations do not occur. Equation (3.107) is modified so that $g_r = g'_r$ as defined in Equation (3.99). For real polymer chains this is likely to be the case, and also the intramolecular term in Equation (3.108) will probably extend over only a few units of the chain adjacent to the reference unit. Thus the application of the Fröhlich theory to a particular model for a liquid polymer gives a relation very similar to that obtained for associating small molecules in the liquid state, i.e. Equations (3.85) and (3.87) above.

3.4g   Effect of Molecular-Weight Distribution

In Section 3.4e, Equations (3.96) and (3.97) were derived for a monodisperse polymer. In the case of a distribution of molecular weights, the system envisaged in Section 3.4e can be treated as a mixture of species, and the static dielectric constant of the mixture is given by (Fröhlich, 1949)

$$(\varepsilon_R - \varepsilon_U) = \frac{3\varepsilon_R}{(2\varepsilon_R + \varepsilon_U)} \frac{4\pi}{3kT} \sum_{s=1}^{z} N_s \overline{\mathbf{m}_s \cdot \mathbf{m^*}_s} \tag{3.109}$$

m^*_s is the average moment of a spherical region of the substance embedded in its own medium if one unit of the sth kind is kept in a configuration corresponding to a moment m_s. There are z different species (polymer molecules say) and N_s is the number of the sth kind per unit volume. From Equations (3.89), (3.92) and (3.97) we may write

$$N_s \cdot \overline{m_s \cdot m^*_s} = (N_r)_s \cdot (g_r)_s p^2 \qquad (3.110)$$

Here $(N_r)_s$ is the number of repeat units per unit volume for molecules of the s kind, $(g_r)_s$ is the effective orientation correlation function for a repeat unit as defined by Equation (3.97) and p is the moment of one repeat unit for polymers of high molecular weight. For polymers of high molecular weight $(g_r)_s$ will be largely determined by short-range correlations, therefore $(g_r)_s$ is essentially the same for all molecular weights. Since $\sum_s (N_r)_s = N_r$ the total number of repeat units per unit volume, Equation (3.109) and (3.110) correspond to Equation (3.94). Thus a molecular-weight distribution for a high molecular weight polymer should not affect the dielectric constant of our model system.

On the other hand, a molecular-weight distribution can be considered in relation to the model of a liquid polymer given in Section 3.4f. Since the basic unit is defined as a repeat unit of the chain, and long-range orientation correlations within a chain are excluded, it follows that a molecular-weight distribution for a high average molecular weight polymer has no effect in our model on the calculated value of the static dielectric constant.

3.5   CALCULATION OF THE MEAN SQUARE DIPOLE MOMENT OF POLYMER MOLECULES

In the previous section we have given equations by means of which $\overline{P^2}$ for a polymer molecule can be evaluated from the measured values of ε_R and ε_U. It is also possible to derive theoretical expressions for $\overline{P^2}$ in terms of the dipole moment of the repeat unit (p) and geometrical parameters such as the main-chain valence angles (α) and averages such as $\eta = \overline{\cos\phi}$ and $\varepsilon = \overline{\sin\phi}$. Here ϕ is the angle, measured from the *trans* conformation, which describes the rotation around main-chain bonds (see Figure 2.3). Knowing ϕ and α, quantities such as η or ε can be obtained from a comparison of theoretical and experimental values of $\overline{P^2}$. These quantities provide some insight into the equilibrium conformation of the polymer molecule.

Calculations of the above kind involve the vector addition of the individual chain dipoles and proceed from Equation (3.89). Eyring (1932) was the first to consider these calculations in a general way and he obtained a result for chains with fixed valence angles and free internal rotation ($\eta = \varepsilon = 0$) and for which the bond vector and the dipole moment vector coincided. Subsequently Fuoss and Kirkwood (1941) derived an equation for the dipole moment of polyvinyl chloride (considering also the case of free rotation) in which the dipole vectors p_i have different directions from the bond vectors. However, their calculations have been criticized by Debye and Bueche (1951) and by Volkenstein (1963). Later studies of this kind include those of Kuhn (1948) who considered a freely jointed Gaussian chain and Debye and Bueche (1951) who studied poly-p-chlorostyrene and took into account the restriction to rotation by assuming that the hindering potential could be approximated by a rectangular well.

Most recent calculations of $\overline{P^2}$ have involved matrix methods as described in detail by Volkenstein (1963). Results of such calculations are conveniently illustrated by reference to two classes of polymers, namely (a) the oxide polymers of general formula $[-(CH_2)_x-O-]_n$ and (b) polymers of general formula $[-CH_2-CHR-]_n$, such as polyvinyl chloride and polyhalostyrenes.

The oxide polymers are of interest because the C—O bond dipole vectors are situated directly within the chain backbone. The C—C bonds are, of course, non-polar and do not contribute to $\overline{P^2}$. Since the C—O—C and O—C—O valence angles are each close to tetrahedral, we may assume that they each equal α. It is also reasonable to assume that the hindering potential is a symmetrical function of ϕ, such that $\overline{\sin\phi} = \varepsilon = 0$, since the carbon atoms are symmetrically substituted with hydrogen. If the rotational state is the same for both C—O and C—C bonds, and rotations around each bond are assumed to be independent, we have obtained (Read, 1965) the following results (for $n \gg 1$ and $|\eta| < 1$).

$x = 1$, *polymethylene oxide*

$$\frac{\overline{P^2}}{n} = 2k^2 \frac{(1+\cos\alpha)}{(1-\cos\alpha)} \frac{(1+\eta)}{(1-\eta)} \tag{3.111}$$

$x = 2$, *polyethylene oxide*

$$\frac{\overline{P^2}}{n} = 2k^2 \frac{(1+\cos\alpha)(1-\eta^2)(1-\cos\alpha)}{3\eta + (1-\eta)^2(1-\cos\alpha+\cos^2\alpha)} \tag{3.112}$$

$x = 3$, *polytrimethylene oxide*

$$\frac{\overline{P^2}}{n} = 2k^2(1+\cos\alpha)\frac{(1+\eta)}{(1-\eta)}\frac{(1+\eta^2+2\eta\cos\alpha+(1-\eta)^2\cos^3\alpha)}{[(1+\eta)^2-4\eta\cos^2\alpha-(1-\eta)^2\cos^4\alpha]} \quad (3.113)$$

$x = 4$, *polytetramethylene oxide*

$$\frac{\overline{P^2}}{n} = 2k^2(1+\cos\alpha)(1-\eta^2)(1-\eta) \times$$

$$\left(\frac{1+\eta+\eta^2-3\eta\cos^2\alpha-(1-\eta)^2\cos^4\alpha}{1-\eta^5+5\eta^2(1-\eta)\cos\alpha+5\eta(1-\eta)^3\cos^3\alpha+(1-\eta)^5\cos^5\alpha}\right)(3.114)$$

In these equations k is the dipole moment of the C—O bond. Each of these equations is analogous to Equation (3.41) which relates the mean square end-to-end distance r^2 to n, l, $\cos\alpha$ and η. Since $p^2 = 2k^2(1+\cos\alpha)$, it follows that Equations (3.111) to (3.114) may each be rewritten in the form (cf. Equations 3.89 and 3.97)

$$\overline{P^2} = ng_r p^2 \quad (3.115)$$

where the correlation parameter g_r is now related to α and η.

Figure 3.7 shows plots of $\overline{P^2}/2nk^2$ against η for $x = 1, 2, 3$ and 4. It will be noted that as η tends to unity (*trans* conformation), $\overline{P^2}/2nk^2$ tends to a very large value for $x = 1$ and 3 and to zero for $x = 2$ and 4. This result follows from the fact that in the extended *trans* conformation, the net

group dipoles (p) add when x is odd and cancel for even x. Equation (3.112) was earlier derived by Marchal and Benoit (1955) and employed in an experimental study of polyethylene oxide which gave $\eta = 0.3$. Equations (3.111) and (3.112) have also been used by Williams (1963a, 1965) in an investigation of the dielectric properties of amorphous polyacetaldehyde and polypropylene oxide. From the values of η obtained in these studies it was concluded that the *trans* conformation had a lower energy than the gauche for both these polymers, the energy differences being about 1 and 0.6 kcal/mole, respectively. More recently Wetton and Williams (1965) have studied amorphous polytetramethylene oxide, and from their experimental value of $\overline{P^2}/n$ they obtained $\eta = 0.54$ to 0.63 at $20°$c. From

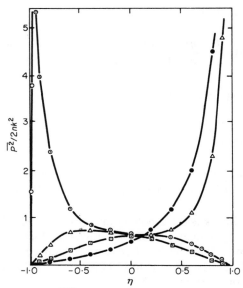

Figure 3.7. Plots of $\overline{P^2}/2nk^2$ against η for $x = 1$ ($\bullet$), $x = 2$ ($\square$), $x = 3$ ($\triangle$) and $x = 4$ ($\odot$) (After Read, 1965).

this result it was estimated that the *trans* conformation was more favourable than the *gauche* by about 0·8 to 1·0 kcal/mole. It should be emphasized that Equations (3.111) to (3.114) are valid only when rotations around neighbouring bonds are independent. This is probably not so for polymethylene oxide ($x = 1$) which forms a predominantly helical conformation (Uchida, Kurita and Kubo, 1956) suggestive of correlations between neighbouring internal rotations.

Polymers of the type $[\text{---}CH_2\text{---}CHR\text{---}]_n$ are interesting in view of the fact that they may exist in different stereoregular forms. Since alternate chain carbon atoms are asymmetrically substituted it cannot be assumed that $\sin \phi = 0$ for these polymers. Volkenstein (1963) has derived expressions for $\overline{P^2}$ for these polymers, for both the isotactic and syndiotactic configurations, in terms of n, p, η and ε. He has also described in detail the calculations of Ptitsyn and Sharonov (1957) in which the effect of the correlation of internal rotations is examined. By taking these correlations into account, better agreement was obtained with experimental data on poly-*p*-halostyrenes.

It should finally be remarked that the above calculations of $\overline{P^2}$ refer to isolated polymer molecules. For comparison of theory and experiment, it is therefore preferable that the experimental determinations of $\overline{P^2}$ be made on dilute polymer solutions in a non-polar solvent. Also, in order that the chain conformations are imperturbed by interactions with the solvent, a so-called θ solvent (Flory, 1953) should be employed. Unfortunately, most dipole moment studies have not so far been performed in a θ solvent. In the case of amorphous bulk polymers, the chain molecules are not isolated but are packed together in the (disordered) condensed system. In the rubber-like state, however, the chains are thought, according to one point of view, to assume their unperturbed conformations (Flory, 1953), and on this basis it would seem reasonable to compare the theoretical expressions for $\overline{P^2}$ with the experimental values determined from the dielectric glass–rubber relaxation magnitude.

4

Phenomenological Theories of
Mechanical and Dielectric Relaxation

The object of a phenomenological theory is not to discover the relaxation properties of a solid in an altogether *a priori* fashion, but rather to predict behaviour under certain circumstances, having observed it under others. For instance, to predict the relationship between a creep curve, $J(t)$, and the frequency dependence of the real part of the dynamic compliance, $J'(\omega)$. This was done for the single relaxation time model in Chapter 1. In Section 4.1 the relationships between the other mechanical parameters and between the dielectric parameters are developed for the single relaxation time model. This section also includes a description of several analytical methods as well as the generalization of the single relaxation time theory to account for the broad relaxations normally observed in polymeric solids. Section 4.2 describes ways in which relaxation parameters may be correlated when they are determined at different temperatures. For instance, to predict the relationship between a creep curve $J^T(t)$ measured at temperature T, and the creep curve $J^{T_0}(t)$ for another temperature T_0. Section 4.3 contains a brief account of the analysis of the mechanical relaxation magnitude in crystalline polymers.

4.1 TIME AND FREQUENCY AS VARIABLES: THE THEORY OF LINEAR RELAXATION BEHAVIOUR

4.1a Superposition Principle

The superposition principle was first used by Boltzmann (1874) in the analysis of creep and stress relaxation properties of solids. Hopkinson (1877) was the first to apply the principle to the problem of residual charge in dielectrics. It turns out that the vast majority of solids when studied at small enough strains (or electric displacements) conform to the superposition principle. For a polymer such as polyethylene at room temperature the strain in a mechanical experiment would have to be below ca. 10^{-3}. For aluminium at room temperature the strain would have to be somewhat lower. There is, nevertheless, for most solids, a strain below which the superposition principle holds, and we shall restrict our attention solely to this region.

Consider a specimen to which is applied a step stress σ_0 at time $t = 0$. The resulting strain can be written (see Chapter 1)

$$\gamma(t) = \sigma_0[J_U + (J_R - J_U)\psi_\sigma(t)] \tag{4.1a}$$

By definition of J_U and J_R

$$\psi_\sigma(t) = 0 \quad \text{for} \quad t = 0 \tag{4.1b}$$

$$\psi_\sigma(t) = 1 \quad \text{for} \quad t = \infty \tag{4.1c}$$

$\psi_\sigma(t)$ is known as the normalized creep function.

For the single relaxation time model $\psi_\sigma(t)$ has the form (see Equation 1.30)

$$\psi_\sigma(t) = (1 - e^{-t/\tau_\sigma}) \tag{4.2}$$

The time constant τ in Equation (1.30) is replaced here by τ_σ (the subscript σ implies that this is the time constant for an experiment in which the stress is constant) and is known as the retardation time. The analogous time constant for a stress relaxation experiment in which the strain is maintained constant is the relaxation time τ_y. The necessity for the distinction between τ_σ and τ_y is that normally $\tau_\sigma \neq \tau_y$, as will appear below (Equation 4.12d).

What can we say about the strain produced at time t by a stress applied at $t = 0$ but which at time $u\,(0 < u < t)$ is not a constant as in Equation (4.1a); that is, for a variable stress $\sigma(u)$? This question can be answered by means of the superposition principle.

Suppose that at time u the stress is increased by $d\sigma$ and that this stress increment causes a strain increment $d\gamma$ at the later time t, given by Equation (4.1a).

$$d\gamma = J_U\,d\sigma + (J_R - J_U)\psi_\sigma(t - u)\,d\sigma \tag{4.3}$$

According to the superposition principle the total strain at time t is a superposition of all the increments $d\gamma$ so that

$$\gamma(t) = J_U\sigma(t) + (J_R - J_U)\int_{-\infty}^{t} \frac{d\sigma(u)}{du}\,\psi_\sigma(t - u)\,du \tag{4.4}$$

Integration by parts gives

$$\gamma(t) = J_U \sigma(t) + (J_R - J_U) \int_{-\infty}^{t} \sigma(u)\alpha_\sigma(t-u)\,\mathrm{d}u \qquad (4.5a)$$

where

$$\alpha_\sigma(t) = \frac{\mathrm{d}\psi_\sigma(t)}{\mathrm{d}t} \qquad (4.5b)$$

In a similar way the dielectric behaviour can be expressed in terms of the superposition principle. The dielectric displacement $D(t)$ at any instant due to a complex electric field history is the sum of displacements arising from the incremental fields applied at times $u \leqslant t$. By a precisely analogous argument to that used above for the mechanical case, we have

$$D(t) = \varepsilon_U E(t) + (\varepsilon_R - \varepsilon_U) \int_{-\infty}^{t} E(u)\alpha_E(t-u)\,\mathrm{d}u \qquad (4.6)$$

$\alpha_E(t) = \mathrm{d}\psi_E(t)/\mathrm{d}t$, and is called the dielectric decay function. $\psi_E(t)$ is the dielectric function which describes the approach of $D(t)$ to equilibrium in the presence of a constant field, and is analogous to $\psi_\sigma(t)$, Equation (4.2). In Equation (4.6) the term $\varepsilon_U E(t)$ corresponds to the instantaneous displacement arising from the electronic distortion and atomic polarization of molecules. The second term in (4.6) is the time-dependent displacement due to the rotation of dipoles.

4.1b Single Relaxation Time

For a single relaxation time model it follows from Equations (4.5b) and (4.2) that

$$\alpha_\sigma(t) = \frac{1}{\tau_\sigma} e^{-t/\tau_\sigma} \qquad (4.7)$$

Differentiating Equation (4.5a) with respect to t

$$\frac{\mathrm{d}\gamma(t)}{\mathrm{d}t} = J_U \frac{\mathrm{d}\sigma(t)}{\mathrm{d}t} + (J_R - J_U)\frac{\mathrm{d}}{\mathrm{d}t}\int_{-\infty}^{t} \sigma(u)\alpha_\sigma(t-u)\,\mathrm{d}u \qquad (4.8)$$

From Equation (4.7) it follows that

$$\frac{\mathrm{d}}{\mathrm{d}t}\int_{-\infty}^{t} \sigma(u)\alpha_\sigma(t-u)\,\mathrm{d}u = \sigma(t)\alpha_\sigma(0) - \int_{-\infty}^{t} \frac{1}{\tau_\sigma}\sigma(u)\alpha_\sigma(t-u)\,\mathrm{d}u \qquad (4.9)$$

Since, from Equation (4.7), $\alpha_\sigma(0) = 1/\tau_\sigma$, using Equations (4.5a), (4.8) and (4.9), we have

$$\tau_\sigma \frac{d\gamma(t)}{dt} + \gamma(t) = \tau_\sigma J_U \frac{d\sigma(t)}{dt} + J_R \sigma(t) \tag{4.10}$$

Equation (4.10) is the differential equation which relates $\gamma(t)$ and $\sigma(t)$ for the special case of a single relaxation time.

Equation (4.10) may be solved for three special circumstances.

(a) The stress level is constant at σ_0, $d\sigma/dt = 0$, and the solution is

$$J(t) = \frac{\gamma(t)}{\sigma_0} = J_U + (J_R - J_U)\left(1 - \exp -\frac{t}{\tau_\sigma}\right) \tag{4.11}$$

(b) The strain level is constant at γ_0, $d\gamma/dt = 0$, and Equation (4.10) becomes for this case

$$\gamma_0 = \tau_\sigma J_U \frac{d\sigma(t)}{dt} + J_R \sigma(t)$$

The solution is

$$\frac{\sigma(t)}{\gamma_0} = G(t) = G_R + (G_U - G_R)\exp -\frac{t}{\tau_\gamma} \tag{4.12a}$$

in which

$$G_R = J_R^{-1} \tag{4.12b}$$

$$G_U = J_U^{-1} \tag{4.12c}$$

$$\frac{\tau_\sigma}{\tau_\gamma} = \frac{G_U}{G_R} \tag{4.12d}$$

These equations give the relationship between τ_σ and τ_γ. Note that τ_σ exceeds τ_γ. Figure 4.1 shows $G(t)$ and $J(t)$ for a model with $G_U = 1.0 \times 10^{10}$ dyn/cm^2, $G_U/G_R = 10$ and $\tau_\gamma = 10$ seconds. Note the inflections in the $G(t)$ and $J(t)$ curves occur at τ_γ and τ_σ respectively.

(c) For an applied alternating stress $\sigma(t) = \sigma_0 \exp i\omega t$ the accompanying strain will be $\gamma(t) = \gamma_0 \exp i(\omega t - \delta_G)$. Substitution into Equation (4.10) gives

$$J^* = \frac{\gamma(t)}{\sigma(t)} = J_U + \frac{(J_R - J_U)}{1 + i\omega\tau_\sigma} \tag{4.13}$$

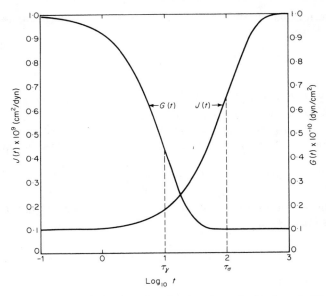

Figure 4.1. $G(t)$ and $J(t)$ plotted against log t for a single relaxation time
model with $G_U/G_R = 10$ and $\tau_\gamma = 10$ sec.

$$G^* = \frac{\sigma(t)}{\gamma(t)} = G_U - \frac{G_U - G_R}{1 + i\omega\tau_\gamma} \tag{4.14}$$

Equations (4.13) and (4.14) are the single relaxation time expressions for
J^* and G^*. Since $J^* = J' - iJ''$ and $G^* = G' + iG''$, we have

$$J' = J_U + \frac{(J_R - J_U)}{1 + \omega^2\tau_\sigma^2} \tag{4.15a}$$

$$J'' = \frac{(J_R - J_U)\omega\tau_\sigma}{1 + \omega^2\tau_\sigma^2} \tag{4.15b}$$

$$G' = G_R + \frac{(G_U - G_R)\omega^2\tau_\gamma^2}{1 + \omega^2\tau_\gamma^2} \tag{4.16a}$$

$$G'' = \frac{(G_U - G_R)\omega\tau_\gamma}{1 + \omega^2\tau_\gamma^2} \tag{4.16b}$$

In Figure 4.2, G' and G'' are plotted against log ω for a model with
$G_U = 1\cdot0 \times 10^{10}$ dyn/cm^2, $G_U/G_R = 10$ and $\tau_\gamma = 10$ seconds. The G''

curve reaches a maximum at $\omega_{max}\tau_y = 1$, and at this angular frequency

$$G' = \frac{(G_U + G_R)}{2} \qquad G''_{max} = \frac{(G_U - G_R)}{2}$$

Also shown in Figure 4.2 are plots of J' and J''.

The loss tangent $\tan \delta_G = J''/J' = G''/G'$ is obtained either from Equations (4.15) or Equations (4.16).

$$\tan \delta_G = \frac{(J_R - J_U)\omega\tau_\sigma}{J_R + J_U\omega^2\tau_\sigma^2} = \frac{(G_U - G_R)\omega\tau_y}{G_R + G_U\omega^2\tau_y^2} \qquad (4.17)$$

If we define an additional time constant $\bar{\tau}_G = (\tau_\sigma\tau_y)^{1/2}$ we obtain using Equation (4.15) (Zener, 1948)

$$\tan \delta_G = \frac{(J_R - J_U)\omega\bar{\tau}_G}{(J_R J_U)^{1/2}(1 + \omega^2\bar{\tau}_G^2)} = \frac{(G_U - G_R)}{(G_U G_R)^{1/2}} \frac{\omega\bar{\tau}_G}{(1 + \omega^2\bar{\tau}_G^2)} \qquad (4.18)$$

Taking $d(\tan \delta_G)/d\omega = 0$ we see that the maximum value of $\tan \delta_G$ (see Figure 4.2) occurs at a frequency given by

$$\omega_{max} = \frac{1}{\bar{\tau}_G} = \left(\frac{J_R}{J_U}\right)^{1/2}\frac{1}{\tau_\sigma} = \left(\frac{G_R}{G_U}\right)^{1/2}\frac{1}{\tau_y}$$

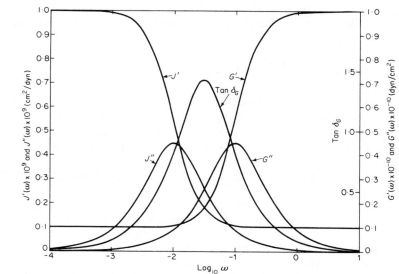

Figure 4.2. $G'(\omega)$, $G''(\omega)$, $J'(\omega)$, $J''(\omega)$ and $\tan \delta_G$ plotted against $\log \omega$ for a single relaxation time model with $G_U/G_R = 10$ and $\tau_y = 10$ sec.

The maximum value of tan δ_G is equal to

$$\frac{(J_R - J_U)}{2(J_U J_R)^{1/2}} \quad \text{or} \quad \frac{(G_U - G_R)}{2(G_U G_R)^{1/2}}$$

It should be emphasized that the maxima for the J'', G'' and tan δ_G curves occur at different frequencies (see Figure 4.2). If J'', G'' and tan δ_G are plotted as a function of temperature at constant frequency (See Section 1.4) then the maxima occur at different temperatures. The G'' maximum occurs at the lowest temperature and the J'' at the highest.

The dynamic properties can be obtained from the superposition Equation (4.5a), by an alternative and more direct method. For example, making the substitution $x = (t - u)$, Equation (4.5a) becomes

$$\gamma(t) = J_U \sigma(t) + (J_R - J_U) \int_0^\infty \sigma(t - x)\alpha_\sigma(x)\, dx \qquad (4.19)$$

For an alternating stress $\sigma(t) = \sigma_0 \exp i\omega t$ we have

$$\frac{\gamma(t)}{\sigma(t)} = J^* = J_U + (J_R - J_U) \int_0^\infty \exp(-i\omega x)\alpha_\sigma(x)\, dx \qquad (4.20)$$

For a single relaxation time $\alpha_\sigma(x)$ is given by Equation (4.7) and Equation (4.20) yields Equation (4.13).

For dielectric relaxation the analogue of Equation (4.10) is obtained from Equation (4.6).

$$\tau_E \frac{dD(t)}{dt} + D(t) = \tau_E \varepsilon_U \frac{dE(t)}{dt} + \varepsilon_R E(t) \qquad (4.21)$$

Equation (4.21) may now be used to study the approach to equilibrium of a condenser when either (a) the field (voltage) or (b) the displacement (true charge, Section 1.1) is held constant. It may also be solved to give (c) the frequency dependence of the complex dielectric constant when an a.c. voltage is applied to the condenser plates.

Case (a). We have $E(t) = E_0$ and $dE(t)/dt = 0$. The solution of Equation (4.21) is

$$D(t) = E_0 \{\varepsilon_U + (\varepsilon_R - \varepsilon_U)(1 - \exp - t/\tau_E)\} \qquad (4.22)$$

Case (b). At constant charge we have $D(t) = D_0$, $dD(t)/dt = 0$. The solution to Equation (4.21) is

$$E(t) = D_0 \{\varepsilon_R^{-1} + (\varepsilon_U^{-1} - \varepsilon_R^{-1}) \exp - t/\tau_D\} \qquad (4.23)$$

where

$$\tau_D = \frac{\varepsilon_U}{\varepsilon_R} \tau_E \tag{4.24}$$

Note that different time constants τ_E and τ_D are required to describe the variation of displacement at constant field and the variation of field at constant displacement respectively. Since, however, the field is nearly always prescribed in dielectric experiments, the most useful time constant is τ_E, known generally as the dielectric relaxation time. Thus the dielectric relaxation time is analogous to the mechanical retardation time (i.e. the first governs the constant voltage experiment, the second the constant stress experiment). The dielectric time constant τ_D has not a specific name but is the analogue of τ_γ, the mechanical relaxation time.

Case (c). An a.c. voltage applied to the condenser may be written $E(t) = E_0 \exp(i\omega t)$ and the resulting displacement as $D(t) = D_0 \exp i(\omega t - \delta_\varepsilon)$. Introducing these expressions into Equations (4.21) leads to

$$\varepsilon^* = \varepsilon_U + \frac{(\varepsilon_R - \varepsilon_U)}{1 + i\omega\tau_E} \tag{4.25}$$

Separating the real imaginary components of $\varepsilon^* = \varepsilon' - i\varepsilon''$ we obtain

$$\varepsilon' = \varepsilon_U + \frac{(\varepsilon_R - \varepsilon_U)}{1 + \omega^2\tau_E^2} \tag{4.26}$$

$$\varepsilon'' = (\varepsilon_R - \varepsilon_U)\frac{\omega\tau_E}{1 + \omega^2\tau_E^2} \tag{4.27}$$

ε' and ε'' are shown as a function of $\log \omega$ in Figure 4.3 for a model with $\varepsilon_R = 10$, $\varepsilon_R/\varepsilon_U = 5$ and $\tau_E = 10^{-4}$ seconds. ε'' shows a maximum ($d\varepsilon''/d\omega = 0$) at a frequency given by $\omega_{max} = 1/\tau_E$ and at this frequency $\varepsilon''_{max} = (\varepsilon_R - \varepsilon_U)/2$ and $\varepsilon' = (\varepsilon_R + \varepsilon_U)/2$. The dielectric loss tangent is given by

$$\tan \delta_\varepsilon = \frac{\varepsilon''}{\varepsilon'} = \frac{(\varepsilon_R - \varepsilon_U)\omega\tau_E}{\varepsilon_R + \varepsilon_U\omega^2\tau_E^2} \tag{4.28}$$

or from (4.26) and (4.27)

$$\tan \delta_\varepsilon = \frac{(\varepsilon_R - \varepsilon_U)}{(\varepsilon_R\varepsilon_U)^{1/2}} \frac{\omega\bar{\tau}_\varepsilon}{1 + \omega^2\bar{\tau}_\varepsilon^2} \tag{4.29}$$

where $\bar{\tau}_\varepsilon = (\tau_E\tau_D)^{1/2} = \tau_E(\varepsilon_U/\varepsilon_R)^{1/2}$.

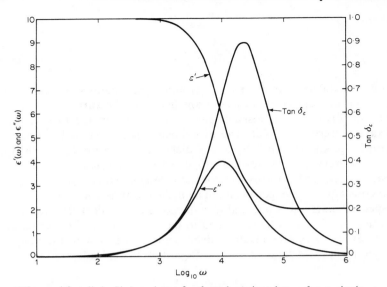

Figure 4.3. $\varepsilon'(\omega)$, $\varepsilon''(\omega)$ and tan δ_ε plotted against log ω for a single relaxation time model with $\varepsilon_R/\varepsilon_U = 5$ and $\tau_E = 10^{-4}$ sec.

According to Equation (4.29) tan δ_ε exhibits a maximum (Figure 4.3) at

$$\omega_{max} = \frac{1}{\bar{\tau}_\varepsilon} = \left(\frac{\varepsilon_R}{\varepsilon_U}\right)^{1/2} \frac{1}{\tau_E}$$

At ω_{max}, tan δ_E has a value

$$\frac{\varepsilon_R - \varepsilon_U}{2(\varepsilon_R \varepsilon_U)^{1/2}}$$

Equations (4.25) to (4.27) are known as the *Debye* equations. This name is somewhat misleading however, since the equations derived by Debye (1929) are similar in form to those above but involve the *intrinsic* relaxation time τ^*_E rather than τ_E. Since τ^*_E has greater molecular significance than τ_E its discussion will be reserved for the section concerning molecular theories (see Chapter 5).

In comparing dielectric and mechanical data it should be noted that the electric field E is analogous to the mechanical stress σ and that the dielectric displacement D is analogous to the mechanical strain γ. It follows that ε^* and τ_E are analogous to J^* and τ_σ respectively. Comparisons between dielectric and mechanical relaxation data might therefore be made, for example, by noting the correlation which exists between the frequencies of maximum ε'' and J''. Alternatively, we may note that G^* is analogous to $(\varepsilon^{-1})^* = E^*/D^*$, which may be termed the complex *reciprocal* dielectric constant. Writing $(\varepsilon^{-1})^*$ in terms of its real and imaginary components,

$$(\varepsilon^{-1})^* = (\varepsilon^{-1})' + i(\varepsilon^{-1})'' \tag{4.30}$$

and noting that

$$(\varepsilon^{-1})^* = (\varepsilon^*)^{-1} = (\varepsilon' - i\varepsilon'')^{-1} \tag{4.31}$$

it follows that

$$(\varepsilon^{-1})' = \varepsilon'/(\varepsilon'^2 + \varepsilon''^2) \tag{4.32a}$$

and

$$(\varepsilon^{-1})'' = \varepsilon''/(\varepsilon'^2 + \varepsilon''^2) \tag{4.32b}$$

By procedures analogous to those used in deriving Equation (4.16b) it may be shown that, for a single relaxation time τ_D, $(\varepsilon^{-1})''$ is given by

$$(\varepsilon^{-1})'' = [(\varepsilon^{-1})_U - (\varepsilon^{-1})_R] \frac{\omega\tau_D}{1 + \omega^2\tau_D{}^2} \tag{4.33}$$

Since τ_D is analogous to τ_y, comparisons between mechanical and dielectric relaxation behaviour would therefore be made by noting the correlation between the frequencies of maximum G'' and $(\varepsilon^{-1})''$. Such comparisons are rare, though a good correlation has been noted in the case of polyvinyl chloride (see Section 11.1a).

4.1c The Cole–Cole Diagram

Cole and Cole (1941) proposed a method for plotting and checking equations (4.26) and (4.27). The procedure consists in constructing an Argand diagram in which ε'' is plotted against ε', each point corresponding to one frequency. From (4.26) and (4.27) we have

$$\left\{\varepsilon' - \frac{(\varepsilon_R + \varepsilon_U)}{2}\right\}^2 + (\varepsilon'')^2 = \left|\frac{\varepsilon_R - \varepsilon_U}{2}\right|^2 \tag{4.34}$$

From this equation the Cole–Cole plot is predicted to give a semicircle of radius $(\varepsilon_R - \varepsilon_U)/2$, its centre lying on the abscissa at a distance $(\varepsilon_R + \varepsilon_U)/2$ from the origin. For given values of ε' and ε'' the curve should be completely defined and uninfluenced by the frequency range in which the relaxation appears (i.e. independent of τ_E). The Cole–Cole diagram corresponding to the plots of Figure 4.3 is given in Figure 4.4. When a single relaxation time does not suffice to describe the relaxation (as in the case of polymers for example) Equations (4.26), (4.27) and (4.34) are inapplicable and a semicircular arc is not obtained. In these cases a distribution of relaxation times is necessary to interpret the data, and for specific forms of the distribution a modified Cole–Cole arc is predicted (see Section 4.1e below).

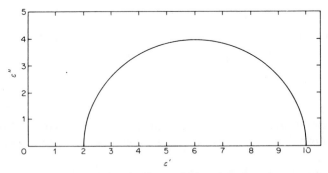

Figure 4.4. Cole–Cole plot for a single relaxation time model with $\varepsilon_R/\varepsilon_U = 5$.

From Equations (4.15) and (4.16) it is clear that the Cole–Cole method of analysis is equally applicable to the analysis of dynamic mechanical data. This method has been little used, probably because of the lack of mechanical data covering a sufficiently wide range of frequencies.

4.1d   Distribution of Relaxation Times

The single relaxation time model predicts that in a stress relaxation experiment a proportion $(G_U - G_R)\gamma_0$ of the initial stress, $G_U\gamma_0$, will relax exponentially to zero with the constant τ_y. For most relaxations this model must be modified to secure agreement between theory and experiment. The modification is straightforward and consists in attributing to a fraction of the stress to be relaxed $(G_U - G_R)\gamma_0\phi_G(\ln \tau)\,d\ln \tau$, a time constant τ. The quantity $\phi_G(\ln \tau)\,d\ln \tau$ can be regarded as the fraction of relaxation processes with relaxation times between $\ln \tau$ and $(\ln \tau + d\ln \tau)$. $\phi_G(\ln \tau)$ satisfies the condition

$$\int_{-\infty}^{\infty} \phi_G(\ln \tau)\,d\ln \tau = 1 \qquad (4.35)$$

and is known as the normalized distribution of mechanical relaxation times. Creep experiments are governed by the distribution of retardation times $\phi_J(\ln \tau) \, d \ln \tau$, which satisfies

$$\int_{-\infty}^{\infty} \phi_J(\ln \tau) \, d \ln \tau = 1 \tag{4.36}$$

The quantity $\phi_J(\ln \tau) \, d \ln \tau$ can be regarded as the fraction of retardation processes with retardation times between $\ln \tau$ and $(\ln \tau + d \ln \tau)$.

We shall now derive the quantities $J(t)$, $J^*(\omega)$, $G(t)$ and $G^*(\omega)$ for a solid with normalized distributions $\phi_J(\ln \tau)$ and $\phi_G(\ln \tau)$; the equations derived will be entirely analogous to the equations for the single relaxation time. For the latter model it is only necessary to specify J_R, J_U and τ_σ and then all the other viscoelastic parameters are specified. This comment holds for the model with a distribution of relaxation times with one addition: instead of τ_σ we must specify $\phi_J(\ln \tau)$ together with J_R and J_U. Having specified these quantities we have also specified G_R, G_U and $\phi_G(\ln \tau)$. The form of the relation between $\phi_J(\ln \tau)$ and $\phi_G(\ln \tau)$ (Gross, 1953) is much more complex than the relation between τ_σ and τ_γ Equation (4.12).

The decay function for the relaxation processes with relaxation times between $\ln \tau$ and $\ln \tau + d \ln \tau$ is clearly, by analogy with Equation (4.7),

$$\frac{\phi_G(\ln \tau)}{\tau} \exp\left(\frac{-t}{\tau}\right) d \ln \tau \tag{4.37}$$

Hence the total decay function for all relaxation centres is

$$\alpha_\gamma(t) = \int_{-\infty}^{\infty} \frac{\phi_G(\ln \tau)}{\tau} e^{-t/\tau} \, d \ln \tau \tag{4.38}$$

Similarly, the total decay function for all retardation centres is

$$\alpha_\sigma(t) = \int_{-\infty}^{\infty} \frac{\phi_J(\ln \tau)}{\tau} e^{-t/\tau} \, d \ln \tau \tag{4.39}$$

From Equation (4.5) we have for the strain $\gamma(t)$ at time t due to the application of a constant stress σ_0 at time $u = 0$,

$$\gamma(t) = J_U \sigma_0 + (J_R - J_U)\sigma_0 \int_0^t \int_{-\infty}^{\infty} \frac{\phi_J(\ln \tau)}{\tau} e^{-(t-u)/\tau} \, d \ln \tau \, du \tag{4.40}$$

That is,

$$J(t) = J_U + (J_R - J_U) \int_{-\infty}^{\infty} \phi_J(\ln \tau)(1 - e^{-t/\tau}) \, d \ln \tau \tag{4.41}$$

Similarly, we have for the dynamic compliances, from Equation (4.20),

$$J'(\omega) = J_U + (J_R - J_U) \int_{-\infty}^{\infty} \frac{\phi_J(\ln \tau) \, d \ln \tau}{1 + \omega^2 \tau^2} \tag{4.42a}$$

$$J''(\omega) = (J_R - J_U) \int_{-\infty}^{\infty} \frac{\phi_J(\ln \tau) \omega \tau \, d \ln \tau}{1 + \omega^2 \tau^2} \tag{4.42b}$$

For the stress relaxation modulus $G(t)$ and the dynamic moduli we may obtain in like manner, using the decay function given by Equation (4.38),

$$G(t) = G_R + (G_U - G_R) \int_{-\infty}^{\infty} \phi_G(\ln \tau) e^{-t/\tau} \, d \ln \tau \tag{4.43}$$

$$G'(\omega) = G_R + (G_U - G_R) \int_{-\infty}^{\infty} \frac{\phi_G(\ln \tau) \omega^2 \tau^2 \, d \ln \tau}{1 + \omega^2 \tau^2} \tag{4.44}$$

$$G''(\omega) = (G_U - G_R) \int_{-\infty}^{\infty} \frac{\phi_G(\ln \tau) \, \omega \tau}{1 + \omega^2 \tau^2} \, d \ln \tau \tag{4.45}$$

The dielectric parameters $\varepsilon(t)$, $\varepsilon'(\omega)$ and $\varepsilon''(\omega)$ may be obtained using the normalized distribution of dielectric relaxation times which satisfies the condition

$$\int_{-\infty}^{\infty} \phi_\varepsilon(\ln \tau) \, d \ln \tau = 1 \tag{4.46}$$

Using the superposition principle, as in the preceding paragraphs, we obtain

$$\varepsilon(t) = \varepsilon_U + (\varepsilon_R - \varepsilon_U) \int_{-\infty}^{\infty} \phi_\varepsilon(\ln \tau)(1 - e^{-t/\tau}) \, d \ln \tau \tag{4.47}$$

$$\varepsilon'(\omega) = \varepsilon_U + (\varepsilon_R - \varepsilon_U) \int_{-\infty}^{\infty} \frac{\phi_\varepsilon(\ln \tau) d \ln \tau}{1 + \omega^2 \tau^2} \tag{4.48}$$

$$\varepsilon''(\omega) = (\varepsilon_R - \varepsilon_U) \int_{-\infty}^{\infty} \frac{\phi_\varepsilon(\ln \tau) \omega \tau \, d \ln \tau}{1 + \omega^2 \tau^2} \tag{4.49}$$

The distribution functions are always positive. It will be seen from Equations (4.42), (4.45) and (4.49) that the $J''(\omega)$, $G''(\omega)$ and $\varepsilon''(\omega)$ curves are formed from the superposition of many single relaxation time curves. The effect of the distribution functions is to broaden the theoretical curves to achieve agreement with experiment. The distribution functions may of course be obtained from the experimental measurements.

The normalized distribution functions ϕ_J and ϕ_G are not used so frequently in the analysis of mechanical experiments as the related functions $L_J(\ln \tau)$ and $H_G(\ln \tau)$

$$(J_R - J_U) = \int_{-\infty}^{\infty} L_J(\ln \tau) \, d \ln \tau \qquad (4.50)$$

$$(G_U - G_R) = \int_{-\infty}^{\infty} H_G(\ln \tau) \, d \ln \tau \qquad (4.51)$$

The functions $L_J(\ln \tau)$ and $H_G(\ln \tau)$ are known as the distribution of retardation and relaxation times respectively and have the dimensions of compliance and modulus.

There are several approximate methods of obtaining $L_J(\ln \tau)$ and $H_G(\ln \tau)$ from experimental measurements of the viscoelastic functions $J(t)$, $J^*(\omega)$ and $G(t)$, $G^*(\omega)$. The results of different authors are summarized in Table 4.1 (Leaderman, 1958). This topic, and the obtaining of viscoelastic functions from $L_J(\ln \tau)$ and $H_G(\ln \tau)$, is discussed by Tobolsky (1960) and Ferry (1961). According to Zener (1948), since the distribution functions are extremely sensitive to small errors in observation it is more useful to express experimental results directly in terms of measured quantities, e.g. tan δ_G, J^*, G^*. This is, in fact, exactly what the majority of workers do.

Table 4.1. Approximations for retardation and relaxation spectra
(After Leaderman, 1958)

Type of test	Order of approximation	Approximation for $L(t)$, $L(1/\omega)$, $H(t)$, $H(1/\omega)$
Creep	1	$L_1(t) = (d/d \ln t)[J(t) - t/\eta]$
	2	$L_2(t/2) = [d/d \ln t - d^2/d(\ln t)^2] \, A(t)$
Dynamic	0	$L''_0(1/\omega) = (2/\pi)[J''(\omega) - 1/\omega\eta]$
	1	$L'_1(1/\omega) = -(d/d \ln \omega)J'(\omega)$
	2	$L''_2(1/\omega) = (2/\pi)[1 - d^2/d(\ln \omega)^2]B(\omega)$
	3	$L'_3(1/\omega) = -|d/d \ln \omega - (\frac{1}{4})d^3/d(\ln \omega)^3|J'(\omega)$
Stress Relaxation	1	$H_1(t) = -(d/d \ln t)G(t)$
	2	$H_2(t/2) = -[d/d \ln t - d^2/d(\ln t)^2]G(t)$
Dynamic	0	$H''_0(1/\omega) = (2/\pi)G''(\omega)$
	1	$H'_1(1/\omega) = (d/d \ln \omega)G'(\omega)$
	2	$H''_2(1/\omega) = (2/\pi)[1 - d^2/d(\ln \omega)^2]G''(\omega)$
	3	$H'_3(1/\omega) = [d/d \ln \omega - (\frac{1}{4})d^3(\ln \omega)^3]G'(\omega)$

$A(t) = J(t)$ or $J(t) - t/\eta$.
$B(\omega) = J''(\omega)$ or $J''(\omega) - 1/\omega\eta$.

4.1e Empirical Distribution Functions

One method of fitting the experimental dynamic data is to express the complex G^*, J^* or ε^* in terms of empirical expressions involving empirical parameters. The empirical Cole–Cole (1941), Fuoss–Kirkwood (1941) and Davidson–Cole (1950, 1951) expressions have been widely used in dielectric studies.

The Cole–Cole Equation

Perhaps the best known empirical distribution is that due to Cole and Cole (1941). They modified the single relaxation time expressions by replacing the term $(1+i\omega\tau)$ in Equation (4.25) by $1+(i\omega\tau_0)^{\bar\beta}$, where $\bar\beta$ is a parameter, $0 < \bar\beta \leqslant 1$. Hence

$$\varepsilon^*(\omega) = \varepsilon'(\omega) - i\varepsilon''(\omega) = \varepsilon_U + \frac{(\varepsilon_R - \varepsilon_U)}{1 + (i\omega\tau_0)^{\bar\beta}} \tag{4.52}$$

giving

$$\frac{\varepsilon'(\omega) - \varepsilon_U}{\varepsilon_R - \varepsilon_U} = \frac{(1 + (\omega\tau_0)^{\bar\beta}\cos\bar\beta\pi/2)}{1 + 2(\omega\tau_0)^{\bar\beta}\cos\bar\beta\pi/2 + (\omega\tau_0)^{2\bar\beta}} \tag{4.53}$$

$$\frac{\varepsilon''(\omega)}{\varepsilon_R - \varepsilon_U} = \frac{(\omega\tau_0)^{\bar\beta}\sin\bar\beta\pi/2}{1 + 2(\omega\tau_0)^{\bar\beta}\cos\bar\beta\pi/2 + (\omega\tau_0)^{2\bar\beta}} \tag{4.54}$$

The dispersion curve from (4.53) is broader than that for a single relaxation time, but is symmetrical about $\omega\tau_0 = 1$. The loss curve from (4.54) is lower in amplitude and broader than that for a single relaxation time. At the loss maximum $d\varepsilon''/d\omega = 0$ we obtain $\omega_{max}\tau_0 = 1$ and

$$\varepsilon''_{max} = \frac{(\varepsilon_R - \varepsilon_U)}{2}\tan\frac{\bar\beta\pi}{4} \tag{4.55}$$

It can be shown that the distribution function, $\phi_\varepsilon(\ln\tau)$, is given by

$$\phi_\varepsilon(\ln\tau) = \frac{1}{2\pi}\frac{\sin\bar\beta\pi}{\{\cosh\bar\beta\ln\tau/\tau_0 + \cos\bar\beta\pi\}} \tag{4.56}$$

The maximum of $\phi_\varepsilon(\ln\tau)$ occurs at $\tau = \tau_0$. Combining Equation (4.53) and (4.54) to eliminate $\omega\tau_0$ yields.

$$\left[\varepsilon' - \frac{(\varepsilon_R + \varepsilon_U)}{2}\right]^2 + \left[\varepsilon'' + \frac{(\varepsilon_R - \varepsilon_U)}{2}\cotan\frac{\bar\beta\pi}{2}\right]^2 = \left[\frac{(\varepsilon_R - \varepsilon_U)}{2}\operatorname{cosec}\frac{\bar\beta\pi}{2}\right]^2 \tag{4.57}$$

This equation represents a circle with centre

$$\left(\frac{(\varepsilon_R + \varepsilon_U)}{2}, \ -\frac{(\varepsilon_R - \varepsilon_U)}{2} \cot \frac{\beta\pi}{2}\right)$$

and radius

$$\frac{(\varepsilon_R - \varepsilon_U)}{2} \csc \frac{\beta\pi}{2}$$

Figure 4.5 shows the plot for $\beta = 1$, $0\cdot6$ and $0\cdot2$ with $\varepsilon_R = 10$, $\varepsilon_U = 2$. It is seen that the Cole–Cole equation gives a depressed semicircle. Many experimental systems conform with Equations (4.53), (4.54) and (4.57). It is to be noted from (4.54) that the plot log ε'' versus log ω has slope β at low frequencies and slope $-\beta$ at high frequencies.

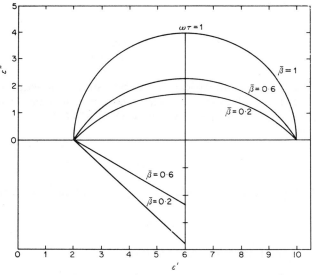

Figure 4.5. Cole–Cole plot for models with $\varepsilon_R/\varepsilon_U = 5$ (i) $\beta = 1$ (single relaxation time), (ii) $\beta = 0.6$ and (iii) $\beta = 0\cdot2$.

The Fuoss–Kirkwood Equation

Fuoss and Kirkwood (1941) noted that the single relaxation time expression for dielectric loss could be rearranged into the form

$$\varepsilon''(\omega) = \varepsilon''_{max} \operatorname{sech} \ln \omega\tau \tag{4.58}$$

Fuoss–Kirkwood introduced a parameter m,

$$\varepsilon''(\omega) = \varepsilon''_{max} \operatorname{sech} m \ln \omega\tau \qquad 0 < m \leqslant 1 \qquad (4.59)$$

or

$$\varepsilon''(\omega) = 2\varepsilon''_{max} \frac{(\omega\tau)^m}{1 + (\omega\tau)^{2m}} \qquad (4.60)$$

It is readily shown that ε''_{max} is given by

$$\varepsilon''_{max} = \frac{(\varepsilon_R - \varepsilon_U)}{2} m \qquad (4.61)$$

Thus (4.59) becomes

$$\varepsilon''(\omega) = \frac{(\varepsilon_R - \varepsilon_U)}{2} m \operatorname{sech} m \ln \omega\tau \qquad (4.62)$$

or

$$\varepsilon''(\omega) = (\varepsilon_R - \varepsilon_U) m \frac{(\omega\tau)^m}{1 + (\omega\tau)^{2m}} \qquad (4.63)$$

Equation (4.63) gives symmetrical loss curves similar in shape but not identical with those of Cole and Cole. The differences between (4.54) and (4.59) are readily seen since (a) from (4.55) and (4.61)

$$m \simeq \tan \frac{\bar{\beta}\pi}{4} \qquad (4.64)$$

(b) from the log ε'' versus log ω plots at very low and high frequencies

$$m \simeq \bar{\beta}$$

(c) by rearranging Equation (4.54) into the form

$$\varepsilon''(\omega) = \frac{\varepsilon''_{max}(1 + \cos \bar{\beta}\pi/2)}{\cosh \bar{\beta} \ln \omega/\omega_{max} + \cos \bar{\beta}\pi/2} \qquad (4.65)$$

and comparing with Equation (4.59), Poley (see Böttcher, 1956) has shown that near the maximum loss

$$m\sqrt{2} \simeq \frac{\bar{\beta}}{\cos \bar{\beta}\pi/4} \qquad (4.66)$$

Evidently m and $\bar{\beta}$ are approximately related. Cole (1955) has discussed the correlation between the two empirical representations of dielectric data.

The reader will have noted that both Cole–Cole and Fuoss–Kirkwood equations, which were derived for dielectric relaxation, can be used also for mechanical data to express J^* or G^* as a function of frequency at constant temperature. As described in Chapter 1, however, because of experimental circumstances J^* and G^* are usually determined as a function of temperature at constant frequency. The Cole–Cole and Fuoss–Kirkwood equations can still be employed if it is assumed (a) that the shape of the distribution is independent of T (i.e. $\bar{\beta}$ or m independent of T), (b) the temperature dependence of τ is given by

$$\tau = \tau_0 \exp \frac{H}{RT} \qquad (4.67)$$

in which the activation energy H is independent of T and (c) the increment $(J_R - J_U)$ or $(G_U - G_R)$ is independent of temperature. These assumptions, the latter especially, are usually of limited validity. If they are acceptable then both Cole–Cole and Fuoss–Kirkwood expressions can be rearranged into a form suitable to fit the data. For example, the Fuoss–Kirkwood expression gives

$$\varepsilon'' = \varepsilon''_{\text{max}} \operatorname{sech} m \frac{H}{R} \left[\frac{1}{T} - \frac{1}{T_{\text{max}}} \right] \qquad (4.68)$$

T_{max} is the value of T at which $\varepsilon'' = \varepsilon''_{\text{max}}$.

It will be shown below that H may be obtained from the area beneath the loss factor $-1/T$ plot under the conditions specified in the previous paragraph. The value of m can be obtained from the half-width of the loss factor $-1/T$ plot. Knowing $\varepsilon''_{\text{max}}$, H and m, the observed loss curve may then be fitted. Alternatively knowing $\varepsilon''_{\text{max}}$ the curve may be fitted by trial and error with H and m as parameters.

The Davidson–Cole Equation

Both the Cole–Cole and Fuoss–Kirkwood empirical equations relate only to dispersion and absorption curves symmetrical about the position $\omega\tau = 1$. It is often found that dielectric loss curves have a high-frequency broadening and the Cole–Cole (1941) arcs for such systems are said to be 'skewed'. Davidson and Cole (1950) attempted to fit the experimental results to the following function

$$\frac{\varepsilon^*(\omega) - \varepsilon_U}{(\varepsilon_R - \varepsilon_U)} = \frac{1}{(1 + i\omega\tau_1)^\gamma} \qquad 0 < \gamma \leqslant 1 \qquad (4.69)$$

$$\frac{\varepsilon'(\omega) - \varepsilon_U}{(\varepsilon_R - \varepsilon_U)} = (\cos\phi)^\gamma \cos\gamma\phi \qquad (4.70)$$

$$\frac{\varepsilon''(\omega)}{(\varepsilon_R - \varepsilon_U)} = (\cos\phi)^\gamma \sin\gamma\phi \qquad (4.71)$$

where

$$\tan\phi = \omega\tau_1 \qquad (4.72)$$

At the maximum loss $\omega\tau_1 \neq 1$, but is given by

$$\omega_{max}\tau_1 = \tan\left\{\frac{1}{(\gamma+1)} \cdot \frac{\pi}{2}\right\} \qquad (4.73)$$

Here ω_{max} is the angular frequency of maximum loss.
In Figure 4.6 the normalized Cole–Cole arc plot is shown for $\gamma = 0.50$.
At low frequencies the arc is circular, but at high frequencies the curve approaches the abscissa along a straight line. The angle between this line and the abscissa is $\gamma\pi/2$.

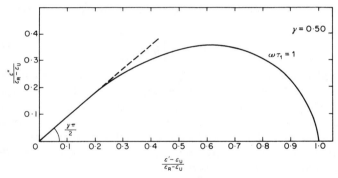

Figure 4.6. Davidson–Cole plot for $\gamma = 0.50$.

The distribution function $\phi(\ln\tau)$ for Equation (4.69) is

$$\phi\left(\ln\frac{\tau}{\tau_1}\right) = \frac{\sin\gamma\pi}{\pi}\left(\frac{\tau}{\tau_1 - \tau}\right)^\gamma \qquad \tau \leqslant \tau_1 \qquad (4.74)$$

$$= 0 \qquad \tau > \tau_1$$

τ_1 is the relaxation time in Equation (4.69).
The Davidson–Cole function describes many systems adequately. However, it seems likely that in a given system two or more discrete

relaxations may be present resulting in a broadening of the loss curve at higher frequencies. When two or more relaxation regions overlap, it can be shown by calculation that curves result which may be fitted to the Davidson–Cole equation within experimental error. Thus, the best test of the reasonable applicability of Equation (4.69) is to see if the experimental dispersion and absorption curves are resolved into the component curves by changing temperature over a wide range. Glarum (1960) has derived the Davidson–Cole equation using a diffusion model.

Wedge and Box Distributions

Tobolsky (1960) has described in detail the 'box' and 'wedge' empirical distributions which have been used in the analysis of mechanical relaxation data. The 'box' distribution is defined by the following equations

$$\left. \begin{array}{ll} H(\ln \tau) = \text{constant} & \text{for} \quad \tau_1 \leqslant \tau \leqslant \tau_2 \\ H(\ln \tau) = 0 & \text{for} \quad \tau > \tau_2 ; \tau < \tau_1 \end{array} \right\} \qquad (4.75)$$

The 'wedge' distribution is given by

$$\left. \begin{array}{ll} H(\ln \tau) = \dfrac{\text{constant}}{\tau^{1/2}} & \text{for} \quad \tau_1 \leqslant \tau \leqslant \tau_2 \\ H(\ln \tau) = 0 & \text{for} \quad \tau > \tau_2 ; \tau < \tau_1 \end{array} \right\} \qquad (4.76)$$

Tobolsky and coworkers used a combination of the wedge and box distributions in an attempt to interpret the relaxation spectrum for polyisobutylene. The 'box' distribution was used for the rubbery flow region, and the 'wedge' distribution was used for the glass–rubber region. The 'box' distribution corresponds formally to the 'rectangular' distribution function used in the Fröhlich molecular theory of relaxation (see Chapter 5).

4.2   TEMPERATURE AS A VARIABLE

The importance of the temperature variation experiment in the study of mechanical relaxation is illustrated in Chapter 1 by Heijboer's study of the cyclohexyl relaxation in polycyclohexyl methacrylate. The theory in Section 4.1 gives the loss factor as a function of time or frequency. It is necessary to consider the extension of this theory to interpret measurements in which the loss factor measured at a constant frequency is obtained as a function of temperature. This discussion is of considerable significance since, in practice, the vast majority of experimentalists employ the temperature variation experiment in studying mechanical relaxation.

This is not true for dielectric relaxation in which a large frequency range may be obtained with relative ease.

In a plot of loss factor versus $1/T$ three important quantities are obtained (a) the temperature of maximum loss at the measuring frequency, (b) the area below the loss factor versus $1/T$ plot (c) the shape of the loss curve, which may be measured roughly as a half-width. The significance of (a) has been discussed in Chapter 1 for a relaxation with a single relaxation time. When there is a distribution of relaxation times the temperature and frequency of maximum loss yield information about the maximum of the appropriate distribution function. The significance of (b) the area, and (c) the half-width of the loss factor versus $1/T$ plot is discussed in the following sections.

4.2a   Area of Loss Factor versus $1/T$ Plots

The theory for areas below loss factor versus $1/T$ plots was first given by Read and Williams (1961b) and a similar procedure has been outlined by Adam and Müller (1962).

Single Relaxation Time

Taking the single relaxation time, expression (4.27), for the dielectric loss factor, and writing τ according to $\tau = \tau_0 \exp H/RT$, the temperature dependence of ε'' is given by

$$\varepsilon''(T) = (\varepsilon_R{}^T - \varepsilon_U{}^T) \frac{\omega\tau_0 \exp H/RT}{1 + \omega^2 \tau_0{}^2 \exp 2H/RT} \tag{4.77}$$

$\varepsilon_R{}^T$ and $\varepsilon_U{}^T$ are the relaxed and unrelaxed dielectric constants respectively at temperature T.

For the special case $(\varepsilon_R{}^T - \varepsilon_U{}^T)$ independent of temperature, from (4.77) we have for constant ω,

$$\int_0^\infty \varepsilon'' \, d\left(\frac{1}{T}\right) = (\varepsilon_R - \varepsilon_U) \frac{R}{H} \left[\frac{\pi}{2} - \tan^{-1} \omega\tau_0 \right] \tag{4.78}$$

Experimental values for τ_0 are usually less than 10^{-11} sec, so that, for experimental frequencies normally encountered, (4.78) may be written as

$$H = (\varepsilon_R - \varepsilon_U) \frac{R\pi}{2} \frac{1}{\int_0^\infty \varepsilon'' \, d(1/T)} \tag{4.79}$$

Thus the area below the ε'' versus $1/T$ plot is simply related to H and $(\varepsilon_R - \varepsilon_U)$, and ω is not involved.

The theories of Onsager, Fröhlich and Debye (see Chapter 3) for polar media lead to a good approximation to the relation,

$$\varepsilon_R{}^T - \varepsilon_U{}^T = \frac{A}{T} \tag{4.80}$$

where A is a constant. In this case the evaluation of the area beneath the ε'' versus $1/T$ curve is more complicated, but the relation has been shown to be (Read and Williams, 1961b)

$$\int_0^\infty \varepsilon'' \, d\left(\frac{1}{T}\right) = \frac{AR^2}{H^2} \left[-\frac{\pi \ln \omega\tau_0}{2} + \sum_{r=0}^\infty \frac{(-1)^r (\omega\tau_0)^{2r+1}}{(2r+1)^2} \right] \tag{4.81}$$

At the maximum loss corresponding to a temperature T_{max} we have

$$A = (\varepsilon_R^{T_{max}} - \varepsilon_U^{T_{max}}) \, T_{max}$$

$$\tau_{T_{max}} = \tau_0 \exp H/RT_{max} \tag{4.82}$$

Schallamach (1946) has shown that $\omega\tau_{T_{max}}$ differs only little from unity, and also $\omega\tau_0 \ll 1$ for frequencies normally encountered, thus from (4.81) and (4.82) we have

$$H = (\varepsilon_R^{T_{max}} - \varepsilon_U^{T_{max}}) \frac{R\pi}{2 \int_0^\infty \varepsilon'' \, d(1/T)} \tag{4.83}$$

Equation (4.83) is seen to differ from (4.79) only in the $(\varepsilon_R - \varepsilon_U)$ term.

Distribution of Relaxation Times

The loss factor ε'' is related generally to ω and τ by the expression

$$\varepsilon'' = (\varepsilon_R - \varepsilon_U) \int_0^\infty \phi_\varepsilon(\tau) \frac{\omega\tau}{1 + \omega^2\tau^2} \, d\tau \tag{4.84}$$

If each τ has the form $\tau = \tau_0 \exp(H/RT)$, a distribution of τ values can arise in three ways.

(A) From a distribution of τ_0, H being equal for every process.
(B) From a distribution of H values, τ_0 being equal for every process.
(C) From a distribution of both τ_0 and H parameters.

In the more general case (C) we may define a distribution function $\phi(\tau_0, H)$ which represents the fraction of processes having τ_0 values between τ_0 and $\tau_0 + d\tau_0$ and *in addition* activation energies lying between H and $H + dH$. The distribution function is normalized,

$$\int_0^\infty \phi(\tau) \, d\tau = \int_0^\infty \int_0^\infty \phi(\tau_0, H) \, d\tau_0 \, dH = 1. \tag{4.85}$$

The loss factor is given by

$$\varepsilon'' = (\varepsilon_R{}^T - \varepsilon_U{}^T) \int_0^\infty \int_0^\infty \frac{\phi(\tau_0, H)\omega\tau_0 \exp(H/RT)}{1 + \omega^2 \tau_0{}^2 \exp(2H/RT)} \, d\tau_0 \, dH \quad (4.86)$$

Assuming $(\varepsilon_R{}^T - \varepsilon_U{}^T)$ to be independent of temperature we obtain, from (4.86),

$$\int_0^\infty \varepsilon'' \, d(1/T) = (\varepsilon_R - \varepsilon_U) \int_0^\infty \int_0^\infty \phi(\tau_0, H) \int_0^\infty \frac{\omega\tau_0 \exp(H/RT) \, d(1/T)}{1 + \omega^2 \tau_0{}^2 \exp(2H/RT)} \, d\tau_0 \, dH$$

$$(4.87)$$

$$= (\varepsilon_R - \varepsilon_U)R \int_0^\infty \left[\frac{\pi}{2} \int_0^\infty \phi(\tau_0, H) \, d\tau_0 \right.$$

$$\left. - \int_0^\infty \phi(\tau_0, H) \tan^{-1} \omega\tau_0 \, d\tau_0 \right] \frac{dH}{H}. \quad (4.88)$$

If we consider that experimentally $\phi(\tau_0, H)$ is only significant for small values of τ_0, the second integral in brackets is negligible with respect to the first integral and (4.88) becomes

$$\int_0^\infty \varepsilon'' \, d(1/T) = (\varepsilon_R - \varepsilon_U) \frac{R\pi}{2} \int_0^\infty \int_0^\infty \phi(\tau_0, H) \frac{1}{H} \, d\tau_0 \, dH$$

$$= (\varepsilon_R - \varepsilon_U) \frac{R\pi}{2} \left\langle \frac{1}{H} \right\rangle_{av} \quad (4.89)$$

The average activation energy is therefore given by

$$\left\langle \frac{1}{H} \right\rangle_{av}^{-1} = (\varepsilon_R - \varepsilon_U) \frac{R\pi}{2} \frac{1}{\int_0^\infty \varepsilon'' \, d(1/T)} \quad (4.90)$$

By an analogous treatment we may show for the mechanical cases

$$\left\langle \frac{1}{H} \right\rangle_{av}^{-1} = (G_U - G_R) \frac{R\pi}{2} \frac{1}{\int_0^\infty G'' \, d(1/T)} \quad (4.91)$$

$$\left\langle \frac{1}{H} \right\rangle_{av}^{-1} = (J_R - J_U) \frac{R\pi}{2} \frac{1}{\int_0^\infty J'' \, d(1/T)} \quad (4.92)$$

Equations (4.90) to (4.92) are of the same form as (4.79). According to this general analysis the area below the loss factor–reciprocal temperature plot is related to $(\varepsilon_R - \varepsilon_U)$ and an average activation energy of the process. The result is independent of the frequency of measurement, and the form of the distribution function $\phi(\tau_0, H)$.

Two special cases of the general result (4.89) are (A) and (B), page 123. For case (A) we have τ_0 distributed, H constant (see Figure 4.9),

$$\left\langle \frac{1}{H} \right\rangle_{av} = \frac{1}{H} \int_0^\infty \phi(\tau_0) \, d\tau_0 = \frac{1}{H} \tag{4.93}$$

where $\phi(\tau_0)$ is equivalent to $\int_0^\infty \phi(\tau_0, H) \, dH$. Thus in this case (4.89) and (4.93) yield an expression identical with (4.79), the single relaxation time result.

For case (B) we have H distributed, τ_0 constant

$$\left\langle \frac{1}{H} \right\rangle_{av} = \int_0^\infty \psi(H) \frac{1}{H} \, dH \tag{4.94}$$

Here $\psi(H)$ is equivalent to $\int_0^\infty \phi(\tau_0, H) \, d\tau_0$ and is identical with the function introduced by Gevers (1946).

It is to be noted that empirical distributions commonly used also give Equation (4.90) for the area, if the form of the distribution is independent of temperature. For example, the Cole–Cole empirical Equation (4.52) yields, for $\varepsilon_R - \varepsilon_U$ and $\bar\beta$ independent of temperature,

$$\int_0^\infty \varepsilon'' \, d\left(\frac{1}{T}\right) = \frac{(\varepsilon_R - \varepsilon_U)}{\bar\beta H} R\left[\frac{\pi}{2} - \tan^{-1}\left\{\frac{(\omega\tau_0)^{\bar\beta} + \cos \bar\beta\pi/2}{\sin \bar\beta\pi/2}\right\}\right] \tag{4.95}$$

which for $\omega\tau_0 \ll 1$ reduces to (4.90). Similarly, the Fuoss–Kirkwood and Davidson–Cole empirical equations give (4.90) if the distribution parameter is independent of temperature. The condition that the empirical distribution parameter is independent of temperature is equivalent to case (A) above for the general distribution, i.e. τ_0 distributed, H constant.

We have mentioned that H may not be independent of temperature, a condition found experimentally in amorphous polymers above the glass-transition temperature. This imposes a restriction on the foregoing analysis which is not too severe. Since the loss curve for typical relaxation processes occurs over a rather narrow range of $(1/T)$, the integral may be cut off on either side of the loss region and the activation energy thus derived is an average value for the temperature range in which the loss occurs.

4.2b The Widths of Curves of Loss Factor against $1/T$

Single Relaxation Time

We have

$$\varepsilon'' = 2\varepsilon''_{\text{max}} \frac{\omega\tau}{1+\omega^2\tau^2} \qquad (4.96)$$

Putting $\varepsilon''/\varepsilon''_{\text{max}} = r^{-1}$ we have, solving (4.96) for $\omega\tau$

$$\omega\tau = r \pm \sqrt{r^2 - 1} \qquad (4.97)$$

Considering τ as the variable and $\tau = \tau_0 \exp(H/RT)$ as before, at $\varepsilon''_{\text{max}}$ we have $T = T_{\text{max}}$ and $\omega\tau_{T_{\text{max}}} = \omega\tau_0 \exp(H/RT_{\text{max}}) = 1$, hence

$$\omega\tau_0 \exp H/RT_r = r \pm \sqrt{r^2 - 1} \qquad (4.98)$$

T_r is the temperature at which $\varepsilon'' = \varepsilon''_{\text{max}}/r$.

Hence

$$H = \frac{2 \cdot 303 R \log\left[r \pm \sqrt{r^2 - 1}\right]}{\delta_{1/r}} \qquad (4.99)$$

$$\delta_{1/r} = \left[\frac{1}{T_r} - \frac{1}{T_{\text{max}}}\right]$$

If $r = 2$, then $\delta_{1/r}$ is one-half of the total half-width of the loss factor against $1/T$ plot, and (4.99) becomes

$$H = \frac{2 \cdot 63}{\delta_{1/2}} = \frac{5 \cdot 26}{\Delta_{1/2}} \qquad (4.100)$$

$\Delta_{1/2}(= 2\delta_{1/2})$ is the total half-width of the curve.

It is seen from Equation (4.99) that for a single relaxation time, the loss factor versus $1/T$ plot is symmetrical about $1/T_{\text{max}}$. The loss factor versus *temperature* plot is not symmetrical being broader at higher temperatures.

Distribution of Relaxation Times

It does not appear that a simple expression for the width of the loss curve can be obtained for a general distribution. This is to be expected since a general distribution would include such cases as the summation of distinct relaxation processes. However, one might expect to obtain a reasonable result for symmetrical distribution functions which are independent of temperature.

We considered above that the distribution might arise from a distribution of τ_0 and H values. If we consider the special case that there is a distribution of τ_0 values, H being equal for every process, the resulting loss factor versus $1/T$ curves will be of the same shape, independent of the frequency of measurement. Distribution functions of the Cole–Cole type Equations (4.52) and (4.56) with the empirical parameter $\bar{\beta}$ independent of temperature can be regarded as a distribution of τ_0 and we can evaluate the half-width in this case. By a treatment similar to the single relaxation time case above, we may show that for $\bar{\beta}$ independent of temperature, the half-width $\delta_{1/2}$ is given by

$$ H = \frac{2 \cdot 303R \log \left[\alpha \pm \sqrt{\alpha^2 - 1}\right]}{\bar{\beta} \delta_{1/2}} \tag{4.101} $$

where $\alpha = \lceil 2 + \cos \bar{\beta}\pi/2 \rceil$.

Similarly, if we consider the Fuoss–Kirkwood distribution, with the distribution parameter m independent of temperature, we have

$$ H = \frac{2 \cdot 303R \log \left[r \pm \sqrt{r^2 - 1}\right]}{m \, \delta_{1/r}} \tag{4.102} $$

For $m = 1 = \bar{\beta}$, (4.101) and (4.102), reduce to (4.99).

It is readily noted that in Equation (4.102) the smaller m becomes (broadening distribution) the larger the half-width becomes.

4.2c Time-Temperature Superposition

Leaderman (1943) was the first to recognize the consequences of the similarity between creep curves measured at closely separated temperatures. As an example, consider the creep curves for polymethyl methacrylate taken between 20°c and −50°c shown in Figure 4.7(a) (McCrum and Morris, 1964). A superficial inspection shows that if each curve is displaced horizontally to the left it will more or less coincide with the curve at the next highest temperature. Thus, if the curve at 10°C is moved horizontally to the left to superpose on top of the 20°C curve and if the 0°C curve is then moved until it superposes on the combined 20°C and 10°C curves and so on, we ultimately obtain one curve, sometimes known as the master curve.

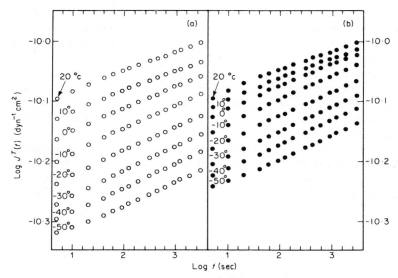

Figure 4.7. (a) Measured values of $J^T(t)$ for polymethyl methacrylate; (b) Values of the reduced compliance $J_p{}^T(t)$ obtained from $J^T(t)$ using Equation (4.112) with $T_0 = 20°\text{C}$. (McCrum and Morris, 1964).

In Figure 4.8 measurements are given of $J'(\omega)$ for polyisobutylene (Fitzgerald, Grandine and Ferry, 1953). These results are clearly also capable of being superposed to form a master curve. Now the relaxations shown in Figures 4.7 and 4.8 are of entirely different origin; that in polymethyl methacrylate is the β relaxation due to the oxycarbonyl side-group (see Chapters 1 and 8) and in polyisobutylene the α or glass–rubber relaxation (see Chapter 10). It is found that this striking similarity between viscoelastic parameters when measured at closely spaced temperatures is a common occurrence and is not confined to a few relaxations. The procedure of superposing curves at different times and temperatures has come to be known as time–temperature superposition.

We turn now to formulate a phenomenological theory to account for time–temperature superposition. The question to be answered is: does the shifting of creep curves have physical significance? An answer (which is yes, under certain circumstances) is drafted in the following paragraphs.

McCrum–Morris Reduction Equations

Let the creep at temperature T be governed by the distribution of retardation times $\phi_J{}^T(\ln \tau)$. Consider $\phi_J{}^T(\ln \tau)$ to obey the following hypothesis:

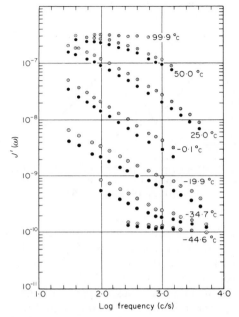

Figure 4.8. Measurements of $J'(\omega)$ versus log frequency (open circles) at different temperatures for polyisobutylene (Fitzgerald, Grandine and Ferry, 1953). The closed circles are the reduced values ($T_0 = 99.9°C$).

all retardation times in the distribution are displaced along the $\ln \tau$ axis by $\ln a_T$ when the temperature is changed from T_0 to T (Figure 4.9). The distribution $\phi_J^{T_0}(\ln \tau)$ is related to the distribution $\phi_J^T(\ln \tau)$

$$\phi_J^T(\ln \tau) = \phi_J^{T_0}(\ln \tau/a_T) \qquad (4.103)$$

It is found that for the great majority of relaxations $\ln a_T$ is a simple function of T being given either by the Arrhenius equation,

$$\ln a_T = \frac{H}{R}\left(\frac{1}{T} - \frac{1}{T_0}\right) \qquad (4.104)$$

or the equation of Williams, Landel and Ferry (1955)

$$\ln a_T = -\frac{C_1(T - T_0)}{C_2 + (T - T_0)} \qquad (4.105)$$

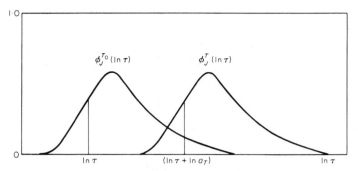

Figure 4.9. Graphical description of the hypothesis that the retardation spectra at T_0 and T are displaced by $\ln a_T$ but are identical in shape. (After, Alfrey 1948, Figure 62).

In Equation (4.104) H is the activation energy of the relaxation and in Equation (4.105) C_1 and C_2 are constants. The equation of Williams, Landel and Ferry (WLF) is nearly always applicable to the glass–rubber relaxation of amorphous polymers. The Arrhenius equation is usually applicable to relaxations in amorphous polymers occurring at temperatures below the glass transition (β, γ relaxations) and for relaxations in crystalline polymers.

Equations (4.103) to (4.105) account for the effect of temperature on the kinetics of the relaxation. It is also necessary to account for temperature induced changes in J_U and J_R. This can be done by means of parameters b_T, c_T and d_T (McCrum and Morris, 1964),

$$J_R{}^T - J_U{}^T = b_T(J_R{}^{T_0} - J_U{}^{T_0}) \tag{4.106}$$

$$J_U{}^T = c_T J_U{}^{T_0} \tag{4.107}$$

$$J_R{}^T = d_T J_R{}^{T_0} \tag{4.108}$$

$J_R{}^T$ and $J_R{}^{T_0}$ are the relaxed compliances at T and T_0. $J_U{}^T$ and $J_U{}^{T_0}$ are the unrelaxed compliances at T and T_0. Only two of the three parameters b_T, c_T and d_T are needed to describe the temperature variation of the relaxation.

It follows from Equations (4.41), (4.103) and (4.106) that the creep compliance at T and time t, $J^T(t)$, is given by

$$J^T(t) = c_T J_U{}^{T_0} + b_T(J_R{}^{T_0} - J_U{}^{T_0}) \int_{-\infty}^{\infty} \phi_J{}^{T_0}\left(\ln\frac{\tau}{a_T}\right)(1 - e^{-t/\tau})\,d\ln\tau \tag{4.109}$$

But the creep compliance at T_0 and time t/a_T is given by,

$$J^{T_0}\left(\frac{t}{a_T}\right) = J_U{}^{T_0} + (J_R{}^{T_0} - J_U{}^{T_0})\int_{-\infty}^{\infty} \phi_J{}^{T_0}(\ln\tau)(1 - e^{-(t/a_T)/\tau})\, d\ln\tau$$

$$= J_U{}^{T_0} + (J_R{}^{T_0} - J_U{}^{T_0})\int_{-\infty}^{\infty} \phi_J{}^{T_0}\left(\ln\frac{\tau}{a_T}\right)(1 - e^{-t/\tau})\, d\ln\tau$$

$$= \frac{1}{b_T}J^T(t) + J_U{}^{T_0}\left(1 - \frac{c_T}{b_T}\right) \tag{4.110}$$

Since experiment usually gives c_T and d_T, we have, from Equations (4.106)–(4.108),

$$b_T = \frac{d_T J_R{}^{T_0} - c_T J_U{}^{T_0}}{J_R{}^{T_0} - J_U{}^{T_0}} \tag{4.111}$$

$$\therefore \quad J^{T_0}\left(\frac{t}{a_T}\right) = J^T(t)\frac{J_R{}^{T_0} - J_U{}^{T_0}}{d_T J_R{}^{T_0} - c_T J_U{}^{T_0}} + J_U{}^{T_0} J_R{}^{T_0}\frac{d_T - c_T}{d_T J_R{}^{T_0} - c_T J_U{}^{T_0}} \tag{4.112}$$

The other viscoelastic parameters may be derived by an analogous argument.

$$J''^{T_0}(a_T\omega) = J''^T(\omega)\frac{J_R{}^{T_0} - J_U{}^{T_0}}{d_T J_R{}^{T_0} - c_T J_U{}^{T_0}} \tag{4.113}$$

$$J'^{T_0}(a_T\omega) = J'^T(\omega)\frac{J_R{}^{T_0} - J_U{}^{T_0}}{d_T J_R{}^{T_0} - c_T J_U{}^{T_0}} + J_U{}^{T_0} J_R{}^{T_0}\frac{d_T - c_T}{d_T J_R{}^{T_0} - c_T J_U{}^{T_0}} \tag{4.114}$$

$$G^{T_0}\left(\frac{t}{a_T}\right) = G^T(t)\frac{G_U{}^{T_0} - G_R{}^{T_0}}{c_T{}^{-1}G_U{}^{T_0} - d_T{}^{-1}G_R{}^{T_0}} + G_U{}^{T_0}G_R{}^{T_0}\frac{c_T{}^{-1} - d_T{}^{-1}}{c_T{}^{-1}G_U{}^{T_0} - d_T{}^{-1}G_R{}^{T_0}} \tag{4.115}$$

$$G''^{T_0}(a_T\omega) = G''^T(\omega)\frac{G_U{}^{T_0} - G_R{}^{T_0}}{c_T{}^{-1}G_U{}^{T_0} - d_T{}^{-1}G_R{}^{T_0}} \tag{4.116}$$

$$G'^{T_0}(a_T\omega) = G'^T(\omega)\frac{(G_U{}^{T_0} - G_R{}^{T_0})}{c_T{}^{-1}G_U{}^{T_0} - d_T{}^{-1}G_R{}^{T_0}} + G_U{}^{T_0}G_R{}^{T_0}\frac{(c_T{}^{-1} - d_T{}^{-1})}{c_T{}^{-1}G_U{}^{T_0} - d_T{}^{-1}G_R{}^{T_0}} \tag{4.117}$$

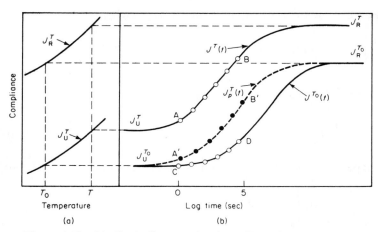

Figure 4.10. Idealized diagram showing effect of the temperature dependencies of $J_U{}^T$ and $J_R{}^T$ on the creep compliance. (a) Plot of $J_R{}^T$ and $J_U{}^T$ against temperature; (b) Dependence of $J^T(t)$ and $J^{T_0}(t)$ on log time. AB and CD represent the measurable portions of the curves (i.e. for times between ca.1 and 10^5 sec). A′B′ represents AB after reduction.

The implications of these equations are described graphically in Figure 4.10. The temperature dependence of $J_U{}^T$ and $J_R{}^T$ is shown in Figure 4.10(a).* The curves $J^{T_0}(t)$ and $J^T(t)$ for temperatures T_0 and T are plotted by full lines in Figure 4.10(b). $J^T(t)$ cannot be superposed by a horizontal shift on $J^{T_0}(t)$ since the limiting compliances are unequal. According to Equation (4.112) if $J^T(t)$ is modified, or reduced, then the reduced $J^T(t)$ (which is written $J_p{}^T(t)$ and is identical to the right-hand side of Equation 4.112) is given in terms of the creep curve at T_0,

$$J^{T_0}\left(\frac{t}{a_T}\right) = J_p{}^T(t) \tag{4.118}$$

Graphically this means that the reduced curve at T can be superposed upon the curve at T_0 by a horizontal shift along the log t axis equal to $\log a_T$.

* The temperature coefficients of $J_U{}^T$ and $J_R{}^T$ are both taken in Figure 4.10(a) to be positive. This is the usual behaviour in solids and is due to thermal expansion. The argument of this section still holds when the temperature coefficients have different signs, as, for instance, at the α or glass–rubber relaxation. For this relaxation the temperature coefficient of $J_U{}^T$ is positive and that of $J_R{}^T$ is negative.

In principle, $J^T(t)$ and $J^{T_0}(t)$ can be determined over their whole length. In fact, measurements can only be made between say 1 sec and approximately 10^5 sec. The shorter time is limited by the inertia of the creep apparatus (see Chapter 5) and the longer time by the patience of the observer. Suppose that only the portions of the curves AB and CD are measurable, Figure 4.10(b). Now, by inspection, AB, may be superposed, apparently, by a horizontal shift upon $J^{T_0}(t)$ of which only CD has been determined. But this of course is a delusion due to the measurements of $J^T(t)$ and $J^{T_0}(t)$ not including the portions of the curves close to the limiting compliances. The superposition of curves such as AB upon CD gives information (a master curve and shift factors) which are without physical meaning. But the superposition of the reduced curves A'B' upon CD is meaningful. The master curve is $J^{T_0}(t)$: each reduced curve $J_p{}^T(t)$ must be shifted to superpose upon the master curve by log a_T which is equal to the shift of the retardation spectra at T and T_0 (Equation 4.103, Figure 4.9).

What are the magnitudes of c_T and d_T? For the α relaxation of an amorphous polymer d_T is known theoretically from the theory of rubber elasticity (Chapter 3)

$$d_T = \frac{\rho_0 T_0}{\rho T} \tag{4.119}$$

in which ρ_0 and ρ are the densities of the specimen at T_0 and T. To obtain an estimate of the expected range of d_T take $T_0 = 300°$K, $T = 400°$K, thermal expansion coefficient $= 3 \times 10^{-4}°$K^{-1} giving

$$d_{400°\text{K}} = 0.78$$

There is unfortunately no reliable theory for the temperature dependence of the other limiting compliances. Estimates of c_T can, however, be made from ultrasonic data when mechanical damping is low; the temperature dependence of the high-frequency modulus is taken to be equal to the temperature dependence of the unrelaxed modulus. Using the experimental data of Wada and others (1959) for polymethyl methacrylate McCrum and Morris (1964), Table 4.2, obtain (taking $T = -80°$c and $T_0 = 20°$c)

$$c_{-80°\text{C}} = 0.817$$

The value is in good agreement with values of temperature coefficient of high-frequency modulus above and below the glass transition quoted by Work (1956). For the usual range of superposition temperatures c_T and

d_T are of the same order of magnitude, being normally found in the range from 0·7 to 1·5.

In the β range of polymethyl methacrylate McCrum and Morris (1964) obtained d_T using the hypothesis $c_T = d_T$. Measurements of d_T for the β relaxation cannot be made directly by measuring the temperature dependence of the modulus at temperatures above the relaxation region since the β and α mechanisms overlap. The assumption $c_T = d_T$ is equivalent to the assumption that the fraction of G_U relaxed by the β mechanism does not vary with temperature.

The effect of the quantities c_T and d_T given in Table 4.2 in the reduction equation is demonstrated in Figure 4.7 (b). The effect of the reduction is to squash the curves noticeably together. This leads to a very significant difference in the value of H calculated according to Equation (4.104). The temperature dependence of the shift factors obtained by horizontal displacements of the unreduced curves (Figure 4.7a) yields $H = 28$ cal/mole. The shift factor for the reduced curves (Figure 4.7b) yield $H = 17$ kcal/mole (McCrum and Morris, 1964). The latter value is in good agreement with the value obtained from internal friction, $H = 18$ kcal/mole (Deutsch, Hoff and Reddish, 1954; Heijboer, 1956). Using the hypothesis $c_T = d_T$, it may be shown that the activation energy obtained from internal friction is the correct one (McCrum and Morris, 1964).

Table 4.2. A table of log c_T values
(McCrum & Morris, 1964)

Temperature (°C)	Wada, Hirose, Asano and Fukitomi (1959) ($3·3 \times 10^4$ c/s)
−80	−0·088
−70	−0·084
−60	−0·080
−50	−0·076
−40	−0·071
−30	−0·064
−20	−0·056
−10	−0·047
0	−0·036
10	−0·019
20	0·000
30	0·018
40	0·039

The reduction equations can, under certain circumstances, be simplified greatly. For the α relaxation of an amorphous polymer, $J_R/J_U \sim 10^3$. (Note that, for instance, in Figure 4.8 the values of $J'(\omega)$ at $-44 \cdot 6$ and $99 \cdot 9°C$, which are approximately equal to J_U and J_R, give $J_R/J_U \sim 3 \times 10^3$). Therefore, neglecting terms in J_U, we have, from Equation (4.112),

$$J^{T_0}\left(\frac{t}{a_T}\right) = \frac{1}{d_T}(J^T(t) - c_T J_U{}^{T_0}) + J_U{}^{T_0} \qquad (4.120a)$$

The analogous equation for $J'(\omega)$ is

$$J'^{T_0}(a_T\omega) = \frac{1}{d_T}(J^T(\omega) - c_T J_U{}^{T_0}) + J_U{}^{T_0} \qquad (4.121)$$

Equations of this form were first derived and used (with the approximation $c_T = 1$) by Ferry and Fitzgerald (1953) to reduce dynamic compliance measurements at temperatures such that $(J^T(\omega) - c_T J_U{}^{T_0})$ is of the order of $J_U{}^{T_0}$. For most temperatures in the α region of an amorphous polymer $J^T(t) \gg c_T J_U{}^{T_0}$. For these temperatures Equation (4.120a) becomes

$$J^{T_0}\left(\frac{t}{a_T}\right) = \frac{1}{d_T} J^T(t) \qquad (4.122a)$$

Take $J(t)$ to rise through three decades as the relaxation progresses, see Figure 4.10b: for the first decade the reduction should be performed using Equation (4.120a); for the second and third decades, to within the usual experimental error, Equation (4.122a) is indistinguishable from Equation (4.120a) (Ferry, 1961).

An example of the use of Equation (4.121) is shown in Figure 4.8. The open circles are measured values of $J'^T(\omega)$ (Fitzgerald, Grandine and Ferry, 1953). The full circles are the reduced values: $T_0 = 99 \cdot 9°C$, thermal expansion coefficient $= 5 \cdot 8 \times 10^{-4}$ and $J_U{}^{T_0} = 0 \cdot 11 \times 10^{-9}$ cm^2 dyn^{-1}. The reduction is made using the approximation $c_T = 1$.

For the stress relaxation modulus the equations exactly analogous to Equations (120a) and (122a) for $G_U{}^{T_0}/G_R{}^{T_0} \sim 10^3$ are

$$G^{T_0}\left(\frac{t}{a_T}\right) = c_T\left[G^T(t) - \frac{G_R{}^{T_0}}{d_T}\right] + G_R{}^{T_0} \qquad (4.120b)$$

$$G^{T_0}\left(\frac{t}{a_T}\right) = c_T G^T(t) \qquad (4.122b)$$

Only for approximately the last decade of the relaxation, that is for

$$\left(10 > \frac{G^T(t)}{G_R{}^T} > 1 \right)$$

is Equation (4.120b) distinguishable from Equation (4.122b). Equations (4.120b) and (4.122b) [as opposed to Equations (4.120a) and (4.122a)] have not been used for polymeric relaxations.

For the α relaxation of an amorphous polymer d_T is known from theory, Equation (4.119), whereas c_T is not. Rewriting Equation (4.120b) in the form

$$G^{T_0}\left(\frac{t}{a_T}\right) = d_T G^T(t) + (G^T(t) - G_R{}^T)(c_T - d_T)$$

using Equation (4.108) we see that, to within an experimental error of 1 %, the equation

$$G^{T_0}\left(\frac{t}{a_T}\right) = d_T G^T(t) \tag{4.123}$$

is valid for

$$d_T G^T(t) > 10^2 (G^T(t) - G_R{}^T)(c_T - d_T)$$

In general d_T is of the order 1 and $(c_T - d_T)$ of the order of $0 \cdot 1$. Hence Equation (4.123) is a valid approximation to Equation (4.122b) for

$$G^T(t) > 10(G^T(t) - G_R{}^T)$$

$$\frac{G^T(t) - G_R{}^T}{G^T(t)} < \frac{1}{10}$$

That is, for the α relaxation of an amorphous polymer Equation (4.123) holds in approximately the last tenth of the last decade of modulus drop. A major virtue of Equation (4.123) is that it brings coincidence of the curves to be superposed as $G^T(t) \rightarrow G_R{}^T$.

4.3 RELAXATION MAGNITUDES IN CRYSTALLINE POLYMERS

It is sometimes difficult to assign a relaxation in a partially crystalline polymer to either the crystalline or the amorphous phase (see Chapter 2). If the polymer can be prepared in the completely amorphous state then the procedure is straightforward. For instance, partially crystalline polymethyl methacrylate is a two-phase solid containing crystals of iso- or syndiotactic molecules and amorphous regions of atactic molecules. As expected, partially crystalline polymethyl methacrylate exhibits a relaxation whose origin is obviously the same as the glass–rubber relaxation in the completely amorphous polymer (Chapter 10). The relaxations of the partially crystalline polymer can be compared to those of the completely amorphous polymer, and in this way relaxations are assigned to crystalline or amorphous regions. This method of assignment can be used only with crystalline polymers which may also be obtained in the amorphous state.

There are crystalline polymers, however, which cannot be obtained in the amorphous state (e.g. polyethylene, polytetrafluoroethylene, polyoxymethylene). For these polymers the physical structure can be varied by changing the rate of cooling from the melt or by annealing the solid after crystallization at temperatures below the melting point. These treatments affect the density (i.e. crystalline content), spherulite size, lamellar thickness, etc. The experimentalist can then observe the correlation between physical structure and characteristics of the relaxations (most typically temperature and frequency of the loss peak and relaxation magnitude). A good example of this method of analysis is shown in Figure 11.27(b). This graph shows the dependence of the logarithmic decrement on temperature for four specimens of polytetrafluoroethylene. The specimens were prepared by different heat treatments which caused the specimens to vary in crystalline content between 48 % and 92 %.

Consider the peaks in the logarithmic decrement marked α and γ, Figure 11.27(b). They both decrease in magnitude with increasing crystallinity (Illers and Jenckel, 1958a ; McCrum, 1958, 1959). Does this mean that both relaxations occur in the amorphous regions of the solid? This question can be answered in one of two ways. Firstly, we may set up an equivalent model (Takayanagi, 1963) and work out the consequences. Secondly, we may proceed by hypothesis. These alternative analytical methods will be considered in order.

4.3a Takayanagi's Model

Takayanagi has proposed a model for crystalline polymers applicable when the mechanical properties of the crystalline and amorphous regions are known. The model is illustrated in Figure 4.11. The mechanical properties of the specimen are considered to be duplicated by a unit cube of crystalline (C) and amorphous (A) material with complex moduli E^*_C and E^*_A. The amorphous region is of volume $\phi\lambda$ and the crystalline

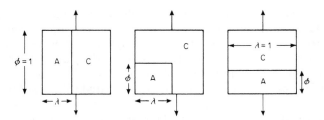

Figure 4.11. Takayanagi models. The unit cubes of amorphous (A) and crystalline (C) material are deformed in tension. Parallel model to the left, series model right.

region $(1 - \phi\lambda)$. The complex modulus of the model is

$$E^* = \left[\frac{\phi}{\lambda E^*_A + (1 - \lambda)E^*_C} + \frac{1 - \phi}{E^*_C} \right]^{-1} \qquad (4.124)$$

The limiting cases at the left ($\phi = 1$) and right ($\lambda = 1$) are sometimes known as the parallel model and the series model. In the parallel model the strain is everywhere constant but not the stress. In the series model the stress is everywhere constant but not the strain (neglecting the trivial case $E^*_A = E^*_C$). For $\lambda = 1$ we have, from Equation (4.124),

$$D^* = \phi D^*_A + (1 - \phi)D^*_C$$

and for $\phi = 1$,

$$E^* = \lambda E^*_A + (1 - \lambda)E^*_C$$

For an amorphous relaxation, the compliance increment in the series model ($\lambda = 1$) is

$$D_R - D_U = \phi(D_{AR} - D_{AU}) \qquad (4.125)$$

D_{AR} and D_{AU} are the relaxed and unrelaxed compliances of the amorphous region. The crystalline region is assumed not to relax. The compliance increment $(D_R - D_U)$ is thus a linear function of the amorphous fraction ϕ. Similarly for the parallel model ($\phi = 1$) the modulus increment for an amorphous relaxation is

$$E_U - E_R = \lambda(E_{AU} - E_{AR}) \qquad (4.126)$$

For this model the modulus increment $(E_U - E_R)$ is a linear function of λ, the amorphous fraction. Therefore, for the series model D''_{max} (proportional to $D_R - D_U$, see Equation 4.42b) is a measure of the amorphous content. Consequently, if there is *a priori* evidence that a polymer can be represented by a series model, D'' is the best parameter to use for analytical purposes. For the parallel model E'' is the best parameter since E''_{max} is proportional to $E_U - E_R$ (see Equation 4.45). It is not to be expected, however, that many crystalline polymers will be represented by either extreme model, Figure 4.11. *A priori* it is to be expected that an intermediate model will be more typical. This presents an obvious problem: how can we decide which of the many intermediate models (many values of ϕ and λ) is the best?

A second problem with the Takayanagi model is that the moduli of the amorphous and the crystalline portions of the solid are often unknown. This is particularly true of polymers which cannot ever be obtained in the completely amorphous or completely crystalline state. For these polymers there is no alternative other than to assign relaxations to the crystalline or amorphous regions by means of a hypothesis.

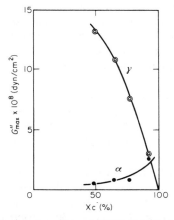

Figure 4.12. Maximum values of $G''(G''_{max})$ at the α and γ relaxations in polytetrafluoroethylene plotted against crystallinity, X_C, in per cent (After Ohzawa and Wada, 1964).

4.3b Assignment by Hypothesis

The hypothesis is that a particular loss parameter (say Λ or G'') is proportional to the volume fraction of the lossy part of the solid. According to the hypothesis of Illers and Jenckel (1958a) and McCrum (1958, 1959) the value of Λ at the peak, Λ_{max} is proportional to the volume fraction of lossy material in polytetrafluoroethylene. That is to say, the specific loss (see Equation 1.22b) is proportional to the volume fraction of lossy material. The α and γ relaxations, (Figure 11.27b), occur therefore in the amorphous regions. The hypothesis of Ohzawa and Wada (1964), however, is that G''_{max} for polytetrafluoroethylene is proportional to the volume fraction of lossy material. Figure 4.12 shows plots due to Ohzawa and Wada of the data of McCrum in which G''_{max} is plotted against crystallinity for the α and γ relaxations in polytetrafluoroethylene. For the γ relaxation both G''_{max} and Λ_{max} (see Figure 11.27b) decrease with increasing crystallinity. But for the α relaxation Λ_{max} decreases but G''_{max} increases with increasing crystallinity. Therefore, according to Ohzawa and Wada (1964) the α relaxation cannot occur in the amorphous fraction of polytetrafluoroethylene so that the original assignment of Illers and Jenckel (1958a) and McCrum (1959) is incorrect.

The only way out of this impasse is to accept the use of Λ_{max} or G''_{max} as a working hypothesis and to seek to confirm or reject the derived conclusions by other evidence (McCrum and Morris, 1965). For polytetrafluoroethylene there is, in fact, evidence of a glassy transition in the region of 130°c (see Chapter 11). This would seem to confirm the hypothesis of Illers and Jenckel (1958a) and McCrum (1958, 1959) that the α relaxation in polytetrafluoroethylene occurs in the amorphous regions of the solid and that for this polymer the equivalent model is probably not the parallel model.

Of course, the fundamental question is how are stress and strain distributed throughout the solid? This problem also occurs in the calculation of the elastic properties of a metal made up of small single crystals disposed at all possible orientations (Hearmon, 1956). Different solutions are obtained by assuming, (a) that strain is everywhere uniform but not the stress (Voigt, 1910), (b) that stress is everywhere uniform but not strain (Reuss, 1929). The experimental values fall in between solutions (a) and (b) (Hearmon, 1956). Clearly (a) and (b) are equivalent to the parallel and series models of Takayanagi. For a crystalline polymer the probability of a model somewhere between the extreme models is *a priori* more likely than either the parallel or the series models themselves, and the parameter Λ is, for this reason, to be preferred to G'' or J''. The use of Λ is nevertheless a hypothesis (albeit the best available) to be confirmed or rejected by the validity of the conclusions to which it leads.

5

Molecular Theories of Relaxation

5.1 SURVEY OF PROPOSED MOLECULAR MECHANISMS

Unlike the phenomenological theories, the molecular theories attempt to explain the relaxation behaviour of polymers in terms of proposed mechanisms for the motions of polymer molecules. Since most polymers exhibit more than one mechanical or dielectric relaxation region, no single molecular mechanism is adequate to describe the behaviour over a wide frequency and temperature range. Hence, before reviewing the molecular theories, we will survey generally the different types of relaxation regions observed for solid polymers. To aid this discussion it is convenient to introduce a system of nomenclature for labelling the different loss regions.

In order to identify and compare the locations of loss peaks for different polymers, the peaks are often labelled with the Greek letters α, β, γ, etc. According to this nomenclature, which was first suggested by Deutsch, Hoff and Reddish (1954), the α loss peak corresponds to the relaxation observed at the highest temperature (at a given frequency) or the lowest frequency (at a given temperature). The β and γ symbols then apply to the other relaxation regions in order of decreasing temperature or increasing frequency.

For polymers in the amorphous state a relaxation region associated with the glass transition is observed at temperatures around and above T_g (see Chapter 2). This relaxation is usually labelled α and is referred to as the primary or glass–rubber relaxation. It is by far the most pronounced mechanical relaxation, the shear modulus increasing with increasing frequency from a rubber-like value of $G_R \approx 10^7$ dyn/cm^2 to a glass-like value of $G_U \approx 10^{10.5}$ dyn/cm^2. Frequently, the glass–rubber relaxation is also the most prominent dielectrically. From a molecular point of view it has been widely accepted for many years that the glass–rubber relaxation results from large-scale conformational rearrangements of the polymer chain backbone. Such rearrangements occur by a mechanism of hindered rotation around main-chain bonds. For dipolar chains, Mikhailov (1955) has termed these long-range motions 'dipole-elastic' processes. At present the majority of molecular theories of mechanical and dielectric relaxation are concerned with this mechanism (Section 5.3 below).

In addition to the glass–rubber relaxation, amorphous polymers usually exhibit at least one secondary relaxation region. The secondary loss regions (β, γ, δ relaxations) result from motions within the polymer in the *glass-like* state. In this state the main chains are effectively 'frozen in' so that these relaxations cannot be due to large-scale rearrangements of the main polymer chain. Since the molecules of most amorphous polymers contain side-groups capable of undergoing hindered rotations independently of the chain backbone, the secondary relaxations are often ascribed to such rotations. For example, the dielectric and mechanical β peaks in the methacrylate polymer series are generally thought to involve rotations of the —COOR side-group (Chapter 8). Within this category fall polycyclohexyl methacrylate and related polymers, which show loss peaks ascribed to the 'chair-to-chair' motions of the cyclohexyl groups in the side-chains (Heijboer, 1956; see Chapter 1). However, certain linear polymers considered to be largely amorphous, for example, polyvinyl chloride (Chapter 11) and the polyvinyl acetals (Chapter 9), show dielectric β peaks which cannot result from side-group rotations. Hence, some limited local motions of the chain backbone may be possible in the glassy state. The local motions responsible for secondary dielectric loss peaks have been termed 'dipole radical' processes by Mikhailov (1955). Proposed theories of these local motions are discussed in Sections 5.4 and 5.5 below.

An understanding of molecular mechanisms of relaxation in crystalline polymers is complicated by the complexity and uncertainty of their physical structure. As discussed in Chapter 2, the so-called 'fringed-micelle' model regards crystalline polymers as two-phase systems containing both ordered crystalline regions and amorphous regions in which the arrangement of chain segments is completely disordered. On the other hand, according to the one-phase model the molecules of a crystalline polymer are considered to form a crystal lattice containing defects of various kinds. Takayanagi (1963) has proposed a third model based on the hypothesis that disordered regions occur both between and within the crystal lamellae.

The α, β, γ nomenclature is also used in labelling relaxations in crystalline polymers. These symbols do not imply, in the case of crystalline polymers, the same molecular mechanisms with which they are often associated in amorphous polymers. For example, the α label is occasionally given to the region of high mechanical tan δ at temperatures very close to the melting point of a crystalline polymer. Most frequently, the α symbol is given to a loss peak associated with the *crystalline* regions often observed some 50 to 100°C below the melting point. An example in the latter category is the well-known α peak in polyethylene which occurs at about 50°C at 1 c/s (Chapter 10). Various molecular interpretations have been proposed for this loss region. Takayanagi (1963) has suggested that the molecular motions occur in the chain-folded lamellae and are of two kinds involving, respectively, translational motions along the chain axis and the local torsional oscillations around the chain axis. These torsional motions, which are termed 'local mode' processes by Japanese workers (Yamafuji and Ishida, 1962; Saitô and others, 1963) will be discussed in Section 5.5b below. On the other hand, Pechold, Blasenbrey and Woerner (1963) have proposed that the α peak in polyethylene is associated with motions of lattice defects, where the defects take the form of 'kinks' in the predominantly planar zig-zag chain. The dielectric α peak of (oxidized) polyethylene has been interpreted by Tuijnman (1963a, b) in terms of the rotation of the complete planar zig-zag chain about the chain (c) axis in the crystal. Tuijnman's theory, which is based on the concept of diffusion over a potential barrier, is considered in Chapter 10. Many other crystalline polymers, notably polyvinyl alcohol (Chapter 9), polypropylene (Chapter 10), polychlorotrifluoroethylene (Chapter 11), polyethylene terephthalate (Chapter 13) and polyoxymethylene (Chapter 14) exhibit crystalline (α) loss peaks, and in a few of these cases molecular mechanisms have been proposed. Polytetrafluoroethylene (Chapter 11) also shows a crystalline relaxation thought to be associated with oscillations around the helical chain axis. This relaxation, however, occurs at about 20°C (280°C below the melting point) and is usually labelled β.

In addition to the relaxation mechanisms occurring in the crystalline phase, crystalline polymers often show two loss peaks associated with the disordered or amorphous regions. As shown in Chapter 10, polyethylene exhibits such peaks at about -20°C (β peak) and -120°C (γ peak), respectively, at 1 c/s. Other crystalline polymers showing two amorphous peaks include polyethylene terephthalate (β peak at 82°C, γ peak at -65°C), polyhexamethylene adipamide (α peak at 65°C, γ peak at -123°C), polytetrafluoroethylene (α peak at 127°C, γ peak at -97°C) and polychlorotrifluoroethylene (β peak at 100°C, γ peak at -20°C). The higher-temperature peaks are often attributed to movements of relatively large sections of chain in the amorphous regions, and these peaks are often clearly related to the glass transition. The lower-temperature amorphous peaks, which are often labelled γ, are thought to involve limited motions of relatively short chain segments, and 'crankshaft'-type motions have occasionally been proposed (see Section 5.5a). Takayanagi (1963), on the other hand, attributes these low-temperature peaks to the 'local mode' mechanism outlined in Section 5.5b below. It is appropriate at this stage to mention the amorphous dielectric relaxation found for polyvinyl alcohol (Chapter 9). Kurosaki and Furumaya (1960) have interpreted their data for this relaxation in terms of a chain mechanism, proposed by Sack (1952), involving successive dipole rotations within linear hydrogen-bond chains. An outline of Sack's theory is given in Chapter 9.

To complete this general survey it should finally be added that relaxation effects in polymers can also arise from the presence of impurities of low molecular weight. Notable examples are polymethyl methacrylate and the 'nylons,' each of which show relaxations in the presence of absorbed water. The precise mechanisms of such relaxations are not clearly understood.

From the above remarks it is clear that a wide variety of molecular mechanisms have been proposed to account for relaxation effects in solid polymers. Many of the suggested processes have not been conclusively verified and are not widely accepted. Nevertheless, it seems appropriate at this time to review several molecular theories in the hope that they may form the foundation for future developments. Several of the molecular theories apply to specific polymer systems and will be considered in Chapters 8 to 14. In this chapter we propose to discuss those theories which are considered to be of more general applicability. This discussion is conveniently divided into three sections. In Section 5.3 we will outline several theories which have been applied to the glass–rubber relaxation of amorphous polymers. Such theories generally consider the large-scale motions of a polymer chain suspended in a viscous continuum. The so-called 'barrier theories' are considered in Section 5.4. The barrier models specify the interactions of a moving unit with neighbouring units by means of a potential energy barrier. Relaxation is then envisaged as the 'jumping' of units across the barrier from one position of equilibrium to another. In Section 5.5 we outline the proposed 'crankshaft' mechanism and also some recent 'local mode' theories, which consider that secondary relaxations can result from damped oscillations of molecular groups close to their equilibrium positions. Before discussing the theories of polymer relaxations, however, it is appropriate to present the results of Debye's theory for small molecules.

5.2 THEORY OF DEBYE FOR SMALL RIGID MOLECULES

Debye (1929) considered the dielectric behaviour of small molecules in the liquid state by treating the molecules as rigid spheres rotating in a viscous continuum. From the equations of Brownian motion both in the absence and in the presence of an electric field he obtained the result

$$\frac{\varepsilon^* - 1}{\varepsilon^* + 2} = \frac{4}{3}\pi N \left(\alpha_e + \frac{\mu_0^2}{3kT} \frac{1}{1 + i\omega\tau^*} \right) \tag{5.1}$$

where N is the number of dipoles per cm^3, α_e is the electronic polarizability, μ_0 is the dipole moment for the molecule and τ^*, the *intrinsic* relaxation time, given by

$$\tau^* = \frac{3V\eta}{kT} \tag{5.2}$$

where V is the volume of a molecule and η the viscosity of the medium in which the dipole turns.

Equation (5.1) may be rearranged with the aid of the Mosotti–Clausius relation

$$\frac{\varepsilon_U - 1}{\varepsilon_U + 2} = \frac{4}{3}\pi N \alpha_e \tag{5.3}$$

giving

$$\varepsilon^* = \varepsilon_U + \frac{\varepsilon_R - \varepsilon_U}{1 + i\omega\tau} \tag{5.4}$$

where

$$\tau = \tau^* \frac{\varepsilon_R + 2}{\varepsilon_U + 2} \tag{5.5}$$

Thus, according to the Debye theory, a medium consisting of small dipoles will be characterized by a single relaxation time τ which is related to the molecular relaxation time τ^* by means of Equation (5.5).

Debye's theory has been modified by Onsager (1936) to include the reaction field giving the relation between ε_R and μ_0^2 (see Chapter 3). Hill (1961) generalized the Onsager theory to alternating fields and obtained a relation which, although complex in form, is numerically equivalent to Equation (5.4) for reasonable values of ε_R and ε_U. However, in Hill's theory, Equation (5.5) is replaced by $\tau \approx \tau^*$.

The available molecular theories for simple molecules have been reviewed by Fröhlich (1949), Böttcher (1952) and Smyth (1955) among others. With the exception of the theory of Perrin (1934) for ellipsoidal molecules, these theories generally predict a single relaxation time expression.

5.3 THEORIES OF THE GLASS–RUBBER RELAXATION

As mentioned above, the glass–rubber relaxation has for many years been thought to result from large-scale conformational rearrangements of the chain backbone. These rearrangements involve cooperative thermal motions of individual chain segments. The hindrance to these 'micro-Brownian' motions can be described in terms of viscous or frictional forces which result from interactions of the moving segment with neighbouring molecules or with segments within the same chain (internal viscosity). Relaxation arises from the diffusion of chains to a new equilibrium distribution of conformations subsequent to the application of an external stress or electric field. It should be added that the distribution of conformations and the rates of conformational rearrangements are only slightly perturbed by the applied stress or electric field if the latter are sufficiently small to correspond to linear behaviour. Theories based on this mechanism involve the solution of diffusion equations in multidimensional chain space and yield results involving distributions of relaxation times instead of the single relaxation time predicted for small molecules. Several early attempts were made to describe mechanical relaxation in polymers in terms of this type of diffusion process (Alfrey, 1944; Kirkwood, 1946; Frenkel, 1946; Kuhn and Kuhn, 1945, 1946). Kirkwood (1949, 1954) also considered the problem of chain diffusion in the most general way by specifying chain conformations by the coordinates of each atom in the chain. The earliest dielectric theories of this kind were presented by Kirkwood and Fuoss (1941) and also Kuhn (1950) who introduced the effect of internal viscosity. Kuhn's theory has been reviewed by Brouckère and Mandel (1958).

Our discussion of current molecular theories of the glass–rubber relaxation is conveniently divided into three sections depending on the details of the models employed. In Section 5.3a we will discuss those theories based on the model of a random Gaussian chain which is arbitrarily divided into a number of segments or 'submolecules' each of which is sufficiently long that its end-to-end distance may also be approximated by the Gaussian distribution. The motion of this chain is then described as the superposition of a number of normal modes after transforming the coordinates of the end of each submolecule into a set of normal co-ordinates. We may note immediately that this model excludes from consideration motions of chain segments which are shorter than the hypothetical submolecule. However, the neglect of these short-range motions greatly simplifies the mathematical treatment. In Section 5.3b we will outline some theoretical attempts to treat these local motions by considering more detailed models of polymer chains, and in Section 5.3c we shall describe some theories which relate the average relaxation time to the so-called 'free volume'.

5.3a Normal Mode Theory Based on the Gaussian Submolecular Model

Normal mode theories of the types outlined above were advanced inde-pendently by Kargin and Slonimskii (1948, 1949), Rouse (1953) and Bueche (1954) to account for mechanical relaxation in polymers. These mechanical theories were subsequently discussed and extended by Gotlib and Volkenstein (1953), Nakada (1955, 1960), Cerf (1955, 1959), Zimm (1956, 1960), Mooney (1959), Pao (1962) and Kästner (1962a, 1963). Reviews of these theories have been given by Tobolsky (1960), Ferry (1961) and Saitô and others (1963). The submolecular model has also been applied to the analysis of dielectric relaxation by Zimm (1956), Van Beek and Hermans (1957), Kästner (1961a, b; 1962b) and Stockmayer and Baur (1964).

Before discussing the normal mode theories in some detail, it should be emphasized that the many versions so far presented are all closely related with regard to the 'mechanism' of relaxation. Detailed differences between the various treatments are centred largely on the two main problems, namely, (1) diffusion and (2) the charge or dipole distribution along the polymer chain. The diffusion problem naturally enters into both the mechanical and dielectric theories, whilst the problem of the dipole distribution concerns only the dielectric theory. Regarding the problem of diffusion, the sole question concerns the effect of hydrodynamic inter-actions on the motional processes. This may be illustrated as follows. When a polymer chain segment moves relative to its liquid environment it will exert frictional forces on the liquid. These forces will modify, to a varying extent, the velocity distribution of the liquid medium in the vicinity of the polymer chain. This effect will, in turn, influence the motions of other segments of the same polymer coil. In the so-called 'free-draining' approximation the velocity of the liquid medium is assumed to be unaffected by the moving polymer molecules, i.e. the effect of hydro-dynamic interaction is neglected. We may note here, for example, that the original Rouse (1953) theory applies to the free-draining case, whereas in the later treatment of Zimm (1956) the effects of hydrodynamic interaction are introduced in an approximate way into the Rouse model. Concerning the question of charge distribution, it should be noted that the manner in which the dipoles are arranged within the polymer chain will have an important bearing on the dielectric behaviour of the polymer. In the dielectric theory of Zimm (1956) the two ends of the chain only were assumed to carry electric charges. Subsequently, Van Beek and Hermans (1957) considered the case in which charges of $+e$ and $-e$ alternated along the chain at the submolecule junction points. In the most recent treatment, due to Stockmayer and Baur (1964), several different charge distributions are considered. This latest work may be regarded as a generalization of all the dielectric normal mode theories to date, including those of Zimm, Van Beek and Hermans, and also the several models treated by Kästner (1961a, b; 1962b).

It should also be emphasized that many of the normal mode theories were basically designed to account for relaxation in dilute polymer solutions in which the polymer molecules are isolated and the solvent acts as the viscous continuum. In applying such theories to undiluted polymers it is usually assumed that polymer molecules themselves constitute the viscous environment and that the friction forces may be suitably modified (Bueche, 1952; Ferry, 1961).

Treatment of Rouse

The model chain adopted by Rouse consists of a large number (n) of freely jointed links, each link being statistically equivalent to a few (typically from about 1 to 10) monomer units and having a length l. The length, l, of each 'random' link will be related to bond lengths, valence angles and energy differences between stable rotational conformations of the real chain. This idealized random chain has been successfully employed to account for many features of the equilibrium elasticity of rubbers (Chapter 3). Its success is largely due to the fact that the conformation of both the real chain and the idealized chain can be approximated by a Gaussian probability distribution if the number of atoms in the chain exceeds about fifty and we are concerned only with small deformations (Flory, 1953; Treloar, 1958). Following Rouse's procedure we arbitrarily divide the random chain into v 'submolecules' each of which contains a sufficient number ($z = n/v$) of links so that its end-to-end distance can also be approximated by a Gaussian distribution (Figure 5.1). If one end of the ith submolecule is placed at the origin of a private coordinate system $OXYZ$ then, at equilibrium, the probability that the other end will lie at

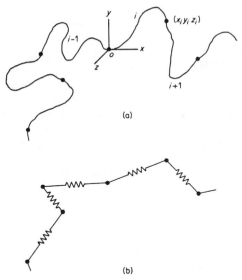

Figure 5.1. The Rouse model. (a) Subdivision of a chain molecule into submolecules; (b) The equivalent spring-bead representation.

the point x_i, y_i, z_i in the volume element $dx_i\, dy_i\, dz_i$ is (cf. Equation 3.33)

$$P_i(x_i y_i z_i)\, dx_i\, dy_i\, dz_i = \frac{b^3}{\pi^{3/2}} \exp[-b^2(x_i^2 + y_i^2 + z_i^2)]dx_i\, dy_i\, dz_i \quad (5.6)$$

where $b^2 = 3/(2zl^2)$.

The conformational probability of the entire chain can now be represented by a point in $3v$ dimensional space. The probability that this point lies at the point $x_1, y_1, \ldots, z_v$ in the volume element $dx_1\, dy_1 \ldots dz_v$ is given by

$$P_v\, dx_1 \ldots dz_v = \prod_{i=1}^{v} P_i(x_i y_i z_i)\, dx_i dy_i\, dz_i =$$

$$\left(\frac{b^3}{\pi^{3/2}}\right)^v \exp\left[-b^2 \sum_{i=1}^{v} x_i^2 + y_i^2 + z_i^2\right] dx_1 \ldots dz_v \quad (5.7)$$

Equation (5.7) shows that, at equilibrium, the most probable values of the x_i, y_i and z_i coordinates are zero. Hence, in the absence of external forces each flexible submolecule will exist for most of the time in a coiled-up conformation. If an external force is suddenly applied to the polymer, the submolecule end-to-end separations will initially be displaced from their equilibrium values to a less probable distribution of conformations. This will result in an overall decrease in entropy, ΔS, and hence an increase in Helmholtz free energy ΔA. Since all conformations of the model chain are assumed to have the same internal energy we may write $\Delta A = -T\Delta S$. The tendency for the free energy of the system to seek a minimum value causes the conformations to change gradually toward a new equilibrium distribution subsequent to the application of the external force.

A diffusion of the end of each submolecule will occur along each space coordinate, the velocity of flow along each coordinate being proportional to the driving force represented by the negative gradient of free energy. The rate of change of x_i due to motions of the junction point between submolecules i and $i-1$ may be written

$$\xi(\dot{x}_i)_{i-1} = \left(T\,\frac{\partial S_v}{\partial x_i} - \frac{\partial S_v}{\partial x_{i-1}}\right) \tag{5.8}$$

where ξ is a friction factor defined as the ratio of the net moving force to the velocity of a junction point. S_v is the entropy of a molecule of conformation $x_1, y_1, \ldots, z_v$ and is given by

$$S_v = k\ln P_v \tag{5.9}$$

where k is Boltzmann's constant.

Similarly the rate of change of x_i due to motions of the $i, i+1$ junction point is given by

$$\xi(\dot{x}_i)_{i+1} = \left(T\,\frac{\partial S_v}{\partial x_i} - \frac{\partial S_v}{\partial x_{i+1}}\right) \tag{5.10}$$

Combining Equations (5.8) and (5.10) we obtain for the net rate of change of length x_i,

$$\xi\dot{x}_i = T\left(2\frac{\partial S_v}{\partial x_i} - \frac{\partial S_v}{\partial x_{i-1}} - \frac{\partial S_v}{\partial x_{i+1}}\right) \tag{5.11}$$

From Equations (5.7), (5.9) and (5.11) we obtain

$$\xi\dot{x}_i + \frac{3kT}{zl^2}(2x_i - x_{i-1} - x_{i+1}) = 0 \tag{5.12}$$

Analogous equations exist for $\dot{y}_i$ and $\dot{z}_i$. Neglecting inertial forces, which are generally much smaller than the viscous and elastic forces, the $3v$ equations such as (5.12) are the equations of motion for a single random chain. The above model is thus equivalent to a series of Hooke's law springs of zero unstretched length joined flexibly together by beads and immersed in a viscous medium (see Figure 5.1b). The constant $3kT/zl^2$ is the 'entropy elastic' force constant for the submolecular springs. The mathematical details for solving these equations have been given many times (see references quoted above) and need not be reproduced here. The procedure involves a matrix method for transforming the x_i, y_i, z_i coordinates into a system of normal coordinates. The 3ν equations such as (5.12) are thus transformed into 3ν new equations each of which is a function of a single normal coordinate. The diffusion equation is subsequently obtained via the equation of continuity. For stress relaxation and dynamic mechanical experiments, respectively, solution of the diffusion equation yields results of the form

$$G(t) = NkT \sum_{p=1}^{\nu} e^{-t/\tau_p} \tag{5.13}$$

$$G' = NkT \sum_{p=1}^{\nu} \frac{\omega^2 \tau_p^2}{1+\omega^2 \tau_p^2} \tag{5.14}$$

where N is the number of molecules per cm^3 and τ_p, the relaxation time governing motions of the pth mode, is given by

$$\tau_p = zl^2 \zeta [24kT \sin^2 \{p\pi/2(\nu+1)\}]^{-1} \tag{5.15}$$

$$p = 1, 2, \dots, \nu$$

From Equations (5.13) and (5.14) we see that $G(t)$ and G' are determined by a discrete number of relaxation times each of which characterizes a given normal mode of motion. The longest and shortest relaxation times correspond to $p = 1$ and $p = \nu$ respectively. The physical nature of these normal modes has been discussed by Bueche (1954) and Zimm (1960). The mode corresponding to $p = 1$ is one in which the ends of the molecule move in opposite directions whilst the centre of the molecule remains stationary. In general, the pth mode corresponds to motions in $p + 1$ segments, a node existing between each segment. The first three normal modes are illustrated in Figure 5.2.

According to the Rouse model the submolecule is the shortest length of chain which can undergo relaxation. The motions of very short segments within the submolecules are ignored. At frequencies for which $\omega\tau_\nu > 0.1$ these excluded processes should begin to contribute to G'. This frequency therefore marks an upper limit of applicability of the theory. On this basis Rouse has shown that relaxation times for which

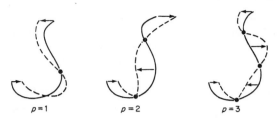

Figure 5.2. Illustration of the first three normal modes of a chain molecule (Zimm, 1960).

$p < v/5$ account for the dispersion of G' up to frequencies within about one decade of the range in which excluded processes are operative. In this case Equation (5.15) becomes (for $v \gg 1$)

$$\tau_p \approx \frac{v^2 z l^2 \xi}{6\pi^2 p^2 kT} \approx \frac{n^2 l^2 \xi_0}{6\pi^2 p^2 kT} \qquad (5.16)$$

where $\xi_0 = \xi/z$ is the friction coefficient appropriate to a single link.

Equation (5.16) illustrates that values of τ_p depend only on the parameters of the link, ξ_0 and l, and the number of links in the chain. They are not basically related to properties of the arbitrarily chosen submolecule. This equation also shows that three factors govern the temperature dependence of τ_p. First, there is the factor $1/T$. Second, the factor nl^2, which equals the equilibrium mean square separation of the chain ends, may vary slightly with temperature if there are strong energy differences between potential minima for rotations around bonds in the real chain. Third, the friction coefficient ξ_0 will decrease rapidly with increasing temperature. This last effect, which may be interpreted in terms of a variation of viscosity or 'free volume' (Section 5.3c) with temperature, is the predominant factor influencing the temperature dependence of τ_p. Since each τ_p is predicted to have the same temperature dependence, the molecular theory provides support for the empirical time–temperature superposition procedure outlined in Chapter 4. The horizontal shift factors (a_T) required in the construction of reduced master curves, are determined largely by the temperature dependence of ξ_0.

Treatment of Zimm

In the theory of Zimm (1956), hydrodynamic interaction between the moving submolecules is considered by a method introduced by Kirkwood and Riseman (1948) for calculating the viscosity of dilute polymer solutions. A parameter h, equal to $v^{1/2}\xi/(12\pi^3)^{1/2}\sigma\eta$, is introduced to specify the magnitude of the hydrodynamic interaction. η is the solvent viscosity and $\sigma^2 = zl^2$ is the mean square end-to-end distance for a submolecule. Zimm's theory, like the Rouse theory, predicts a discrete number of relaxation times, and $G(t)$ and $G'(\omega)$ may be represented by equations such as (5.13) and (5.14). For vanishing hydrodynamic interaction (free draining, $h \ll 1$), Zimm's expression for τ_p becomes identical with that of Rouse (Equation 5.15) (Tschoegl, 1963). However, for dominant hydrodynamic interaction (non-free draining, $h \gg 1$) the Zimm theory yields the following equation for τ_p,

$$\tau_p = \frac{\sigma^2 v^2 \xi}{24h\lambda'_p kT} \tag{5.17}$$

where the λ'_p are eigenvalues which, for $p = 1, 2, 3, 4, 5, 6$ and 7, have values of 4·04, 12·79, 24·2, 37·9, 53·5, 70·7 and 89·4 respectively (Zimm, Roe and Epstein, 1956). Zimm has also shown that τ_p is given by

$$\tau_p = \frac{6M\eta[\eta]}{\pi^2 p^2 RT} \tag{5.18}$$

for free-draining chains, and

$$\tau_p = \frac{M\eta[\eta]}{0\cdot586\lambda'_p RT} \tag{5.19}$$

for the non-free-draining case, where M is the molecular weight and $[\eta]$ the intrinsic viscosity.

Zimm calculated specifically the mean extension of a chain, $\langle x_N - x_0 \rangle$, when equal and opposite forces of magnitude f are applied to the ends of the chain in the x direction. For both free-draining and non-free-draining chains his result becomes

$$\langle x_N - x_0 \rangle \exp(i\omega t) = \frac{\sigma^2 f}{2kT} \sum_{p \text{ odd}} \frac{8v}{\pi^2 p^2} \cdot \frac{1}{1 + i\omega\tau'_p} \tag{5.20}$$

where x_N and x_0 are the x coordinates of the chain ends and $\tau'_p = 2\tau_p$. The factor $1/p^2$ inside the summation shows that the predominant contribution to the *compliance* comes from the longer relaxation time processes, particularly the process with time constant τ'_1. We may note for comparison that each relaxation process contributes the *same* amount (NkT)

to the shear *modulus* (Equation 5.14). Hence, the compliance relaxation should occur at lower frequencies than the relaxation of the modulus, a result found for the glass–rubber relaxation.

In applying his theory to the case of dielectric relaxation, Zimm assumed that the electric moment of a submolecule was proportional to its mean extension. This condition is fulfilled if each bond in the real chain has a dipole moment along the direction of the chain contour. The model would not be expected to apply to polymers with dipoles in the side-chains such as polyvinyl chloride and polyvinyl acetate. In the case specifically worked out by Zimm, only the two ends of the chain were assumed to carry electric charges. He obtained for the mean polarization per molecule per unit field (P_ω), for both free-draining and non-free-draining chains,

$$P_\omega = \frac{\mu^2 \sigma^2}{3kT} \sum_{p \text{ odd}} \frac{8v}{\pi^2 p^2} \frac{1}{1 + i\omega \tau'_p} \tag{5.21}$$

where $\mu^2 \sigma^2$ is the mean square moment of a submolecule. Equation (5.21) is clearly analogous to Equation (5.20) for the relaxation of the mechanical compliance.

At zero frequency Equation (5.21) becomes

$$P_0 = v \left(\frac{\mu^2 \sigma^2}{3kT} \right) \tag{5.22}$$

Thus

$$\frac{P_\omega}{P_0} = \sum_{p \text{ odd}} \frac{8}{\pi^2 p^2} \frac{1}{1 + i\omega \tau'_p} \tag{5.23}$$

As in the mechanical case the maximum contribution to the dielectric dispersion magnitude is predicted to come from the process with time constant τ'_1. Hence, Zimm's theory predicts a single non-symmetrical loss peak with a relaxation time essentially equal to τ'_1. It follows from Equations (5.18) and (5.19) that τ'_1 is a strong function of molecular weight for both free-draining and non-free-draining chains.

Treatment of Van Beek and Hermans

Van Beek and Hermans (1957) calculated the dielectric properties of a system of free-draining chains each of which was divided into $v - 1$ submolecules the ends of which carry charges $-e$ and $+e$ respectively. Alternatively their chain model may be regarded as a string of v beads

(v even) in which the kth bead carries a charge $(-1)^k e$. The polarization P_ω was found to be

$$P_\omega = \frac{\mu^2}{3kT} \sum_{p\,\text{odd}}^{v-1} \frac{a_p}{1 + i\omega\tau_p} \tag{5.24}$$

where

$$\tau_p = \tau_0 \sin^{-2}\left(\frac{p\pi}{2v}\right) \qquad a_p = \left(\frac{2}{v}\right)\sin^{-2}\left(\frac{\pi p}{v}\right) \tag{5.25}$$

and

$$\tau_0 = \frac{\xi_0 \sigma^2}{12kT} = \frac{\xi_0 n^2 l^2}{12 v^2 kT} \tag{5.26}$$

From these equations Van Beek and Hermans showed that two dielectric dispersions should exist, at frequencies in the region of

$$\omega_{\text{max},1} = \tau_0^{-1} = \frac{12 v^2 kT}{\xi_0 n^2 l^2} \tag{5.27}$$

and

$$\omega_{\text{max},2} = \frac{\pi^2}{4v^2 \tau_0} = \frac{3\pi^2 kT}{\xi_0 n^2 l^2} \tag{5.28}$$

respectively.

Van Beek and Hermans showed that the high-frequency dispersion (Equation 5.27) can be described to a good approximation by the simple Debye formula for a single relaxation time τ_0. However, since τ_0 depends on the value of v, which can be chosen arbitrarily, it has no molecular significance.

The low-frequency relaxation (Equation 5.28) has an effective relaxation time $(1/\omega_{\text{max},2})$ which is independent of any parameters describing the hypothetical submolecule. Furthermore this relaxation time depends on n^2 and is thus a strong function of molecular weight. We may finally add that τ_0 is essentially the same as τ'_N, in Zimm's terminology, and the long relaxation time $(4v^2 \tau_0/\pi^2)$ is equivalent to Zimm's τ'_1.

Treatment of Stockmayer and Baur

Using Zimm's model, Stockmayer and Baur (1964) have analysed the dielectric relaxation behaviour to be expected from polymers having different charge distributions along the chain contour. The theories of Zimm, and Van Beek and Hermans are thus specific cases of this more general treatment. According to Stockmayer and Baur's theory, dielectric relaxations are specified by the Fourier components of the electric charge distribution along the chain. Low-frequency dispersions of the type predicted by Zimm, and Van Beek and Hermans are shown to require chains whose units have a component of dipole moment in the direction of the chain contour. If the direction of this component is the same for all repeat units, the appropriate relaxation time is essentially τ'_1, which corresponds approximately to a rotation of the entire polymer chain (see Figure 5.2). As noted above τ'_1 is a strong function of molecular weight, and a low-frequency molecular weight dependent peak in polypropylene oxide (Baur and Stockmayer, 1959) has been attributed to this mechanism.

Comparison with Experimental Data

For dilute polymer solutions the Rouse and Zimm theories have successfully accounted for the frequency dependence of G' and G'' (Tschoegl and Ferry, 1963; Lamb and Matheson, 1964). Most applications of Equation (5.14) to solid polymers (Ferry, 1961) have involved replacing the summations by integrals, assuming that the discrete distribution of relaxation times may be approximated by a continuous spectrum. Ferry (1956) has shown that this procedure is valid to a good approximation if the three longest relaxation times ($\tau_1 \rightarrow \tau_3$), which are spaced too far apart, are ignored. The distribution of relaxation times $H_G(\ln \tau)$ is then given by

$$H_G(\ln \tau)\, \mathrm{d} \ln \tau = -NkT(\mathrm{d}p/\mathrm{d}\tau)\, \mathrm{d}\tau \qquad (5.29)$$

From Equations (5.16) and (5.29) we obtain

$$H_G(\ln \tau) = \left(\frac{Nnl}{2\pi}\right)\left(\frac{kT\zeta_0}{6}\right)^{1/2} \tau^{-1/2} \qquad (5.30)$$

Hence the Rouse theory predicts that a plot of $\log H_G(\ln \tau)$ against $\log \tau$ should have a slope of $-\frac{1}{2}$. For dominant hydrodynamic interaction, however, the Zimm theory predicts that this slope should be $-\frac{2}{3}$ (Ferry, 1961). A slope between $-\frac{1}{2}$ and $-\frac{2}{3}$ would therefore be expected depending on the amount of hydrodynamic interaction. It can likewise be shown that a double logarithmic plot of the retardation spectrum $L_J(\ln \tau)$ should have a theoretical slope between $\frac{1}{2}$ and $\frac{2}{3}$ (Ferry, 1961).

Figure 5.3 shows plots of $\log H_G(\ln \tau)$ against $\log \tau$, reduced to 100°C, for five methacrylate polymers in the glass–rubber relaxation region. The shapes of these spectra are typical of those obtained for the glass–rubber relaxation of amorphous polymers. The slope of each of these plots tends

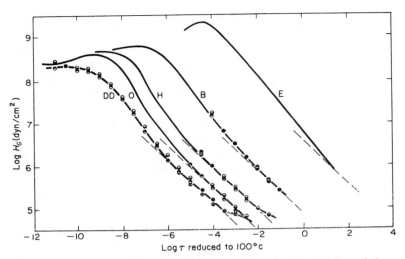

Figure 5.3. Mechanical relaxation spectra, reduced to 100°C, of the glass–rubber relaxation for E polyethyl methacrylate, B poly-n-butyl methacrylate, H poly-n-hexyl methacrylate, O poly-n-octyl methacrylate and DD poly-n-dodecyl methacrylate. Dashed lines are of slope $-\frac{1}{2}$ according to Equation (5.30) (Ferry, 1961).

to a value of $-\frac{1}{2}$ in the regions of long τ or low $H_G(\ln \tau)$. With decreasing τ the absolute values of the slopes increase, exceeding even the value of $\frac{2}{3}$, and then the curves pass through maxima. Thus the molecular theory appears to be valid only in the long τ region of this relaxation, covering about 2 to 3 decades of τ or about 1 to 1·5 decades of $H_G(\ln \tau)$. In this region the curves of Figure 5.3 can be used to evaluate the friction coefficient ξ_0. Rearranging Equation (5.30) we obtain

$$\log \xi_0 = 2 \log H_G + \log \tau + \log \left(\frac{6}{kT}\right) + 2 \log \left(\frac{2\pi}{Nnl}\right)$$

$$= 2 \log H_G + \log \tau + \log \left(\frac{6}{kT}\right) + 2 \log \left(\frac{2\pi M_0}{\rho N_0 l}\right) \quad (5.31)$$

where ρ is the density, N_0 is Avogadro's number and M_0 the molecular weight of the statistical link. Ferry (1961) has calculated ξ_0 values from this equation for a wide variety of polymers. In these calculations M_0 was taken as the monomer molecular weight and l as the root mean square end-to-end distance per square root of the number of monomer units.

The latter information was obtained from dilute solution measurements in a θ solvent (Flory, 1953), assuming that chain conformations in the bulk polymer are unperturbed by long-range 'excluded volume' effects. For the methacrylate polymers referred to in Figure 5.3, values of $\log \xi_0$ at 100°C ranged from -0.21 (ethyl) to -6.57 (n-dodecyl). This decrease reflects the increased chain mobility effected by increasing the length of the side-chain, an effect known as 'internal plasticization' (Chapter 8). So far, no *a priori* method has apparently been devised to calculate ξ_0 from molecular parameters. ξ_0 is probably related to factors such as free volume, intermolecular forces and chain flexibility which together determine the 'potential barriers' (see Section 5.4 below) hindering segmental motion.

The failure of the molecular theory in the region of short τ (Figure 5.3) is not surprising in view of the fact that the submolecule was taken as the shortest relaxing unit. The processes of short relaxation time probably involve motions of smaller segments within the submolecules. These local motions would be mainly responsible for $H_G(\ln \tau)$ values above 10^6 (Figure 5.3) and for the corresponding high-frequency moduli between about 10^7 and $10^{10.5}$ dyn/cm². In fact, the above theory, based on the concept of entropy elasticity, cannot reasonably account for moduli above about 10^8 dyn/cm². This fact can be seen qualitatively from Equation (5.14) which predicts that the limiting value of G' at high frequencies ($\omega\tau_p \gg 1$, for all p) is given by

$$G'_U = vNkT = \frac{\rho RT}{M_0 z} \tag{5.32}$$

If we let the molecular weight of the statistical link $M_0 = 50$, the number of links per submolecule $z = 5$, and assume that $\rho = 1$ g/cm³ and that $T = 300°$K, we obtain

$$G'_U = \frac{1 \times 8.314 \times 10^7 \times 300}{50 \times 5} \approx 1 \times 10^8 \text{ dyn/cm}^2 \tag{5.33}$$

A more rigorous calculation for polyisobutylene (Williams, 1962a) suggests that the Rouse theory cannot be expected to predict the experimental results when G' exceeds about 5.0×10^6 to 1.55×10^7 dyn/cm². These values compare with the rubber elastic value of 3.3×10^6 dyn/cm².

We thus conclude that the Rouse theory, at best, can only partially account for the glass–rubber relaxation behaviour. A theory is clearly required which considers local motions within the submolecule and which is based on energy elasticity in addition to entropy elasticity. Recently Tobolsky and Aklonis (1964) have proposed a theory similar to that of

Rouse but in which the entropy elastic force constant $(3kT/zl^2)$ in Equation (5.12) is replaced by a force constant c which characterizes local torsional motions along the chain. The normal mode analysis again gives rise to a discrete number of relaxation times, but τ_p is now related to c. This theory was applied to stress relaxation data for polyisobutylene in the glass–rubber relaxation region, and reasonable agreement was obtained between theory and experiment. The dynamic birefringence studies of Read (1962a, 1964a, b) also suggest that the Rouse theory is not strictly valid in the glass–rubber relaxation region. These studies indicate further that an interpretation of this relaxation requires the simultaneous consideration of both the long-range and local 'distortional' modes of chain motion.

Tobolsky and Aklonis (1964) have proposed that the original Rouse theory should be applied entirely to the 'terminal zone,' where $G(t)$ and $G'(\omega)$ decrease to zero at very long times or low frequencies, and the location of which is strongly molecular-weight dependent. On the other hand, according to Ferry (1961), the behaviour in this terminal zone is determined predominantly by the longest relaxation time, τ_1, of the Rouse theory. The problem therefore remains concerning the exact mechanical region in which the Rouse theory is applicable. At low molecular weights the problem is eliminated to some extent since the rubbery plateau region essentially disappears, and the glass–rubber and terminal regions merge. This merging may be explained in terms of the disappearance of chain entanglements responsible for the rubber-like plateau, or in terms of the strong molecular-weight dependence of the longest relaxation times.

Very few experimental tests of the above theories have been made in the case of dielectric relaxation. As mentioned above, however, a low-frequency molecular weight dependent loss peak in polypropylene oxide has been attributed by Stockmayer and Baur (1964) to the mechanism with relaxation time τ'_1. This process would seem to be analogous to the mechanical terminal zone or, at low molecular weights, to the low-frequency tail of the glass–rubber relaxation.

The experimental results discussed here serve to emphasize that the Gaussian submolecular model can, at best, account only for a few of the longer relaxation times in the glass–rubber region. These relaxation times correspond to the lowest few normal modes, involving cooperative motions of relatively long chain segments. The failure of these theories to account for the shorter relaxation time region results from the fact that motions of units shorter than the submolecule are ignored. A description of these shorter-range motions clearly requires theories based on more detailed models of polymer chains.

5.3b   Theories Based on More Detailed Chain Models

Only a few theoretical attempts have been made to describe relaxation in terms of fairly realistic chain models. This is, of course, not surprising in view of the mathematical complexities involved. Existing theories of this kind have been developed mainly for the case of dielectric relaxation and are due largely to Kirkwood and coworkers and Yamafuji and Ishida. Before outlining these theories it should be emphasized that the mathematical developments involve several approximations which tend to obscure the details of the models and cast doubt on the validity of the final expressions. These mathematical approximations are avoided in the approach of Bueche (1961) who considered the motions of a single chain bond, assuming that the cooperation between distant points along the chain could be largely ignored.

Theory of Kirkwood and Coworkers

Kirkwood and Fuoss (1941) considered generally the motions of a polymer chain in a medium of viscosity η, neglecting the effects of hydrodynamic interaction. Their procedure involved a solution of the diffusion equation in which the time-dependent probability distribution function of the chain was expressed in a very general way. This required the specification of the coordinates of each atom in the chain in terms of bond lengths, valence bond angles, and angles of internal rotation. Such a specification is, of course, necessary if relaxation is to be described in terms of the elementary processes of rotation around individual bonds. Kirkwood and Fuoss treated, in particular, the special case of the [—CH$_2$—CH(X)—] chain, exemplified by polyvinyl chloride (X = Cl). We may note that the C—X dipoles are rigidly attached to the main chain, and are perpendicular to the chain axis, so that relaxation must involve motions of the chain backbone. After making many mathematical approximations, Kirkwood and Fuoss obtained the following result,

$$\frac{\varepsilon^* - \varepsilon_U}{\varepsilon_R - \varepsilon_U} = \int_0^\infty \frac{\phi(\tau)\,d\tau}{1 + i\omega\tau} \qquad \int_0^\infty \phi(\tau)\,d\tau = 1 \qquad (5.34)$$

$$\phi(\tau) = \frac{\tau_m}{(\tau + \tau_m)^2} \quad \text{when} \quad \frac{4\tau_m}{n\pi^2} \leqslant \tau \leqslant \frac{n\tau_m}{6} \qquad (5.35)$$

$$\phi(\tau) = 0 \quad \text{when} \quad \frac{4\tau_m}{n\pi^2} > \tau > \frac{n\tau_m}{6} \qquad (5.36)$$

Here $\phi(\tau)$ is the normalized distribution of relaxation times which, according to Equation (5.36), is zero outside the stated limits. n is the degree of polymerization and $\tau_{m} = 3/(2nD_{0})$ where D_{0} is the diffusion coefficient. The theory also yields $D_{0} = kT/2\pi n^{2}\eta\,(a^{2}b)$ where a is the C—C bond length and $b = (3v_{m}/8\pi)^{1/3}$, v_{m} being the volume of a monomer unit. Hence $\tau_{m} = 3\pi a^{2}b\eta n/kT.$

In evaluating the dispersion and absorption curves from Equations (5.34) to (5.36), Kirkwood and Fuoss considered a molecular-weight distribution of the form $\phi(n) = [\exp(-n/\bar{n})]/\bar{n}$ where n is the degree of polymerization and $\bar{n}$ the average degree of polymerization. Such a distribution often applies for polymers formed by free-radical polymerization without subsequent fractionation (Mark, 1940). They obtained

$$J_{\omega} = \frac{\varepsilon' - \varepsilon_{U}}{\varepsilon_{R} - \varepsilon_{U}} = \int_{0}^{\infty} \frac{[1 + u + u(u+2)e^{u}\mathrm{Ei}(-u)]\,du}{1 + x^{2}u^{2}} \qquad (5.37)$$

and

$$H_{\omega} = \frac{\varepsilon''}{\varepsilon_{R} - \varepsilon_{U}} = \int_{0}^{\infty} \frac{xu[1 + u + u(u+2)e^{u}\mathrm{Ei}(-u)]\,du}{1 + x^{2}u^{2}} \qquad (5.38)$$

where $x = \omega\bar{\tau}_{m}$, $u = n/\bar{n}$ and Ei is the exponential integral. Here $\bar{\tau}_{m} = \bar{n}\tau^{*}$ where $\tau^{*}(= \tau_{m}/n)$ is a relaxation time of magnitude appropriate to one monomer unit. Equations (5.37) and (5.38) predict that J_{ω} and H_{ω} are symmetrical functions about $x = 0.63$, at which H_{ω} has a maximum value $H_{\omega,\,max} = 0.27$. The frequency of maximum loss ω_{max} is equal to $0.63/\bar{\tau}_{m}$. An interesting feature of (5.37) and (5.38) is that normalized plots of J_{ω} and H_{ω} against $\log(\omega/\omega_{max})$ are independent of any arbitrary distribution parameter and also independent of temperature. The latter result is consistent with the time–temperature superposition principle. Figure 5.4 shows these plots, calculated by numerical integration of (5.37) and (5.38), together with the experimental master curve for the glass–rubber relaxation of polyacetaldehyde. The experimental and theoretical loss peaks have about the same height and half-width (2·2 decades). However, whereas the theoretical loss peak is symmetrical, the experimental peak is slightly non-symmetrical, being broader on the high-frequency side. This non-symmetry is a typical feature of α peaks of amorphous polymers, and in some cases (e.g. polymethyl acrylate, polypropylene oxide; see Figure 5.5) it is more marked than in the case of polyacetaldehyde.

The Kirkwood–Fuoss theory also predicts that the 'average' relaxation time, $\bar{\tau}_{m}$, should be proportional to the average degree of polymerization $\bar{n}$. This result is not observed for bulk amorphous polymers. In fact the average relaxation time for the primary relaxation is generally found to

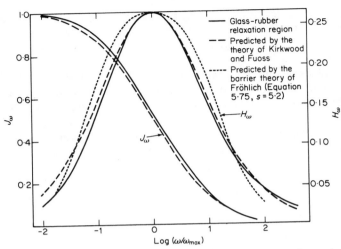

Figure 5.4. Dielectric dispersion and absorption curves for polyacetaldehyde (Williams, 1963a).

become independent of molecular weight for values of n above about 200
(see results for polyvinyl acetate, Chapter 9, and polyisobutylene, Chapter
10).

Hammerle and Kirkwood (1955) have modified the above theory by
taking into account the effect of hydrodynamic interaction between
monomer units. This interaction was predicted to have little effect on the
shape of the loss curves but, for Gaussian chains, the average relaxation
time was found to be proportional to $n^{1/2}$. Again this result is not consistent with experimental observations on the glass–rubber relaxation.
Owing to the considerable mathematical approximations introduced in
the Kirkwood theories the final results are certainly doubtful (Bueche,
1961; Stockmayer and Baur, 1964).

Theory of Yamafuji and Ishida

Yamafuji and Ishida (1962) proposed a theory for the dielectric relaxation of a linear polymer chain which also takes into account local chain
motions. In order to avoid the complications of the general Kirkwood–
Fuoss formulation, they neglected translational motions of the chain
dipoles on the grounds that these do not alter the potential energy with
regard to an external electric field. They further assumed that the rotational motion of a 'dipole' was the same as the rotational motion of the
'motional unit'.

The rotation of each dipolar unit along the chain is considered (approximately) to occur in $(X-Y)$ plans, an origin of coordinates being chosen at the centre of each unit and the polar axis in the direction of motion. If the most stable equilibrium direction of the unit is taken as the origin of the rotational angle ϕ then the instantaneous rotation angle of the lth dipole (ϕ_l) is given by

$$\phi_l = \phi_l{}^0(j) + \delta\phi_l(j) \tag{5.39}$$

where $\phi_l{}^0(j)$ is the azimuthal angle of the jth equilibrium direction for the lth dipole and $\delta\phi_l(j)$ is the instantaneous deviation of the lth unit from the jth equilibrium direction.

The motion of a chain unit is regarded in terms of (a) a rotation of the unit between the j equilibrium directions, and (b) a torsional motion of the unit about a given equilibrium direction. Motions of the type (a) are considered to give rise to the glass–rubber or α relaxation. Type (b) motions are governed by local energy elastic restoring forces. If the latter motions are highly damped, as might be the case in glassy polymers, they may give rise to a (β) relaxation rather than a resonance effect. The α and β relaxations may roughly be described in terms of changes in the mean values of $\phi_l{}^0$ and $\delta\phi_l$, respectively, subsequent to the application of an external field.

The time average for the lth dipole $(\overline{\phi_l})$ is written as follows,

$$\overline{\phi_l} = \phi_l{}^0 + \Delta_\alpha\phi_l + n_\beta\Delta_\beta\phi_l \tag{5.40}$$

where

$$\phi_l{}^0 = \sum_{j=1}^{M} \omega_l{}^0(j)\phi_l{}^0(j)$$

$$\Delta_\alpha\phi_l = \sum_{j=1}^{M} \Delta\omega_l{}^0(j)\phi_l{}^0(j) \tag{5.41}$$

$$\Delta_\beta\phi_l = \sum_{j=1}^{M} \Delta\omega_l{}^0(j)\delta\overline{\phi}_l(j)$$

n_β is a factor, defined by the authors, which denotes the mean probability that such local relaxations can occur. M is the number of equilibrium directions for the lth dipole; $\omega_l{}^0(j)$ is the mean probability for the jth direction of the lth dipole in the absence of an applied field, and $\Delta\omega_l{}^0(j)$ is the deviation of the mean probability from $\omega_l{}^0(j)$ when a field is applied.

The terms $\Delta_\alpha \phi_l$ and $\Delta_\beta \phi_l$ in Equation (5.41) are related to the α and β relaxations respectively. We will reserve our consideration of the β process until Section 5.5b below. In Yamafuji and Ishida's terminology, the diffusion equation for the micro-Brownian motion responsible for the α relaxation is

$$\frac{\partial f^\alpha}{\partial t} = \sum_{l=1}^{n} \left[\frac{kT}{\zeta_l^\alpha} \frac{\partial^2 f^\alpha}{\partial \phi_l^2} + \frac{1}{\zeta_l^\alpha} \frac{\partial}{\partial \phi_l} \left\{ f^\alpha \frac{\partial V^\alpha}{\partial \phi_l} \right\} \right] \tag{5.42}$$

where $f^\alpha = f^\alpha(\phi_1, \phi_2, \dots, \phi_n)$ is the probability distribution function for the chain conformation, ζ_l^α is the rotational friction coefficient for the lth dipole appropriate to the α process, n is the total number of dipole units in the chain and V^α, the total potential of the chain in the diffusion process pertaining to the α process, is given by

$$V^\alpha = V_0^\alpha + V_E^\alpha \tag{5.43}$$

where

$$V_0^\alpha = \frac{1}{2} \sum_{l=2}^{n} A^\alpha (\phi_l - \phi_{l-1})^2 + \frac{1}{2} \sum_{l=1}^{n} (B^\alpha + C^\alpha) \phi_l^2 + \frac{1}{2} \sum D^\alpha (\phi_l - \langle \phi_l - z/2 \rangle)^2 \tag{5.44}$$

and

$$V_E^\alpha = -p_0 \sin \theta_l \cos (\Phi_l - \phi_l) E_r \tag{5.45}$$

A^α is a force constant of the order of the energy differences between internal rotational isomers. B^α is a restoring force constant due to local interactions between nearest neighbour chains. C^α is a force constant arising from middle- and long-range dipole–dipole interactions and D^α is the entropy elastic force constant. The entropy elastic term (i.e. the third term in Equation (5.44) is an approximation based effectively on the submolecular model of Section 5.3a above. In deriving this term Yamafuji and Ishida arbitrarily divided the chain into 'segments' each containing an even number (z) of dipoles. They denoted the average value of ϕ_l for dipoles in the lth group as $\langle \phi_l \rangle$. V_E^α is the potential due to the local electric field, E_r, p_0 is the dipole moment of the motional unit, and θ_l ...nd Φ_l are the angles which the lth dipole makes with the electric field.

Yamafuji and Ishida obtained an approximate solution of Equation (5.42) with the aid of (5.44) and (5.45). Assuming that $\zeta_l^\alpha = \zeta_0^\alpha$ is the same for all dipoles they obtained an expression for ε_ω^* involving a discrete number of relaxation times as follows,

$$\tau_\alpha = \zeta_0^\alpha [4A^\alpha \sin^2 (k_2 \pi/z) + B^\alpha + C^\alpha + 4D^\alpha \sin^2 (k_1 \pi/2(v-1))]^{-1} \tag{5.46}$$

where k_1 and k_2 are integers which range from $k_1 = 1$ to $v - 2$ and $k_2 = -z/2 + 1$ to $z/2$. Here v ($= n/z$) is the number of 'segments' per molecule ($v \gg 1$). According to Equation (5.46) each elementary relaxation time has the same dependence on temperature through ξ_0^α, A^α, B^α, C^α and D^α. Hence the shape of the distribution function of $\log \tau$ should be independent of temperature.

For the dispersion magnitude ($\omega = 0$) Yamafuji and Ishida obtain

$$(\varepsilon_R - \varepsilon_U)_\alpha = \frac{4\pi Nn}{3} \left(\frac{3\varepsilon_R}{2\varepsilon_R + \varepsilon_U} \right)_\alpha \left(\frac{\varepsilon_U + 2}{3} \right)^2 \left(\frac{P_\alpha^2}{\Gamma_\alpha} \right) \tag{5.47}$$

where N is the number of chains per unit volume, P_α^2 is the effective mean square dipole moment per dipole unit and Γ_α is a complex function of the force constants A^α, B^α, C^α and D^α. Yamafuji and Ishida derived approximate expressions for Γ_α at temperatures both above and below T_g. For $T \gg T_g$ they predict that Γ_α^{-1} is proportional to $(kT)^{-1}$ for an amorphous polymer and proportional to $(kT)^{-1/2}$ for a semi-crystalline polymer. This also expresses the temperature dependence of $(\varepsilon_R - \varepsilon_U)_\alpha$ according to Equation (5.47). At temperatures below T_g they predict that $(\varepsilon_R - \varepsilon_U)_\alpha$ will decrease with decreasing temperature for both amorphous and semi-crystalline polymers. These predictions were considered to be consistent with experimental observations.

Yamafuji and Ishida were able to reduce their expression for ε^*_ω to the following form,

$$J_\omega - iH_\omega = \left(\frac{\varepsilon^*_\omega - \varepsilon_U}{\varepsilon_R - \varepsilon_U} \right)_\alpha = [(1 + i\omega\tau_0 y)(1 + i\omega\tau_0 y^{-1})]^{-1/2} \tag{5.48}$$

where

$$\tau_0 = \xi_0^\alpha [(4A^\alpha + B^\alpha + C^\alpha + 4D^\alpha)(B^\alpha + C^\alpha + 4D^\alpha)]^{-1/2} \tag{5.49}$$

and

$$y = \left(\frac{4A^\alpha + B^\alpha + C^\alpha + 4D^\alpha}{B^\alpha + C^\alpha + 4D^\alpha} \right)^{1/2} \tag{5.50}$$

It does not seem easy to calculate the condition of maximum loss from Equation (5.48). The authors have given plots of $\varepsilon''/(\varepsilon_R - \varepsilon_U)$ against $\log(\omega/\omega_{max})$ for various y values. For $y = 1$ Equation (5.48) becomes

$$J_\omega - iH_\omega = \frac{1}{1 + i\omega\tau_0} \tag{5.51}$$

which is the equation for a single relaxation time. In the limit $y \to \infty$ Equation (5.48) becomes

$$J_\omega - iH_\omega = \frac{1}{[(1 - \omega^2\tau_0{}^2) + i\omega\tau_0 y]^{1/2}} \tag{5.52}$$

For the special case $\omega^2\tau_0{}^2 \ll 1$ this equation is identical with the empirical equation of Davidson and Cole (1951). Hence plots of J_ω and H_ω against $\log \omega$ should be non-symmetrical, corresponding to 'skewed' Cole–Cole arcs (see Chapter 4). More specifically the H_ω curves should be broader for $\omega/\omega_{max} > 1$ than for $\omega/\omega_{max} < 1$. This is usually found for the glass–rubber relaxation of amorphous polymers. In Chapter 9 (Figure 9.2) it is seen that Equation (5.48) provides a good fit to the experimental results for polyvinyl acetate and polyvinyl benzoate using y values equal to 20 and 5 respectively. Figure 5.5 shows normalized master curves of $H_\omega/H_{\omega, max}$ and J_ω against $\log(\omega/\omega_{max})$ for polymethyl acrylate, poly-acetaldehyde and polypropylene oxide, each of these polymers being in the amorphous state. The limiting theoretical curve according to Equation (5.52) is also shown. Although the experimental and theoretical curves are

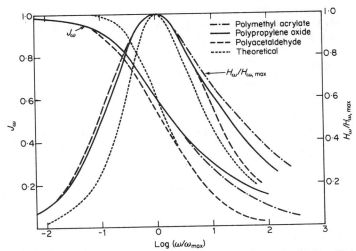

Figure 5.5. Dielectric dispersion and absorption curves for polymethyl acrylate, polypropylene oxide and polyacetaldehyde in the glass–rubber relaxation region (Williams, 1965). The theoretical curves correspond to Equation (5.52) of the Yamafuji and Ishida theory.

both non-symmetrical, the limiting theoretical half-width (1·6 decades) is considerably below the observed half-widths of 2·15 to 2·5 decades for these polymers.

In the case of partially crystalline polymers Yamafuji and Ishida argue that the α relaxation may not be characterized by a single ξ^α value. In this case they propose that a distribution of τ_0 values should be introduced as follows,

$$\left(\frac{\varepsilon^*_\omega - \varepsilon_U}{\varepsilon_R - \varepsilon_U}\right)_\alpha = \int_{-\infty}^{+\infty} \frac{\phi(\ln \tau_0)\, d \ln \tau_0}{[(1 + i\omega\tau_0 y)(1 + i\omega\tau_0 y^{-1})]^{1/2}} \qquad (5.53)$$

Both $\phi(\ln \tau_0)$ and y may be chosen arbitrarily leading to ε^*_ω curves of any desired form.

To summarize, the model of Yamafuji and Ishida is more detailed than the submolecular models considered in Section 5.3a above, the local motions of individual chain dipoles being considered. The theory makes reasonable predictions concerning the temperature dependence of the dispersion magnitude and, unlike the Kirkwood–Fuoss theory, predicts the non-symmetrical shape of the loss peaks observed for the glass–rubber relaxation. It also differs from the Kirkwood–Fuoss theory in predicting that the average relaxation time is independent of molecular weight. However, the friction coefficient $\xi_0{}^\alpha$ and the force constants A^α, B^α, C^α and D^α are parameters difficult to evaluate in an *a priori* fashion. Also, the very complex and approximate mathematical procedures make the validity of this theory difficult to assess.

Theory of Bueche (1961)

Before concluding our discussion of the molecular theories of the glass–rubber relaxation, mention should be made of a paper by Bueche (1961) in which dielectric dispersions in polar polymers are discussed from a relatively simple viewpoint. Bueche suggests that the major portion of the primary dielectric dispersion might be explained by ignoring co-operative motions of distant portions of a chain and considering only the motions of a single chain bond. Bueche considers a model chain whose bond angle is 90°. For simplicity he takes the first bond in the chain to point in the $+x$ direction and successive bonds to point in the directions $\pm y$, $\pm z$, $\pm x$ and so on. He then calculates the 'jump frequency' P which, for this model chain, is the number of times per second that a given bond reverses its direction. For chains in which the dipoles are rigidly attached to the main-chain bonds he obtains a simple Debye dielectric dispersion in which the relaxation time is given by

$$\tau = P/2 = \frac{l^2 \xi_0}{kT} \qquad (5.54)$$

where, as usual, l is the length of a bond and ξ_0 the friction factor associated with the bond. Experimentally the glass–rubber relaxation is, as we have seen, considerably broader than that for a single relaxation time. Bueche has attributed this broadening to local effects rather than the long-range cooperative nature of chain motions treated in the above theories. In particular, he considers that the local liquid or crystalline order in the polymer may give rise to several equilibrium positions for each dipole separated by energy barriers of varying height. According to the barrier theory of Hoffmann and coworkers (Section 5.4b below) such a range of barrier heights can give rise to a broadening of the distribution. Furthermore, Bueche has suggested that a broadening of the dispersion can arise from a variation of the rotational frequency with the *direction* of rotation. In particular, rotations of a chain segment around the local chain axis will usually be much freer than end-to-end rotations of the same segment. In this sense the jump frequency P and the relaxation time τ in Equation (5.54) are average values for these different rotational frequencies. It seems unlikely that Bueche's analysis is applicable to the low-frequency tail of the dielectric glass–rubber relaxation, particularly the molecular weight dependent regions of the type studied by Stockmayer and Baur (Section 5.3a). These regions must clearly involve the longer range cooperative chain motions. However, Bueche's approach is certainly consistent with observations on the main glass–rubber peak which suggest that the average relaxation time becomes independent of molecular weight at high molecular weights. It also has the advantage of relative simplicity, avoiding the mathematical approximations necessary in the above theories. Future developments could conceivably yield an interpretation of the width and shape of primary dielectric loss peaks in terms of the local effects discussed above.

The theories outlined in Sections 5.3a and 5.3b have described the glass–rubber relaxation in terms of relaxation times related to parameters such as the segmental friction coefficient, the diffusion coefficient or the steady flow viscosity. Such coefficients are, however, macroscopic in nature and enter into the theories on the assumption that the medium in which a polymer chain moves may be regarded as a continuum. In this sense the above theories are not truly molecular, but phenomenological, and friction or diffusion coefficients cannot yet be predicted quantitatively from the chemical structure of polymers. However, one promising approach, which has been fairly extensively pursued, involves the relationship between the relaxation time or diffusion coefficient and the so-called free volume.

5.3c Free-Volume Theories

Although the 'free volume' has no unique or rigorous definition, it is a useful semiquantitative concept which has been employed in statistical thermodynamical theories of the liquid state (Lennard-Jones and Devonshire, 1939; Glasstone, Laidler and Eyring, 1941; Frenkel, 1946; Fowler and Guggenheim, 1956). According to Glasstone, Laidler and Eyring (1941) the free volume may be regarded as the volume in which each molecule of a liquid moves in an average potential field due to its neighbours. However, theoretical estimates of free volume depend on postulates regarding the compressibilities of the molecules and the nature of their packing in the liquid state. A definition of free volume often used in polymer studies is that employed by Doolittle (1951, 1952)

$$v_f = v - v_0 \tag{5.55}$$

where v_f is the free volume per gm and v the measured specific volume of the polymer at temperature T. v_0 has been termed the 'occupied volume'. In Doolittle's studies v_0 was taken as the value of v extrapolated to $0°K$ and was therefore regarded as a constant independent of temperature. This definition assumes that v_f must tend to zero as the temperature tends to absolute zero, and that the increase of v with temperature, due to thermal expansion, is associated entirely with an increase in v_f.

The free volume is a time-average quantity which can be determined from equilibrium experiments. In a liquid-like system, however, the local free volume is continually being redistributed throughout the medium, the redistribution occurring simultaneously with the random thermal motions of the molecules. The basic idea underlying the free-volume approach to relaxation phenomena is that the molecular mobility at any temperature is dependent on the available free volume at that temperature. As temperature increases the free volume increases and molecular motions become more rapid. A few molecular theories based on this free-volume concept have been proposed, with the ultimate aim of relating dynamic quantities such as the diffusion coefficient, viscosity or relaxation time to the equilibrium quantity, free volume. These theories are applicable to the liquid-like state and can therefore be applied to amorphous polymers at temperatures of the order of and above T_g. Before discussing these molecular theories it is instructive to consider some relevant empirical expressions.

Doolittle (1951, 1952) found that the flow viscosity of low molecular weight hydrocarbon liquids could be represented by an empirical equation which, with the aid of Equation (5.55), can be written in the form

$$\eta = a \exp (b/f) \tag{5.56}$$

where a and b are constants and $f = v_f/v$ is the *fractional* free volume. Like v_f, f must depend on variables such as temperature and external pressure.

As already noted in Chapter 4, Williams, Landel and Ferry (1955) have proposed an empirical equation which describes the temperature dependence of relaxation times in the glass-transition region. This equation is now well known as the WLF equation and may be written generally as follows,

$$\log a_T = \log \frac{\tau(T)}{\tau(T_0)} = \frac{-c_1(T-T_0)}{c_2 + (T-T_0)} \tag{5.57}$$

in which T_0 is an arbitrary reference temperature and c_1 and c_2 are material constants dependent on the value chosen for T_0. $\tau(T)$ and $\tau(T_0)$ may be regarded as average relaxation times at temperatures T and T_0 respectively. To a rough approximation it was first found that the values of c_1 and c_2 were the same for a wide variety of glass-forming materials if T_0 was suitably chosen for each substance. In particular, if the dilatometric glass-transition temperature T_g was selected for T_0 then Equation (5.57) could be written in the 'universal' form

$$\log a_T = \frac{-c_1{}^g(T-T_g)}{c_2{}^g + (T-T_g)} \tag{5.58}$$

the 'universal' constants $c_1{}^g$ and $c_2{}^g$ having approximate values of 17·44 and 51·6 respectively. Equation (5.58) was found to apply in the temperature range T_g to about $T_g + 100°C$.

As pointed out by Williams, Landel and Ferry (1955) the WLF-type equation can be derived on the basis of Doolittle's empirical free-volume equation (5.56). Taking T_g as the reference temperature we may write (Ferry, 1961)

$$\log a_T = \log \frac{\tau(T)}{\tau(T_g)} \approx \log \frac{\eta(T)}{\eta(T_g)} = \frac{b}{2\cdot303}\left(\frac{1}{f(T)} - \frac{1}{f(T_g)}\right) \tag{5.59}$$

Assuming that $f(T)$ increases linearly with temperature according to

$$f(T) = f(T_g) + \alpha_f(T - T_g) \tag{5.60}$$

where α_f has the dimensions of a thermal expansion coefficient, then it follows from (5.59) and (5.60) that

$$\log a_T = -\frac{(b/2\cdot303f(T_g))(T-T_g)}{f(T_g)/\alpha_f + (T-T_g)} \tag{5.61}$$

Equation (5.61) can now be identified with Equation (5.58) if $c_1{}^g = b/2\cdot303f(T_g)$ and $c_2{}^g = f(T_g)/\alpha_f$. Choosing $b = 1$, in accord with Doolittle's data, Williams, Landel and Ferry obtained from the 'universal' values of $c_1{}^g$ and $c_2{}^g$

$$\alpha_f = 4\cdot8 \times 10^{-4} \text{ deg}^{-1} \tag{5.62}$$

and

$$f(T_g) = 0\cdot025 \tag{5.63}$$

Equation (5.63) suggests that all polymers have approximately the same fractional free volume at T_g, a condition first proposed by Fox and Flory (1950). A close examination of experimental data by Ferry (1961) has revealed variations in the values of the 'universal' constants, particularly $c_1{}^g$, $c_2{}^g$ and α_f, among different polymers. This could partly be due to uncertainties in the T_g values, measurements of which depend on such factors as cooling rate and the presence of residual monomer and moisture. However, these variations do not seem to seriously undermine the significance of the above relationships.

The Doolittle- and WLF-type empirical equations have been derived on the basis of molecular theories by Bueche (1956, 1959) and Cohen and Turnbull (1959). The basic idea underlying the theory of Bueche (1956) is that, at temperatures close to T_g, molecules or polymer segments do not move as individuals requiring a high activation relative to kT, but that several segments must cooperate in a movement. Accordingly, Bueche considered as a model a group of n segments and postulated that this group could move cooperatively only when its local fractional free volume exceeded some critical value f_c. From considerations of fluctuations in local free volume, and assuming the validity of Equation (5.60), Bueche derived the WLF equation (5.61), the constant b being equated to nf_c. Bueche's theory also gave $f_c \approx 0\cdot2$. Since $b \approx 1$ it follows that n is about 5, a reasonable value. In a later theory Bueche (1959) considered the temperature dependence of relaxation times on the basis of fluctuations in the vibrations of concentric shells surrounding a given molecule or segment. This theory resulted in an equation of the WLF form at low temperatures and an Arrhenius equation at high temperatures, a result consistent with experimental data.

The theory of Cohen and Turnbull (1959) was based on the hypothesis that molecular transport in liquids occurs by the movement of molecules into voids having a volume greater than some critical volume v^* formed by the redistribution of free volume. They considered a model of a liquid consisting of hard spheres and assumed that no energy change was associated with the redistribution of free volume. For the self-diffusion coefficient D they obtained a relationship of the form

$$D = D_0 \exp(-\gamma v^*/v_f) \qquad (5.64)$$

where D_0 is proportional to the molecular diameter and the gas kinetic velocity, and γ is a constant ($0 \cdot 5 \leqslant \gamma \leqslant 1$) accounting for the overlap of free volume. The Doolittle-type equation follows readily from (5.64) assuming that D is inversely proportional to η and that the fractional free volume is identified with $v_f/\gamma v^*$.

It should be emphasized that in Cohen and Turnbull's theory the free volume v_f corresponds to that part of the excess volume $v-v_0$ which can be redistributed *without a change in energy*. This concept of free volume was subsequently clarified by Turnbull and Cohen (1961). They assumed a Lennard-Jones function for the potential energy of a molecule in a 'cage' formed by its neighbours as a function of the cage radius. For small cage radii they argued that a large energy would be required for the redistribution of the excess volume. However, at a sufficiently large radius, corresponding to the linear region of the potential energy curve, the excess volume could be redistributed with no energy change. In accordance with this idea they wrote

$$v = v_0 + \Delta v_c + v_f \qquad (5.65)$$

where Δv_c is that part of the excess volume $v-v_0$ which requires an energy for redistribution. At temperatures below some critical temperature T_∞, where the cage radii are small, the redistribution energy will be large and thus $v_f \approx 0$ and $v-v_0 = \Delta v_c$. At temperatures above T_∞, where the cage radii have increased to values corresponding to the linear region of the potential energy curve, most of the volume added by thermal expansion will be 'free' for redistribution. According to this picture, free volume is introduced into the system at some temperature well above absolute zero, in contrast to Doolittle's definition (Equation 5.55). Accordingly, Cohen and Turnbull have written

$$v_f = \alpha \bar{v}_m(T-T_\infty) \quad \text{for} \quad T \geqslant T_\infty$$
$$v_f = 0 \qquad\qquad\quad \text{for} \quad T < T_\infty \qquad (5.66)$$

where α is the average expansion coefficient of the medium and $\bar{v}_m$ the average value of the 'molecular' volume v_0 in the temperature range T_∞ to T. Substituting Equation (5.66) into (5.64) and comparing with self-diffusion data on simple van der Waals' liquids, Cohen and Turnbull found that γv^* had values close to the molecular volume v_0, consistent with their model. An alternative approximate expression for v_f, also consistent with Cohen and Turnbull's picture (Saitô and others, 1963) is the following,

$$v_f = v_f(T_0) + v_g\Delta\alpha(T - T_0) \tag{5.67}$$

where $\Delta\alpha$ is the difference in the thermal expansion coefficients above and below T_g, and v_g is the specific volume of the polymer at T_g. Putting the reference temperature T_0 equal to T_∞ and $v_f(T_\infty) = 0$, Equation (5.67) takes on a form similar to (5.66). Combining Equations (5.64) and (5.66) and assuming D to be inversely proportional to the average relaxation time, τ, we also obtain an equation of the form

$$\tau = \tau_0 \exp[B/(T - T_\infty)] \tag{5.68}$$

where τ_0 and B are constants. Equations of this form have been found to fit experimental relaxation data on many systems (Tammann and Hesse, 1926; Davidson and Cole, 1951; Williams, Landel and Ferry, 1955; Williams, 1963a; Miller, 1963; Kovacs, 1964). Hence, Cohen and Turnbull's picture, as represented by Equation (5.68), predicts that the relaxation time will tend to infinity as the temperature decreases toward T_∞ owing to the disappearance of free volume. The 'transition' thus predicted at T_∞ is, however, essentially unattainable since the *volume* relaxation time will also tend to infinity as the free volume decreases to zero. An infinitely slow cooling rate would consequently be required for the establishment of volume equilibrium at T_∞. The dilatometric transition actually observed at T_g (several degrees above T_∞) is an apparent one, and measured values of T_g will depend on the rate of cooling into the glass-transition region, as observed, for example, by Kovacs (1958). At temperatures in the glass-transition region, relaxation quantities such as G', $\tan\delta$, ε'' and the apparent activation energy also become dependent on the cooling rate owing to the slow establishment of volume equilibrium (see results for polyvinyl acetate, Chapter 9 and polyvinylidene chloride, Chapter 11).

In this section we have attempted to give only a brief outline of the free-volume approach to the glass–rubber relaxation and, in particular, no attention has been given to relevant studies involving the application of hydrostatic external pressures. For a more extensive account of the free-volume approach the reader is referred to Ferry (1961), Saitô and others (1963), Williams (1964) and Kovacs (1964).

5.4 BARRIER THEORIES

In a condensed system the forces of interaction between a molecule and its neighbouring molecules can be represented by a plot of the potential energy of the molecule against its angular orientation or separation from its neighbours. The slopes of this plot give the intermolecular force field. In general, potential energy diagrams consist of minima, corresponding to the equilibrium positions of the molecule, separated by energy 'barriers' or maxima. The simplest type of potential energy diagram comprises two minima separated by a single barrier. This case has already been discussed in Chapter 1 and is exemplified schematically in Figure 5.6(a) which shows a plot of potential energy against angular orientation for a molecule having two non-equivalent equilibrium orientations 180° apart. Thermal fluctuations in the system may enable the molecule to occasionally gain sufficient kinetic energy to surmount the barrier and rotate from one equilibrium position to the other. In terms of the energy differences W and V indicated in Figure 5.6(a), the transition probabilities for jumps in the two directions are given by

$$\Gamma_{12} = \Gamma = A\,e^{-(W+V)/RT}$$
$$\Gamma_{21} = \Gamma' = A\,e^{-W/RT}$$

(5.69)

where A is a constant and the subscripts 12 and 21 represent jumps from site 1 to 2 and vice versa. As shown by Debye (1945) and later by Fröhlich (1949) and Hoffman (1959) the single barrier model yields a single relaxation time given by

$$\tau_2 = [2(\Gamma + \Gamma')]^{-1}$$

(5.70)

We have already seen that a single relaxation time is inadequate to explain observed relaxation processes in polymers. In terms of the barrier model a distribution of relaxation times can be envisaged as arising in two ways. Firstly, if the arrangement of chains in the polymer is heterogeneous, as is the case with a partially crystalline polymer, then the barrier height may vary from molecule to molecule. A description of the relaxation behaviour of such systems will require a consideration of the distribution of barrier heights throughout the assembly of molecules. In Section 5.4a below

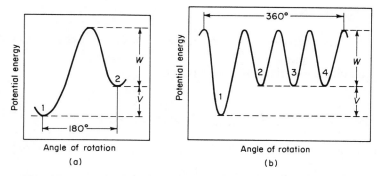

Figure 5.6. (a) The single barrier model having two non-equivalent rotational sites 180° apart and numbered 1 and 2 respectively; (b) The barrier model of Hoffman (1959).

we will outline the theory of Fröhlich (1949) which yields an expression for the complex dielectric constant arising from an arbitrary rectangular distribution of barrier heights. Secondly, a distribution of relaxation times can also arise if the force field experienced by a particular molecule is such that it can adopt a number of equilibrium positions separated by a range of barrier heights. This situation is thought to exist, for example, in certain molecular lattices owing to the anisotropy of the crystalline field in which the molecules reorient. The latter problem has been discussed in a series of papers by Hoffman and coworkers (Hoffman, 1954, 1955, 1959; Hoffman and Pfeiffer, 1954; Hoffman and Axilrod, 1955). Hoffman's theory is considered in Section 5.4b below.

5.4a   Theory of Fröhlich

Fröhlich (1949) described a theory of dielectric relaxation which considers a rectangular distribution of barrier heights. Each relaxation process in a given system is assumed to conform to the Arrhenius relation

$$\tau_i = \tau_0 \exp\left(\frac{Hi}{RT}\right) \tag{5.71}$$

where τ_0 is the relaxation time at infinite temperature and is the same for every relaxation process in the medium. He considers that H_i is distributed between H_1 and H_2, there being an equal probability of occurrence for every process in this range, and zero probability outside this range. Thus, if $\phi(H)\,dH$ is the fraction of processes with H between H and $H + dH$ we have $\phi(H) = $ constant independent of H in the range H_1 to

H_2 and $\phi(H) = 0$ for $H_1 > H > H_2$

$$\int_0^\infty \phi(H)\,\mathrm{d}H = \int_{H_2}^{H_1} \phi(H)\,\mathrm{d}H = \phi(H)(H_2 - H_1) = 1 \qquad (5.72)$$

The complex dielectric constant is given by

$$\frac{\varepsilon^*_\omega}{\varepsilon_R - \varepsilon_U} = \int_{H_1}^{H_2} \frac{\phi(H)\,\mathrm{d}H}{1 + i\omega\tau_0 \exp H/RT} \qquad (5.73)$$

which gives

$$J_\omega = \frac{\varepsilon'_\omega - \varepsilon_U}{\varepsilon_R - \varepsilon_U} = 1 - \frac{1}{2s} \ln \frac{[1 + x^2 \exp s]}{[1 + x^2 \exp(-s)]} \qquad (5.74)$$

$$H_\omega = \frac{\varepsilon''_\omega}{\varepsilon_R - \varepsilon_U} = \frac{1}{s}\left\{ \tan^{-1}\left[x \exp \frac{s}{2} \right] - \tan^{-1}\left[x \exp\left(-\frac{s}{2} \right) \right] \right\} \qquad (5.75)$$

and

$$\frac{\varepsilon''_{max}}{\varepsilon_R - \varepsilon_U} = \frac{1}{s}\left[\tan^{-1} \exp \frac{s}{2} - \tan^{-1} \exp\left(-\frac{s}{2} \right) \right] \qquad (5.76)$$

where

$$x = \omega/\omega_{max} \quad \text{and} \quad s = (H_2 - H_1)/RT$$

The above equations predict symmetrical absorption and dispersion curves, and for $s = 0$ they reduce to the equations for a single relaxation time. Figure 5.4 shows a plot of Equation (5.75) for $s = 5\cdot2$ for comparison with the experimental curve for amorphous polyacetaldehyde in the glass–rubber relaxation region. It is seen that the value $s = 5\cdot2$ yields a theoretical curve which is broader around the peak but narrower in the wings than the experimental loss curve. Thus the simple rectangular distribution of relaxation times is not a good approximation in the case of the glass–rubber relaxation. Also, the assumption that each relaxation time is given by the Arrhenius equation (5.71) is invalid in the case of the glass–rubber relaxation, since H_i is found experimentally to be a strong function of temperature. Furthermore, since s tends to zero with increasing temperature, the distribution of barrier heights predicts that the loss peak must narrow with increasing temperature tending toward the width for a single relaxation time. This prediction is inconsistent with the time–temperature superposition principle. The two latter objections seem less serious for partially crystalline polymers since these materials often exhibit relaxations for which the Arrhenius equation is a good approximation and for which the loss peak narrows as the temperature increases. The Fröhlich equations can be shown to resemble the empirical Cole–Cole and Fuoss–Kirkwood equations (Chapter 4) when s is near zero. Higasi (1961) has discussed the distribution function of Fröhlich and has considered the relationship between the parameters of the Fröhlich and Cole–Cole equations. We may finally add that Fröhlich's theory is itself semi-empirical since, although the barrier concept has molecular significance, the assumption of a rectangular distribution of barrier heights is arbitrary.

5.4b Theory of Hoffman

The barrier theory of Hoffman (1954, 1955, 1959) was proposed to account for dielectric relaxation in molecular crystals. This theory proceeds briefly as follows. The complex dielectric constant for a discrete set of relaxation modes is given by

$$\varepsilon^*_\omega = \sum_{\beta=1}^{\Omega} \frac{\Delta\varepsilon_\beta}{1+i\omega\tau_\beta} \tag{5.77}$$

Here τ_β is a set of relaxation times where $\beta = 1, 2, \ldots, \Omega$. $\Delta\varepsilon_\beta$ is the dielectric constant increment associated with the βth mode of relaxation.

Let the dipole be a single-axis rotator which may occupy one of four orientation sites as shown in Figure 5.6(b) all 90° apart. Site 1 is of depth $(W+V)$ and the other wells of depth W. The barrier for jumps from 2 to 3, 3 to 4, their inverse jumps and from 4 to 1 and 2 to 1 are all equal at value W. The jumps 1 to 2 and 1 to 4 have a barrier $(W+V)$. The transition probabilities are

$$\Gamma_{12} = \Gamma_{14} \equiv \Gamma = A \exp -(W+V)/RT \tag{5.78}$$

$$\Gamma_{41} = \Gamma_{21} = \Gamma_{23} = \Gamma_{32} = \Gamma_{34} = \Gamma_{43} \equiv \Gamma' = A \exp(-W/RT) \tag{5.79}$$

If N_i is the population of site i then the rate equations (if only single jumps are allowed) are

$$\frac{dN_1}{dt} = -2\Gamma N_1 + \Gamma' N_2 + \Gamma' N_4$$

$$\frac{dN_2}{dt} = \Gamma N_1 - 2\Gamma' N_2 + \Gamma' N_3$$

$$\tag{5.80}$$

$$\frac{\mathrm{d}N_3}{\mathrm{d}t} = \Gamma'N_2 - 2\Gamma'N_3 + \Gamma'N_4$$

$$\frac{\mathrm{d}N_4}{\mathrm{d}t} = \Gamma N_1 + \Gamma'N_3 - 2\Gamma'N_4$$

The eigenvalues of the operator $(\mathrm{d}/\mathrm{d}t)$ in the characteristic determinant of Equations (5.80) are 0, $-2\Gamma'$, $-(2\Gamma' + \Gamma - \sqrt{Q})$ and $-(2\Gamma' + \Gamma + \sqrt{Q})$, where Q is $\Gamma'^2 + (\Gamma' - \Gamma)^2$. Hoffman (1954) has thus shown that the relaxation times associated with this model are

$$\tau_2 = \frac{1}{2\Gamma'}$$

$$\tau_3 = \frac{1}{2\Gamma' + \Gamma - \sqrt{Q}} \qquad (5.81)$$

$$\tau_4 = \frac{1}{2\Gamma' + \Gamma + \sqrt{Q}}$$

The model thus yields multiple relaxation times. The general feature of Hoffman's work is that multimodal relaxation curves are obtained which tend to a single relaxation time at sufficiently high temperatures. The individual relaxation processes are not independent of each other but are each functions of Γ and Γ' of the model. An analogous treatment for the mechanical case has been given by Beyer (1957). For polymers, Hoffman's barrier model would seem most applicable to relaxation processes occurring in the crystalline regions, and Illers (1964) has recently discussed Hoffman's model in relation to mechanical relaxation occurring in paraffins and polyethylene.

By an extension of Hoffman's model Ishida and Yamafuji (1961) have developed a theory which illustrates the possible nature of the α and β relaxations in the methacrylate polymers. Although (in the amorphous state) each dipole will be surrounded by a slightly different environment and will, therefore, experience different energy barriers to rotation, the model is based on the so-called 'one-body approximation' in which the energy barriers to rotation are considered for a single side-group dipole only. This simplified model is illustrated in Figure 5.7 where $\otimes$ represents the chain backbone which is perpendicular to the plane of the paper and R is the ester alkyl group. Rotations around the R—O and O—C bonds are assumed to have set in at temperatures below that of the β relaxation so that a net average dipole moment is associated with the entire side-group. In addition to sites 1 and 2 (Figure 5.7) sites 3 and 4 are also

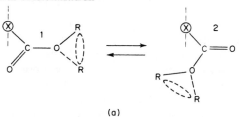

(a)

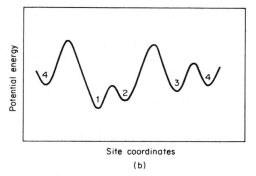

Site coordinates

(b)

Figure 5.7. Illustration of the barrier model of Ishida and Yamafuji (1961) for the dielectric α and β relaxations of polymers containing —COOR side-chains. (a) Conformations of the dipolar side-group in the proposed sites; (b) Schematic diagram of the barrier system.

proposed which are the symmetrical configurations of 2 and 1 with respect to the dashed lines in Figure 5.7. The rotations between sites 1 and 2 and between 3 and 4 give rise to the β relaxation, whereas transitions between 1 and 4 and between 2 and 3, which involve motions of the chain segments (and which also lead to a directional change of the net dipole moment), lead to the α relaxation. Of the four different barrier heights those for the transitions $1 \rightleftarrows 2$ and $3 \rightleftarrows 4$ are assumed to be lower than for $2 \rightleftarrows 3$ and $1 \rightleftarrows 4$. An analysis of this barrier model by Hoffman's method yields the following expression for the complex dielectric constant,

$$\varepsilon^* - \varepsilon_U = \frac{4\pi N p_0{}^2 n}{3kT} \left(\frac{3\varepsilon_R}{2\varepsilon_R + \varepsilon_U}\right) \left(\frac{\varepsilon_U + 2}{3}\right)^2 \sum_{l=1}^{III} \frac{\Delta\varepsilon_l}{1 + i\omega\tau_l} \quad (5.82)$$

where N is the number of chains per cm³, n the degree of polymerization and p_0 the dipole moment of the side-group. Hence, three discrete relaxation processes are predicted having relaxation times τ_I, τ_{II} and τ_{III}

and magnitudes $\Delta\varepsilon_I$, $\Delta\varepsilon_{II}$ and $\Delta\varepsilon_{III}$ respectively. The τ_i's are related to the transition probabilities (or heights) of the various barriers and the $\Delta\varepsilon_i$'s are related both to these transition probabilities and also to the vertical angle of the cone traced out by the composite dipole of the ester group in rotating between sites 1 and 2. Processes I and III contribute to the β absorption whilst process II determines the α relaxation. If the different barrier heights become of a similar order of magnitude, which occurs by a lowering of barriers 2 ⇌ 3 and 1 ⇌ 4 by an increase in the n-alkyl group length, the α and β absorptions will of course merge. Clearly the one-body approximation cannot yield a distribution of relaxation times for both the α and β relaxations. It does, however, predict expressions for the temperature dependence of $\varepsilon_R - \varepsilon_U$, and predicts the fact that $\varepsilon_R - \varepsilon_U$ passes through a maximum near T_g. It also predicts qualitatively the *total* magnitudes of the $\alpha + \beta$ absorptions above T_g for the various methacrylate polymers, and also the ratio of the magnitudes of the α and β absorptions in several polymer systems.

5.5 THEORIES FOR SECONDARY RELAXATIONS

In Section 5.1 above we have noted that secondary relaxations in amorphous or partially crystalline polymers can result from rotations of side-groups attached to the main chain or from limited motions within the chain backbone. Side-group rotations have been considered briefly in connection with the barrier theories in Section 5.4. In this section we propose to discuss those mechanisms which involve local in-chain movements. These are largely of two kinds which are termed, respectively, the crankshaft and the local mode mechanisms.

5.5a The Crankshaft Mechanism

A few authors, notably Schatzki (1962, 1965) and Boyer (1963) have recently proposed that a so-called 'crankshaft' mechanism may be responsible for many relaxations in polymers at temperatures below T_g. Schatzki's proposals are limited mainly to the relaxation region at about $-120°$C (1 c/s) which is observed for polymers containing linear $(-CH_2-)_n$ sequences, where $n \geqslant 3$ or 4. This relaxation is usually labelled γ and is observed, for example, for linear polyethylene (Chapter 10), polyamides (Chapter 12), polyesters (Chapter 13) and the oxide polymers (Chapter 14), as well as for some higher methacrylate polymers (Chapter 8) containing linear methylene sequences in the side-chains. The mechanism proposed by Schatzki, and illustrated in Figure 5.8(a),

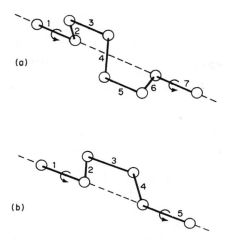

Figure 5.8. Illustration of the crankshaft mechanism. Models proposed by (a) Schatzki (1965) and (b) Boyer (1963).

involves the simultaneous rotation about the two bonds 1 and 7, such that the intervening carbon atoms move as a crankshaft. An essential criterion for the mechanism is that two bonds 1 and 7 are *colinear*. The chain bonds on either side may then remain frozen and only a relatively small volume is required for the movement. If all bond valence angles are tetrahedral, the bond lengths are equal, and the stable rotational angles have the usual 0° (*trans*), +120° (*gauche*$^+$) and −120° (*gauche*$^-$) values then, according to Schatzki, colinear bonds can occur if there are four intervening carbon atoms (Figure 5.8a). Boyer (1963) has proposed the conformation shown in Figure 5.8(b) in which there are only two invervening carbon atoms. However, this structure involves the energetically unfavourable *cis* conformation about bond 3, and is far less probable than Schatzki's structure. Wunderlich (1962) has also suggested that colinear bonds separated by three carbon atoms could give rise to a crankshaft motion. The latter structure appears to be a low-energy, tight-helical, conformation.

Several experimental facts are consistent with the crankshaft mechanism. Firstly, as mentioned above, ($-CH_2-$)$_n$ sequences ($n \geqslant 3$ or 4) separated from each other by immobile groups give rise to the γ relaxation in a variety of polymers (Willbourn, 1958). Available evidence suggests that this relaxation occurs in the amorphous regions of these polymers. This is a requirement of the crankshaft mechanism owing to the condition that two bonds must be colinear. Such a condition is not fulfilled in the crystalline regions where the $(-CH_2-)_n$ sequences are exclusively in the *trans* conformation. Secondly, the observed activation energy for the γ relaxation (12–15 kcal/mole) agrees well with the value of about 13 kcal/mole estimated by Schatzki. This estimation is based on twice the butane potential barrier (7·5 kcal/mole) plus a 'van der Waals' barrier of 5–6 kcal/mole determined from the cohesive energy density. Thirdly, the predicted free volume of activation, about four times the molar volume of a $-CH_2$ unit, is in good agreement with experimental estimates.

From an extension of the above ideas, particularly by Boyer (1963), it has been suggested that a crankshaft motion may be responsible for secondary relaxations in a wide variety of polymers. However, more experimental tests are required to establish whether or not the proposed crankshaft mechanism is valid, and no theory yet exists which predicts, on the basis of this mechanism, either the magnitude or the breadth of secondary dielectric or mechanical relaxations.

5.5b   Local Mode Theories

Most molecular theories of relaxation in polymers, including all those discussed above, are based on mechanisms which fundamentally require hindered rotations around chemical bonds. These processes are thermally activated and, in terms of the barrier concepts, they involve the passage of atoms across a potential barrier from one position of equilibrium to another. An alternative type of relaxation mechanism has recently been proposed to account for secondary loss regions in glassy polymers (Yamafuji and Ishida, 1962; Saitô and others, 1963; Takayanagi, 1963; Gotlib and Salikhov, 1962). This mechanism involves the damped oscillation of chain segments about their equilibrium positions and has been termed the 'local mode' process by Japanese workers. A similar mechanism has been suggested by Hill (1963) to account for a possible dielectric dispersion in polar liquids at frequencies above the Debye frequency.

Saitô and others (1963) have considered in some detail the vibrational motions of a molecule in the neighbourhood of a local equilibrium conformation. In this analysis the vibrations were considered in the usual manner as the superposition of a set of normal modes. The characteristic frequency of the μth normal mode (ω_μ) is given by

$$\omega_\mu = (C_\mu/m)^{1/2} \qquad (5.83)$$

where C_μ is the appropriate (energy elastic) force constant and m the mass of a vibrating unit. The root mean square of the amplitude X_μ of the μth mode is given by

$$\{\overline{(X_\mu)^2}\}^{1/2} = \left\{ \frac{2kT}{m\omega_\mu^2} \cdot \frac{\hbar\omega_\mu/kT}{\exp(\hbar\omega_\mu/kT) - 1} \right\}^{1/2} \tag{5.84}$$

The characteristic frequencies, and thus the average amplitudes, of the normal vibrations are distributed over a wide range. Since the force constant for torsional oscillations is very small compared with the force constants for other types of vibration (e.g. bond stretching and valence angle deformation), the torsional oscillations should account largely for the modes of lowest frequency and largest amplitude. For polyethylene in the planar zig-zag conformation, Saitô and others estimate that the lowest-frequency torsional mode has a root mean square amplitude of about 21°. They argue that torsional modes of this amplitude will be strongly damped in the highly viscous glass-like state and will give rise to relaxation effects. They have treated this relaxation mechanism by introducing the appropriate friction constant ξ_l into the Kramers–Chandrasekhar diffusion equation (Kramers, 1940). The relaxation time (τ_λ) for this local mode process is given by

$$\tau_\lambda^{-1} = \frac{C_\lambda}{\xi_l} \ll \omega_\lambda \tag{5.85}$$

where C_λ is the torsional force constant for mode λ and ω_λ the corresponding eigenfrequency. Equation (5.85) states that for large ξ_l the relaxation frequency, τ_λ^{-1}, will be much less than the characteristic frequency of the original normal vibration. The latter frequency usually lies within the infrared region. The order of magnitude of the activation energy associated with ξ_l was estimated by Saitô and others assuming that the moving group must be surrounded by a void of volume greater than some critical volume. The critical volume was arbitrarily selected on the basis of an intermolecular potential of the Lennard-Jones form. This calculation yielded an activation energy of about 10 kcal/mole. Saitô and others have given the following expression for the magnitude of the local mode dielectric relaxation,

$$\varepsilon_R - \varepsilon_U = \frac{4\pi n N g p_0^2}{3C_\lambda} \left(\frac{3\varepsilon_R}{2\varepsilon_R + \varepsilon_U} \right) \left(\frac{\varepsilon_U + 2}{3} \right)^2 \tag{5.86}$$

in which n is the number of dipolar units in the chain, p_0 the dipole moment of the repeat unit and g the correlation parameter (Chapter 3).

Yamafuji and Ishida's analysis of the local mode relaxation has been introduced in Section 5.3b above. As already noted they labelled this relaxation β. Their diffusion equation for the β process is identical with Equation (5.42) except that the superscript α is replaced by β. The total potential of the chain (V^β) for small displacements of the mean directions of dipoles is written

$$V^\beta = V_0^\beta + V_E^\beta \qquad (5.87)$$

where

$$V_0^\beta = \tfrac{1}{2} \sum_{l=2}^{n} A^\beta(\phi_l - \phi_{l-1})^2 + \tfrac{1}{2} \sum_{l=1}^{n} (B^\beta + C^\beta)\phi_l^2 \qquad (5.88)$$

and

$$V_E^\beta = -n_\beta p_0 \sin \theta_l \cos (\Phi_l - \phi_l)E_r \qquad (5.89)$$

where n_β, the mean probability that the local mode relaxation will occur, equals unity for $T \ll T_g$. A^β is an intramolecular force constant appropriate to chain distortions and B^β is a force constant related to interchain forces. At zero frequency they obtain for the dispersion magnitude

$$\varepsilon_R - \varepsilon_U = n_\beta \frac{4\pi n N p_0^2}{3\Gamma_\beta}\left(\frac{3\varepsilon_R}{2\varepsilon_R + \varepsilon_U}\right)\left(\frac{\varepsilon_U + 2}{3}\right)^2 \qquad (5.90)$$

in which

$$\Gamma_\beta = [(B^\beta + C^\beta)(4A^\beta + B^\beta + C^\beta)]^{-1/2} \qquad (5.91)$$

Assuming reasonable values for A^β, B^β and C^β, Yamafuji and Ishida obtain $\Gamma_\beta \simeq 8 \times 10^{-13}$ erg. We note that for $n_\beta = 1$ and $g = 1$ Equations (5.86) and (5.90) become identical if Γ_β is equated with the torsional force constant C. In fact the local mode theories of Saitô and others and Yamafuji and Ishida become virtually identical as soon as the former apply the inequality of (5.85) which eliminates the inertial terms.

Yamafuji (1960) had earlier proposed the following alternative expression for Γ_β,

$$\Gamma_\beta \simeq G'(\omega_{max}^\beta)a(nN)^{-2/3} \qquad (5.92)$$

assuming that the polar groups may be regarded as spheres of effective radius a. Here $G'(\omega_{max}^\beta)$ is the value of G' at the frequency of maximum G'' for the local mode relaxation. For a variety of polymers, including polyvinyl chloride, polyethylene terephthalate and polyoxymethylene, good agreement was obtained between the observed values of $\varepsilon_R - \varepsilon_U$ for the low-temperature relaxations and values calculated from Equation (5.90) with Γ_β estimated from either (5.91) or (5.92).

For the complex dielectric constant Yamafuji and Ishida have given the following equation,

$$\frac{\varepsilon^* - \varepsilon_U}{\varepsilon_R - \varepsilon_U} = \int_{-\infty}^{\infty} \frac{\phi_\beta(\ln \tau_0) \, d \ln \tau_0}{[(1 + i\omega\tau_0 y)(1 + i\omega\tau_0 y^{-1})]^{1/2}} \qquad (5.93)$$

where

$$\tau_0 = \xi_0^\beta \left[(B^\beta + C^\beta)(4A^\beta + B^\beta + C^\beta \right]^{-1/2} \qquad (5.94)$$

and

$$y = \left[\frac{4A^\beta + B^\beta + C^\beta}{B^\beta + C^\beta} \right]^{1/2} \qquad (5.95)$$

Equation (5.93) involves effectively a 'double distribution' since we may arbitrarily select both a distribution of τ_0 values, $\phi_\beta(\ln \tau_0)$, and also a y value. Choosing $y^2 = 300$ and a symmetrical function for $\phi_\beta(\ln \tau_0)$ of half-width about ln 12, Yamafuji and Ishida were able to fit the non-symmetrical curve of ε'' against $\log \omega/\omega_{max}$ for the low-temperature relaxation of polyethylene terephthalate.

To summarize, it seems that the local mode theory offers a plausible explanation of many secondary relaxation phenomena in polymers, and it reasonably predicts the magnitude of several such dielectric dispersions. However, like the theories considered in Sections 5.3a and 5.3b, the local mode theory relates relaxation times to unpredictable friction coefficients, the significance of which are difficult to understand in terms of the basic molecular mechanism.

6

Experimental Methods (1): Mechanical

Section 6.1 of this chapter contains a brief account of the distribution of stress in the types of specimen geometry normally used for viscoelastic measurements on polymeric solids. Several of the important experimental techniques are described in detail in Section 6.2.

6.1 STRESS DISTRIBUTION IN EXPERIMENTAL SPECIMENS

Elastic compliances and moduli are usually determined by applying forces to a body and measuring the resulting deformation. If a rod of length L with a circular cross-section of radius a is subject to a load W uniformly distributed over the ends (Figure 6.1a) then the tensile stress in the rod is

$$t_z = \frac{W}{\pi a^2} \qquad (6.1)$$

and the tensile strain is given by

$$e_z = D t_z$$

where D is the tensile compliance. The total elongation is

$$\Delta L = L e_z$$

Hence the compliance is given by

$$D = \frac{\Delta L}{L} \frac{\pi a^2}{W} \qquad (6.2)$$

D can therefore be determined by measuring the dependence of ΔL on W.

When the tensile compliance is time dependent, $D(t)$, as for a viscoelastic rod, we have, for the case of a load W applied at $t = 0$,

$$D(t) = \frac{\pi a^2}{LW} \Delta L(t) \qquad (6.3)$$

$D(t)$ may therefore be determined by measuring the time dependence of the rod extension, $\Delta L(t)$.

In order to measure the complex tensile compliance

$$D^*(\omega) = D'(\omega) - iD''(\omega) \qquad (6.4)$$

it is necessary to apply a sinusoidal load to the specimen. Let the load be,

$$W^* = W_0 \exp i\omega t \qquad (6.5a)$$

so that,

$$t_z^* = \frac{W_0}{\pi a^2} \exp i\omega t \qquad (6.5b)$$

The strain is given

$$e_z^* = D^* t_z^*$$

The elongation of the specimen is therefore

$$\Delta L^* = L e_z^*$$

$$= \frac{L W_0}{\pi a^2}(D' - iD'') \exp i\omega t$$

The component of the elongation in phase with stress is

$$\Delta L' = \frac{L W_0 D'}{\pi a^2} \qquad (6.6a)$$

and the component of the elongation $\pi/2$ out of phase with the stress is

$$\Delta L'' = \frac{L W_0 D''}{\pi a^2} \qquad (6.6b)$$

D' and D'' can be obtained therefore by measuring $\Delta L'$ and $\Delta L''$. This is often done by measuring $|\Delta L|$,

$$|\Delta L| = \sqrt{(\Delta L')^2 + (\Delta L'')^2} \qquad (6.7a)$$

and the phase angle δ_E between stress and strain

$$\tan \delta_E = \frac{\Delta L''}{\Delta L'} \qquad (6.7b)$$

This type of experiment is known as the forced oscillation experiment.

Equation 6.3 does not, of course, hold for times in step-function experiments less than the time required for the stress in all parts of the rod to come to equilibrium at the value given by Equation (6.1). This time, which is of the order of L/v, where v is the velocity of sound, is usually below a millisecond. In forced oscillation experiments the measuring frequency must be sufficiently low so that there is no phase lag between the load W^* and the stress t_z^*. In Equations 6.5(a) and (b) load and stress are taken to be absolutely in phase. It is found that forced oscillation (usually in shear) can be used at frequencies up to the kilocycle range. Frequencies above the kilocycle range are studied with resonance or ultrasonic techniques.

In the tensile experiment the stress t_z is constant throughout the volume of the rod. The experiment is therefore valuable even when the relationship between stress and strain is not linear. Another technical advantage of the tensile experiment is that Youngs modulus is probably the most valuable of all moduli for the engineering designer. The experiment is also, in principle, extremely simple. Yet tensile experiments are not invariably used to study viscoelastic relaxation. There are several reasons for this. Firstly, for an elastic solid the tensile compliance D is given in terms of the shear and bulk compliances, J and B

$$D = \tfrac{1}{3}J + \tfrac{1}{9}B$$

For a viscoelastic solid the equation

$$D(t) = \tfrac{1}{3}J(t) + \tfrac{1}{9}B(t) \tag{6.8}$$

is normally assumed to hold.* Whatever the precision of Equation (6.8), viscoelastic deformation in tension is clearly a mixture of shear deformation and volume deformation. This would not matter if both types of deformation relaxed in the same way. However, according to Kono (1961) the temperatures of the maxima in G'' and K'' (the imaginary part of K^*, the complex bulk modulus) at the α relaxation of at least one family of polymers, the methacrylates, differ by as much as 20°C. Furthermore, according to Kono (1961) the activation energy for shear deformation is less than that for volume deformation. One reasonable interpretation of this is that for the methacrylate polymers at the α relaxation and at the frequencies used by Kono ($\sim 10^6$ c/s) one type of molecular motion leads to shear relaxation and another to volume relaxation. Clearly, if this is so, tensile experiments are more difficult to interpret than experiments in which the relaxation of the shear modulus alone is observed.

Another reason why tensile experiments are not invariably used in practice is that shear experiments are fairly simple to perform. Consider

---

* See Staverman and Schwarzl (1956) and Turner (1964). According to Turner the experimental justification for Equation (6.8) is not strong.

the torsion of an elastic rod of length L with a circular cross-section of radius a under the action of equal and opposite torques Γ applied at each end, Figure 6.1(b). It may be shown (see, for instance, Timoshenko and Goodier, 1951) that the tangential shear stress σ acting on the planes of the cross-section is given by

$$\sigma = \frac{G\phi r}{L} \qquad (6.9)$$

where G is the shear modulus and ϕ the relative angle of twist of the two ends of the rod (Figure 6.1b). Taking coordinates x, y, z as shown in Figure 6.1(c) we have

$$\sigma_{yz} = \sigma\frac{x}{r} = \frac{G\phi x}{L} \qquad (6.9a)$$

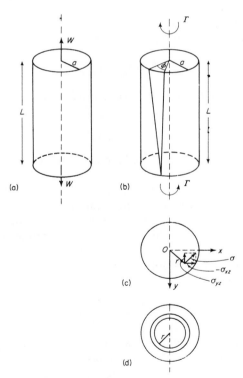

Figure 6.1. (a) Cylindrical rod under tension; (b) Cylindrical rod under torque; (c) and (d) show cross-sections of (b).

$$\sigma_{xz} = -\frac{\sigma y}{r} = -\frac{G\phi y}{L} \tag{6.9b}$$

All other stresses are taken to be zero,

$$t_x = t_y = t_z = \sigma_{xy} = 0 \tag{6.9c}$$

In order for this stress distribution to hold, Equations (6.9) must satisfy (A) the equations of compatibility, (B) the equations of equilibrium and (C) the boundary conditions. Equations (6.9) satisfy (A) and (B) (see Timoshenko and Goodier, 1951). The boundary conditions are (a) the curved sides of the rod must be free from forces, (b) the ends of the rod must be subject to forces, the sum of whose moments equal the applied torque Γ. It is a simple matter to show that the effect of stresses (6.9) is to cause zero force on the curved sides of the rod (Timoshenko and Goodier, 1951).

Concerning boundary condition (b), it follows from Equation (6.9) that the torque on an elemental tube of thickness dr and radius r (Figure 6.1d) is $2\pi r^2 \sigma \, dr$ so that the total torque on the shaft is

$$\Gamma = \int_0^a 2\pi r^2 \sigma \, dr \tag{6.10}$$

Inserting Equation (6.9) into (6.10) and integrating,

$$G = \frac{2\Gamma L}{\pi \phi a^4} \tag{6.11}$$

Consequently, G may be measured for an elastic rod by applying a torque Γ and measuring ϕ.

The foregoing theory applies exactly to the viscoelastic case so long as the relationship between stress and strain is linear. Consider a typical creep experiment in which a constant torque Γ_0 is applied at $t = 0$ and ϕ is measured as a function of time. With increasing time the linear relationship between stress and strain ensures that the spatial distribution of stress within the rod does not change. Therefore the compliance at time t, $J(t)$, is given by

$$J(t) = \frac{\pi a^4}{2L\Gamma_0} \phi(t) \tag{6.12a}$$

where $\phi(t)$ is the time-dependent angle of twist of the rod. Similarly, the stress relaxation modulus $G(t)$ may be obtained,

$$G(t) = \frac{2L}{\pi a^4 \phi_0} \Gamma(t) \tag{6.12b}$$

where $\Gamma(t)$ is the torque necessary at time t to maintain the constant angle of twist ϕ_0.

Strictly speaking, the shearing stresses should be applied at the ends of the rod as specified in Equation (6.9). This particular stress distribution cannot be obtained with normal clamping methods, e.g. a pin chuck. According to Saint-Venant's principle, however, if forces acting on a small part of the surface of a body of dimension $2a$ are replaced by statically equivalent forces, the stress state in the body is negligibly changed at distances large compared to $2a$. Consequently, in the region near the pin chuck, the stress will not be given by Equation (6.9). But if L is much larger than $2a$, Equation (6.9) will hold for the greater part of the rod. The error involved in measuring $J(t)$ or $G(t)$ is thus reduced by increasing the ratio L/a. For other geometries also there is usually a small experimental error due to the perturbation of the clamp. It is extremely difficult to allow for this theoretically. Normal practice is to minimize the error by a judicious choice of dimensions.

This theory for a twisted rod is only valid when the relationship between stress and strain is linear. If the relationship is not linear then the stress distribution is not given by Equation (6.9) nor do any of the other results hold. The shear experiment under these circumstances is therefore valueless. The level of strain at which the relationship between stress and strain ceases to be linear depends on the polymer, the temperature and also the time (see, for instance, Turner, 1964a, b). Usually, the lower the experimental error the lower is the strain at which departures from linearity are observed. Normally, measurements are essentially linear for strains below 10^{-3}.

The real and imaginary parts of the shear compliance can be measured by forced oscillation by a method analogous to the forced oscillation experiment in tension. Two important examples of this, due to Morrison, Zapas and DeWitt (1955) and Fitzgerald and Ferry (1953), are described in the next section.

The elastic and viscoelastic properties of the solid have been assumed to be isotropic in the foregoing theory. For an amorphous polymer this is clearly reasonable since there is no *a priori* reason for the elastic properties measured in one direction to differ from those measured in another direction. It is known, however, that the elastic properties of polymeric crystals are anisotropic. The assumption of isotropic elastic properties is found, nevertheless, to be applicable to crystalline polymers prepared by moulding. The reasons for this are: (1) the crystal size ($\sim 10^{-6}$ cm, see Chapter 2) is considerably less than the size of the specimen, which is usually of the order of 1 cm, (2) the crystals are randomly oriented in space.

The stress distribution in a twisted rod, for instance, is given by Equation (6.9) and a shear compliance may be obtained. The exact theoretical significance of the measured compliance is, however, a matter on which there is at present little understanding. The great complication is not so much the random orientation of the crystals but the existence between the crystals of amorphous polymer. The problem of the distribution of stress between the crystalline and amorphous regions has been treated empirically by Takayanagi (1963) (see Chapter 4).

6.2 EXPERIMENTAL TECHNIQUES

Only a selection of the many methods of measuring viscoelastic properties are included in this section. For a complete account the reader is referred to standard texts on physical acoustics. The techniques described here are those most commonly used to study polymeric solids.

6.2a Torsion Pendulum

The experimental arrangements which have been described differ in detail but the principle is identical to that described in Figure 6.2(a) (Heijboer, Dekking and Staverman, 1954). The lower end of the specimen is clamped rigidly. The upper end is clamped to a vertical rod supported by a counterbalance. The vertical rod carries a horizontal rod, known as the inertia rod, with weights at the end. The inertia rod is displaced slightly, twisting the specimen, and released. The decay of the oscillations is then detected, most simply, by galvanometer lamp and scale. For frequencies above ca. 2 c/s a Kymograph (Schmieder and Wolf, 1952) or differential transformer (Nielsen, 1951) may be used. In some equipment the upper clamp is held rigidly and the lower attached to the inertia arm, Figure 6.2(b). This arrangement causes a tensile stress in the specimen which may cause elongation at high temperatures.

The logarithmic decrement is obtained by measuring the nth amplitude A_n and the $(n+1)$th amplitude A_{n+1} (Equation 1.23). In practice it is usual to measure the decay of amplitude over as many swings as possible to reduce random error,

$$\Lambda = \frac{1}{k} \ln\left(\frac{A_n}{A_{n+k}}\right) \tag{6.13}$$

where k is the number of swings. The relationships between Λ and other measures of mechanical loss are given in Chapter 1, (Equations 1.22a–d).

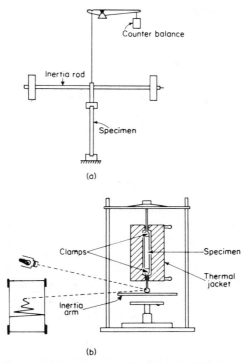

Figure 6.2. (a) The torsion pendulum of Heijboer, Dekking and Staverman (1954); (b) The torsion pendulum of Schmieder and Wolf (1952) showing the thermal jacket.

The shear modulus G is obtained by observing the period of the oscillation. It is also necessary to know the specimen geometry and the inertia of the system. It may be shown (Inoue and Kobatake, 1958) that when a tensile stress t exists in a specimen of rectangular cross-section the torque Γ produced by a twist through an angle θ is

$$\Gamma = \frac{1}{3} G \frac{\theta}{L} ab^3 \left[\left(1 - 0{\cdot}63\frac{b}{a}\right) + \frac{1}{120}\left(\frac{E}{G}\right)\frac{a^3}{b}\left(\frac{\theta}{L}\right)^2 + \frac{1}{4}\left(\frac{t}{G}\right)\left(\frac{a}{b}\right)^2 \right] \qquad (6.14)$$

where G is the shear modulus, E is Young's modulus, a is the breadth, b the thickness ($a > 3b$) and L the length. In practice the angle of deflection θ is usually less than 10^{-2} and consequently the term in θ^3 can be neglected.

If the moment of inertia of the pendulum around the central axis be M, the torque per unit deflection $\Gamma(1)$, then the period T is given by

$$T = 2\pi \sqrt{\frac{M}{\Gamma(1)}} \sqrt{1 + \Lambda^2/4\pi^2} \qquad (6.15)$$

The shear modulus is then obtained from Equations (6.14) and (6.15).

$$G = \left(\frac{4\pi^2 ML}{NT^2}\right)\left(1 + \frac{\Lambda^2}{4\pi^2}\right) - \frac{mga^2}{12N} \qquad (6.16)$$

where

$$N = \frac{1}{3}ab^3\left(1 - 0.63\frac{b}{a}\right) \qquad (6.17)$$

The term containing Λ is usually negligible. In this equation m is the mass of the pendulum supported by the specimen (Figure 6.2b) and g the gravitational constant. N is a form factor which depends only on the specimen geometry.

The most usual specimen geometry is a thin rectangular strip with dimensions of the order of $6 \times 1 \times 0.1$ cm. The reason for this geometry is partly a question of convenience. Sheets of this thickness may be easily compression moulded. The specimen may be quenched from the melt if necessary to give a cross-section of fairly uniform density. The weight of polymer (ca. 1 gram) is low and consequently the torsion pendulum method can be used to study experimental samples which are in short supply. In addition, the rectangular cross-section possesses a larger surface for heat exchange with the surrounding gas than say a circular specimen of equal cross-sectional area. A further advantage is that a specimen with a rectangular cross-section develops a lower torsional rigidity for a given cross-sectional area than, for instance, circular or elliptic specimens. (For a detailed discussion, see Timoshenko and Goodier, 1951.)

The maximum strain, γ_{max}, occurs in the centre of the long side of the specimen and is given (Timoshenko and Goodier, 1951) by

$$\gamma_{max} = \frac{\theta_{max}b}{L} \qquad (6.18)$$

where θ_{max} is the maximum value of θ.

The specimen is usually placed in a thermal jacket. The arrangement of Schmieder and Wolf (1952) is shown in Figure 6.2(b). Many other thermostatting devices are described in the literature.

The upper range of frequency obtainable with a torsion pendulum depends on the limit of detection of the recording device and the onset of apparatus resonance. For instance, it is necessary to stay well below the natural resonance of the supporting rods. Below ca. 0·1 c/s the oscillation of the pendulum can be materially affected by disturbances such as air currents.

The accuracy of the measurement of the logarithmic decrement depends on the resolution of the trace on the recording paper and the care taken in measuring amplitudes. If the trace is observed directly with a lamp and scale then several measurements should be taken to reduce random error. Systematic errors can occur if the pendulum is not vertical or if the axis of the specimen is off the axis of rotation. The method of excitation should only induce oscillatory motions of the pendulum. The greatest error in the measurement of the shear modulus lies in the determination of b^3, Equations (6.16) and (6.17). This quantity is usually not known to better than 3 %. The period of oscillation can be determined to better than 0·5 % either with a stop watch or from the trace.

The torsion pendulum cannot be used to measure extremely low values of internal friction since the background losses are not so low as in the resonance methods. Only at liquid helium temperatures, however, does the internal friction of a high polymer approach the background level. The great advantage enjoyed by the torsion pendulum over resonance methods is at high loss levels. Values of $\Lambda > 0·1$ can be easily and reproducibly measured. Highly damped vibrations require more sophisticated analysis than is inherent in Equation (6.13) (Westphal, 1947; Kohlrausch, 1955).

Losses of the magnitude observed at the α relaxation of an amorphous polymer ($\Lambda > 1$) are studied most accurately using forced oscillations (see below).

6.2b Resonant Rod

Three types of vibration occur in rods: longitudinal, torsional and flexural. In longitudinal vibration the elements of the rod expand and contract parallel to the axis, which is parallel to the direction of wave propagation. In torsional vibration the elements oscillate around the axis of the rod. In flexural vibration the elements are translated at right angles to the axis of the rod. The internal friction is measured by determining either the half-width of the resonant peak (Equation 1.22c) and the resonant frequency or by observing the number of oscillations required for the amplitude to decay between certain limits (Equation 6.13). The modulus is calculated from the resonant frequency. Internal friction and modulus can be measured similarly at overtone frequencies. A typical experimental arrangement is illustrated in Figure 6.3 (Heijboer, Dekking and Staverman, 1954). The earlier work on metals is reviewed by Zener (1948). The rod, usually with a circular or rectangular cross-section, is supported at the nodes. One method of doing this is illustrated in Figure 6.3. The type of vibration generated in the specimen depends critically on the method of suspension. In general, flexural vibrations are easier to generate than either longitudinal or torsional vibrations (Lord Rayleigh, 1877). Pickett (1945) gives a comprehensive list of equations for computing elastic constants from flexural and torsional resonant frequencies for specimens in the shape of prisms and cylinders.

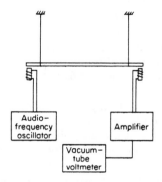

Figure 6.3. Resonant rod apparatus. The rod is suspended by fine threads at the nodes. The vibration is stimulated by an audio-frequency oscillator and the response, after amplification, detected by a vacuum-tube voltmeter. Small pieces of magnetic steel, glued to the specimen close to the transducer and detector, couple the specimen magnetically to the oscillator and vacuum-tube voltmeter (After Heijboer, Dekking and Staverman, 1954).

Stimulation of a horizontal rod into oscillation is achieved by vibrating one of the supporting wires or by applying an oscillatory force to one end of the rod. If a supporting wire is vibrated then the other wire is attached to a detector to measure the amplitude of vibration (Kline, 1956). An oscillating force may be applied to the end of the rod directly by means of a fine wire attached to a piezoelectric transducer. The oscillation is detected at the other end of the rod by means of another fine wire attached to a similar piezoelectric transducer. The force may also be applied and detected magnetically by attaching small pieces of steel to the ends of the rod (Heijboer, Dekking and Staverman, 1954, Figure 6.3). If the rod is held in a clamp at one end it may be driven by bolting the clamp directly to a transducer (Robinson, 1955). The detection may be magnetic, electrostatic or photoelectric. An important resonant rod apparatus is described by Foerster (1937, 1955).

Resonant rod techniques are capable of great accuracy for low values of internal friction but are not easily used for values of tan $\delta > 0.2$. They demand considerable experimental technique especially in mounting and driving the specimen and in avoiding spurious resonances. Subsidiary parts of the apparatus frequently have resonances in the same range as that of the specimen, i.e. 50 to 5×10^4 c/s. It is therefore necessary to design in such a way that apparatus resonance does not interfere with the resonance of the rod. The support wires (Figure 6.3) can cause large errors (a), if they resonate at a frequency close to that of the specimen, (b) if they are not placed exactly at the nodes since they have the effect of adding additional mass and stiffness to the vibrating bar (Wachtman and Lam, 1959).

The specimen must be machined accurately to prevent double resonance peaks lying close together. Voids and other inhomogeneities can also lead to double resonance peaks. The resonant rod technique cannot be used if the specimen distorts under its own weight. This limitation is particularly critical if the bar is supported horizontally. A frequency range of several decades may be observed, either by varying the geometry of the specimen or by using harmonics, though the resonant rod technique is normally employed for (essentially) constant frequency experiments.

6.2c Ultrasonic Methods

The wave equation for the propagation of a wave in the x direction of frequency ω in an extended lossy medium is

$$\mu = A \exp i\omega \left(t - \frac{x}{v^*} \right) \tag{6.21}$$

where v^* is the complex velocity. For longitudinal waves

$$v^*_l = \sqrt{\frac{K^* + \frac{4}{3}G^*}{\rho}} \tag{6.22}$$

where K^* and G^* are the complex bulk modulus and torsion modulus respectively at frequency ω. For transverse waves

$$v^*_t = \sqrt{\frac{G^*}{\rho}} \tag{6.23}$$

In terms of the experimentally determined attenuation coefficient α_t and sound velocity v_t,

$$v^*{}_t = \frac{\omega}{\omega/v_t - i\alpha_t} \qquad (6.24)$$

A similar equation gives $v^*{}_1$ in terms of v_1 and α_1. Consequently both K^* and G^* can be determined by measuring the four quantities α_t, v_t, α_1 and v_1.

The several experimental methods all use pulse techniques and differ largely in the method by which the sound is transmitted into the specimen. In the early experiments of Nolle and Mowry (1948) longitudinal pulses of about 2 microseconds were transmitted from a piezoelectric crystal through a liquid medium to a reflection block where they were reflected back to the crystal. The experiment was then repeated with a thin polymer film immersed in the liquid so that the pulse traversed the film. The sound velocity and attenuation of the film were then calculated from the difference in amplitude and phase of the two pulses.

Since liquids will not transmit transverse waves it is necessary to use the rotating plate method to determine v_t and α_t. The specimen is placed at such an angle to the direction of propagation that longitudinal waves are totally reflected (Kono, 1960).

McSkimin (1951) has described an elegant technique in which the piezoelectric crystals are attached to the specimen by means of quartz rods. This eliminates the liquid bath so that the method is very suited to cryogenic work (Mason, 1958). Both K^* and G^* may be measured in high loss materials over a frequency range of $5-50 \times 10^6$ c/s.

McSkimin's Method

The specimen, in the form of a thin film of thickness ~ 1 mm, is mounted between two similar rods of fused silica, Figure 6.4. Piezoelectric crystals are attached to the ends of the rods. Waves generated by each of the piezoelectric crystals strike the specimen and are transmitted and reflected as shown in Figure 6.5. The displacements of the incoming waves at the surface of the specimen μ_1 and μ_5 are then adjusted so that μ_8 is zero, allowing calculation of the elastic constants.

Consider the displacements μ of waves travelling in the x direction,

$$\mu = A \exp i(\omega t - k^* x) \qquad (6.25)$$

where k^* is the complex propagation constant. The pressure P is given by the equation of state

$$P = \rho(C^*)^2 \frac{\partial \mu}{\partial x}$$

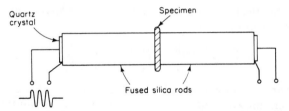

Figure 6.4. Showing the specimen and fused silica rods used in the McSkimin's ultrasonic apparatus (McSkimin, 1951).

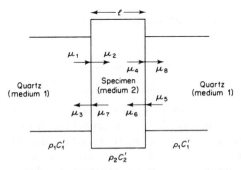

Figure 6.5. Detail of wave amplitudes (McSkimin, 1951). The measurement consists essentially in adjusting μ_1 and μ_5 to make μ_8 zero.

where C^* is the complex velocity and ρ the density. Let the amplitudes μ_1 and μ_5 be adjusted so that μ_8 is zero. The following standard boundary conditions then hold,

$$\mu_1 + \mu_3 = \mu_2 + \mu_7$$

$$\mu_4 + \mu_6 = \mu_5$$

$$P_1 + P_3 = P_2 + P_7$$

$$P_4 + P_6 = P_5$$

In addition we have in medium 2 (the specimen, Figure 6.5),

$$\mu_4 = \mu_2 \exp{-ik^*_2 l}$$

$$\mu_7 = \mu_6 \exp{-ik^*_2 l}$$

in which l is the thickness of the specimen. These equations may then be solved for the ratio P_1/P_5,

$$\frac{P_1}{P_5} = \frac{\sinh(ik^*{}_2 l)}{2}\left(\frac{\rho_1 C^*{}_1}{\rho_2 C^*{}_2} - \frac{\rho_2 C^*{}_2}{\rho_1 C^*{}_1}\right) \tag{6.26}$$

The density and velocity of sound in quartz are both well known. Therefore if ρ_2 is determined we have an equation relating a measurable quantity (P_1/P_5) to $k^*{}_2$ and $C^*{}_2$. But

$$k^*{}_2 = \frac{\omega}{C_2} - i\alpha_2$$

where C_2 is the velocity of sound in medium 2, and α_2 is the attenuation coefficient. Also,

$$C^*{}_2 = \frac{\omega}{k^*} = \frac{\omega}{\omega/C_2 - i\alpha_2} \tag{6.27}$$

Therefore Equation (6.26) gives the complex quantity (P_1/P_5) in terms of the unknowns C_2 and α_2. In the experiment ω is varied until (P_1/P_5) is real. The amplitudes of P_1 and P_5 are then varied until P_8 is zero so that Equation (6.26) holds. The reader is referred to the original paper for details of the solution for C_2 and α_2 (McSkimin, 1951).

6.2d Forced Oscillation

The forced oscillation technique has been used for a continuous range of frequencies from 10^{-4} up to 10^4 c/s. Of the many designs which have been used, those described by Morrison, Zapas and DeWitt (1955) and Fitzgerald and Ferry (1953) have proved amongst the most versatile and accurate.

The Apparatus of Morrison, Zapas and DeWitt

The design of this apparatus is illustrated in Figure 6.6. The frequency range is from $\sim 3 \times 10^{-4}$ to 10 c/s. An oscillating torque is applied to a strip specimen by passing an oscillating current through a coil in a radial magnetic field. The coil former must be non-conducting. The deflection of the system is observed by a light beam reflected from a mirror glued to the coil. When the current producing the torque is observed with a high speed galvanometer internal friction may be determined by measuring the phase angle between the two light beams reflected from the mirrors on the galvanometer and coil.

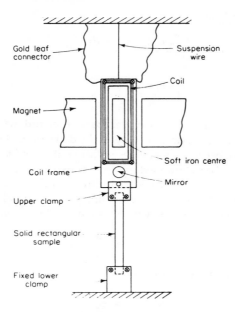

Figure 6.6. Forced oscillation apparatus of Morrison, Zapas and DeWitt (1955).

The differential equation governing the system is

$$M\ddot{\theta} + \frac{N}{L}\eta'\dot{\theta} + \left(\frac{NG'}{L} + \frac{ba^3 t_z}{12L} + k_0\right)\theta = k_i i_0 \exp i\omega t \qquad (6.28)$$

in which M is the moment inertia of the moving parts, θ is the angle of deflection from the equilibrium, N is a geometric factor given by Equation (6.17), L is the length of the specimen, η' is the dynamic viscosity, G' is the dynamic shear modulus, a is the width of the specimen, b is the thickness of the specimen, t_z is the tensile stress in the specimen, k_0 is the torsion constant of the suspension wire, k_i, is the coil constant (torque per unit current), ω is the angular frequency of the current and i_0 is the maximum value of the current.

The origin of the terms is fairly obvious except for the term containing the stress t_z. This term occurs because the tension in the specimen has a horizontal component when the specimen is rotated from the equilibrium (see Equation 6.14).

Solution of Equation (6.28) gives

$$G' = \frac{L}{N}\left[\frac{k_i i_0 \cos\phi}{\theta_0} - \frac{ba^3 t_z}{12L} - k_0 + M\omega^2\right] \tag{6.29}$$

$$\eta' = \frac{L}{N}\left[\frac{k_i i_0 \sin\phi}{\omega\theta_0}\right] \tag{6.30}$$

in which ϕ is the phase angle between the current and the deflection of the system and θ_0 is the maximum value of the deflection. The measurement of ϕ and θ_0 therefore leads to a determination of G' and η'. The apparatus constants have to be determined in separate experiments.

Equation (6.28) is analogous to the differential equation governing the flow of charge q in a circuit containing electrical inductance L_e, capacitance C and resistance R under an applied sinusoidal voltage V.

$$L_e\ddot{q} + R\dot{q} + \frac{q}{C} = V \tag{6.31}$$

From network theory we know that

$$\dot{q} = \frac{V}{Z_e^*} \tag{6.32}$$

where the electrical impedance Z_e^* is given by

$$Z_e^* = R + i\left(\omega L_e - \frac{1}{\omega C}\right) \tag{6.33}$$

The mechanical impedance of the torsion apparatus is governed by Equation (6.28) is therefore defined by

$$Z^* = \frac{\Gamma}{\theta} \tag{6.34}$$

where Γ is the applied torque and Z is given by

$$Z^* = \frac{N}{L}\eta' + i\left[\omega M - \frac{1}{\omega}\left(\frac{NG'}{L} + \frac{ba^3 t_z}{12L} + k_0\right)\right] \tag{6.35}$$

The imaginary part of the impedance disappears at the resonant frequency ω_0,

$$\omega_0 = \left[\frac{1}{M}\left(\frac{NG'}{L} + \frac{ba^3 t_z}{12L} + k_0\right)\right]^{1/2}$$

Note that at frequency ω_0 the total impedance to the applied torque is due to the mechanical loss in the specimen. Measurements in the frequency range around ω_0 are difficult. By adjusting (N/L) according to the value of G', ω_0 can be placed outside the desired measuring range.

The Apparatus of Fitzgerald and Ferry

The principle of the Fitzgerald apparatus is shown in Figure 6.7. The driving tube is vibrated horizontally by means of an oscillating current in the driving coil which lies in a radial magnetic field. The movement of the driving tube is resisted by the inertia of the system and also by the rigidity of the two specimens which are wedged between the tube and a large suspended mass. A pick-up coil is wound onto the driving tube and also lies in a radial magnetic field. The complex ratio of the current in the driving coil and the induced voltage in the pick-up coil is a measure of the internal friction and modulus.

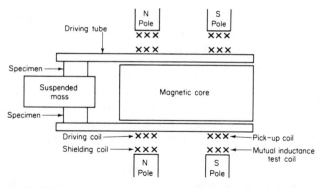

Figure 6.7. Schematic diagram of the apparatus of Fitzgerald and Ferry (1953). The driving tube is held in position by a balanced set of tightly strung wires.

Consider first the movement of the driving tube with no specimen. If an alternating current $i_1 = i_0 \sin \omega t$ is passed through the driving coil it will exert a force $f = f_0 \sin \omega t$ on the driving tube; or f(newton) $= B_1 l_1 i_1$ where B_1 is the radial flux (weber/m^2), l_1 the length of wire in the driving coil (metres) and i_1 is given in amperes. The instantaneous velocity of the tube will be v(metre/sec) $= f/Z_0^*$ in which Z_0^* is the mechanical impedance of the driving tube. The open circuit voltage generated in the pick-up coil is e_2 (volt) $= B_2 l_2 v$ where B_2 is the radial flux (weber/m^2) and l_2 the length of wire in the pick-up coil (metres). The mechanical impedance of the driving tube is, therefore,

$$Z_0^* = \frac{f}{v} = B_1 B_2 l_1 l_2 \left(\frac{i_1}{e_2}\right)$$

When two specimens are placed between the driving coil and the suspended mass as shown in Figure 6.7, the total mechanical impedance becomes $Z_t^* = Z_0^* + Z^*$ where Z^* is the impedance of the specimens. Thus Z^* can be determined by measuring Z_t^* and Z_0^*, i.e. by determining i_1/e_2 with and without the specimens. The magnetic fluxes and wire lengths are previously determined. The complex shear modulus is then given for a thin disc-shaped specimen of thickness b and area A,

$$G^* = i\omega Z^* \frac{b}{A}$$

The method of Fitzgerald and Ferry therefore measures the complex shear modulus in terms of electrical circuit elements. The measurements may be made over a frequency range 25 to 5000 c/s. The temperature range of the apparatus is at present limited by the sensitivity of the coil insulation to temperatures in excess of 150°C.

Mutual inductance between the driving coil and the pick-up coil is eliminated by a shielding coil, Figure 6.7. The shielding coil is glued to the inside of the pole piece close to the driving coil. It carries a current of frequency equal to that in the driving coil but with a phase and amplitude sufficiently different to cause the directly induced voltage in the pick-up coil to be zero. This is checked by means of the mutual inductance test coil which is glued to the pole piece close to the pick-up coil. Before each determination the phase and amplitude of the current in the shielding coil are adjusted to give zero voltage in the mutual inductance test coil. It is then assumed that the voltage e_2 induced in the pick-up coil is entirely due to the movement of the driving tube.

Forced oscillation techniques have two principal assets; they can be used to measure high values of mechanical loss (tan $\delta > 1$) and have a wide and continuous frequency range. The frequency is varied by twisting the knob of a sine wave generator, not, as in the resonance methods, by changing the specimen or its method of support. Forced oscillation techniques are not generally suitable for the measurement of low loss (tan $\delta < 0.01$) and cannot easily be used over a very wide temperature range.

There are many other forced oscillation techniques including those of Alexandrov and Lazurkin (1940), Maxwell (1956), Payne (1961) and Takayanagi (1963). The instrument described by Takayanagi is of particular interest owing to its capability of measuring low losses (tan δ_E range 0·001 → 1·7). McKinney and coworkers (McKinney, Edelman and Marvin, 1956; McKinney and Bowyer, 1960) have described a forced oscillation method for determining the components of the complex *bulk* modulus or compliance.

6.2e Creep and Stress Relaxation

Creep and stress relaxation are, in principle, simpler than the dynamic experiments described in the preceding sections. In creep, a constant stress is applied to a specimen and the change of strain measured as a function of time. In stress relaxation the stress needed to maintain a specimen at a constant strain is measured as a function of time.

Both creep and stress relaxation may be measured in shear or tension. The apparatus of Morrison, Zapas and DeWitt (1955), Figure 6.6, is suitable for the measurement of creep and stress relaxation in shear. For creep, a constant current is passed through the coil and the deflection measured as a function of time. For stress relaxation the necessary current to maintain a constant deflection is determined. If it is required to maintain a constant shear stress in all parts of the specimen (for instance, when studying non-linear effects) then the specimen should be a thin-walled cylinder (Lethersich, 1947, 1950). At strains below 10^{-3} the modulus is essentially independent of strain, so that specimens with circular or rectangular cross-sections are frequently used instead of cylinders.

In tension, the simplest method of measuring creep is to hang a weight on a specimen, usually of rectangular or circular cross-section, and follow the elongation as a function of time using a cathetometer (Hoff, Clegg and Sherrard-Smith, 1958; Gohn and Cummings, 1960), or an extensometer (Findley, 1962; Jackson and McMillan, 1963; Dunn, Mills and Turner, 1964). If the strain becomes large then the stress is increased owing to reduction in the cross-section of the specimen. Under these circumstances constant stress can be maintained to within $\pm 1\%$ using Andrade–Chalmers-type lever arms (Andrade and Chalmers, 1932; Sherby and Dorn, 1958). Strain gauges (Nagamatsu, Yoshitomi and Takemoto, 1958) or balances (Tobolsky, 1960) are used to measure the force in stress relaxation experiments in tension. Creep and stress relaxation in tension can also be measured with commerical force-elongation equipment.

In order to measure creep right 'through' a relaxation, it is necessary to make measurements over at least five and preferably ten decades of time. If the first measurements of the deflection are made at $t \sim 10^{-3}$ seconds then the tenth decade of time elapses after $\sim \frac{1}{2}$ year. Lethersich (1950) first performed an experiment of this duration on polymethyl methacrylate at 30°C (see Figure 8.4). His results show the presence of two superposed relaxations, one in the short time region and the other at long times. Reliable measurements for $t \sim 10^{-3}$ seconds were achieved by applying a torque abruptly to a specimen of low inertia. Usually, experiments are designed for the initial reading to lie between 0·1 and 10 seconds. Lack of patience causes experiments to be discontinued after times of the order of 10^6 seconds (~ 1 week). Consequently, most workers measure creep (and stress relaxation) over approximately five decades of time. Frequently, attempts are made to extend the time scale by the analytical procedure known as time–temperature superposition, which is described in Chapter 4.

7

Experimental Methods (2): Dielectric

7.1 INTRODUCTION

It is a feature of the dielectric technique that measurements can be performed nearly continuously over the frequency range 10^{-4} to 3×10^{10} c/s. Table 7.1 summarizes the methods which are employed in particular regions of this large frequency range. The frequency range of a particular method is indicated in the table. This should be taken only as a guide since the different techniques may be extended to slightly higher or lower frequencies than those indicated, often accompanied by loss of accuracy in the measurement.

The methods may be classified into two general groups, namely 'lumped circuit' and 'distributed circuit' methods. In the 'lumped circuit' range, 10^{-4} to 10^8 c/s (approximately), the experimental technique is designed to measure the equivalent capacitance and resistance at a given frequency. At higher frequencies, the effect of residual inductance in the measuring assembly makes it difficult to regard the sample as a resistance–capacitance arrangement. Thus, in the frequency range 10^8 to 3×10^{10} c/s, which is the 'distributed circuit' range, the experimental technique is designed to measure the attenuation factor α and the phase factor β at a given frequency. We shall show below how the experimental quantities, resistance and capacitance, and attenuation and phase factor, are related to the complex dielectric constant ε^*.

A large part of the dielectric work on polymers has been confined to the frequency range 10^2 to 10^5 c/s. It is, however, essential that as large a frequency range as possible should be covered, since dielectric relaxation curves for polymers are broad and very sensitive to temperature variation. Special emphasis will be given below to the direct current transient method since its value and ease of operation have not been fully appreciated by polymer workers. Extensive reviews of experimental dielectric methods have been given by Westphal (1954), De Vos (1958) and Smyth (1955).

Table 7.1. Experimental methods

Frequency range	Method	Remarks
10^{-4} to 10^{-1}	d.c. Transient measurements	Analogous to creep effect
10^{-2} to 10^{2}	Ultra low frequency bridge	Precise determination of $\varepsilon' - i\varepsilon''$
10 to 10^{7}	Schering bridge Transformer bridge	} Precise determination of $\varepsilon' - i\varepsilon''$
10^{5} to 10^{8}	Resonance circuits	Upper limit of lumped circuit methods
10^{8} to 10^{9}	Coaxial line	Good only for medium and large ε''
	Re-entrant cavity	Good only for low ε'' values; poor for temperature variation
10^{9} to 3×10^{10}	H_{01n} cavity resonator	Same as above.
	Coaxial lines and waveguides	Good for medium and high ε'' only

7.2 LUMPED CIRCUITS

In the frequency range 10^{-4} to 10^{8} c/s, it is convenient to regard a polymer sample as being electrically equivalent to a capacitance C_x in parallel with a resistance R_x, at a particular frequency. Both C_x and R_x will in general be frequency dependent. We must now find a relationship between C_x, R_x and the complex dielectric constant ε^*.

Consider a parallel plate capacitor, of plate area A, and let d be the plate separation. If the space between the electrodes is vacuum filled, the capacitance (in farads) C_0 of the capacitor is given by

$$C_0 = \frac{A\varepsilon_0}{d} \tag{7.1}$$

$\varepsilon_0 = 10^{-12}/3 \cdot 6\pi$ (F/cm), and in Equation (7.1) the ratio (A/d) has the dimensions of cm. For details of the electrical units, see Panofsky and Phillips (1955).

If a sample replaces the vacuum between the plates, the new capacitance, C^*, is complex and is given by (see Von Hippel, 1954)

$$C^* = \varepsilon^* C_0 \tag{7.2}$$

If a voltage $V(t) = V_0 \exp i\omega t$ is applied to the sample filled capacitor, the total current $I(t)$ traversing the capacitor at the steady state is given by

$$I(t) = \frac{dq(t)}{dt} = \frac{d[C^*V(t)]}{dt} \qquad (7.3)$$

$q(t)$ is the charge on the capacitor plates at time t.

From (7.2) and (7.3), $I(t)$ is given by

$$I(t) = iI_c(t) + I_L(t) = [i\omega\varepsilon' + \omega\varepsilon'']C_0V(t)$$

$$I_C = \omega C_0 \varepsilon' V(t) \qquad I_L = \omega C_0 \varepsilon'' V(t) \qquad (7.4)$$

Figure 7.1 shows $I(t)$ and its components, in relation to $V(t)$. Here $I_c(t)$ is the capacitance component of the current, and $I_L(t)$ is the loss (or resistive) component of the current. It is seen that $I_L(t)$ is in phase with $V(t)$ and a power loss will occur in the medium.

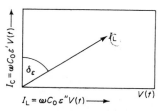

Figure 7.1. The currents in a capacitor containing a lossy dielectric.

From Figure 7.1, the dielectric loss tangent $\tan \delta_\varepsilon$ is given by

$$\tan \delta_\varepsilon = \frac{\varepsilon''}{\varepsilon'} \qquad (7.5)$$

Figure 7.1 is consistent with an equivalent circuit of a capacitance C_x in parallel with a resistance R_x (Figure 7.2). The electrical admittance Y_x of this simple circuit is given by

$$Y_x = \frac{1}{R_x} + i\omega C_x \qquad (7.6)$$

$I(t)$ is now given by

$$I(t) = Y_x \cdot V(t) \qquad (7.7)$$

Comparing (7.4) and (7.7) together with (7.6) gives

$$\varepsilon' = \frac{C_x}{C_0} \qquad (7.8a)$$

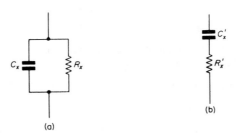

Figure 7.2. (a) Equivalent parallel (R_x, C_x) circuit for a dielectric;
(b) Equivalent series (R'_x, C'_x) circuit for a dielectric.

$$\varepsilon'' = \frac{1}{R_x \omega C_0} \tag{7.8b}$$

$$\tan \delta_\varepsilon = \frac{1}{R_x C_x \omega} \tag{7.8c}$$

The experimental techniques to be described below are mainly concerned with the measurement of C_x and R_x. Equation (7.8) gives the relationships between C_x and R_x, and the components of ε^*.

It is also possible to regard the sample as equivalent to a capacitance C'_x in *series* with a resistance R'_x (Figure 7.2b). In this case the impedance Z'_x is given by

$$Z'_x = R'_x + \frac{1}{i\omega C'_x} \tag{7.9}$$

Also

$$I(t) = \frac{V(t)}{Z'_x} \tag{7.10}$$

Comparing (7.4) and (7.10) together with (7.9) gives

$$\varepsilon' = \frac{C'_x}{C_0(1 + \tan^2 \delta_x)} \tag{7.11a}$$

$$\varepsilon'' = \frac{R'_x (C'_x)^2 \omega}{C_0(1 + \tan^2 \delta_x)} \tag{7.11b}$$

$$\tan \delta_\varepsilon = R'_x C'_x \omega \tag{7.11c}$$

Hence, if the experimental technique evaluates the equivalent *series* capacitance and resistance, C'_x and R'_x respectively, it is possible to evaluate ε', ε'' and $\tan \delta_x$ for the sample using Equation (7.11).
Comparison of Equations (7.8) and (7.11) gives

$$C_x = \frac{C'_x}{(1 + \tan^2 \delta_x)} \tag{7.12a}$$

$$R_x C_x = \frac{1}{R'_x C'_x \omega^2} \tag{7.12b}$$

It is important at this stage to distinguish between dielectric loss arising from d.c. conductivity in a specimen, and loss arising from other sources, notably dipole relaxation.

If a sample is considered equivalent to a capacitance C_x in parallel with a resistance R_x, we allot a portion R_0 to R_x as arising from d.c. conductivity of the specimen, and a portion R_1 as arising from non-d.c. resistance.

Since R_0 and R_1 are in parallel, $(R_x)^{-1} = (R_0)^{-1} + (R_1)^{-1}$. By its definition, R_0 will be a constant independent of frequency at a given temperature. R_1 will in general be a function of frequency.

According to Equation (7.8b) the total loss factor ε'' is related to R_0 and R_1 in the following way

$$\varepsilon''(\omega) = \frac{1}{R_x \omega C_0} = \frac{1}{R_0 \omega C_0} + \frac{1}{R_1 \omega C_0}$$

$$= \varepsilon''_0(\omega) + \varepsilon''_1(\omega) \tag{7.13}$$

The term $\varepsilon''_0(\omega)$ due to d.c. conductivity of the specimen is given by

$$\varepsilon''_0(\omega) = \frac{1}{R_0 \omega C_0} = \frac{G_0}{\omega C_0} \tag{7.14}$$

G_0 is the d.c. conductivity of the sample (ohm^{-1}).

Since R_0 is independent of frequency, the product $(\varepsilon''_0 \times \omega)$ is a constant for the sample, independent of frequency. For a given R_0 value, (ε''_0) increases rapidly with decreasing frequency.

In view of Equation (7.8a), where R_x is not involved, it follows that a d.c. conductivity process contributes to the total loss in a manner prescribed by Equation (7.14), but it cannot contribute to the dielectric constant ε'. For a dielectric relaxation process, ε' and ε'' are related to each other (Fröhlich 1949). It is thus easy to distinguish between loss arising from a d.c. conductivity process and loss arising from other sources.

Equation (7.14) can be written in an alternative form. Since R_0 is proportional to the separation of the plates, d, for a parallel plate capacitor, and is inversely proportional to the area, A, of the plates, we have

$$R_0 = [R_{spec}]\frac{d}{A} \text{(ohm)} \qquad (7.15)$$

$[R_{spec}]$ is the resistance of a sample of unit area (1 cm^2) and unit thickness (1 cm). $[R_{spec}]$ is known as the specific resistance of a given sample, in (ohm cm).

Combination of (7.14), (7.15) with the aid of (7.1) gives

$$\varepsilon''_0(\omega) = \frac{1 \cdot 8 \times 10^{12}}{[R_{spec}]f} = \frac{1 \cdot 8 \times 10^{12}[G_{spec}]}{f} \qquad (7.16)$$

f is in c/s.

$[G_{spec}]$ is the reciprocal of $[R_{spec}]$ and is known as the specific conductivity (ohm^{-1} cm^{-1}) of a given sample.

The lumped circuit techniques to be described below are mainly designed to measure R_x and C_x, at a particular frequency. C_x and R_x may be converted to ε' and ε'' values, respectively, knowing C_0, and using Equation (7.8). The contribution to the total loss arising from a d.c. conductivity process is given by Equation (7.14) or its alternative form (7.16). $\varepsilon''_0(\omega)$ can be evaluated in practice by measuring R_0 or $[R_{spec}]$ by a simple direct current method, and using Equation (7.14) to obtain $\varepsilon''_0(\omega)$ at any desired value of ω.

7.3 SAMPLE AND SAMPLE HOLDER CONSIDERATIONS FOR LUMPED CIRCUIT METHODS OF MEASUREMENT

In the frequency range 10^{-4} to 10^8 c/s, it has become common practice to measure a polymer sample in the form of a flat circular disc, of area A and thickness d. The vacuum capacitance of this geometry is given by Equation (7.1).

The bridge methods and the transient current method to be described below are usually constructed to measure C_x values in the range 10 to 100 $\mu\mu$F. It is necessary therefore to arrange the geometry of the sample (A and d) so that $C_x = \varepsilon'C_0$, is within the range of the instrument. The convenient procedure is to design a sample holder to accommodate a fixed area of sample, and make samples of a suitable thickness for study.

It is essential that the area and thickness of a sample is known accurately. The faces of the disc should be accurately parallel to each other, so that Equation (7.1) is obeyed. The accuracy of the measurement of C_x, R_x may be far greater than the accuracy to which C_0 can be determined, in which case, the accuracy of the derived ε' and ε'' values will be determined wholly by the measured C_0 value.

The sample (in disc form) is mounted between metal electrodes. The electrical contact between the sample faces and electrodes is best ensured by evaporating a metal film on the sample surfaces. A less satisfactory arrangement is to attach metal foil to the surface with the aid of a thin film of silicone grease. If an air gap occurred between the sample face and the electrode, a capacitance would be introduced in series with the sample, leading to appreciable error in the measurement of C_x, R_x.

In order to eliminate fringing fields and also surface conduction across the edge of a sample, a guard ring may be incorporated into the sample assembly, and this is shown schematically in Figure 7.3(a). The sample geometry of this arrangement corresponds to a thickness d and area $A = \pi(r + \delta r/2)^2$ for a circular disc. The guard-ring assembly of Figure 7.3(a) can be expressed as the equivalent circuit shown in Figure 7.3(b).

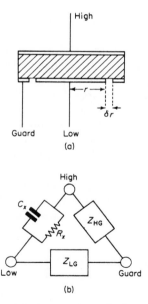

Figure 7.3. (a) Three-terminal arrangement for lumped circuit measurements, incorporating a guard ring; (b) Equivalent circuit for a three-terminal arrangement.

The technique of measurement in the frequency range 10^{-4} to 10^7 c/s must be designed in such a way that C_x and R_x can be measured independent of the impedances Z_{HG} and Z_{LG}. In the Schering bridge described below, these impedances may be eliminated using a Wagner earthing device. The transformer ratio-arm bridge and the ultra low frequency (Scheiber) bridge are designed so that Z_{HG} and Z_{LG} do not enter the balance conditions. If a two-terminal system is employed (high to low), the fringing field can be calculated theoretically, and used to correct the capacitance of the sample (Hartshorn and Ward, 1936).

The electrode assembly is mounted in a metal case, (normally at earth potential) and the resulting cell may be immersed in a thermostat bath for temperature variation. Detailed designs of cells which could be used in the range 10^{-4} to 10^5 c/s have been given by Rushton and Pratt (1940), Baker (1949) and by Scott and others (1962). The electrode assembly for measurements in the frequency range 10^5 to 10^8 c/s using a resonance technique has been described by Hartshorn and Ward (1936).

7.4 D.C. TRANSIENT CURRENT METHOD 10^{-4} TO 10^{-1} c/s

Figure 7.4 gives a simple circuit to illustrate the method. Switch (S_1, a) is closed, the sample C_x responds to the step voltage V, giving rise to a transient charging current through C_x which is measured by the amplifier circuit. After charging equilibrium has been attained, switch (S_1, b) is closed [opening (S_1, a)], and the transient discharge current is measured. Figure 7.5 shows a typical charge-discharge cycle for a polymer sample. The charging current decays to a time-independent value I_0 which is the steady conduction current for the sample. Discharging the sample produces the negative discharge current. If we subtract I_0 from the observed charging current we find that the net charge and discharge currents are identical in size and shape.

If the sample consisted of a pure capacitance only, there would be no transient current. Since transients are obtained in practice, the dielectric must be considered as having a time-dependent resistance associated with it.

A general theory assuming only that the superposition principle holds (Von Schweidler, 1907; Manning and Bell, 1940) related the time-dependent charge–discharge currents to the frequency dependence of the complex dielectric constant. The following relation was obtained

$$\varepsilon^*(\omega) = \varepsilon'(\omega) - i\varepsilon''(\omega) = \varepsilon_U + \left\{ \int_0^\infty \phi(t)e^{-i\omega t}\, dt \right\} - i\frac{G_0}{\omega C_0} \qquad (7.17)$$

$\varepsilon^*(\omega)$ is the complex dielectric constant at a frequency ω (radian). ε_U is the high-frequency (unrelaxed) dielectric constant of the medium and is due to electronic and atomic polarization. $\phi(t)$ is derived from the observed reversible charge–discharge currents $I_{(t)}$ using the relation $\phi(t) = I_{(t)}/C_0V$. Here V is the applied step voltage and C_0 is the capacitance of the electrodes when the sample is replaced by air. t is the time which has elapsed after the application (or withdrawal) of the step voltage. G_0 is the steady d.c. conductance of the sample (see Equation 7.14).

It is important to note that $\phi(t)$ does not contain the contribution to the observed changing current from the steady d.c. conductivity of the medium. The loss factor of the sample is thus separated into two components experimentally. The loss $\varepsilon''_0(\omega)$ arising from the d.c. conductivity component is given by Equation (7.14).

The relaxation component of the complex dielectric constant from (7.17) is

$$\varepsilon^*(\omega) = \varepsilon_U + \int_0^\infty \phi(t)e^{-i\omega t}\,\mathrm{d}t \tag{7.18}$$

Thus we can obtain $\varepsilon^*(\omega)$ at any chosen value of ω if $\phi(t)$ is known over the range $t = 0$ to ∞. In the simplest case where $\phi(t)$ is given by $ke^{-t/\tau}$, k and τ are constants, we evaluate (7.18) directly as

$$\varepsilon^*(\omega) = \varepsilon_U + \frac{(\varepsilon_R - \varepsilon_U)}{1 + i\omega\tau} \tag{7.19}$$

This is the familiar equation for a single dipolar relaxation process (see Chapter 4). k is equal to $(\varepsilon_R - \varepsilon_U)/\tau$. Thus if k and τ are observed experimentally, we have $\varepsilon^*(\omega)$ over all values of ω.

In practice, however, $\phi(t)$ may be a complicated function of time, and the integration of Equation (7.18) cannot be performed analytically. In these circumstances, the integration must be performed numerically at a given ω value. This may be done by hand or with the aid of a computer. If $\phi(t)$ is known over its entire range of time, numerical integration of Equation (7.18) will yield exact values for $\varepsilon'(\omega)$ and $\varepsilon''(\omega)$.

A less satisfactory procedure which avoids exact numerical integration has been widely used in practice. It is assumed that $\phi(t)$ may be expressed as

$$\phi(t) = At^{-n} \tag{7.20}$$

A and n are constants for the dielectric at a given temperature. Experimental results for transient currents in polymer solids have been found to be consistent with the form of Equation (7.20) *over a limited time scale.*

Insertion of Equation (7.20) into (7.18) gives

$$\varepsilon'(\omega) = \varepsilon_U + \omega^{n-1} A\Gamma(1-n)\sin\frac{n\pi}{2} \qquad 0 < n < 1 \qquad (7.21a)$$

$$\varepsilon''(\omega) = \omega^{n-1} A\Gamma(1-n)\cos\frac{n\pi}{2} \qquad 0 < n < 2 \qquad (7.21b)$$

Γ denotes the gamma function.

If n and A are known experimentally, $\varepsilon^*(\omega)$ can be derived from (7.21a) and (7.21b) at any desired ω value. The obvious limitation of this method is that (7.21b) does not predict a maximum loss which is so characteristic of the higher frequency dielectric relaxation data.

A significant advance was made by Hamon (1952) who assumed relation (7.20) and rewrote (7.21b) as

$$\varepsilon''(\omega) = \frac{\phi(t)}{\omega}\left[\frac{\Gamma(1-n)\cos n\pi/2}{(\omega t)^{-n}}\right] \qquad (7.22)$$

Arranging ω and t combinations as

$$\omega t = [\Gamma(1-n)\cos n\pi/2]^{-1/n} \qquad (7.23)$$

Hamon found that for $0\cdot1 < n < 1\cdot2$, the right-hand side of (7.23) was approximately constant, at a value $\pi/5$, to about three per cent accuracy. Thus $\varepsilon''(\omega)$ may be evaluated at any chosen ω value using the relations

$$\varepsilon''(\omega) = \frac{\phi(t)}{\omega} \qquad \omega t = \frac{\pi}{5} \qquad (7.24)$$

The main difference between (7.21b) and (7.24) is that (7.24) does not involve the parameters A and n, also the relationship between frequency and time is specified. Rewriting $\omega t = \pi/5$ as $f = 0\cdot1/t$ we see that the measurement of current at 1 second and 100 seconds corresponds to $0\cdot1$ and 1×10^{-3} c/s respectively. The other important feature of Hamon's work was that he showed that relations (7.24) would still give a reasonably accurate conversion of d.c. to a.c. data even if relation (7.20) were not exactly true. The real part $\varepsilon'(\omega)$ is obtained from (7.21a) and (7.24) as

$$\varepsilon'(\omega) = \varepsilon_U + \varepsilon''\tan\frac{n\pi}{2} \qquad \omega t = \frac{\pi}{5} \qquad 0\cdot1 < n < 1 \qquad (7.25)$$

Unfortunately the Hamon treatment is not rigorous since (7.20) will not hold in practice over the complete time range. However, it has been shown (Williams, 1962b) that the Hamon-type equations may be derived if a

Cole–Cole distribution of relaxation times (Cole and Cole, 1941) describes the dielectric properties at a given temperature. The analysis is as follows. For a Cole–Cole distribution, $\varepsilon^*(\omega)$ is given by

$$\varepsilon^*(\omega) = \varepsilon'(\omega) - i\varepsilon''(\omega) = \varepsilon_U + \frac{\varepsilon_R - \varepsilon_U}{1 + (i\omega\tau_0)^{\bar{\beta}}} \tag{7.26}$$

$\bar{\beta}$ is the distribution parameter and τ_0 is the average relaxation time. From (7.18) we have

$$\phi(t) = \frac{1}{\pi}\int_0^\infty \varepsilon^*(i\omega)e^{i\omega t}\,d\omega \tag{7.27}$$

Cole and Cole (1942) combined (7.26) and (7.27) and evaluated the integral as two equivalent series which reduce for $t/\tau_0 \ll 1$ and $t/\tau_0 \gg 1$ to

$$\phi(t) = \frac{(\varepsilon_R - \varepsilon_U)}{\tau_0}\frac{1}{\Gamma(\bar{\beta})}\left(\frac{t}{\tau_0}\right)^{-(1-\bar{\beta})} \qquad \frac{t}{\tau_0} \ll 1 \tag{7.28a}$$

$$\phi(t) = \frac{(\varepsilon_R - \varepsilon_U)}{\tau_0}\frac{\bar{\beta}}{\Gamma(1-\bar{\beta})}\left(\frac{t}{\tau_0}\right)^{-(1+\bar{\beta})} \qquad \frac{t}{\tau_0} \gg 1 \tag{7.28b}$$

From (7.26) we have

$$\varepsilon''(\omega) = (\varepsilon_R - \varepsilon_U)(\omega\tau_0)^{-\bar{\beta}}\sin\frac{\bar{\beta}\pi}{2} \qquad \omega\tau_0 \gg 1 \tag{7.29a}$$

$$\varepsilon''(\omega) = (\varepsilon_R - \varepsilon_U)(\omega\tau_0)^{\bar{\beta}}\sin\frac{\bar{\beta}\pi}{2} \qquad \omega\tau_0 \ll 1 \tag{7.29b}$$

Combining short times and high frequencies, i.e. (7.28a) and (7.29a) and also long times and low frequencies, i.e. (7.28b) and (7.29b) we obtain

$$\varepsilon''(\omega) = \frac{\phi(t)}{\omega} \tag{7.30a}$$

$$\omega t = \left[\Gamma(1-z)\cos\frac{z\pi}{2}\right]^{-1/z} \tag{7.30b}$$

Here $z = 1 - \bar{\beta}$ for $t/\tau_0 \ll 1$; $\omega\tau_0 \gg 1$ and $z = 1 + \bar{\beta}$ for $t/\tau_0 \gg 1$; $\omega\tau_0 \ll 1$. Thus $0 \leq z < 1$ for short times–high frequencies and $1 < z \leq 2$ for long times–low frequencies. It was shown that $\omega t \simeq \pi/5$ for $0 < z \leq 1.2$, so (7.30) are identical with the Hamon equations (7.24). It was also shown that Equations (7.30a) and (7.30b) could be used when the current–time curve passes through the relaxation region ($t \cong \tau_0$) at which the curve shows a sharp change in slope. This method avoids the incorrect assumption of

(7.20), and gives meaning to the observed slopes of the current–time curve. The Hamon equations (7.24) are most accurate when the double logarithmic current–time curve has a slope less than unity, and care must be taken for slopes greater than unity.

Experimentally, if a polymer sample having a capacitance of 10 to 100 $\mu\mu$F is subjected to a step voltage in the range 1 to 500 volts, the transient current is in the range 10^{-9} to 10^{-15} amp. Various methods have been used. The most convenient arrangement is shown in Figure 7.4. The step voltage V applied to the three-terminal sample produces the transient current which is measured as a voltage appearing across a standard resistor R_1 by means of a d.c. amplifier which incorporates an electrometer valve. The voltage drop from point A to ground is made zero by a negative feedback in the amplifier which produces a voltage across R_2 which is equal and opposite to that across R_1. Currents are recorded in the time range 1 second to 1000 seconds. Figure 7.5 shows the schematic current–time curve after the application of a step voltage at $t = 0$, and its subsequent withdrawal at $t = 63$ seconds; i_0 is the steady d.c. component of the charging current.

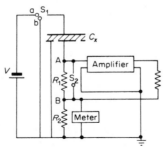

Figure 7.4. Schematic diagram of a circuit used for the d.c. transient experiment (Williams 1963b).

It has become customary to use the Hamon approximation to transform relation (7.18) in order to obtain ε''_ω from transient current measurements. It should be emphasized that exact numerical integration procedures should really be used in order to obtain precise values of ε''_ω. Good agreement has been obtained between the bridge and d.c. measurements for several polymer systems, using the Hamon approximation (Reddish, 1958, 1959). Davidson and others (1951) have used a modification of the d.c. method in which the applied voltage is increased linearly with time. This method has no apparent advantage over the simpler method outlined above.

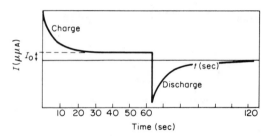

Figure 7.5. Charge–Discharge currents (schematic) obtained using the d.c. transient technique.

7.5 ULTRA-LOW FREQUENCY BRIDGE 10^{-2} TO 10^2 c/s

Scheiber (1961) has described a bridge which will measure very accurately in the range 10^{-2} to 10^2 c/s. The main difficulty in the lower region of this frequency range is that the generator cannot be coupled via a transformer to the bridge but must be coupled directly. This was achieved in Scheiber's design, and, also, a Wagner earth was not required to balance the bridge, which is a great time consumer at low frequencies. The actual bridge works on the Schering bridge principle (see below) and using a substitution method very precise ε^*_ω measurements are possible on polymer compounds. Other bridges used in this region have been reviewed by Scheiber.

The schematic circuit diagram of the Scheiber bridge is shown in Figure 7.6. The generator is directly coupled to the bridge circuit. Z_1, Z_2 and Z_L are stray impedances within the generator, J_1, J_2, J_3 and J_4 are 15 kΩ resistors in the generator. C_G is shorted for bridge measurements. The sample C_x, R_x is measured by balancing the bridge with sample 'in' using R_1 and R_2, and C_S. Here R_1 is a 100 Ω decade resistor, $R_2 = 10^6$, 10^7, 10^8 or 10^9 Ω interchangeable calibrated resistors. C_S and C_B are precision three-terminal variable capacitors (10 to 110 $\mu\mu$F). C_3 and C_4 are 1000 $\mu\mu$F precision two-terminal capacitors. R_3 and R_4 are matched precision 10^5 Ω resistors.

The balance condition is obtained as a null reading on the detector. For 30 c/s to 100 c/s, a cathode follower preceding a microvolt tuned amplifier is used as the detector assembly. For 30 c/s down to 10^{-2} c/s, a d.c. electrometer is used to visually follow the a.c. signal, and again the balance sensitivity can be in terms of microvolts.

If the sample is 'in', the balance readings are $(C_S)_{in}$, $(R_1)_{in}$ and $(R_2)_{in}$. The sample is removed from the arm by shorting its low-potential end (see Figure 7.6) to ground. This effectively places the sample in parallel

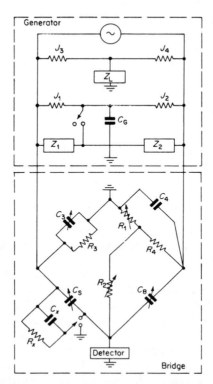

Figure 7.6. Schematic diagram of the ultra low frequency Bridge, 10^{-2} to 10^{2} c/s (Scheiber 1961).

with C_3 and R_3; thus in the 'out' condition, C_3 must be decreased to compensate for C_x. The balance condition for sample 'out' is obtained by changing C_S and R_1, keeping $(R_2)_{in}$ fixed. If $(C_S)_{out}$ and $(R_1)_{out}$ denote the out reading, the sample capacitance C_x and $\tan \delta_x = 1/(R_x C_x)\omega$ are given by

$$C_x = (C_S)_{out} - (C_S)_{in} \tag{7.31}$$

$$\tan \delta_x = \frac{(R_1)_{in} - (R_1)_{out}}{\omega C_x (R_2)_{out} R_4} \tag{7.32}$$

Equations (7.31) and (7.32) are good approximations of the exact equations given by Scheiber (1961).

The bridge is extremely accurate, capacitances may be measured to better than $10^{-3}\ \mu\mu$F accuracy. Large and small losses can be measured with great precision.

One advantage of the Scheiber bridge is that d.c. conductivity is easily measured down to the lowest frequencies. For d.c. conductivity, the loss factor $\varepsilon''_0(\omega)$ is related to frequency via Equation (7.16). From Equations (7.16) and (7.32)

$$[G_{\text{spec}}] = \frac{(R_1)_{\text{in}} - (R_1)_{\text{out}}}{C_0 (R_2)_{\text{out}} R_4 3 \cdot 6\pi 10^{12}} \tag{7.33}$$

The specific conductivity $[G_{\text{spec}}]$ is a constant, hence for a given sample (C_0 = constant), the same value of $(R_1)_{\text{in}} - (R_1)_{\text{out}}$ is obtained at each frequency, i.e. ω is not involved in Equation (7.33). This property of the Scheiber bridge is quite different from that for a Schering bridge, where it is found that the rising loss at low frequencies due to d.c. conductivity rapidly exhausts the balancing range of the bridge.

7.6 SCHERING BRIDGE 10 TO 10^6 c/s

This is the most common method for the measurement of ε^* particularly in polymer work. Various designs differing in detail are used, but the basic principle of one of the most commonly used is illustrated in Figure 7.7. This is a simple capacitance bridge.

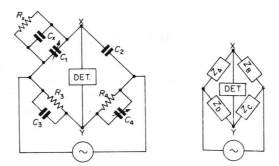

Figure 7.7. Schematic diagram of a modified Schering bridge.

For sample in arm A, at balance we have $(Z_A Z_C)_{\text{in}} = (Z_B Z_D)_{\text{in}}$. Here Z_A is the total impedance of arm A, etc. For sample out, only C_1 and C_4 need be adjusted to rebalance the bridge giving $(Z_A Z_C)_{\text{out}} = (Z_B Z_D)_{\text{out}}$.

Since arms B and D were not adjusted on going from the sample in to out measurements we have $(Z_A Z_C)_{in} = (Z_A Z_C)_{out}$. Solution of this equation gives the following result for the sample capacitance C_x and its loss tangent $\tan \delta_x$

$$C_x = \frac{[(C_1)_{out} - (C_1)_{in} - R_4^2 \omega^2 (C_4)_{in} \{(C_4)_{in}(C_1)_{in} - (C_4)_{out}(C_1)_{out}\}]}{1 + [R_4 \omega (C_4)_{in}]^2}$$

(7.34)

$$\tan \delta_x = \frac{R_4 \omega (C_1)_{out} [(C_4)_{in} - (C_4)_{out}]}{C_x [1 + \{R_4 \omega (C_4)_{in}\}^2]}$$

(7.35)

For low loss ($\tan \delta_x < 10^{-2}$) Equations (7.34) and (7.35) reduce to

$$C_x = (C_1)_{out} - (C_1)_{in}$$

(7.36a)

$$\tan \delta_x = \frac{R_4 \omega (C_1)_{out} [(C_4)_{in} - (C_4)_{out}]}{C_x}$$

(7.36b)

A Wagner earth is sometimes used in order to ensure that at balance the potentials X and Y are not merely equal but are at earth potential so that noise and stray pick-up are at a minimum at balance.

The Schering bridge is capable of very high accuracy for ε' and ε'' and the uncertainty of measurement is often due to the fact that the sample dimensions are not known to the accuracy that C_1 and C_4 changes can be determined.

7.7 TRANSFORMER RATIO-ARM BRIDGE 10 TO 10^6 c/s

The conventional bridge methods, e.g. Schering bridge, are based on the Wheatstone bridge principle, where the impedances of the ratio arms correspond to capacitance–resistance networks. Cole and Gross (1949), Thompson (1956) and Lynch (1959) have described bridges in which the ratio arms comprise closely coupled transformers. The advantages of this technique over the conventional methods are that residual impedance and guard circuit impedance are essentially eliminated from the bridge balance conditions. Also, transformer arms can be wound to give accuracy of a few parts in 10^6 in transformer ratios. Thus the transformer bridge is capable of high accuracy over a wide range of capacitance and loss.

Figure 7.8 shows a very simple form of transformer bridge. A toroidal core carries a primary transformer L_P, which activates the secondary transformer (L_S, L_U), which is also wound on the toroid which carries L_P.

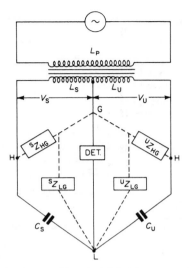

Figure 7.8. A simple capacitance bridge incorporating transformer ratio arms.

The secondary transformer is tapped at point G, and the voltage ratio V_U/V_S is given accurately by n_U/n_S where n_U and n_S are the number of turns of L_U and L_S respectively. The primary transformer activates the bridge circuit, but is not involved in the bridge balance conditions. The simple bridge circuit contains a standard capacitor C_S, with which is associated stray impedances $^SZ_{HG}$ and $^SZ_{LG}$. The other arm contains the unknown C_U, and stray impedances $^UZ_{HG}$ and $^UZ_{LG}$. The balance condition is attained for zero current through the detector arm. The impedances $^SZ_{LG}$ and $^UZ_{LG}$ shunt the detector, and thus are not included in the balancing conditions of the bridge. The impedances $^SZ_{HG}$ and $^UZ_{HG}$ shunt the transformer L_S and L_U respectively, and thus do not affect the bridge balance. At balance it can be shown that

$$V_S C_S = V_U C_U \tag{7.37}$$

where $V_U/V_S = n_U/n_S$. Thus C_U can be determined, with high accuracy in terms of the standard C_S.

Figure 7.9(a) shows a schematic diagram of a sophisticated transformer bridge capable of measuring capacitance and loss over very wide ranges with high accuracy (General Radio Expt., 1962). The sample is represented as C_x, R_x. The transformer in the measuring arm has taps along

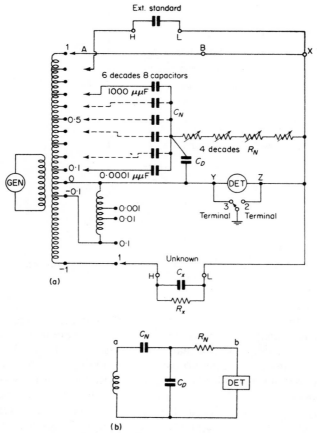

Figure 7.9. (a) Sophisticated transformer ratio-arm bridge (by permission of General Radio Co.); (b) Equivalent T network for the evaluation of $\tan \delta_N$.

its length at $\frac{1}{10}$ intervals. The measuring arm contains eight fixed capacitors (C_N), ranging from 1000 $\mu\mu$F to 0·0001 $\mu\mu$F in decade intervals. The capacitors C_N are connected at a common point to decade resistors R_N, which cover the range 100 Ω to 0·1 Ω in four decades. The standard capacitors C_N may tap the transformer in any desired combination. The bridge is balanced when the current through the detector is zero. This is achieved at a particular value for C_N and R_N, and it is seen that the measurement essentially corresponds to a measurement of the equivalent *series* capacitance and resistance of the sample. Owing to stray capacitance between the common junction point of the capacitors C_N and resistors R_N to the bridge shields, the basic network which provides the loss tangent measurement is a T network shown in Figure 7.9(b). The effective direct impedance of this network between points a and b may be regarded as a capacitance C_N, with a loss tangent $\tan \delta_N$

$$\tan \delta_N = \omega R_N (C_N + C_D) = \tan \delta_x \qquad (7.38)$$

$(C_N + C_D)$ includes all the direct and stray capacitance connected to the junction of C_N and R_N. Although the effective capacitance C_N of the bridge is varied by varying the voltages applied to the several bridge capacitors, $(C_N + C_D)$ can be made a constant. Experimentally, the resistance R_N can be calibrated to read $\tan \delta_N$ directly at a chosen frequency. In the G.R. 1615-A bridge (see General Radio Expt., 1962), the frequency chosen was 10^3 c/s. At other frequencies, the indicated $\tan \delta_N$ on the bridge dials was multiplied by the frequency of measurement in kc/s, to obtain the actual $\tan \delta_N$ of the measuring arm. If we denote the capacitance of the measuring arm at balance as $(C_N)_{\text{balance}}$, this is equal to the equivalent series capacitance C'_x of the sample (see Equation 7.9). The equivalent parallel capacitance C_x of the sample is given by Equation (7.12a)

$$C_x = \frac{C'_x}{1 + \tan^2 \delta_x} = \frac{(C_N)_{\text{balance}}}{1 + \tan^2 \delta_x} \qquad (7.39)$$

The balance conditions given above correspond to the case where the transformer ratio in the bridge circuit is 1 : 1 (measuring arm : sample arm). The range of the bridge can be extended via taps in the sample arm as is shown in Figure 7.9(a).

With the combination of eight fixed capacitors, four decades of R_N and four voltage ratios, it is possible to measure sample capacitances in the range 10^6 $\mu\mu$F to 10^{-5} $\mu\mu$F, and $\tan \delta_x$ values can be obtained in the range 1×10^{-2} to 1×10^{-6} at 10^3 c/s. Additional capacitors may be added to the measuring arm to extend the measurement of $\tan \delta_x$ up to 1.

For high loss specimens it is more convenient to use a resistance network to obtain a variable conductance. This network is obtained by removing R_N in Figure 7.9(a), and connecting the junction point of the standard capacitors C_N to point Z. A standard 10^5 Ω resistor is placed between A and B, and another standard 10^5 Ω resistor is placed between B and X. The junction point B is connected to four decade resistors R'_N in series, which are then connected to Y in the figure. With this arrangement the values of C_N and R'_N at balance yield the equivalent parallel capacitance and resistance of the sample.

$$C_x = (C_N)_{\text{balance}} \tag{7.40}$$

$$\tan \delta_x = \frac{1}{\omega C_x} \cdot \left[\frac{(R'_N)_{\text{balance}} \times 10^{-10}}{1 + 2 \times 10^{-5} (R'_N)_{\text{balance}}} \right] \tag{7.41}$$

$(C_N)_{\text{balance}}$ and $(R'_N)_{\text{balance}}$ are the values of C_N and R_N at the balance condition of the bridge.

R'_N is adjustable over 10^2 to $0.1 \, \Omega$ in four decades, which provides a range of conductance $G = \omega C_x \tan \delta_x$ of $0.1 \, \mu$mho to $10^{-11} \times$ mho. The range can be extended using the transformer taps on the sample arm of the bridge. Equations (7.40) and (7.41) relate to a 1:1 transformer ratio setting between the measuring arm and the sample arm. The dielectric constant and loss factor of the sample follow from Equation (7.8).

The transformer ratio-arm bridge described above has the advantage over other bridge methods in view of its accuracy of measurement over an extremely large range of capacitance and loss tangent.

7.8 RESONANCE CIRCUITS 10^5 TO 10^8 c/s

Above about 10^6 c/s the effects of stray impedance (particularly inductance) become increasingly significant. The bridge methods cannot be used above about 10 Mc/s for this reason. Various methods have been devised for this range and we shall confine ourselves to the conductance variation resonance method which is probably the most widely used in polymer studies and also gives very precise results. The apparatus is illustrated schematically in Figure 7.10 (Hartshorn and Ward, 1936).

The resonance circuit is made up of two precision variable capacitors C_1 and C_2, inductance L, the sample C_x, R_x, and the voltmeter V. The circuit is brought to resonance using a loosely coupled generator circuit of variable frequency. At resonance the half-width δ_{in} of the resonance curve is determined using the micrometer capacitor $C_2 (\simeq 0\text{–}8 \, \mu\mu\text{F})$. The sample is then withdrawn from the resonance circuit and resonance restored by changing C_1 only.

The half-width δ_{out} of the resonance circuit without sample is determined using C_2. The sample capacitance is given by the change in C_1 on going from sample in to sample out resonance. The loss tangent of the sample is given by

$$\tan \delta_x = \frac{\delta_{\text{in}} - \delta_{\text{out}}}{2C_x} \tag{7.42}$$

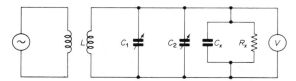

Figure 7.10. Conductance variation circuit, 10^5 to 10^8 c/s.

Frequency variation is achieved by a suitable choice of the primary and secondary inductance values. Measurements can be easily made over the entire frequency range at room temperature. If measurements are required on the sample over a range of temperature special care must be taken to keep the electrical leads from the sample cell to the resonance circuit as short as possible. Even then measurements may be restricted to the range 10^5 to 5×10^6 c/s.

This method gives very precise ε^*_ω values at the lower frequencies but at higher frequencies the accuracy diminishes particularly when an auxiliary cell is used. A two-terminal sample only can be used and the effects of the fringing must be carefully eliminated.

7.9 RE-ENTRANT CAVITY 10^8 TO 10^9 c/s

For low loss ($\tan \delta \simeq 10^{-4}$) a resonant cavity is appropriate in the frequency range 10^8 to 10^9 c/s. Various methods and designs have been employed (Works and others, 1944; Reynolds, 1947; Parry, 1951). A good illustration of a re-entrant cavity technique is that due to Parry (1951). Figure 7.11 shows Parry's apparatus in schematic form. The method is an extension of the Hartshorn–Ward method (1936) to higher frequencies. The sample is placed between the electrodes, and the system is equivalent to a closed coaxial transmission line (see Section 7.10) in which the central conductor is in two parts, separated by the sample. The cavity is brought to resonance using frequency as the variable, the resonance being detected using a silicon crystal and loop. The resonance curve is defined in terms of capacitance using the variable micrometer capacitor C_2. The half-width of the resonance curve is a measure of the total loss in the cavity. The sample is removed and the cavity returned to resonance using the main micrometer C_1. The change in equivalent air capacitance of the main micrometer capacitor C_1 is noted. The half-width of the resonance curve is again obtained by variation of C_2.

The capacitance of the sample is obtained from the change in capacitance of the main micrometer on removal of the sample. The loss tangent of the sample is obtained using Equation (7.42) above.

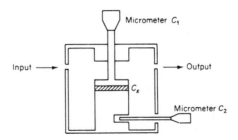

Figure 7.11. Re-entrant cavity, 10^8 to 10^9 c/s.

Using this method ε' is obtained to about one per cent, and ε'' to about five per cent accuracy. Temperature variation of the sample is not easily achieved, and the frequency range of a given cavity is very small.

7.10   DISTRIBUTED CIRCUITS 10^8 TO 3×10^{10} c/s

At frequencies above about 10^8 c/s, it is extremely difficult to make lumped circuit measurements, due to the increasing importance of residual inductance. Methods have been developed which avoid this problem, and are based on the concepts of wave propagation through a dielectric medium. These methods involve the propagation of electromagnetic waves along rectangular or cylindrical waveguides or coaxial transmission lines. A description of the transmission line and cavity resonator techniques will be given below, but first the relationship between the experimental quantity γ_s, known as the propagation factor of the dielectric, and the complex dielectric constant ε^* will be derived.

A plane electromagnetic wave propagated in the x coordinate of an infinite continuum of absolute complex dielectric constant ε^*_s is characterized by the equations

$$\mathbf{E} = \mathbf{E}_0 \exp\left[i\omega t - \gamma_s x\right] \tag{7.43a}$$

$$\mathbf{H} = \mathbf{H}_0 \exp\left[i\omega t - \gamma_s x\right] \tag{7.43b}$$

The derivation of Equation (7.43) has been given by Von Hippel (1954). $\mathbf{E}$ and $\mathbf{H}$ are the electric and magnetic field components of the wave. $\mathbf{E}$ and $\mathbf{H}$ are perpendicular to each other, and are both perpendicular to the direction x. x, E and H form a right-hand coordinate system of sequence $+x \rightarrow E_y \rightarrow H_z$. The plane wave is known as a transverse electromagnetic or TEM wave.

γ_s is the complex propagation factor of the dielectric medium and is given by

$$\gamma_s = i\omega(\varepsilon^*_s\mu^*_s)^{1/2} = \alpha_s + i\beta_s \qquad (7.44)$$

ε^*_s and μ^*_s are the absolute complex dielectric constant and complex permeability of the dielectric. α_s and β_s are the attenuation and phase factor respectively of the dielectric, and $\beta_s = 2\pi/\lambda_s$, where λ_s is the wavelength of the wave in the dielectric. For a vacuum, Equation (7.44) is modified to.

$$\gamma_o = i\omega(\varepsilon_o\mu_o)^{1/2} = i\beta_o \qquad (7.45)$$

Subscripts o refer to vacuum and $\beta_o = 2\pi/\lambda_o$, where λ_o is the wavelength in vacuum. Introducing a refractive index

$$n = \frac{\lambda_o}{\lambda_s} = \frac{\beta_s}{\beta_o}$$

and remembering $\mu^*_s = \mu_o$ for non-magnetic dielectrics, Equations (7.44) and (7.45) give $\varepsilon^* = \varepsilon' - i\varepsilon''$ as

$$\varepsilon' = \left(\frac{\varepsilon'_s}{\varepsilon_o}\right) = n^2\left(1 - \frac{\alpha^2_s}{\beta^2_s}\right) \qquad (7.46a)$$

$$\varepsilon'' = \left(\frac{\varepsilon''_s}{\varepsilon_o}\right) = n^2 2\frac{\alpha_s}{\beta_s} \qquad (7.46b)$$

ε' and ε'' are the dielectric constant and loss factor respectively, relative to vacuum.

Equations (7.46) relate the experimentally determined quantities α_s and β_s to the complex dielectric constant for the special case of a TEM wave propagated in an infinite continuum.

If the wave is confined, restrictions are imposed on the solutions of the wave equations. Figure 7.12 shows a cylindrical and rectangular waveguide and a coaxial transmission line. For a wave propagated down a waveguide Equations (7.46) do not apply, and are replaced for TE or TM waves by (see Von Hippel, 1954)

$$\varepsilon^* = \frac{1/\lambda_c^2 - (\gamma_{sg}/2\pi)^2}{1/\lambda_c^2 + 1/\lambda_{og}^2} \qquad (7.47)$$

λ_c is the cut-off wavelength of the guide. λ_c values are determined by the

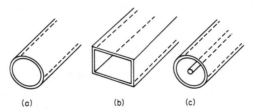

Figure 7.12. (a) Cylindrical waveguide; (b) Rectangular waveguide; (c) Coaxial line.

geometry of the guide. $\lambda_c = \infty$ for a coaxial line, and Equation (7.47) is identical in this case with Equation (7.46). λ_{og} is the wavelength in a vacuum filled guide. The attenuation factor γ_{sg} is given by

$$\gamma_{sg}^2 = \left(\frac{2\pi}{\lambda_c}\right)^2 - \omega^2 \varepsilon^*_s \mu^*_s \qquad (7.48)$$

λ_c and λ_{og} are related according to

$$\frac{1}{\lambda_o^2} = \frac{1}{\lambda_{og}^2} + \frac{1}{\lambda_c^2} \qquad (7.49)$$

The experimental distributed circuit methods are designed to measure γ_{sg} (or γ_s for the case of a coaxial line). Equations (7.46) and (7.47) give the relationship between γ_s and γ_{og} and the complex dielectric constant relative to vacuum of the sample.

7.11   TRANSMISSION LINES 10^8 TO 3×10^{10} c/s

7.11a   Introduction

Figure 7.12(a), (b) and (c) show in schematic form a cylindrical waveguide, rectangular waveguide and a coaxial line. These are constructed in a good conducting material such as copper, and may be silver plated in order to reduce attenuation in the walls of the line. For standing wave and travelling wave methods, the line has a slot along part of its length, and the field in the line is measured using a crystal probe which travels along the slot. Westphal (Von Hippel, 1954), De Vos (1958) and Bos (1958) have reviewed the experimental arrangements and the various methods used for the determination of ε^*. It would be inappropriate to give the details of all the methods here, so a description will be given of the Roberts–

Von Hippel method (1946) as an illustration of the principles involved in the standing wave method using waveguides or a coaxial line.

7.11b Roberts–Von Hippel Method

Figure 7.13 shows a rectangular waveguide, shorted at one end, and partially filled with a dielectric medium which is in contact with the shorted end. Denoting the vacuum region, sample region and short circuit as o, s and e respectively, consider a wave propagated down the line in the x direction. The wave strikes the boundary os at normal incidence and is partially reflected back into region o. The transmitted wave travels through the dielectric and is totally reflected at the boundary se. The resultant wave pattern in region o will be composed of superposed forward and backward travelling waves, and gives a standing wave pattern in region o. The electric and magnetic field vectors in region o are given by

$$(E_y)_o = (E_o)_o \exp\left[i\omega t - \gamma_{og}x\right] + (E_1)_o \exp\left[i\omega t + \gamma_{og}x\right] \qquad (7.50\text{a})$$

$$(H_z)_o = (H_o)_o \exp\left[i\omega t - \gamma_{og}x\right] + (H_1)_o \exp\left[i\omega t + \gamma_{og}x\right] \qquad (7.50\text{b})$$

The first term on the right-hand side of Equations (7.50a) and (7.50b) refers to the forward wave, and the second term refers to the reflected wave at the interface os. $(E_o)_o$ and $(E_1)_o$ are the amplitudes of forward and reflected waves respectively, and similar definition applies to $(H_o)_o$ and $(H_1)_o$.

Introducing a reflection coefficient at the boundary os as r_{os} and $-r_{os}$ for the electric and magnetic components respectively, Equations (7.50a) and (7.50b) may be combined to give

$$Z = \frac{(E_y)_o}{(H_z)_o} = Z_o \frac{\left[\exp - \gamma_{og}x + r_{os} \exp \gamma_{og}x\right]}{\left[\exp - \gamma_{og}x - r_{os} \exp \gamma_{og}x\right]} \qquad (7.51)$$

$$r_{os} = (E_1)_o/(E_o)_o = -\frac{(H_1)_o}{(H_o)_o}$$

$Z_o = (E_o)_o/(H_o)_o$, and Z_o is known as the characteristic impedance. At the boundary $x = 0$

$$Z(0) = Z_o \frac{1 + r_{os}}{1 - r_{os}} \qquad (7.52)$$

The wave in the sample region can be expressed in a similar way to Equation (7.50), but, the reflection coefficient at the short circuit is now -1 for the electric wave component and $+1$ for the magnetic wave component. Using a development similar to that above, the impedance at the boundary $x = 0$ is given by

$$Z(0) = Z_s \tanh \gamma_{sg} d \qquad (7.53)$$

$Z_s = (E_o)_s/(H_o)_s$. $(E_o)_s$ and $(H_o)_s$ are the amplitudes of the forward electric and magnetic components in the sample region. For non-magnetic dielectrics

$$Z_o \gamma_{og} = Z_s \gamma_{sg} \qquad (7.54)$$

Combination of (7.52), (7.53) and (7.54) gives the relationship between the standing wave pattern in the region o, and the propagation factor of the dielectric. Roberts and Von Hippel (1946) used Equation (7.50) to define the standing wave pattern in the o region of the line. It was shown that the electric component of the wave had successive minima in region o, separated by $\lambda_{og}/2$. Figure 7.13 shows the amplitude (E^2) of the standing wave pattern in the line. The first minimum occurs at a distance x_o from the sample surface. The voltage standing wave ratio $(E_{\min}/E_{\max})$ in the region o is a measure of the attenuation of the wave due to the sample. It was shown that r_{os} was a particular function of x_o, d, λ_{og} and $(E_{\min}/E_{\max})$. Combination of Equations (7.52), (7.53) and (7.54) gave the relation

$$\frac{\tanh \gamma_{sg}d}{\gamma_{sg}d} = -\frac{i\lambda_{og}}{2\pi d} \frac{[(E_{\min}/E_{\max}) - i \tan 2\pi x_o/\lambda_{og}]}{[1 - i(E_{\min}/E_{\max}) \cdot \tan 2\pi x_o/\lambda_{og}]} \qquad (7.55)$$

Equation (7.55) expresses the relationship between the experimentally measured quantities x_o, d, λ_{og} and $(E_{\min}/E_{\max})$, and the propagation factor of the dielectric γ_{sg}.

The theory outlined above does not take into account factors such as wall loss and short circuit loss which will contribute to the experimental value of $(E_{\min}/E_{\max})$ determined in the air portion (strictly vacuum) portion of the line. The apparent $(E_{\min}/E_{\max})$ can be measured with the travelling probe by direct observation of the minima and maxima. For low loss samples $(E_{\min}/E_{\max})$ is obtained from measurements around a voltage minimum. In this case

$$\left|\frac{E_{\min}}{E_{\max}}\right| = \frac{\sin \theta}{[(E_x/E_{\min})^2 - \cos^2 \theta]^{1/2}} \qquad (7.56)$$

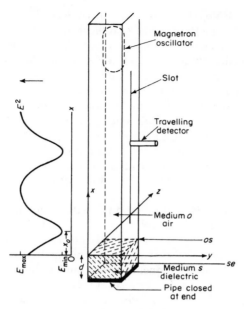

Figure 7.13. Rectangular waveguide, partially filled with a dielectric sample (Roberts and Von Hippel, 1946).

E_x is the field at a distance $\Delta x/2$ from the voltage minimum. $\theta = \pi \Delta x / \lambda_{og}$.

When Δx corresponds to the distance between points of twice minimum power, $(E_x/E_{\min})^2 = 2$, and

$$\left|\frac{E_{\min}}{E_{\max}}\right| = \frac{\pi \Delta x}{\lambda_{og}} \tag{7.57}$$

The observed value of Δx must be corrected for wall loss and short circuit loss. Westphal (1954) has given an extensive account of the correction procedures. De Vos (1958) has described the effect of parasitic reflections of the slot and connections to the cell on the measurement of x_o and $(E_{\min}/E_{\max})$.

x_o is obtained by comparing the minima positions for sample in with those obtained with no sample in the line. Since a minimum occurs at the shorted end for both sample in and out, x_o is easily evaluated. λ_{og} is observed directly, in the air portion of the line.

The solution of the transcendental equation (7.55) presents some difficulty, but a number of procedures have been given. Roberts and Von

Hippel (1946), De Vos (1958), Bos (1958) and Westphal (Von Hippel, 1954) have reviewed the methods. The graphical method of Roberts and Von Hippel (1946) is accurate for medium and high loss dielectrics, whereas for low loss media, the Dakin and Works approximations of Equation (7.55) are more appropriate (Dakin and Works, 1947).

Having solved Equation (7.55), a number of $\gamma_{sg}d$ values are obtained due to the transcendental nature of the function tanh $\gamma_{sg}d/\gamma_{sg}d$. The correct value can be recognized since $\beta_{sg}d \simeq 2\pi d\sqrt{\varepsilon'}/\lambda_{og}$. Knowing an approximate value for ε', $(\beta_{sg}d)$ can be roughly predicted, and the correct value for γ_{sg} value obtained from Equation (7.55). Alternatively the correct value can be obtained from a comparison of the multiple values obtained from measurements at two different depths d_1 and d_2 for the dielectric.

The Roberts–Von Hippel method is used in conjunction with coaxial lines in the frequency range 10^8 to 5×10^9 c/s (De Vos, 1958; Williams, 1959). At higher frequencies cylindrical and rectangular waveguides are used in preference to the coaxial lines. The method is capable of good accuracy over the entire frequency range, and yields ε' and ε'' values to about $\pm 1\%$ and $\pm 5\%$ accuracy respectively. For low loss dielectrics the correction factors for wall loss and short circuit loss are appreciable, and the cavity resonator techniques are more appropriate.

7.12   CAVITY RESONATOR METHODS 10^9 TO 3×10^{10} c/s

Lamont (1942) has summarized the various properties of cavity resonators, and Horner and others (1946) have described methods for the measurement of ε^* using cavity resonator techniques. Figure 7.14 shows an empty cylindrical cavity resonator operating in a particular resonance condition known as the H_{011} mode. The electromagnetic field inside the cavity is described in terms of the field vectors E_θ, H_z and H_r, where (z, θ, r) are the cylindrical coordinates. Resonance occurs in a H_{01n} resonator when the length is equal to $n\lambda_{og}/2$. In Figure 7.14, $n = 1$, and the mode of the resonance is denoted as H_{011}. The cut-off wavelength for a cylindrical cavity resonator is given by $\lambda_c = 1 \cdot 63a$, where a is the radius of the cylinder. The resonance condition is measured using a probe placed at $l/2$, and the cavity is excited by feeding the high-frequency signal in via a coupling hole at $l/2$. Horner and others (1946) have described the experimental arrangement of a H_{01n} resonator for $n = 4$ and $n = 5$, which operates at an energizing wavelength of 3 cm.

The dielectric constant and loss factor measurements are made using a H_{01n} resonator by partially filling the resonator with a dielectric medium.

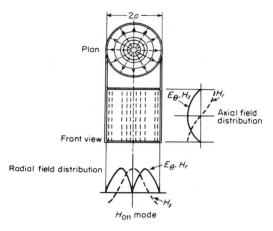

Figure 7.14. H_{01n} cavity resonator resonating in the H_{011} mode (Horner and others, 1946).

Figure 7.15 shows the partially filled resonator and the air filled resonator. The resonant lengths for sample out and sample in are l_o and l_r respectively. The dielectric constant is obtained by solving the relation

$$\frac{\tan \beta_{sg} d}{\beta_{sg}} + \frac{\tan \beta_{og}(l_r - d)}{\beta_{og}} = 0 \qquad (7.58)$$

β_{sg} is related to ε' according to

$$\varepsilon' = \frac{\beta_{sg}^2 + k^2}{\beta_{og}^2 + k^2} \qquad (7.59)$$

$$k = \left(\frac{3 \cdot 832}{a}\right) \qquad \beta_{og} = \frac{2\pi}{\lambda_{og}}$$

where λ_{og} is the wavelength in the empty resonator. The cavity is brought to resonance by changing its length. A non-contact piston is used as the variable of length. Knowing l_r, l_o, λ_{og} d and a, the dielectric constant is evaluated.

The sample loss is measured from the half-width of the resonance curve for sample in and sample out.

$$\tan \delta_s = \beta_{og}^2 \left\{ 2\delta l_s - \frac{s'\delta l_o}{(2a\beta_{og}^2 + l_o k^2)} \right\} \times \frac{1}{p(2d - s)(\beta_{sg}^2 + k^2)} \qquad (7.60)$$

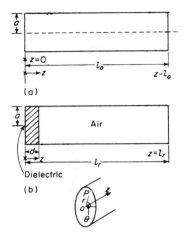

Figure 7.15. (a) Air filled cavity resonator; (b) Cavity resonator containing a dielectric sample at one end (Horner and others, 1946).

δl_s and δl_o are the half-widths for sample in and sample out respectively.

$$p = \frac{\sin^2 \beta_{og}(l_r - d)}{\sin^2 \beta_{sg}d} \qquad (7.61)$$

$$q = \frac{\sin 2\beta_{og}(l_r - d)}{\beta_{og}} \qquad (7.62)$$

$$s = \frac{\sin 2\beta_{sg}d}{\beta_{sg}} \qquad (7.63)$$

$$s' = k^2[p(2b - s) + 2(l_r - d) - q] + 2a[p\beta_{sg}^2 + \beta_{og}^2] \qquad (7.64)$$

Horner and others (1946) have also described a E_{010} resonator for dielectric measurements. In this case the cavity is brought to resonance by varying the frequency inside the cavity. Variation of the axial length of the cavity does not affect the resonance condition.

Cavity resonator methods are used mainly in the study of low loss dielectrics. The low loss polymers such as polyethylene, polystyrene and PTFE are readily measured to a high degree of accuracy, since the Q factor for empty cavities can be made extremely high. For an empty

H_{01n} cavity, Q is given by

$$Q_o = \frac{(\beta_{og}^2 + k^2)}{\beta_{og}^2} \frac{l_o}{\delta l_o} \qquad (7.65)$$

It is difficult to measure high loss dielectrics using the cavity resonator methods. The sample length must be made extremely small, and under these conditions the experimental uncertainty in $(l_r - l_o)$ affects the accuracy in the derived values for β_{sg} and $\tan \delta_s$. Another difficulty with the cavity resonator technique is that temperature variation of the sample is not easily achieved.

8

Methacrylate and Related Polymers

8.1 ALKYL METHACRYLATE POLYMERS

$$\left[\begin{array}{cc} \text{H} & \text{CH}_3 \\ -\text{C} & -\text{C} - \\ \text{H} & \text{O}=\text{C}-\text{OR} \end{array}\right]_n$$

(1)

Alkyl methacrylate polymers (1) have been known since 1877 when Fittig polymerized ethyl methacrylate ($R = CH_2\text{—}CH_3$). The study of their physical properties was rudimentary prior to their commercial production which began with polymethyl methacrylate ($R = CH_3$) in 1931. When prepared by the conventional free-radical methods at elevated temperatures the alkyl methacrylate polymers are invariably amorphous. However, by varying the polymerization temperature, or by the use of stereospecific catalysts, many of them have been prepared in crystallizable forms having a high degree of either syndiotactic or isotactic stereoregularity. Most mechanical and dielectric studies have been made on the conventional polymers and unless indication is given to the contrary it will be understood that conventional polymers are under discussion. It may be noted that the dielectric properties of the alkyl methacrylate polymers are determined largely by the strong electric dipole in the side ester (—COOR) group. We consider first the lowest member of the homologous series, namely polymethyl methacrylate ($R = CH_3$), the study of which has received much attention.

8.2 POLYMETHYL METHACRYLATE

Polymethyl methacrylate (PMMA) can be obtained in different stereoregular forms depending on the polymerization method. Evidence from high-resolution n.m.r. studies suggests that the 'conventional' PMMA prepared by free-radical initiation at elevated temperatures has a fairly high degree ($\approx 60\%$) of syndiotactic character (Bovey and Tiers, 1960; Fox and Schnecko, 1962). These n.m.r. studies also show that if the temperature of the free-radical polymerization is lowered, the proportion of syndiotactic groupings within the chains increases. The above conclusions are supported by the infrared measurements of Baumann, Schreiber and Tessmar (1959) and of Fox and Schnecko (1962). Moreover, Fox and others (1958) have observed that PMMA prepared by free-radical means below about 0°C (and thus having a high degree of syndiotactic character) may be crystallized by treatment with a borderline solvent (e.g. 4-heptanone), whereas the polymers prepared at higher temperatures are non-crystallizable. Crystallizable PMMA samples having a high degree of isotactic character (95%) are also produced by the use of stereospecific catalysts such as phenylmagnesium bromide at 0°C (Fox and others, 1958; Fox and Schnecko, 1962). An isotactic–syndiotactic stereoblock polymer may also be prepared by an anionic polymerization at low temperatures in a medium of moderate solvating power. The properties of the different stereospecific forms of PMMA are summarized in Table 8.1. Both the syndiotactic (Type I) and isotactic (Type II) molecules crystallize in a helical conformation, the former having ten monomer units and four turns and the latter five monomer units and two turns in the repeat unit (Stroupe and Hughes, 1958). The isotactic helix is less stiff than the syndiotactic helix which probably accounts for the lower melting point and glass-transition temperature of the isotactic polymer (Table 8.1). The value of T_g for 'conventional'

Table 8.1. Properties of the various types of stereoregular PMMA (Fox and others, 1958; Stroupe and Hughes, 1958).

Type	Density of amorphous polymer at 30°C (g/cc)	Helix	Volume–Temperature measurements		Stereoregular form
			T_g	T_m	
I	1·19	10_4 (stiff)	115°C	>200°C	Predominantly syndiotactic
II	1·22	5_2 (less stiff)	45°C	160°C	Predominantly isotactic
III	1·20 → 1·22	—	60°C → 95°C	170°C	Stereoblock 'copolymer'
Conventional	1·188	—	105°C	—	Fairly high degree of syndiotactic character

PMMA has been reported variously between 70°C and 105°C. According to Table 8.1 this variation may be due to differences in stereoregularity between different samples, although it could partly be due to the occasional presence of absorbed moisture or monomer. The value of 105°C (Heijboer, 1952; Loshaek, 1955; Rogers and Mandelkern, 1957) is generally accepted for the glass-transition temperature of conventional PMMA. The relaxation properties of conventional PMMA will be discussed in Sections 8.2a to 8.2f and the stereoregular polymers will be considered in Section 8.2g.

8.2a   The α and β Relaxations—Mechanical

The first significant study of mechanical relaxation in PMMA was made by Alexandrov and Lazurkin (1940). They measured the temperature variation of the amplitude of deformation caused by a sinusoidal stress of constant amplitude over a frequency range $1\cdot67 \times 10^{-2}$ to $16\cdot7$ c/s. Their results are shown in Figure 8.1. For the unplasticized polymer, relaxation is first evident at 110°C and is virtually complete at 180°C within the frequency range studied. Alexandrov and Lazurkin also studied PMMA containing 10% and 30% plasticizer and observed that the relaxation region was lowered by approximately 40°C and 80°C respectively (Figure 8.1). These results are clearly consistent with a mechanism involving chain backbone motions and related to the glass transition.

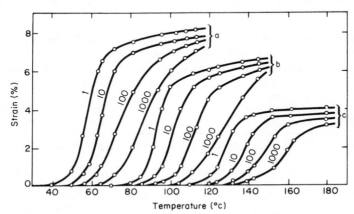

Figure 8.1. Variation of the deformation of PMMA with temperature at frequencies of 1, 10, 100 and 1000 vibrations per minute respectively. (a) 30% plasticizer; (b) 10% plasticizer; (c) Without plasticizer (Alexandrov and Lazurkin, 1940).

The mechanical α relaxation in PMMA was later investigated by McLoughlin and Tobolsky (1952) by means of stress relaxation measurements in the time range 0·001 hours to about 100 hours, and temperature range 40°C to 155°C (Figure 8.2). These authors found the time–temperature superposition principle to hold in this time and temperature region and were thus able to construct a master stress relaxation curve and evaluate the corresponding mechanical relaxation spectrum (Figure 8.3). They also observed that the location of the α relaxation region was independent of (viscosity average) molecular weight in the range 1.5×10^5 to 3.6×10^6. However, the plasticizing effect of small amounts of absorbed water was found to cause a decrease in the temperature of the α relaxation (at constant time).

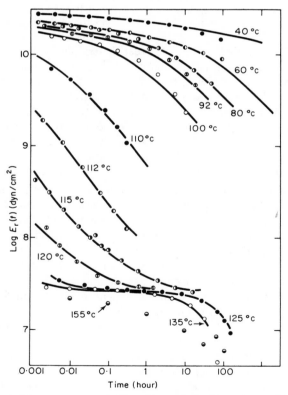

Figure 8.2. Log $E_r(t)$ versus log t for PMMA ($\overline{M}_v = 3.6 \times 10^6$) between 40 and 135°C (McLoughlin and Tobolsky, 1952).

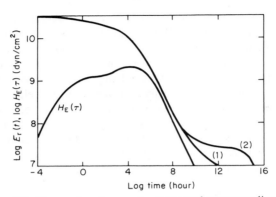

Figure 8.3. Stress relaxation master curve and corresponding mechanical relaxation spectrum (each reduced to 40°C) for PMMA. Sample (1) has $\overline{M}_v = 1.5 \times 10^5$; Sample (2) has $\overline{M}_v = 3.6 \times 10^6$ (Tobolsky, 1960).

Bueche (1955) studied the tensile creep behaviour of PMMA in the region of the α relaxation (40 to 160°C; 10^{-1} to 10^2 minutes) and constructed a master creep curve from his data. From the temperature variation of the shift factors (a_T) used in the formation of the master curve Bueche found that the apparent activation energy ($H_{app} = Rd \ln a_T/d(1/T)$) increased from a value of about 50 kcal/mole at 150°C to a maximum value of 250 kcal/mole at 110°C (i.e. 5°C above the measured T_g value). This observation conforms to the WLF equation (Ferry, 1961). Below 110°C, however, H_{app} was found to decrease with decreasing temperature, an effect also found by McLoughlin and Tobolsky. The observation of a maximum in H_{app} at temperatures close to T_g has been made for several other polymers including polyvinyl chloride (Figure 11.8) and vinylidene chloride–vinyl chloride copolymers (Figure 11.22). Therefore this would appear to be a general phenomenon which might be taken to indicate that the molecular mechanism of the glass–rubber relaxation changes in the vicinity of T_g. On the other hand, at temperatures around and below T_g, observed H_{app} values were found to depend markedly on the rate at which samples are initially cooled into this temperature region (Bueche, 1955; see also Figure 11.22). Hence the phenomenon seems partly related to the fact that, in the glass-transition region, the time scale for the establishment of volume equilibrium becomes comparable to the duration of an experiment (see Sections 5.3c, 9.2a, 11.1a and 11.3).

In addition to the mechanical α relaxation in PMMA many workers have also observed a small mechanical β relaxation which is generally assigned to the rotation of the —COOCH$_3$ side-group. The β relaxation was first detected in the creep measurements of Lethersich (1950) and, as shown in Figure 8.4, the average retardation time at 30°C is about 1 sec. Using dynamic mechanical techniques many workers later observed the β relaxation, and in most cases attributed it to the side-group motions. These workers included Schmieder and Wolf (1952, 1953), Iwayanagi and Hideshima (1953a, b), Heijboer, Dekking and Staverman (1954), Deutsch, Hoff and Reddish (1954), Jenckel and Illers (1954), Jenckel (1954), Fukada (1954), Becker (1955), Maxwell (1956), Heijboer (1956, 1965) and Koppelmann (1958, 1965). The results of Heijboer (1965) for PMMA are included in Figure 8.5 in which G'' at 1 c/s is plotted against temperature for four methacrylate polymers. The β loss peak is seen at 10°C, and the peak at 120°C is due to the α relaxation discussed above.

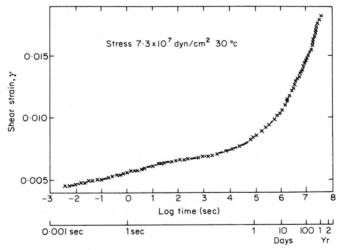

Figure 8.4. Strain versus time PMMA at 30°C and constant load. The 'hump' at about 100 sec is due to the β relaxation. The rapid rise for t about 10^7 sec is due to the α relaxation (Lethersich, 1950).

From their dynamic data Iwayanagi and Hideshima (1953a, b), Deutsch, Hoff and Reddish (1954) and Heijboer (1956) have each evaluated an activation energy for the mechanical β relaxation and quote values of 29, 18 and 18 kcal/mole respectively. From creep measurements Sato and others (1954) obtained a value of 30 kcal/mole by means of a superposition analysis. However, McCrum and Morris (1964) have shown that if corrections are first made for changes in the limiting (unrelaxed and

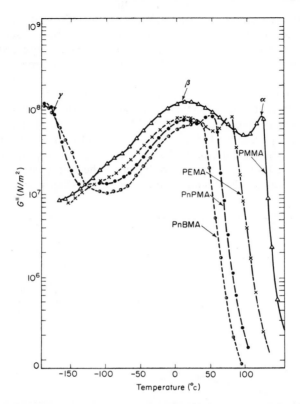

Figure 8.5. Temperature dependence of G'' at 1 c/s for PMMA, poly-ethyl methacrylate (PEMA), poly-n-propyl methacrylate (PnPMA) and poly-n-butyl methacrylate (PnBMA) (Heijboer, 1965).

relaxed) compliances with temperature, then a value of 17 kcal/mole is obtained from the temperature dependence of the shift factors required to produce a master creep curve. This value is in good agreement with the values of 18 kcal/mole obtained from two of the dynamic investigations, and also agrees with values obtained from dielectric measurements (Table 8.2). The values of 29 and 30 kcal/mole therefore seem to be in error. Hideshima and others (see Saitô and others, 1963) have also constructed master creep and stress relaxation curves for the β relaxation in PMMA from measurements below T_g, and have evaluated the corresponding mechanical retardation and relaxation spectra. The apparent success of these master curve constructions implies that when the β relaxation is well resolved from the α relaxation (i.e. at long time scales), then the widths of the relaxation and retardation spectra are either independent of temperature or change only slowly with temperature. The latter suggestion is favoured by the dielectric results for the β relaxation (see Figure 8.21 below).

8.2b The α and β Relaxations—Dielectric

The close relationship between mechanical and dielectric relaxation in PMMA is evident when the above mechanical data are compared with the dielectric results of Mikhailov and others (1956). The temperature dependence of the dielectric loss tangent at 20 c/s for unplasticized PMMA is shown in Figure 8.6, curve 1. Two loss maxima are observed, a small α peak at about 117°c and a large β peak at 35°c, which clearly correspond to the mechanical α and β loss peaks respectively. Upon the addition of plasticizer (curves 2, 3 and 4) the α peak is shifted to lower temperatures

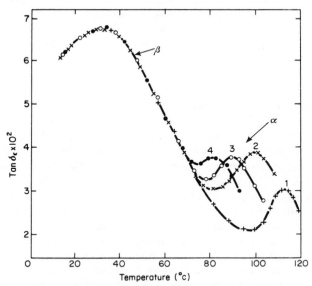

Figure 8.6. The temperature dependence of the dielectric loss tangent for PMMA at 20 c/s. Curve 1, without plasticizer; Curves 2, 3 and 4, PMMA containing, respectively, 5, 10 and 25% dibutyl phthalate plasticizer (Mikhailov and others, 1956).

(cf. the mechanical data of Alexandrov and Lazurkin above) but the β peak remains unchanged. The dielectric relaxation of PMMA has been investigated by many other workers including Mead and Fuoss (1942), Deutsch, Hoff and Reddish (1954), Rushton (1954), Heijboer (1956), Brouckere and Offergeld (1958), Mikhailov and Borisova (1958), Saito and Nakajima (1959b) and Ishida and Yamafuji (1961). These authors obtained values for the activation energy of the dielectric β relaxation between 19 and 23 kcal/mole, and for the dielectric α relaxation Saito and Nakajima quote an activation energy of 110 kcal/mole. These values, which are shown in Table 8.2 for comparison with the higher methacrylate polymers, are in good agreement with the corresponding values for the respective mechanical relaxations. The Cole–Cole distribution parameter $\bar{\beta}$ and the temperature dependence of $\varepsilon_R - \varepsilon_U$ for the dielectric β relaxation in PMMA is discussed in Section 8.3b below.

It is now generally thought that the β relaxation in PMMA arises from the hindered rotation of the —$COOCH_3$ group about the C—C bond which links it to the main chain. An inspection of molecular models shows that the steric hindrance to this rotation comes largely from the main-chain methyl substituents of the two *adjacent* repeat units (Heijboer, 1965). However, the detailed molecular mechanism of the β relaxation is not completely understood and, in particular, it is not clear whether, or to what extent, the main chain is involved locally in the movement (Roetling, 1965; Read, 1965a).

8.2c Relaxation Due to Absorbed Water

Owing to the presence of the hydrophilic ester groups PMMA can absorb up to about 2 % by weight of water. In addition to the plasticizing effect of water in lowering the temperature of the α relaxation (McLoughlin and Tobolsky, 1952), absorbed moisture also gives rise to an additional relaxation region which is not found for the dry polymer. This relaxation was first noticed by Wada and Yamamoto (1956) at 0°C (10^5 c/s). The data of Gall and McCrum (1961), illustrated in Figure 8.7, show that at 1 c/s the loss peak due to absorbed water occurs in the region of -100°C. Hendus and others (1959) studied PMMA containing 0.2 % moisture and observed the 'water relaxation' in dynamic mechanical experiments ($10^3 \rightarrow 10^6$ c/s) and also by a narrowing of the n.m.r. line width in the -10 to -40°C region. From dielectric measurements Scheiber and Mead (1957) have observed the 'water peak' at 5 kc/s (-20°C) which correlates approximately with the location of the mechanical peak (see Figure 8.8).

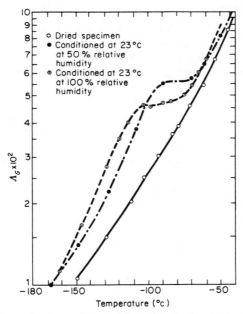

Figure 8.7. The dependence of logarithmic decrement Λ_G on water content for PMMA in the region of $-100°$c (Gall and McCrum, 1961).

8.2d Rotation of the α-Methyl Groups

A small mechanical loss peak has also been observed in PMMA by Hendus and others (1959) and by Sinnott (1960), and is assigned to the rotation of the methyl groups attached directly to the main chains (i.e. the α-methyl groups). In Sinnott's experiments the loss peak was detected at $-173°$c at a frequency of about 1 c/s. Hendus and others observed the relaxation at temperatures ranging from $-20°$c to $-150°$c in the frequency range $10^6 \rightarrow 10^3$ c/s. Since the rotation of methyl groups does not result in a directional change in their net dipole moments these rotations are not observed dielectrically. However, they are most easily observed in n.m.r. line-width and T_1 measurements because of the large magnetic dipolar interactions within the —CH₃ groups. The rotation of the α-methyl groups in PMMA is consistent with a line-width narrowing centred at about $-110°$c (Powles, 1956; Odajima and others, 1957; Slichter and Mandell, 1959; Hendus and others, 1959; Sinnott, 1960; Odajima and others, 1961) and with a minimum in T_1 at about $-10°$c (Kawai, 1961;

Powles, Hunt and Sandiford, 1964). Polymethyl acrylate (PMA), which does not contain α-methyl groups, does not exhibit these n.m.r. effects in similar temperature regions. Hence, this relaxation region, which correlates with the small mechanical relaxation region, is probably due to the rotations of α-methyl groups, *not* the methyl groups in the ester side-chains.

It is difficult to understand how the rotations of a methyl group can give rise to mechanical loss if, as might be expected, the barrier diagram for this rotation has three energetically equivalent minima each separated by 120° (as illustrated in Figure 2.3a). One possibility is that the onset of rotation of the α-methyl groups gives rise to some local cooperative movement of the main chain. Another possibility is that the barrier diagram contains *multiples* of three potential minima such that each of the minima are not energetically equivalent. For example, McCall (1964) has suggested that a barrier of the following kind,

where the ordinate is the potential energy and the abscissa is the rotation angle, might cause the methyl group reorientation to be mechanically active. In the case of the α-CH_3 groups in PMMA, a complex barrier system of this kind is not inconceivable owing to the close proximity of (and interaction with) the main-chain substituents of the two *neighbouring* monomer units. However, the above suggestions must, at present, be regarded as speculative.

8.2e Rotation of the Methyl Groups in the Ester Side-Chains

Both dynamic mechanical and n.m.r. studies have indicated that the rotation of the —CH_3 group about the O—C bond in the —$COOCH_3$ side-group sets in at extremely low temperatures. For example, Sinnott (1959) failed to detect a mechanical loss peak due to this methyl group rotation using torsion pendulum measurements at 1 c/s down to −269°C (4°K). Also, the observed second moment of the n.m.r signal at −196°C for both PMMA and PMA was about equal to that calculated assuming that one methyl group was rotating at a frequency above 10^4 to 10^5 c/s (Sinnott, 1960). Kawai (1961) and Powles and Mansfield (1962) have found a rather shallow minimum in the T_1–temperature plot for PMMA at about −200°C. At this temperature the side-chain —CH_3 groups are therefore rotating at frequencies of about 30 Mc/sec. Since the activation energy for this motion is small (Powles and Mansfield, 1962) this supports Sinnott's mechanical observations which suggest that the —CH$_3$ group reorientation at 1 c/s occurs below 4°κ. By comparison with the previous section it is apparent that the side-chain —CH$_3$ group rotation sets in at much lower temperatures than the rotation of the α-methyl group. Presumably the larger hindrance to rotation of the α-methyl groups arises from the presence of the neighbouring —COOCH$_3$ substituents on the main chain.

8.2f Summary of Data for Conventional PMMA

The temperature–frequency locations of the five different relaxation regions in PMMA are plotted in Figure 8.8. The mechanical results are given by the filled points and the dielectric data are plotted as open points. The temperature regions of n.m.r. line-width narrowings are also indicated and also the temperatures of minimum T_1. The merging of the α and β relaxations at a frequency of about 10^4 c/s is clearly seen. It will also be observed that for the β mechanism the mechanical tan δ maxima tend to lie at lower frequencies than the corresponding dielectric tan δ peaks. Thurn and Wolf (1962) have found, however, that the temperature of maximum E'' correlates closely with the temperature of maximum $ε''$ for the β process.

8.2g Stereoregular PMMA

The properties of the various stereoregular forms of PMMA have been discussed above and are summarized in Table 8.1. The influence of stereoregularity on the dielectric properties of PMMA has been investigated by Mikhailov and Borisova (1961) and also by Nagata, Hikichi, Kaneko and Furuichi (1963). Some results of Mikhailov and Borisova are summarized in Figure 8.9. The isotactic and syndiotactic samples were prepared using organolithium catalysts and were precipitated from benzene solution with methanol. This preparation guaranteed the complete absence of crystallinity, as confirmed by x-ray examination. It will first be noted that the loss-temperature curves for the conventional and syndiotactic polymers are similar, except that the α peak is located at a slightly higher temperature for the syndiotactic specimen. This result supports the evidence from n.m.r and infrared studies (referred to above) which suggests that the conventional polymer, prepared by free-radical initiation, has a high degree of syndiotactic character. For the isotactic specimen, however, the α loss peak occurs about 60°c below the temperature of

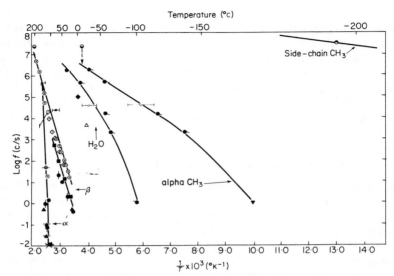

Figure 8.8. Plot of log f against $1/T$ for the mechanical (filled points) and dielectric (open points) loss maxima for PMMA. In the following key (f) refers to the frequency of maximum loss in constant temperature experiments, and (T) refers to the temperature of maximum loss in constant frequency experiments.

● tan δ_G (T) (Schmieder and Wolf, 1952, 1953)
▲ tan δ_G (T) (Iwayanagi and Hide-shima, 1953a)
◖ (Hendus and others, 1959)
◖ tan δ_G (T) (Gall and McCrum, 1961)
▼ tan δ_G (T) (Sinnott, 1960)
× inflection point of double logarithmic stress relaxation curve at 112°c (McLoughlin and Tobolsky, 1952)

■ tan δ_G (f) (Heijboer, Dekking and Staverman, 1954)
◆ tan δ_E (T) (Wada and Yamamoto, 1956)
◓ tan δ_G (T) (Heijboer, 1956)
▲ tan δ_G (f) (Heijboer, 1956)
◈ tan δ_E (f) (Koppelmann, 1965)
◇ tan δ_ε (f) (Heijboer, 1960a)
⊕ tan δ_ε (T) (Koppelmann, 1965)
⊙ tan δ_ε (f) (Mikhailov and Borisova, 1958a)
△ ε'' (f) (Scheiber and Mead, 1957)

Also shown in the diagram are the temperature regions of the n.m.r. line-width narrowings (├────┤) and the temperatures of minimum T_1 (half-filled points). ├── ⊙ ──┤ (Hendus and others, 1959); ◓ (Kawai, 1961); ◑ (Powles and Mansfield, 1962). The sample studied by Hendus and others (1959) contained 0.2% moisture.

the α peak for the conventional polymer, and is partially merged with the β peak. Furthermore, the α peak is of larger magnitude and the β peak of smaller magnitude for the isotactic polymer than for the conventional and syndiotactic samples. Thus, whereas the height of the β peak is

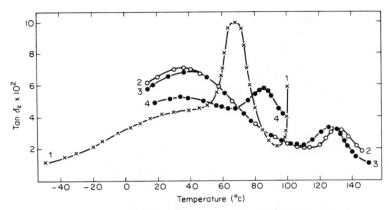

Figure 8.9. Temperature dependence of the dielectric loss tangent at 28 c/s. Curve 1 isotactic; Curve 2 syndiotactic and Curve 3 conventional PMMA; Curve 4 is for a sample of PMMA of intermediate degree of isotacticity (Mikhailov and Borisova, 1961).

about twice that of the α peak for the conventional polymer, the opposite situation exists in the case of the isotactic polymer. The results for the specimen of intermediate degree of isotacticity are intermediate between those for the isotactic and syndiotactic samples. The data for the isotactic polymer are further illustrated in Figure 8.10 which shows plots of the dielectric loss tangent against frequency at various temperatures. Mikhailov and Borisova have attributed the curves between 24°C and 56°C to the β process. At 67°C the merged α and β peaks are both observed, the α region being on the low-frequency side. The curves at temperatures above about 76°C were attributed largely to the α mechanism but with the β relaxation superposed at high frequencies. Figure 8.13 shows the variation of $\log f_{max}$ with $1/T$, and the merging of the α and β relaxation regions at high frequencies and temperatures is clearly observed. It will also be seen from this diagram that at each frequency the β peak is located at somewhat lower temperatures for the isotactic than for the conventional polymer. The activation energy for the β process in the isotactic polymer (24 kcal/mole) was similar to that found for the conventional polymer (21 kcal/mole).

Gall and McCrum (1961) have determined the temperature dependence of the shear modulus and logarithmic decrement for the predominantly isotactic and syndiotactic forms of PMMA. Their data are shown in Figure 8.11 together with the results of similar measurements made on

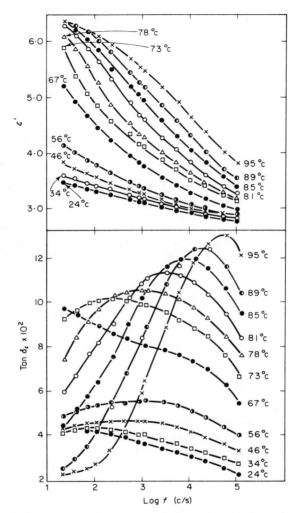

Figure 8.10. Frequency dependence of ε' and the dielectric loss tangent for isotactic PMMA at several temperatures (Mikhailov and Borisova, 1961).

the conventional polymer. As in the case of the dielectric measurements described above, each of the specimens was studied in the *amorphous* state. They were each conditioned at 50% relative humidity prior to the measurements. For the conventional polymer the water peak ($-82°$C,

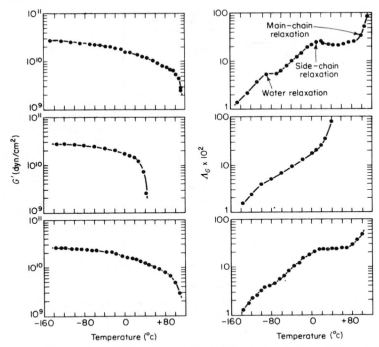

Figure 8.11. Temperature dependence of the shear modulus and logarithmic decrement at 1 c/s for conventional (upper graphs), isotactic (intermediate graphs) and syndiotactic (lower graphs) PMMA (Gall and McCrum, 1961).

0·98 c/s) and the β peak are clearly shown, and the low-temperature side of the α peak is shown by the rise in internal friction and drop in modulus above 80°C. The loss peak due to absorbed water is shown also both by the isotactic ($-95°$C, 1·14 c/s) and syndiotactic ($-95°$C, 1·21 c/s) polymer. The β peak is resolved for the syndiotactic polymer (32°C, 0·90 c/s) but apparently not for the isotactic polymer. A likely explanation for the apparent absence of the mechanical β peak for the isotactic polymer is that it is merged with the much larger α peak, the latter having shifted by a relatively large amount (about 50°C) to lower temperatures. The results of Heijboer (1965) (Figure 8.12) for a 95% isotactic PMMA sample show clearly the presence of a mechanical β peak. This result is consistent with the dielectric results described above.

An investigation of the temperature dependence of the n.m.r. spin-lattice relaxation time (T_1) and the mechanical loss tangent (at about

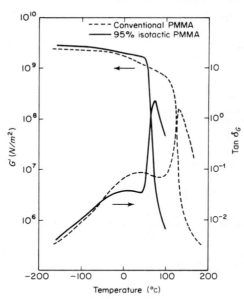

Figure 8.12. Temperature dependence of G' and $\tan \delta_G$ at 1 c/s for conventional and 95% isotactic PMMA (Heijboer, 1965).

200 c/s) for isotactic, syndiotactic and conventional PMMA has been described by Powles, Strange and Sandiford (1963). For the α relaxation, the temperatures of minimum T_1 and maximum $\tan \delta_E$ were some 40°C and 75°C, respectively, lower for the isotactic than for the syndiotactic specimen. In both the mechanical and n.m.r. plots the β relaxation was unresolved owing to the high frequencies employed. From the T_1 measurements the reorientation of the α-CH$_3$ groups were also found to occur at a temperature some 40°C lower for the isotactic than for the conventional and syndiotactic materials. Odajima, Woodward and Sauer (1961) have also found that the temperatures of the n.m.r. line-width narrowings were lower for isotactic than for conventional PMMA, both for main-chain and α-CH$_3$ group reorientations.

A summary of the data for isotactic PMMA is presented in the plot of $\log f$ versus $1/T$ (Figure 8.13). From this plot we conclude that the chain backbone mobility is considerably enhanced, at each temperature, by the isotactic configuration. On the basis of the simple side-group interpretation of the β process, both the dielectric and mechanical data further indicate that the —COOCH$_3$ group motions occur somewhat more

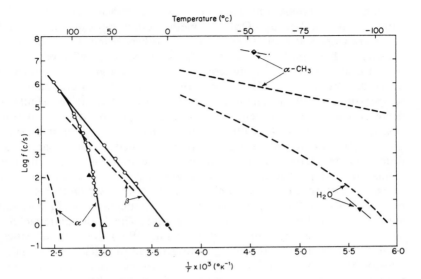

Figure 8.13. Plot of log f against $1/T$ for the dielectric (open points) and mechanical (filled points) loss maxima for isotactic PMMA. In the following key (f) refers to the frequency of maximum loss in constant temperature experiments and (T) refers to the temperature of maximum loss in constant frequency experiments.

○ $\tan \delta_\varepsilon$ (f) (Mikhailov and Borisova, 1961)	▲ $\tan \delta_E$ (T) (Powles, Strange and Sandiford, 1963)
△ ε'' (T) (Nagata, Hikichi, Kaneko and Furuichi, 1963)	▼ $\tan \delta_G$ (T) (Gall and McCrum, 1961)
	● $\tan \delta_G$ (T) (Heijboer, 1965)

The dashed lines indicate the positions of the mechanical or dielectric loss maxima for conventional PMMA. Also shown is the temperature of the n.m.r. T_1 minimum (◆), attributed by Powles, Strange and Sandiford (1963) to the reorientati⋯ ⋯-CH₃ groups in isotactic PMMA.

rapidly in the syndiotactic polymer. Finally, from the n.m.r. results it is seen that the mobility of the α-CH$_3$ groups is also increased by the isotactic chain configuration.

8.3   HIGHER METHACRYLATE POLYMERS

It is of interest to compare the behaviour of PMMA with that of the higher methacrylate polymers in which R = ethyl, n-propyl, isopropyl, n-butyl, etc. (**1**, page 238.) Such a comparison has aided considerably in the assignment of the various loss peaks in PMMA. The properties of the conventional n-alkyl methacrylate polymers depend critically on the length of the alkyl group. Only the methyl, ethyl, n-propyl and n-butyl esters are glasses at room temperature, the others being rubber-like if of sufficiently high molecular weight. This property is reflected by the decrease of T_g with increasing length of the n-alkyl group as shown in Table 8.2 (Rogers and Mandelkern, 1957). As will be seen below, the temperatures of the mechanical and dielectric α relaxations similarly decrease with increasing length of the n-alkyl group. An explanation occasionally given for these observations is that as the length of the side-group increases, neighbouring chains are pushed further apart, thus decreasing the hindrance to chain backbone motions. This effect is similar to that produced by the addition of plasticizer and is usually known as 'internal plasticization'. In support of this hypothesis is the observation that the density shows a systematic decrease (Table 8.2) with increasing length of the n-alkyl group. The 'molecular packing factors', calculated by Ishida and Yamafuji (1961) from the densities, exhibit a similar trend. According to Rogers and Mandelkern (1957) this density decrease is consistent with a similar decrease in the magnitude of the interchain interactions as determined from the cohesive energy densities. (For a definition of cohesive energy density see Hildebrand and Scott, 1950).

It should be emphasized that the effects of internal plasticization discussed above refer only to the n-alkyl methacrylate polymers. If the n-alkyl group is replaced by the corresponding iso- or t-alkyl group then T_g and the α relaxation temperature are generally observed to shift back to higher temperatures. This situation is exemplified by the fact that T_g for polyisobutyl methacrylate (PiBMA) is some 40°C higher than that for poly-n-butyl methacrylate (PnBMA) (Table 8.2). Hence the shorter and more rigid isobutyl group is far less effective at internal plasticization than the linear flexible n-butyl group. Since the densities of the two polymers are apparently very similar (Table 8.2) the interchain cohesive forces are probably also similar. Therefore the effect of internal plasticization may be compensated for in the case of PiBMA· by an increase in steric hindrance to *intra*molecular rotations produced by the more rigid isobutyl substituent. The above effect is further exemplified in Table 8.2 by the fact that T_g is appreciably increased if the n-propyl group is replaced by the isopropyl group or if the n-hexyl group is replaced by the less flexible cyclohexyl group.

Like PMMA, several of the higher methacrylate polymers have been obtained in forms which differ from the conventional polymers, probably on account of differences in stereoregularity (Shetter, 1963). As shown in

Table 8.3 the Type I polymers, which may (by comparison with PMMA) have a high degree of syndiotactic regularity, have similar glass-transition temperatures to the corresponding conventional polymers. However, the Type II polymers, which may be largely isotactic, have considerably lower T_g values. For these Type II polymers T_g also decreases as the n-alkyl side-group increases in length, although this decrease is less than that observed for conventional polymers. As will be seen below (Section 8.3d) the α relaxation temperature is also lower for the isotactic than for the conventional and syndiotactic polymers. Firstly, however, we will discuss the considerable quantity of experimental data available for the conventional polymers.

8.3a   The α and β Relaxations—Mechanical

A systematic study of mechanical relaxation in a series of n-alkyl methacrylate polymers has been made by Ferry and coworkers using the dynamic mechanical technique in the frequency range 20 c/s to 3000 c/s. The polymers studied were polyethyl methacrylate, PEMA (Ferry and others, 1957), poly-n-butyl methacrylate, PnBMA (Child and Ferry, 1957a), poly-n-hexyl methacrylate, PnHMA (Child and Ferry, 1957b), poly-n-octyl methacrylate, PnOMA (Dannhauser, Child and Ferry, 1958) and poly-n-dodecyl methacrylate, PnDDMA (Kurath and others, 1959). By studying a range of temperatures some 10°C to 90°C above T_g for each polymer these workers derived master curves of J' and J'' by the usual shift procedure (Section 4.2c), and shift factors (a_T) were obtained. However, for all polymers up to poly-n-hexyl methacrylate, anomalies were found in a limited temperature region where the determined shift factors did not conform to the WLF equation. These anomalies were attributed to the presence of a small secondary β relaxation which overlapped the main α relaxation. By methods described in the above references the two relaxations were resolved. For the β mechanisms the temperature dependence of relaxation times was found to obey the Arrhenius equation and the activation energy was observed to decrease with increasing length of the side-chain (Table 8.2). According to Heijboer and Schwarzl (1962), however, the above methods may not be sufficiently accurate to give reliable values for the activation energy of the β mechanism. Since the temperature dependence of relaxation times for the α relaxations conforms to the WLF equation, the apparent activation energies (H_{app}) are strongly temperature dependent. Values of H_{app} for the α relaxations at given temperatures are shown in Table 8.2 for later comparison. In Figures 8.14(a), (b) and (c) the mechanical relaxation and retardation spectra for

Table 8.2. Summary of data for the conventional alkyl methacrylate polymers

Methacrylate polymer	Structure of alkyl group R	Density (g/cm³)	Ref.[b]	T_g(°C)	Ref.[b]	Activation energies[a] (kcal/mole) α Relaxation	β Relaxation	Ref.[b]
Methyl (PMMA)	—CH₃	1·170 (25°C)	1	105	1	110 D	23 D	4
						100 D	19 D	5
							21 D	6
							19 D	2
							20 D	7
							20 D	8
							21 D	10
						80 M	18 M	7
							29 M	9
							18 M	10
							30 M	11
							17 M	12
						250 M (100°C)		13
Ethyl (PEMA)	—CH₂CH₃	1·125 (25°C)	1	65	1	92 D	24 D	4
						50 D	21 D	6
							18·5 D	2
						69 M (100°C)	31 D {107·5 (to 133°C)} 31 M	14
n-Propyl (PnPMA)	—(CH₂)₂CH₃	1·077 (25°C)	1	35	1	49 D	21 D	6
n-Butyl (PnBMA)	—(CH₂)₃CH₃	1·053 (25°C)	1	20	1	29 D		2
						30 D β		4
						24 D β	24 M	14
						37 M (100°C)		14
n-Hexyl (PnHMA)	—(CH₂)₅CH₃	1·008 (25°C)	1	−5	1	21 D β	13 M	14
						27 M (100°C)		14

Polymer	R group	Density (Temp)		Temp (°C)		col 7 (α)	col 8 (β)	Refs
n-Octyl (PnOMA)	—(CH₂)₇CH₃	0·971 (25°c)	1	−20	1		18 D β 21 M (100°c)	14 14
n-Nonyl (PnNMA)	—(CH₂)₈CH₃	0·970 (20°c)	2				23 D	2
n-Dodecyl (PnDDMA)	—(CH₂)₁₁CH₃	0·929 (25°c)	1	−65	1			2
IsoPropyl (PiPMA)	CH₃ —CH CH₃			81	3		16 D	6
Isobutyl (PiBMA)	CH₃ —CH₂—CH CH₃	1·041 (20°c)	2	67 53	4 3		45 D 43 D β	2 4
tertiary-Butyl (PtBMA)	CH₃ —C—CH₃ CH₃			95	16	17 D	45 D	15
Cyclohexyl (PCMA)		1·103 (20°c)	2	90	16		53 D	2

ᵃ D = Dielectric, M = Mechanical. Activation energies given in column 7 are assigned to the α relaxation and those in column 8 to the β relaxation. For PnBMA, PiBMA, PnOMA, PnHMA, PnNMA, PnDDMA the complete merging of the dielectric α and β relaxations has resulted in some inconsistencies between the assignments of various authors (see text). In cases where our dielectric α assignment disagrees with that given in the original reference, the original assignment is indicated by the Greek letter β. When activation energies have been determined at a given temperature or over a narrow temperature range the appropriate temperatures are indicated in brackets.

ᵇ 1. Rogers and Mandelkern (1957). 2. Ishida and Yamafuji (1961). 3. Shetter (1963) (see Table 8.3). 4. Brouckere and Offergeld (1958). 5. Saito and Nakajima (1959b). 6. Mikhailov and Borisova (1958a), Mikhailov (1958). 7. Deutsch, Hoff and Reddish (1954). 8. Rushton (1954). 9. Iwayanagi and Hideshima (1953a, b). 10. Heijboer (1956). 11. Sato and others (1954). 12. McCrum and Morris (1964). 13. Bueche (1955). 14. Ferry and Strella (1958). 15. Mikhailov (1965). 16. Heijboer (1965).

Table 8.3. Glass transition temperatures (°c) of methacrylate polymers prepared by different methods (Shetter, 1963).

Methacrylate Polymer	Method of Polymerization		
	Free-radical at 44 to 60°c (conventional)	Free-radical at or below −30°c (Type I)	Anionic in non-polar solvent (Type II)
Methyl	104	115 to 122	43 to 50
Ethyl	66	—	8 to 12
Isopropyl	81	85	27
Butyl	19	—	−24
Isobutyl	53	—	8
Cyclohexyl	66[a]	—	51
Isobornyl	—	111	110

[a] Heijboer (1965) has reported a value of 90°c for the glass-transition temperature of conventional polycyclohexyl methacrylate. According to Heijboer the value of 66°c is certainly too low.

the α relaxations, and the retardation spectra for the β relaxations are shown for the n-alkyl methacrylate polymers at 100°c. For the α relaxation the spectra shift to shorter relaxation times (i.e. the main-chain mobility is increased) as the length of the alkyl group increases. This observation is consistent with the lowering of T_g and the temperature of the mechanical α peak (Figure 8.5) as the side-chain increases in length. It also agrees with the creep observations of Bueche (1955) on a series of methacrylate polymers. In terms of the Rouse theory the monomeric friction coefficient (ξ_0) decreases as the length of the alkyl group increases (Section 5.3a). The β mechanisms are probably associated with hindered rotation of the ester side-groups and according to Figure 8.14(c) the average retardation time for this process (at 100°c) appears to be relatively little affected by the length of the n-alkyl group. A similar conclusion is obtained from Figure 8.5. Although the spectra in Figure 8.14(c) are not normalized it appears that for the β mechanism the retardation spectrum increases in height and sharpens as the alkyl group increases in length. This behaviour has been confirmed by Saitô and others (1963) who have compared the β retardation spectra of PMMA and poly-n-propyl methacrylate derived from creep data below the respective T_g values. These authors have also noted that the width of the spectrum depends on the temperature range of measurement. In particular, spectra derived from measurements below T_g are considerably broader than those obtained above T_g (as applies in Figure 8.14c).

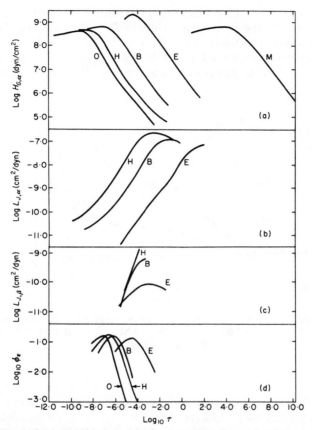

Figure 8.14. Double logarithmic plots of (a) the mechanical relaxation spectra for the α relaxations, (b) the mechanical retardation spectra for the α relaxations and (c) the mechanical retardation spectra for the β relaxations for the n-alkyl methacrylate polymers, (d) the dielectric relaxation spectra (Ferry and Strella, 1958). All spectra reduced to 100°C.

M PMMA (Fujita and Kishimoto, 1958)
E PEMA (Ferry and others, 1957)
B PnBMA (Child and Ferry, 1957a)

H PnHMA (Child and Ferry, 1957b)
O PnOMA (Dannhauser, Child and Ferry, 1958)

Heijboer (1960a) has reported some dynamic mechanical data for PMMA, PEMA, poly-t-butyl methacrylate (PtBMA; R = —C—(CH$_3$)$_3$) and polycyclohexyl methacrylate (PCMA; R = —C$_6$H$_{11}$). As shown in

Figure 8.15 the α and β loss peaks are resolved (at 1 c/s), and the β peak occurs at essentially the same temperature ($\approx 25°$C) for each polymer. A similar observation was made at several frequencies, as illustrated by the frequency–temperature plot in Figure 8.16. Hence, the average mechanical relaxation time at each temperature, and therefore the activation energy,

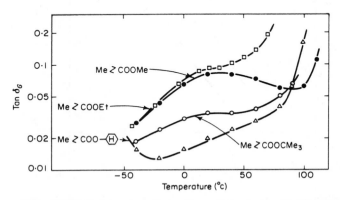

Figure 8.15. Temperature dependence of the mechanical loss tangent at 1 c/s for PMMA ($\bullet$), PEMA ($\square$), PtBMA ($\bigcirc$) and PCMA ($\triangle$) (Heijboer, 1960a).

for the β relaxation appears to be independent of the side-chain alkyl group, *provided that the β and α relaxations are resolved*. At first sight this result seems to contradict the findings of Ferry and Strella (1958) that the activation energy for the β relaxation decreases with an increase in the n-alkyl side-group length (Table 8.2). However, the latter result was obtained within the series PEMA, PnBMA and PnHMA. For each of these polymers the β and α relaxations were merged at the experimental frequencies employed, and it seems likely that the β mechanism was considerably influenced by the α process. In terms of the side-group interpretation of the β relaxation, we suggest that in frequency–temperature regions where the α and β relaxations overlap, the barriers opposing the hindered rotations of side-chains are decreased by an increase in flexibility of the chain backbone. The result of Ferry and Strella can then be understood, since the main-chain mobility increases in the order PEMA, PnBMA, PnHMA.

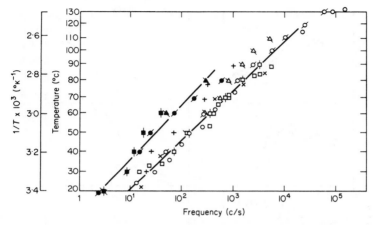

Figure 8.16. Plot of log f against $1/T$ for the mechanical (filled points) and dielectric (open points) β relaxation loss tangent maxima for several alkyl methacrylate polymers (Heijboer, 1960a).

	Mikhailov Dielectric	Heijboer Dielectric	Heijboer Mechanical
PMMA	○	⌀	●
PEMA	□	⊟	■
PnPMA	×		
PiPMA	+		
P*t*BMA		△	▲

8.3b   The α and β Relaxations—Dielectric

Some dielectric data of Mikhailov and Borisova (1958a) at 20 and 100 c/s, are shown in Figure 8.17 for PMMA, PEMA, poly-n-propyl methacrylate (PnPMA), PnBMA, and polyisopropylmethacrylate (PiPMA). Two tan δ_ε peaks are clearly seen for PMMA and for PEMA at 20 c/s. In the case of PEMA at 100 c/s and for PNPMA and PnBMA the two loss peaks are merged owing to the decrease in temperature of the α peak as the length of the n-alkyl group increases. The latter effect is consistent with the corresponding variations in T_g and with the mechanical observations on the α relaxation. In the case of PiPMA (R = —CH—(CH$_3$)$_2$; **1** page 238) the α and β loss peaks are again well resolved. Similarly, Heijboer (1960a) has shown that, as in the mechanical case, the dielectric α and β peaks are resolved at 60 c/s for both P*t*BMA and PCMA (Figure 8.18).

According to Figure 8.16 the frequency–temperature location, and hence the activation energy, of the resolved dielectric β relaxation appears

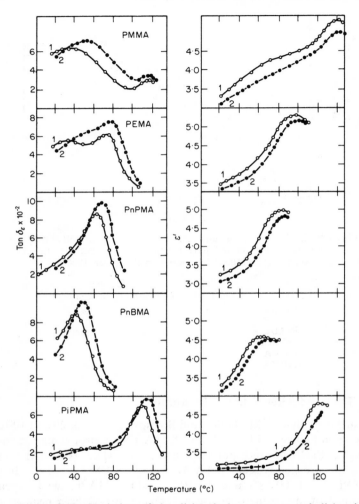

Figure 8.17. Variation of the dielectric loss tangent and dielectric constant with temperature at 20 c/s (curves 1) and 100 c/s (curves 2) for PMMA, PEMA, PnPMA, PnBMA and PiPMA (Mikhailov and Borisova, 1958a).

to be little affected by changes in the side chain alkyl substituent. Mikhailov and Borisova have reported an activation energy of 21 kcal/mole for the β process in PMMA, PEMA and PnPMA, although for PiPMA a somewhat lower value of 16 kcal/mole was obtained. These

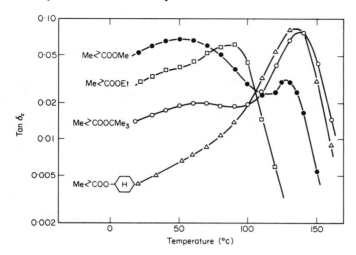

Figure 8.18. Temperature dependence of the dielectric loss tangent at 60 c/s for PMMA (●), PEMA (□), P*t*BMA (○), and PCMA (△) (Heijboer, 1960a).

results are reasonably consistent with Heijboer's mechanical data (Figures 8.5 and 8.15) discussed above.

A further point of interest concerns the relative magnitudes of the dielectric α and β loss peaks. From Figures 8.17 and 8.18 it will be noted that as the side-chain alkyl group increases in size so the α and β peaks increase and decrease in height respectively. Heijboer (1960a, 1965) has attributed this observation to the fact that with the larger side-groups a relatively small number of side-chains can partake in the movement below T_g (owing to some blocking effect). As far as they do partake, they do so with the same ease (same activation energy). Above T_g, on the other hand, movement of these side-groups can occur along with main-chain motions, thus augmenting the magnitude of the α relaxation.

Brouckere and Offergeld (1958) measured the dielectric constants and loss factors of PMMA, PEMA, PnBMA and PiBMA in the frequency range 500 c/s to 500 kc/s. They found a single relaxation region for each of the methacrylate polymers and the normalized $\varepsilon''/\varepsilon''_{max}$ versus frequency plots narrowed as the temperature increased, so that the time–temperature superposition principle was invalid. Also, within the experimental frequency range studied, the Arrhenius equation, and not the WLF equation, was found to describe the temperature dependence of average relaxation times for each polymer. The activation energy was found to increase with increasing length of the n-alkyl group and varied from 23 kcal/mole for PMMA to 30 kcal/mole for PnBMA. Furthermore, the activation energy for the relaxation process in PnBMA (30 kcal/mole) was *lower* than that observed in PiBMA (43 kcal/mole). Since Brouckere and Offergeld regard the single relaxation region which they observed as due to the β process their conclusions are in contradiction to those of Mikhailov and Borisova and of Heijboer.

Ferry and Strella (1958) analysed the dielectric data of Strella and others (1957a, 1957b, 1958) for PEMA, PnBMA, PnHMA, and PnOMA which were obtained in the frequency range 33 c/s to 300 kc/s. Like Brouckere and Offergeld they observed a single loss peak for each of these polymers and found that the time–temperature superposition principle was generally invalid. However, in the highest temperature regions they were able to construct reduced master curves from which the dielectric relaxation spectra were evaluated. The spectra at 100°C are plotted in Figure 8.14(d) for comparison with the mechanical relaxation and retardation spectra. The dielectric spectra are seen to shift to shorter relaxation times (i.e. the mobility is increased) as the length of the n-alkyl group increases. It will also be noticed that the dielectric spectra are located at shorter τ's than the mechanical retardation spectra (for both α and β relaxations) suggesting that the higher-frequency processes make a larger contribution to the dielectric behaviour than the mechanical *retardation* behaviour. However, for PEMA and PnBMA the positions of the maxima of the dielectric spectra agree fairly well with the positions of the maxima of the mechanical *relaxation* spectra for the α *process* (cf. PMA, Section 8.9a). Also, the magnitude of the shift in the position of the dielectric spectra with increasing alkyl group length correlates better with the mechanical spectra for the α relaxation than for the β relaxation. This evidence suggests perhaps that the data of Ferry and Strella are largely influenced by the α (or backbone) relaxation process. However, Ferry and Strella attribute their results to the β mechanism on the grounds that (within a limited temperature region) the temperature dependence of relaxation times conformed to the Arrhenius equation rather than to the WLF equation and the calculated activation energies agreed with those for the mechanical process (Table 8.2). Hence, unlike both Mikhailov and Borisova and also Brouckere and Offergeld, Ferry and Strella reported that for the β process the activation energy decreased with increasing length of the n-alkyl group.

The above inconsistencies arise probably from the fact that for the higher n-alkyl methacrylate polymers the α and β relaxation regions merge into a single region and that over a *wide* temperature or frequency range the Arrhenius equation is not, in fact, obeyed. These facts are clearly illustrated by the dielectric results of Ishida and Yamafuji (1961) obtained over a wide range of frequency (10 to 10^6 c/s) on PMMA, PEMA, PnBMA, PiBMA, poly-n-nonyl methacrylate (PnNMA) and polycyclohexyl methacrylate (PCMA). For PMMA the predominant β relaxation was observed together with some indications of the small α relaxation at the

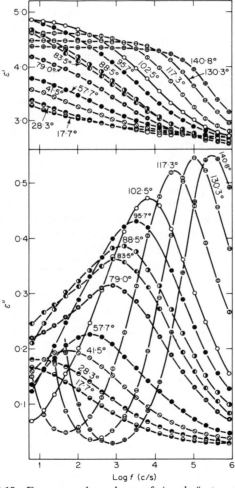

Figure 8.19. Frequency dependence of ε' and ε'' at various temperatures for PEMA (Ishida and Yamafuji, 1961).

highest temperature studied (136·8°C). In the case of PEMA (Figure 8.19) the ε'' against log f curves show only the β peak at temperatures below about 60°C. In the range 79°C to 88·5°C the α absorption appears as a low-frequency shoulder to the main β loss peak, and at higher temperatures the α and β peaks merge and become indistinguishable. For PnBMA, PiBMA and PnNMA single asymmetric loss peaks were found at all temperatures. These were inferred to be predominantly α peaks containing the β peaks in the (high-frequency) tails. For PCMA the β peak appeared as a high-frequency shoulder to the predominant α peak in the temperature range 44·3°C to 103·5°C. The tendency for the ratio of the magnitude of the α relaxation to the magnitude of the β relaxation to increase as the size of the alkyl group increases has already been noted from the data of Mikhailov and Borisova (Figure 8.17). These same data also support the suggestion that for PBMA and higher methacrylates the loss peak is due largely to the α mechanism, and a similar conclusion was arrived at above in comparing the dielectric and mechanical distribution functions of Ferry and Strella.

From the Cole–Cole arc plots Ishida and Yamafuji have calculated values of $\varepsilon_R - \varepsilon_U$ and the Cole–Cole distribution parameters $\bar{\beta}$ and these are plotted in Figures 8.20 and 8.21 respectively. Also the plot of log f_{max} versus $1/T$ is shown in Figure 8.22. According to Figure 8.21 the dielectric relaxation spectrum is seen to narrow slowly with increasing temperature for the β relaxations in PMMA and PEMA. For the (predominantly α) relaxations in PnBMA, PiBMA, PnNMA and PCMA the

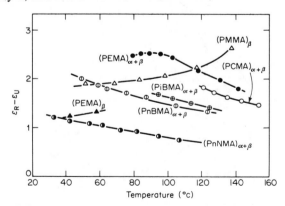

Figure 8.20. $\varepsilon_R - \varepsilon_U$ as a function of temperature for several alkyl methacrylate polymers. The values refer either to the resolved β relaxation or to the merged $\alpha + \beta$ relaxations as indicated (Ishida and Yamafuji, 1961).

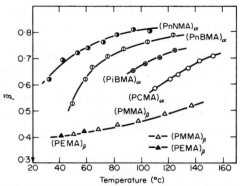

Figure 8.21. Temperature dependence of the Cole–Cole distribution parameter $\bar{\beta}$ for various methacrylate polymers (Ishida and Yamafuji, 1961).

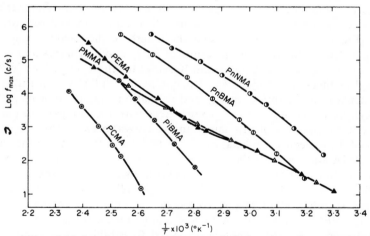

Figure 8.22. Log f_{max} versus $1/T$ for several alkyl methacrylate polymers. The points for PMMA refer to the β relaxation. For the other polymers the points refer to the merged α and β loss peaks as described in the text (Ishida and Yamafuji, 1961).

distribution of relaxation times sharpens rapidly with increasing temperature and tends asymptotically to a limiting width at high temperatures. This is consistent with Ferry and Strella's observation that the time–temperature superposition principle is valid only at high temperatures. Since the distribution parameters for well-resolved α relaxations in amorphous polymers are often independent of temperature (down to T_g) the increase in β with increasing temperature may be due to the merging of the α and β relaxations.

From Figure 8.22 it is seen that at low temperatures the $\log f_{max}$ against $1/T$ plot for PEMA coincides with the plot for the β relaxation in PMMA, providing additional evidence that at low temperatures the single loss peak for PEMA arises solely from the β mechanism. As the temperature increases the plot for PEMA increases in slope fairly abruptly. From the above discussion concerning the relative magnitudes of the two peaks in PEMA it would seem reasonable to regard the slope (or activation energy) in the high-temperature region as due to the β relaxation modified by the α relaxation. For PnBMA, PiBMA, PnNMA and PCMA on the other hand, the above evidence, and the curvature of the $\log f_{max}-1/T$ plots, suggests that these plots are determined largely by the α relaxation but perhaps modified by a superposed β relaxation. Despite the curvature of these frequency–temperature plots Ishida and Yamafuji have evaluated activation energies from the average slopes (Table 8.2). Comparing PnBMA and PnNMA, the decrease in activation energy and relaxation time (Figure 8.22) upon increasing the length of the n-alkyl group is consistent with the mechanical results of Ferry and coworkers and with the dielectric data of Mikhailov and Borisova for the α *relaxations* in the n-alkyl methacrylate polymers. Also, comparing PnBMA with PiBMA, the increase in activation energy and relaxation time on replacing the n-alkyl group by the isoalkyl group agrees with the comparison made above (from the data of Mikhailov and Borisova) between the temperatures of the dielectric α relaxations in PnPMA and PiPMA. In the case of PnBMA and PnNMA the frequency–temperature plots are seen to lie in lower temperature (higher frequency) regions than the Arrhenius plot for the β relaxation in PMMA. Thus, if the relaxation time for the β process is independent of the n-alkyl group length, as in the case of PMMA, PEMA and PnPMA at low temperatures, the loss peaks for PnBMA and PnNMA might be expected to exhibit a broadening or shoulder on the low-frequency side. This effect is not found, however, indicating either that the magnitude of the β peak in these polymers is extremely small, or that once the chain backbone motions have set in the side-group rotations occur automatically. In other words the side-group motions may not be independent of motions within the main chain, and it would seem unlikely that the side-groups could remain unrelaxed above the glass-transition temperature.

Summarizing the dielectric observations, it appears that for PMMA, PEMA and PnPMA at low temperatures, where the α and β relaxations are resolved, the relaxation time for the β process is independent of the n-alkyl group length. For longer n-alkyl groups the situation is complicated by the merging of α and β relaxations. Of course even for PMMA the α and β relaxations merge at high temperatures and frequencies (Figure 8.8) as discussed by Koppelmann (1965). It thus seems unlikely that it is correct to assign the single relaxation region in the higher n-alkyl methacrylate polymers to the β mechanism. Furthermore, since the $\log f_{max}$ versus $1/T$ plots are not linear, the disagreement regarding the variation of activation energy with increasing alkyl group length arises from the fact that these activation energies were derived from different temperature regions.

8.3c   Relaxation of the Side-Chain Alkyl Group (the γ Relaxation)

That the alkyl group R (**1**, page 238) relaxes independently of the oxycarbonyl group and the main chain was first illustrated by the mechanical experiments of Heijboer, Dekking and Staverman (1954) and Hoff, Robinson and Willbourn (1955). The measurements of Hoff and others (1955) for the methyl, ethyl, n-propyl, n-butyl and stearyl methacrylate polymers are shown in Figure 8.23. For the three last polymers (R = $(CH_2)_2CH_3$, $(CH_2)_3CH_3$ and $(CH_2)_{17}CH_3$) a loss peak occurred at temperatures between -100 and $-200°C$ (about 100 c/s) and was attributed to the rotational isomerism of the alkyl group. Later measurements of Sinnott (1959), made with a torsion pendulum, revealed the alkyl loss peak in PEMA at $41°K$ ($-232°C$) at 9.02 c/s (Figure 8.24). As mentioned above (Section 8.2e) Sinnott also inferred from his measurements that in PMMA the hindered rotation of the ester methyl group should lead to a mechanical relaxation below $4.2°K$ ($-268.8°C$) at about 10 c/s. For PEMA Sinnott (1960) also observed a decrease in the n.m.r. second moment between -196 and $-70°C$ which was assigned to the combined motions of the side-chain ethyl group and the α-methyl group (at frequencies of about 10^4 c/s). The T_1 minimum (25 Mc/s) was observed by Kawai (1961) for PEMA and PnBMA at -132 and $-148°C$ respectively.

Low-temperature γ relaxations have also been found for other methacrylate polymers in which the side-chain —R group contains a dipolar unit. For example, Hoff, Robinson and Willbourn (1955) found a mechanical damping maximum at about $-120°C$ (100 c/s) for poly-β-chloroethyl methacrylate (R = —CH_2—CH_2—Cl), and Mikhailov and Borisova (1960) have observed this relaxation by a maximum in the dielectric loss tangent at $-120°C$ (20 c/s). More recently, Mikhailov and Borisova (1964) have reported dielectric data for poly-α-chloromethyl methacrylate

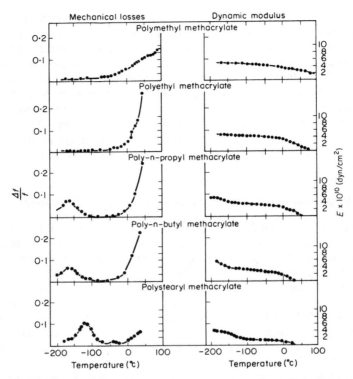

Figure 8.23. Temperature dependence of mechanical losses and Young's modulus for five methacrylate polymers (Hoff, Robinson and Willbourn, 1955).

$(R = -CH_2-Cl)$, poly-β-chloroethyl methacrylate $(R = -(CH_2)_2-Cl)$, poly-γ-chloropropyl methacrylate $(R = -(CH_2)_3-Cl)$ and poly-δ-chlorobutyl methacrylate $(R = -(CH_2)_4-Cl)$. For this polymer series the activation energy of the γ relaxation (about 9 kcal/mole) was essentially independent of the length of the R group. A dielectric γ relaxation has also been observed for poly-β-ethoxyethyl methacrylate $(R = -CH_2-CH_2-O-CH_2-CH_3)$ by Ishida, Amano and Takayanagi (1961). An activation of energy of about 10 kcal/mole was found for this process. The low-temperature relaxation behaviour of poly-β-hydroxyethyl methacrylate $(R = -CH_2-CH_2-OH)$ and related polymers has been studied by Janacek and coworkers using both dielectric (Bares and Janacek, 1965) and mechanical (Ilavski and Janacek, 1965; Janacek and Kolarik,

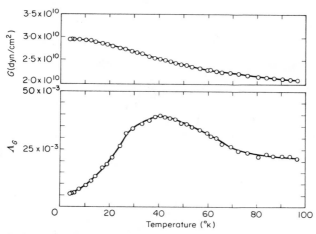

Figure 8.24. Temperature dependence of logarithmic decrement and shear modulus for PEMA at low temperatures (Sinnott, 1959).

1965) methods. Both methods yielded an activation energy of 9 kcal/mole for the low-temperature γ relaxation. The γ mechanism in each of the above polymers was ascribed to motions within the dipolar R group independent of the O—C≡O part of the side-chain.

8.3d Influence of Stereoregularity

Mikhailov and coworkers have studied the effect of stereoregularity on the dielectric behaviour of poly-n-butyl methacrylate (Borisova, Burshtein and Mikhailov, 1962), poly-t-butyl methacrylate (Mikhailov, 1965) and polyphenyl methacrylate (Mikhailov, 1965). Their results for the atactic, isotactic and syndiotactic forms of PnBMA are shown in Figure 8.25. For each specimen a single loss peak is observed. In view of the above discussion concerning the conventional polymer (Section 8.3b), this peak may be regarded predominantly as the (main-chain) α peak with perhaps a small β peak submerged in the low-temperature tail. The tan δ_ε peaks for the syndiotactic and atactic specimens are almost identical both in magnitude and location. The peak for the isotactic sample is somewhat taller and narrower and is located about 40°c lower in temperature. These effects, which are similar to those observed for PMMA (Section 8.2g), suggest that the isotactic chain is considerably more mobile than the syndiotactic and atactic chains.

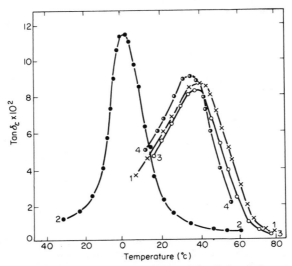

Figure 8.25. Temperature dependence of the dielectric loss tangent at 20 c/s. Curve 1 atactic, Curve 2 isotactic and Curves 3 and 4 syndiotactic PnBMA (Borisova, Burshtein and Mikhailov, 1962).

The results of Mikhailov (1965) for different stereoregular forms of poly-t-butyl methacrylate are presented in Figure 8.26. For each sample the α and β peaks are well resolved, and again the α peak is located some 30°C lower in temperature for the isotactic than for the atactic polymer. The β peak is also shifted to lower temperatures for the isotactic specimen. Thus, both the main-chain and the side-chain (—COOR) mobility appear to be increased by the isotactic configuration, as deduced from the results for PMMA (Section 8.2g). Another observation (Figure 8.26), which is similar to that for PMMA, is that the β peak is reduced and the α peak increased in magnitude by the isotactic configuration. According to Heijboer (1965) this observation suggests that in the isotactic polymer fewer side-chains can move in the glassy region. However, for isotactic poly-t-butyl methacrylate those side-chains which do move in the glassy state apparently do so with greater ease than in the atactic polymer, since the activation energy for the β mechanism decreases from 17 kcal/mole to 11 kcal/mole on going from the atactic to the isotactic configuration (Mikhailov, 1965). For polyphenyl methacrylate the activation energy for the dielectric β mechanism was also lower for the isotactic (10 kcal/ mole) than for the atactic (18 kcal/mole) polymer and, again, the α peak was located at lower temperatures for the isotactic sample.

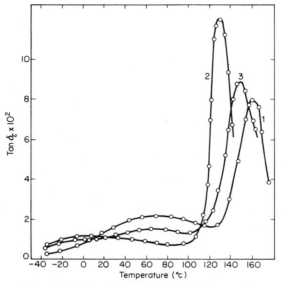

Figure 8.26. Temperature dependence of the dielectric loss tangent at 200 c/s for different stereoregular forms of poly-*t*-butyl methacrylate. Curve 1 Atactic; Curve 2 Isotactic; Curve 3 Intermediate tacticity (Mikhailov, 1965).

8.4 COPOLYMERS OF METHYL METHACRYLATE AND STYRENE

The dielectric relaxation behaviour of random methyl methacrylate–styrene copolymers (see **8.1** and Chapter 10) has been discussed by Mikhailov (1960) (see also Borisova and Mikhailov, 1959). Some results of this investigation are given in Figure 8.27, which shows plots of the dielectric loss tangent against temperature at 20 c/s for various copolymer compositions. With regard to the small α peak at high temperatures three points are worth noting. Firstly, the fact that only a single α peak is observed for each copolymer composition suggests that the relaxing chain unit comprises several repeat units of each type. Secondly, the loss peak shifts to lower temperatures as the proportion of styrene increases. This shift is to be expected on the grounds that T_g for polystyrene is some 20°C to 30°C lower than that for conventional PMMA. Thirdly, the magnitude of the α relaxation, as indicated by the maximum value of $\tan \delta_\varepsilon$, exhibits a maximum at about 60% methyl methacrylate content.

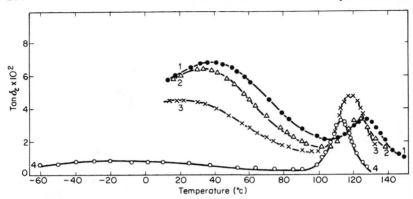

Figure 8.27. Temperature dependence of the dielectric loss tangent at 20 c/s for random methyl methacrylate–styrene copolymers. Curve 1 100% MMA; Curve 2 93·3% MMA; Curve 3 80·5% MMA; Curve 4 24·4% MMA (Mikhailov, 1960).

According to Mikhailov this effect is related to two competing factors; (1) an increase in (non-polar) styrene content lowers the concentration of the dipolar —COOCH$_3$ groups which tends to lower the relaxation magnitude; (2) an increase in the number of styrene units alters the intramolecular interactions such that the change in average chain conformation (in the equilibrium elastic state) results in an increase in the effective dipole moment per statistical segment. The latter effect predominates at low styrene concentrations.

With regard to the β relaxation at low temperatures, it will first be observed that its magnitude decreases considerably as the amount of styrene in the copolymer increases. This probably results from the decreasing concentration of relaxing —COOCH$_3$ side-groups. It is also seen that an increase in styrene content causes the β peak to shift to lower temperatures. This result is obtained over a wide frequency range, as shown by the plots of log f_{max} against $1/T$ (Figure 8.28). From the slopes of these plots Mikhailov (1960) has found that the activation energy for the β mechanism decreases from a value of 18·3 kcal/mole at 0·8% styrene content to a value of 13·4 kcal/mole at 44·5% styrene. Mikhailov attributes these results to a decreased intramolecular steric hindrance to the local motions of the —COOCH$_3$ groups when some of the methacrylate units are replaced by styrene units. This probably results from the replacement of main-chain —CH$_3$ substituents by hydrogen atoms of the styrene units. The mechanical results of Heijboer (1965) on methyl methacrylate–styrene copolymers are in accord with these interpretations.

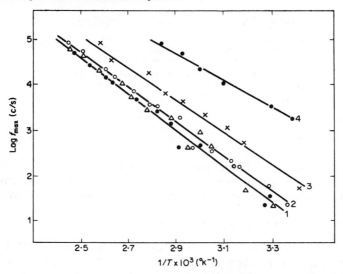

Figure 8.28. Plots of $\log f_{max}$ against $1/T$ for random methyl methacrylate–styrene copolymers in the β relaxation region. Curve 1 99·2% MMA (●) and 93·3% MMA (△); Curve 2 90·2% MMA; Curve 3 80·5% MMA; Curve 4 55.5% MMA (Mikhailov, 1960).

8.5 ALKYL CHLOROACRYLATE POLYMERS

$$\left[\begin{array}{cc} H & Cl \\ | & | \\ -C & -C- \\ | & | \\ H & O=C-OR \end{array}\right]_n$$

(2)

The chloroacrylate polymers (2) differ from the methacrylate polymers in having a chlorine substituent on the α carbon atom in place of the CH_3 group. Since the van der Waals radius of Cl, 1·8 Å, is very close to that of a CH_3 group, the flexibility of the chloroacrylate polymer chain might be expected to resemble that of the methacrylate chain on account of similar steric hindrances to rotation about main-chain bonds. On this basis, the chloroacrylate and methacrylate polymers would be predicted to have similar glass-transition temperatures. However, as shown in Table 8.4, the Vicat softening temperature (which is a rough measure of T_g) is higher for the n-alkyl chloroacrylate polymers than for the corresponding methacrylates (Hoff, Robinson and Willbourn, 1955). This difference is thought

Table 8.4. The Vicat softening point ($T°c$) of the n-alkyl methacrylate and chloroacrylate polymers (Hoff, Robinson and Willbourn, 1955)

Ester	Methacrylate	Chloroacrylate
Methyl	119	140
Ethyl	81	93
n-Propyl	55	71
n-Butyl	30	57

to arise from the large dipole moment of the C—Cl bond (1·86 debye) which produces relatively strong cohesive forces,* due to dipole–dipole interactions, between and within the chloroacrylate polymer molecules.

8.6 POLYMETHYL-α-CHLOROACRYLATE

The dielectric properties of polymethyl-α-chloroacrylate (PMαClA; $R = CH_3$) have been investigated by Mead and Fuoss (1942) and by Deutsch, Hoff and Reddish (1954). Some results of the latter workers are presented in Figure 8.29 which shows the contour diagram of tan δ_ε as a function of temperature and frequency. Both α and β peaks occur, in the region of 150 and 90°c respectively.

The variation of the mechanical modulus and losses with temperature for PMαClA is included in Figure 8.30 (Hoff, Robinson and Willbourn, 1955). These results were obtained from the vibrating cantilever experiment at a frequency of about 200 c/s. Again, both α and β relaxations are observed, by a large increase in loss in the 100 → 150°c range and by a small subsidiary peak at about 90°c respectively. For the lower frequency range (0·01 → 30 c/s) Hoff, Robinson and Willbourn (1955) determined the tensile modulus of PMαClA as a function of temperature and rate of extension. From these measurements they estimated the frequency–temperature location of the α relaxation region.

The correlation between the dielectric and mechanical data is shown on the log f versus $1/T$ plot (Figure 8.31). For the α and β processes Deutsch, Hoff and Reddish (1954) have reported apparent activation energies of 130 and 26 kcal/mole respectively from the dielectric results. From the mechanical data the corresponding values were 105 and 30 kcal/mole.

The α and β mechanisms in PMαClA are attributed, respectively, to segmental motions within the main chain and to the hindered rotation of the —COOCH$_3$ side-groups. These processes are identical with those

---

* See Tobolsky (1960) for a discussion of the effect of cohesive energy density on the glass-transition temperature.

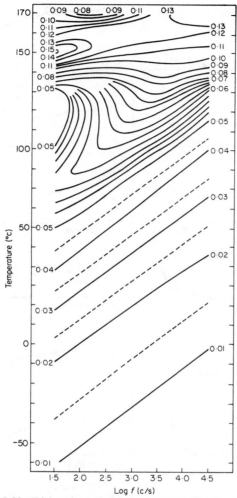

Figure 8.29. Dielectric contour map of tan δ_ε as a function of temperature (°C, ordinate) and log frequency (abscissa) for PMαClA. The tan δ_ε values are indicated at the extremities of the contour lines (Deutsch, Hoff and Reddish, 1954).

proposed for PMMA. However, for PMαClA both the α and β relaxations occur in higher temperature–lower frequency regions than the corresponding relaxations in PMMA. Furthermore, the apparent activation energies for both the α and β relaxations are higher for PMαClA than

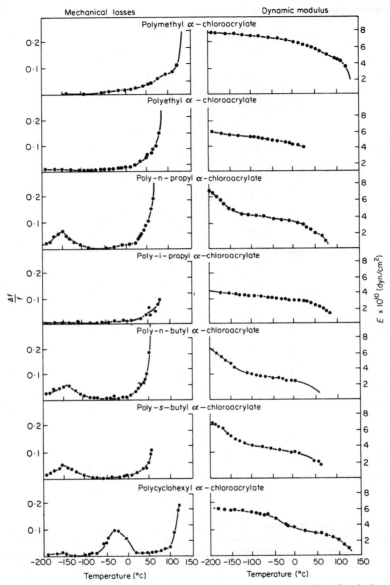

Figure 8.30. Variation of the dynamic Young's modulus and mechanical losses with temperature for various alkyl chloroacrylate polymers. These results were determined at about 2000 c/s using the vibrating cantilever (resonance) experiment (Hoff, Robinson and Willbourn, 1955).

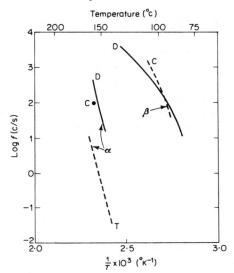

Figure 8.31. Log frequency against reciprocal temperature plot for the loss maxima for PMαClA. Curves constructed from dielectric tan δ_ε maxima (D), and from mechanical cantilever (C) and tensile (T) experiments. The α and β relaxation regions are indicated (Deutsch, Hoff and Reddish, 1954).

for PMMA. Thus, although the —Cl and —CH$_3$ α substituents are approximately equal in size, the —Cl atom apparently provides a larger hindrance to both main-chain and side-group motions. This probably arises from the larger inter- and intramolecular forces resulting from the polarity of the C—Cl bond. The dielectric β relaxation is of a similar magnitude for both PMαClA and PMMA, a result consistent with the side-chain interpretation. However, from Figure 8.29 it is seen that for PMαClA the magnitude of the dielectric α relaxation is greater than that of the β relaxation, which is the reverse situation from conventional PMMA. According to Deutsch, Hoff and Reddish (1954) this is to be expected if the α mechanism involves movements of the main chain since in PMαClA the C—Cl dipole is attached *rigidly* to the chain backbone. The corresponding C—CH$_3$ bond in PMMA is only weakly polar.

8.7 HIGHER ALKYL CHLOROACRYLATE POLYMERS

The mechanical results of Hoff, Robinson and Willbourn (1955) are shown in Figure 8.30 for the methyl, ethyl, n-propyl, isopropyl, n-butyl, s-butyl, and cyclohexyl chloroacrylate polymers.

For each chloroacrylate polymer, the rise in mechanical loss marking the onset of the α peak occurs at a higher temperature than for the corresponding methacrylate polymer (cf. Figure 8.23). This difference is in keeping with that observed for the corresponding glass-transition temperatures (Table 8.4), and probably arises from the decreased chain mobility due to the cohesive dipole–dipole forces in the chloroacrylate polymers. For the n-alkyl chloroacrylate polymers the α relaxation shifts to lower temperatures with increasing alkyl group length. Such a trend, which has already been noted for the methacrylate polymer series, is attributed to the gradual increase in interchain spacing (internal plasticization). However, if the flexibility of the alkyl group is decreased, keeping its size constant (for example by substituting the isopropyl for the n-propyl group or the s-butyl for the n-butyl group), then the α relaxation shifts back to higher temperatures. This trend was also noted within the methacrylate series from both the mechanical and dielectric studies. Apparently the more flexible n-alkyl groups provide less hindrance to the movement of neighbouring chains, possibly on account of the relative ease with which they can undergo conformational changes. In this connection it is also of interest to note that the methyl and cyclohexyl chloroacrylate polymers have similar α relaxation temperatures. Although the cyclohexyl group is far more bulky than the CH_3 group its ability to increase the interchain spacing may be offset by its lack of flexibility.

The β loss peak, which is clearly visible for polymethyl chloroacrylate, becomes less distinct as the n-alkyl group increases in length. Hoff, Robinson and Willbourn (1955) suggest that the length of the alkyl group is unlikely to affect motions of the —COOR group, and that for the higher alkyl chloroacrylates the β peak merges with the α peak as the latter shifts to lower temperatures. As in the case of the methacrylates, therefore, it appears that to a first approximation the —COOR group rotation is unaffected by the length of the n-alkyl group. The β peak does, however, seem to be depressed by the cyclohexyl group in the side-chain.

Mikhailov and Borisova (1960) have investigated the dielectric behaviour of the methyl, ethyl, n-propyl and isopropyl α-chloroacrylate polymers. Their results, shown in Figures 8.32 and 8.33, are in good agreement with the mechanical data of Hoff, Robinson and Willbourn. Owing to the decrease in the α relaxation temperature with increasing n-alkyl group length the dielectric β peak is resolved for the methyl and ethyl but not for the n-propyl chloroacrylate polymer. The β peak is also depressed, however, by the more rigid isopropyl side-group, despite the fact that the α peak has moved back to higher temperatures. Figure 8.33 shows that the frequency–temperature location, and thus the activation

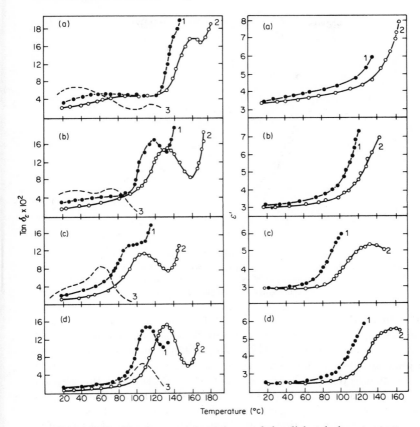

Figure 8.32. Temperature dependence of the dielectric loss tangent and the dielectric constant for the (a) methyl, (b) ethyl, (c) n-propyl and (d) isopropyl α-chloroacrylate polymers. Curves 1 and 2 correspond to frequencies of 20 and 400 c/s respectively. The dashed curves (curves 3) indicate, for comparison, the dielectric loss tangents at 20 c/s for the corresponding methacrylate polymers (Mikhailov and Borisova, 1960).

energy, of the β relaxation is the same for both the methyl and ethyl side-groups. Mikhailov and Borisova quote an activation energy of 22·6 kcal/mole for this β process. Therefore, on the basis of the side-group interpretation, the dielectric results are also consistent with the view that the —COOR group rotations are unaffected by the nature of the —R group.

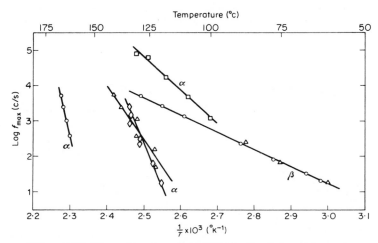

Figure 8.33. Plots of $\log f_{max}$ against $1/T$ for the dielectric loss tangent maxima of the methyl ($\bigcirc$), ethyl ($\triangle$), n-propyl ($\square$) and isopropyl ($\Diamond$) α-chloroacrylate polymers. Both α and β relaxations are indicated (Mikhailov and Borisova, 1960).

It is interesting to note (Figure 8.30) that an additional mechanical γ peak also appears for the n-propyl, n-butyl and s-butyl chloroacrylates at about $-150°$C (200 c/s). This peak, which is also observed for the methacrylate polymers in similar temperature regions (see Figure 8.23), is attributed to the motions of the alkyl group. Presumably the ethyl and methyl chloroacrylate polymers exhibit alkyl relaxations below $-190°$C. Thus the presence of the polar Cl atom in the α position does not affect the motions of the alkyl side-group to any extent. This is also true of the relaxation due to the motions within the cyclohexyl group (see Section 1.5).

8.8 ALKYL ACRYLATE POLYMERS

$$\left[\begin{array}{cc} H & H \\ | & | \\ -C & -C- \\ | & | \\ H & O=C-OR \end{array}\right]_n$$

(3)

The alkyl acrylate polymers (3) differ from the methacrylate polymers in the substitution of a hydrogen atom for the methyl group on the α carbon atom. Several of the acrylate polymers have been prepared by different methods which, by analogy with their application to the preparation of PMMA (see Section 8.2), are thought to yield different stereoregular forms of the polymers. However, a comparison of Tables 8.5 and 8.3 shows that the glass-transition temperatures of the acrylate polymers (unlike the methacrylate polymers) are not significantly dependent on the polymerization method. Furthermore, the T_g values are considerably lower for the acrylate than for the methacrylate polymers, irrespective of the method of preparation. This result probably reflects a decrease in

Table 8.5. Glass-transition temperatures (°C) of several alkyl acrylate polymers as a function of the method of polymerization. For a few of the conventionally prepared polymers the densities are also given (Shetter, 1963; Wolf, 1951; Thurn and Wolf, 1956).

| Acrylate polymer | Structure of alkyl group (—R) | Polymerization Method | | | Density of conventional polymer (room temp.) (g/cm^3) |
		Free-radical at near ambient temp. (44, 60°C) (conventional)	Free-radical at low temp. (below −30°C)	Anionic in non-polar solvent	
Methyl (PMA)	—CH$_3$	3 to 8	—	10	1·22
Ethyl (PEA)	—CH$_2$CH$_3$	−24	−24	−25	1·12
n-Propyl (PnPA)	—(CH$_2$)$_2$CH$_3$	−51·5	—	—	—
Isopropyl (PiPA)	—CH(CH$_3$)CH$_3$	−3 to −6	−2 to 11	−11	—
n-Butyl (PnBA)	—(CH$_2$)$_3$CH$_3$	−70	—	—	1·00
s-Butyl (PsBA)	—CH(CH$_3$)—CH$_2$—CH$_3$	−22	−20	−23	—
t-Butyl (PtBA)	—C(CH$_3$)(CH$_3$)—CH$_3$	43	40	40	—
Cyclohexyl (PCA)	(cyclohexyl)	19	16	12	—
Isobornyl (PiBOA)	(isobornyl: H$_3$C, CH$_3$, CH$_3$)	94	96	90	—

steric hindrance to main-chain motions effected by the replacement of a —CH$_3$ group by a hydrogen atom. To our knowledge, only the conventional acrylate polymers have been studied with respect to their relaxation behaviour.

8.9 POLYMETHYL ACRYLATE

Both mechanical and dielectric studies have revealed the existence of (at least) two relaxation regions in polymethyl acrylate (PMA; R = —CH$_3$). These regions are labelled α and β and are thought to result from main-chain and side-group motions respectively.

8.9a The α Relaxation

The α relaxation in PMA has been studied dielectrically by several workers including Mead and Fuoss (1942), Mikhailov (1951), Thurn and Wolf (1956), Scheiber (1957), Brouckere and Offergeld (1958), Ishida (1961) and Williams (1964). The results of Ishida are shown in Figures 8.34(a) and (b), in which ε' and ε'' are plotted against log frequency in the temperature range 14·5 to 46·5°C. At temperatures above 25·5°C the α absorption maximum is clearly seen within the experimental frequency range. The α loss peaks are slightly non-symmetrical, being broader on the high-frequency side, and between 14·5 and 30°C ε'' increases slightly with frequency in the high-frequency region. The latter observation was attributed to the existence of a β loss maximum (see below) at frequencies above the experimental range. The Cole–Cole arc plots for the α relaxation were also asymmetric, a fact which Ishida attributed entirely to the merging of the β relaxation into the high-frequency tail of the α relaxation. Thus, the calculation of $\varepsilon_R - \varepsilon_U$ and the Cole–Cole distribution parameter $\bar{\beta}$ from the Cole–Cole plots involved an extrapolation, assuming that for the *resolved* α relaxation these plots would be symmetric. However, this procedure may not be strictly valid since the Cole–Cole plots for a few well-resolved α relaxations (e.g. polyvinyl acetate, polyacetaldehyde) are known to be asymmetric. Hence the values of $\varepsilon_R - \varepsilon_U$ and $\bar{\beta}$ for the α relaxation, shown as a function of temperature in Figures 8.35 and 8.36 respectively, may be subject to some error on this account. The decrease in $\varepsilon_R - \varepsilon_U$ with increasing temperature probably arises largely from the predicted $1/T$ dependence (Equation 3.107). The value of $\bar{\beta}$, and therefore the width of the relaxation spectrum, is seen to be independent of temperature, showing that the time–temperature superposition principle should be valid. The plot of $\log f_{\max}$ against $1/T$ (Figure 8.37) shows a slight

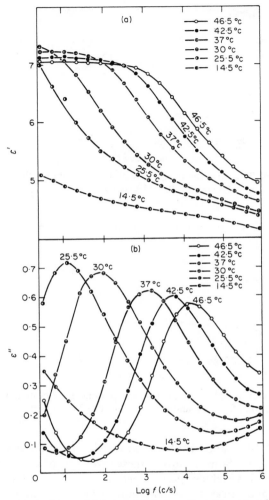

Figure 8.34. Plots of (a) ε' and (b) ε'' against frequency for PMA at several temperatures in the region of the α relaxation (Ishida, 1961).

curvature, consistent with the WLF equation (see also Williams and Ferry, 1955). Nevertheless, Ishida has derived an apparent activation energy of 57 kcal/mole from this plot for the α mechanism.

The α relaxation in PMA has also been studied mechanically by Williams and Ferry (1955), Iwayanagi (1955a), Thurn and Wolf (1956),

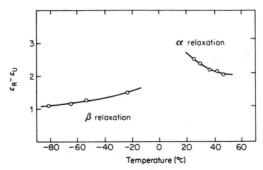

Figure 8.35. The variation of $\varepsilon_R - \varepsilon_U$ with temperature for the α and β relaxations in PMA (Ishida, 1961).

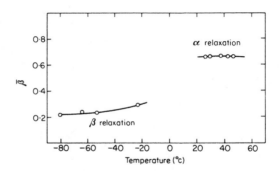

Figure 8.36. The Cole–Cole distribution parameter, $\bar{\beta}$, plotted against temperature for the α and β relaxations in PMA (Ishida, 1961).

Heijboer (1965, 1966) and also by Read (1964a) using the dynamic bire-fringence technique. Some results of these investigations are summarized in Figure 8.37 which includes points corresponding to the location of the G'', E'', J'', D'' and ultrasonic attenuation maxima. Williams and Ferry constructed reduced master curves of the real and imaginary components of the complex shear modulus and compliance and found the temperature dependence of the shift factors to follow the WLF equation. The mechanical relaxation and retardation spectra (at 25°c), derived from these master plots, are reproduced in Figure 8.38 together with the normalized dielectric relaxation spectrum calculated by Ferry, Williams and Fitz-gerald (1955) from the data of Mead and Fuoss (1942). The mechanical relaxation spectrum has a slope of $-\frac{1}{2}$ in the long τ region as predicted by

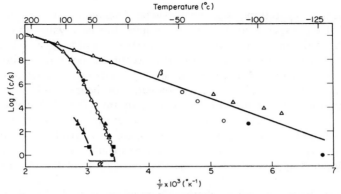

Figure 8.37. Diagram showing the frequency–temperature location of the dielectric (open points) and mechanical (filled points) loss maxima for the α and β absorptions in PMA. In the following key (f) refers to the frequency of maximum loss in constant temperature experiments and (T) denotes the temperature of maximum loss in constant frequency experiments.

△ ε'' (f) (Mikhailov, 1965)
⊙ ε'' (f) (Ishida, 1961)
▲ G'' (f) (Williams and Ferry, 1955)
▲ J'' (f) (Williams and Ferry, 1955)
■ E'' (f) (Read, 1964a)

■ D'' (f) (Read, 1964a)
● G'' (T) (Heijboer, 1966)
● ultrasonic attenuation (T) (Thurn and Wolf, 1956)

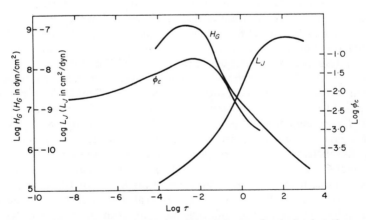

Figure 8.38. Double logarithmic plots of the mechanical relaxation (H_G) and retardation (L_J) spectra against the relaxation time (τ) (Williams and Ferry, 1955). Shown for comparison is the normalized dielectric spectrum (ϕ_ε) calculated by Ferry, Williams and Fitzgerald (1955) from the data of Mead and Fuoss (1942). All spectra are at 25°C.

the Rouse theory (Section 5.3a). It will also be observed that the maximum in the dielectric spectrum occurs at a much shorter τ value than the maximum of the mechanical retardation spectrum, but almost coincides (along the τ axis) with the mechanical relaxation spectrum maximum. Similarly from Figure 8.37 it is seen that at each temperature the frequency of the ε'' maximum coincides with the frequency of the mechanical modulus maximum and not the compliance maximum. These observations indicate that processes with the longer relaxation times, such as the motions of relatively long chain sections, make a relatively larger contribution to $J_R - J_U$ than to $\varepsilon_R - \varepsilon_U$. Also, the local movements of short chain segments apparently determine the frequencies of the ε'' and G'' maxima.

Both the dielectric and mechanical α relaxations in PMA occur in considerably lower temperature–higher frequency regions than the corresponding α relaxation in conventional PMMA. Furthermore, the apparent activation energy of 57 kcal/mole calculated for the dielectric α mechanism in PMA is lower than the corresponding values quoted for PMMA (Table 8.2). These differences reflect the increased chain mobility effected by the substitution of a hydrogen atom for the α-CH$_3$ group.

8.9b The β Relaxation

From dielectric studies a number of authors have reported on the β relaxation in PMA (Mikhailov, 1951; Thurn and Wolf, 1956; Scheiber, 1957: Brouckere and Offergeld, 1958; Ishida, 1961). The data of Ishida (1961) are shown in Figure 8.39(a) and (b). The β loss maximum occurs within the experimental frequency range ($10 \rightarrow 10^6$ c/s) at -53, -64.5 and $-80.5°$c. It will be seen that the β relaxation is of smaller magnitude and broader than the α relaxation. Symmetrical Cole–Cole plots were obtained from the data of Figure 8.39(a) and (b), and the values of $\varepsilon_R - \varepsilon_U$ and $\bar{\beta}$ evaluated from these plots are shown in Figures 8.35 and 8.36 respectively. The small increase in $\bar{\beta}$ with increasing temperature shows that the dielectric relaxation spectrum narrows slightly as the temperature increases. For the β process Ishida calculated an activation energy of 15 kcal/mole from the linear variation of $\log f_{max}$ with $1/T$ (Figure 8.37). This value agrees well with the value of 14 kcal/mole reported by Brouckere and Offergeld (1958), but differs somewhat from the values of 9.6 and 7.4 kcal/mole obtained by Mikhailov (1960) and Scheiber (1957) respectively.

Most of the early mechanical studies of PMA failed to detect a β relaxation in the temperature region corresponding to the β relaxation of

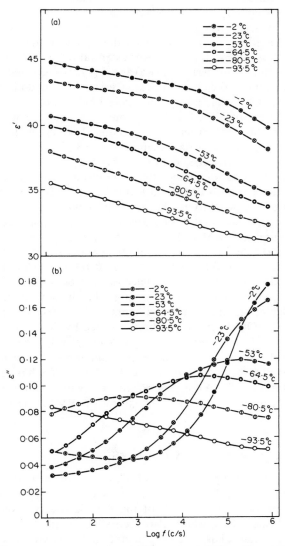

Figure 8.39. Plots of (a) ε' and (b) ε'' versus log frequency for PMA at various temperatures in the region of the β relaxation (Ishida, 1961).

PMMA. Further, from a study of copolymers of methyl methacrylate and
methyl acrylate, Heijboer (1956) observed that the mechanical β peak of
PMMA gradually disappeared as the proportion of methyl acrylate in the
copolymer increased. He concluded from this evidence that steric hindrance provided by the α-CH$_3$ group was essential for the occurrence of
the β loss maximum. Furthermore, Iwayanagi (1955) suggested that the
α-CH$_3$ group maintains the chain backbone essentially rigid at temperatures where the —COOCH$_3$ group undergoes hindered rotation, so that
the α and β peaks are resolved for PMMA but are completely merged for
PMA. Heijboer (1963) has observed a mechanical tan δ_G peak at about
$-120°$c at 1 c/s (Figure 8.42). As shown in Figure 8.37 Heijboer's results
correlate well with the dielectric data at higher frequencies.

The question arises as to whether the same mechanism is responsible
for the β relaxations in both PMA and PMMA. Ishida (1961) has attributed the β relaxation in PMA to the rotation of the —COOCH$_3$ side-
group and suggests, in fact, that this is the same process as that responsible
for the β relaxation in PMMA. However, it will be observed that the β
relaxation for PMA has a lower magnitude and occurs at much lower
temperatures (at a given frequency) than the β relaxation for PMMA.
Also, the activation energy for the dielectric β process in PMA (7 →
15 kcal/mole) is lower than the corresponding value of about 20 kcal/mole
for PMMA. Ishida attributes these differences to a decrease in steric
hindrance to —COOCH$_3$ group rotations effected by the substitution of
a hydrogen atom for the α-CH$_3$ group. Alternatively, they may be
regarded as indicating some fundamental difference between the β
mechanisms for PMA and PMMA. In fact Heijboer (1965) has suggested
that the β relaxation in PMA corresponds to a more limited movement of
the —COOCH$_3$ groups than the β relaxation in PMMA (see Section
8.11).

8.10 HIGHER ACRYLATE POLYMERS

The glass-transition temperatures and densities of several acrylate
polymers are listed in Table 8.5. Both the density and T_g are seen to
decrease with an increase in length of the flexible n-alkyl side-group.
This internal plasticization, which also occurs for the methacrylate and
chloroacrylate polymers, has been attributed to a decreased hindrance to
main-chain motions resulting from an increased interchain spacing. This
effect is not related specifically to the overall size of the side-group but to
its length and flexibility. Bulky side-groups which are relatively short and
rigid, such as isopropyl, t-butyl and cyclohexyl, yield considerably higher

T_g values than the respective linear n-alkyl groups (Table 8.5). A similar effect has been noted for the methacrylate and chloroacrylate polymer series.

Some notable investigations of relaxation in the higher acrylate polymers have been made by Wolf (1951), Thurn and Wolf (1956) and Mikhailov and coworkers (Mikhailov, 1960; Mikhailov, Lobanov and Shevelev, 1961; Mikhailov and Krasner, 1962). The results of these studies are summarized by the plots of $\log f$ against $1/T$ shown in Figure 8.40. It will be noted that both α and β relaxation regions have been observed for most of the acrylate polymers investigated.

8.10a The α Relaxation

For the α relaxation (Figure 8.40) a good correlation is found between the dielectric and mechanical data of the different authors. Also, the effect of internal plasticization is observed by a shift of the α relaxation to lower temperatures (at constant frequency) as the n-alkyl side-group increases in length. It is also of interest to note that the extent of internal plasticization is considerably less for poly-β-chloroethyl acrylate (PβClEA; R = —CH$_2$—CH$_2$—Cl) than for PnPA (R = —CH$_2$—CH$_2$—CH$_3$). Since the CH$_3$ group and the Cl atom have almost identical van der Waals' radii (1·8 Å), it would appear that for PβClEA the effect of internal plasticization is offset by the polarity of the C—Cl bond which may reduce the main-chain mobility through dipole–dipole interactions.

8.10b Low-Temperature Relaxations

Figure 8.40 also summarizes the results of Mikhailov and coworkers for the low-temperature dielectric relaxations in some acrylate polymers. For the low-temperature mechanisms in PMA and PEA, activation energies of 9·6 and 9 kcal/mole respectively were obtained from these results (see references to Figure 8.40). The value of 9 kcal/mole obtained for PEA compares with a value of 7·2 kcal/mole reported by Scheiber (1957). The mechanism of these relaxations is not clearly understood but, as noted in Section 8.9b (see also Section 8.11), they might involve some limited movement of the —COOR side-chains (β relaxation of PMA).

For PnPA and PβClEA, activation energies of 5·7 and 8·6 kcal/mole, respectively, were obtained for the low-temperature processes. The locations of these relaxations are close to those found for the γ relaxations in the corresponding methacrylate polymers (PnPMA and PβClEMA respectively; Section 8.3c). Therefore these processes are probably due largely to motions initiated by the —CH$_2$—CH$_2$—CH$_3$ and the —CH$_2$—CH$_2$—Cl parts of the side-chain.

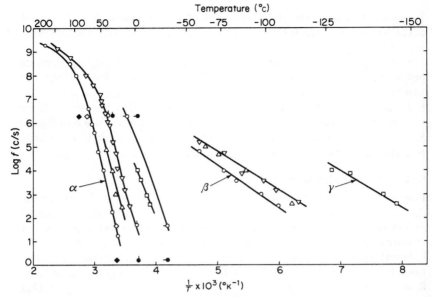

Figure 8.40. Plots of log$f$ against $1/T$ for the dielectric (open points) and mechanical (filled points) loss maxima for the α and β relaxations of several acrylate polymers. In the following key (f) refers to the frequency of maximum loss in constant temperature experiments and (T) denotes the temperature of maximum loss in constant frequency experiments.

○ PMA, tan δ_ε (f) (Mikhailov, Lobanov and Shevelev, 1961)

◇ PMA, tan δ_G (T) (Wolf, 1951; Thurn and Wolf, 1956)

◆ PMA, tan δ_G or ultrasonic attenuation maximum (T) (Wolf, 1951; Thurn and Wolf, 1956)

▽ PEA, tan δ_ε (f) (Mikhailov, Lobanov and Shevelev, 1961)

○̇ PEA tan δ_ε (T) (Wolf, 1951; Thurn and Wolf, 1956)

●̇ PEA, tan δ_G or ultrasonic attenuation maximum (T) (Wolf, 1951; Thurn and Wolf, 1956)

□ PPA, tan δ_ε (f) (Mikhailov, 1960; Mikhailov and Krasner, 1962)

–○ PBA, tan δ_G (T) (Wolf, 1951; Thurn and Wolf, 1956)

–● tan δ_G or ultrasonic attenuation maximum (T) (Wolf, 1951; Thurn and Wolf, 1956)

△ PβClEA, tan δ_ε (f) (Mikhailov, 1960; Mikhailov and Krasner, 1962)

The relaxation mechanism arising from the 'chair-to-chair' motions of the cyclohexyl ring has been studied by Heijboer (1960) using polycyclohexyl acrylate as well as the cyclohexyl methacrylate polymers (see Chapter 1). For the acrylate polymer the mechanical loss peak due to the cyclohexyl group occurs at exactly the same temperature as the corresponding peak for the methacrylate polymer (i.e. about $-23°c$ at 10^3 c/s).

An activation energy of 11.5 kcal/mole was also obtained for the cyclohexyl relaxation in the acrylate polymer, in exact agreement with the value obtained for the methacrylate polymer. The cyclohexyl peak, however, was about 20% higher for the acrylate than for the methacrylate polymer for the same concentration of cyclohexyl groups. Although the reason for this difference was not clear, it was suggested that somewhat more extended movement of the cyclohexyl group may be possible with the more flexible acrylate chain backbone. Copolymers of cyclohexyl acrylate and methyl methacrylate were also investigated and again the height of the loss maximum was found to be proportional to the cyclohexyl concentration.

8.11 METHYL ACRYLATE–METHYL METHACRYLATE COPOLYMERS

The mechanical relaxation behaviour of methyl acrylate–methyl methacrylate random copolymers has been studied by Fujino, Senshu and Kawai (1962) and also by Heijboer (1963). Fujino and coworkers made tensile stress relaxation measurements in the region of the α relaxation, and presented master stress relaxation curves reduced to 100°C. As shown in Figure 8.41, the α relaxation shifts to shorter times as the proportion of methyl acrylate in the copolymer increases. This result was attributed to the increased chain mobility arising from the incorporation of the more flexible acrylate units into the methacrylate polymer chain. A similar conclusion is obtained from the dynamic results of Heijboer (Figure 8.42), since at 1 c/s the α relaxation is seen to move to lower temperatures as the acrylate content increases. Also, for the copolymers containing between 20 and 60% acrylate units, two secondary loss maxima are evident. Labelling these peaks β and γ, in order of decreasing temperature, then it is clear that they correspond, respectively, to the β peaks in PMMA and PMA. They may therefore be assigned, respectively, to the movements of —$COOCH_3$ groups which are hindered (predominantly) by α-CH_3 groups (β peak) and by α-H atoms (γ peak) of the neighbouring repeat units. Figure 8.43 shows the temperature of the β and γ peaks as a function of copolymer composition. Both the β and γ peaks shift to lower temperatures as the proportion of acrylate units is increased, and at intermediate compositions the two peaks remain well separated. The latter result supports the suggestion of Heijboer (1965) that different types of movement of the —$COOCH_3$ groups may be responsible for the β relaxations in PMMA and PMA (Section 8.9b).

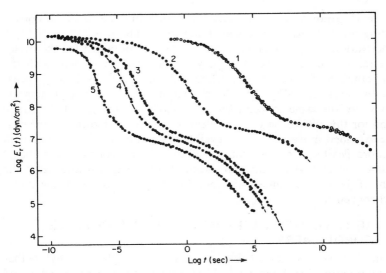

Figure 8.41. Stress–relaxation master curves at 100°C for methyl acrylate–methyl methacrylate copolymers. The mole ratio MA/MMA is as follows: Curve 1 0/100; Curve 2 25/75; Curve 3 39/61; Curve 4 58/42; Curve 5 100/0 (Fujino, Senshu and Kawai, 1962).

8.12 COPOLYMERS OF METHYL ACRYLATE AND STYRENE

Some results of dielectric measurements on random copolymers of methyl acrylate and styrene have been reported by Mikhailov (1960) (see also Mikhailov and Krasner, 1963). These results are summarized in Figures 8.44 and 8.45. Like PMA itself, each copolymer exhibits both an α and a β loss peak.

At constant frequency the α loss peak moves to higher temperatures as the proportion of styrene is increased. This result is ascribed to an increase in steric hindrance to chain backbone motions and shows that the relaxing chain unit must consist both of styrene and acrylate units.

The frequency–temperature location of the β loss peak is seen to be independent of styrene concentration, and from the slope of the $\log f_{max}$ versus $1/T$ plot (Figure 8.45) Mikhailov has evaluated an activation energy of 9·6 kcal/mole. It was concluded that the mobility of the —COOCH$_3$ groups of the acrylate units is unaffected by the introduction of styrene units. Mikhailov has also shown that the maximum value of $\tan \delta_e$, for both the α and β relaxations, increases linearly with an increase in (polar)

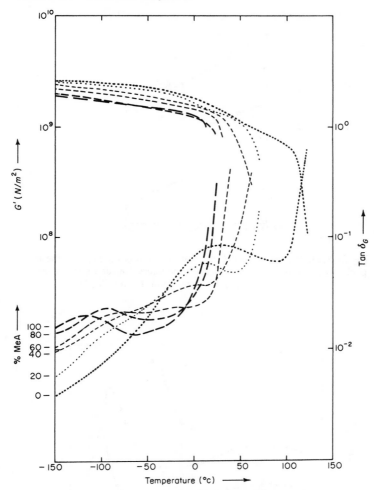

Figure 8.42. The variation of the shear modulus (G') and tan δ_G with temperature at 1 c/s for copolymers of methyl acrylate and methyl methacrylate. The per cent of methyl acrylate in the copolymers is indicated on the ordinate (Heijboer, 1963).

acrylate units. The above evidence suggests that very little interaction occurs between the styrene and acrylate units. These results contrast with those obtained for the styrene–methyl methacrylate copolymers (Section 8.4).

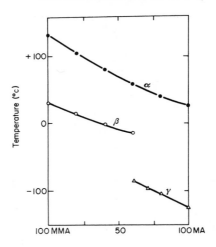

Figure 8.43. Temperatures of the tan δ_G maxima (at 1 c/s) of the loss peaks for methyl acrylate–methyl methacrylate copolymers as a function of copolymer composition (Heijboer, 1965).

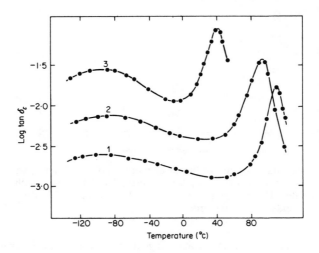

Figure 8.44. Temperature dependence of the dielectric loss tangent at 1000 c/s for styrene–methyl acrylate copolymers. Curve 1 14%, MA; Curve 2 40%, MA; Curve 3 100%, MA (Mikhailov, 1960).

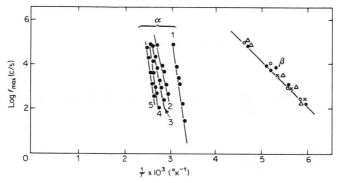

Figure 8.45. Plots of $\log f_{max}$ against $1/T$ for the dielectric loss peaks of styrene–methyl acrylate copolymers. For the α relaxation, Curves 1, 2, 3, 4 and 5 refer, respectively, to 100, 78, 49, 39 and 23 per cent methyl acrylate content (Mikhailov, 1960).

8.13 MIXED (HETEROGENEOUS) POLYMER SYSTEMS

Hughes and Brown (1961) have made an extensive study of the temperature dependence of the shear modulus of several two-component polymer mixtures each of which included either PMA or PEA as one of the components. The mixtures investigated were PMA and PMMA, PMA and polystyrene, PMA and polyvinyl acetate, PEA and PMMA, and PEA and poly-n-butyl methacrylate. Each of these systems consists of polymers of limited compatibility, and the acrylate polymer has, in each case, the lower glass-transition temperature. The results showed that the relaxation temperature of each component was essentially unaffected by the presence of the other component. It was concluded that the systems contain discrete regions of the individual polymers which are sufficiently large for each polymer to retain its characteristic properties. However, in the case of a *graft copolymer* of ethyl acrylate and methyl methacrylate the α relaxation temperature of the methacrylate component was some 40°C lower than that for pure PMMA. This result would indicate that the grafting had produced a greater degree of mixing between the two components.

9

Polyvinyl Esters and Related Polymers

9.1 POLYVINYL ESTERS

$$[-\overset{\overset{\displaystyle H}{|}}{\underset{\underset{\displaystyle H}{|}}{C}}-\overset{\overset{\displaystyle H}{|}}{\underset{\underset{\displaystyle O}{|}}{C}}-]_n$$

$$O=C-R$$

(1)

The relaxation behaviour of the polyvinyl esters (**1**) is of particular interest since these polymers are structurally related to the acrylate and methacrylate polymers. For example, the structural unit of polyvinyl acetate (PVAc; $R = CH_3$) is isomeric with that of polymethyl acrylate and differs only in that the oxygen and carbonyl in the side-chain have been interchanged. Therefore, the fairly close proximity of the glass-transition temperatures of PVAc (28°C) and PMA (3°C) is not surprising. As a further example of the structural similarities, the repeat unit of polyvinyl proprionate ($R = CH_2CH_3$) is isomeric with the repeat units of polyethyl acrylate and polymethyl methacrylate. Consequently, a comparison of the properties of these three polymers may lead to interesting generalizations on the effect of chemical structure on physical properties. Unfortunately, except in the case of PMMA, little is known of the effect on physical properties of changes in stereospecificity.

9.2 POLYVINYL ACETATE

In recent years many publications have appeared concerning the relaxation behaviour of PVAc. The α relaxation in this polymer has been studied in great detail and, in particular, the effects of molecular weight and addition of plasticizer have been investigated. The studies have also revealed the existence of a β mechanism at low temperatures which is thought to arise from motions of the $-OCOCH_3$ side-group.

300

9.2a The α Relaxation

Many authors have presented dielectric results for PVAc in the region of the α relaxation. These authors include Holzmüller (1941), Mead and Fuoss (1941), Würstlin (1951, 1953), Veselovskii and Slusker (1955), Broens and Müller (1955), Thurn and Wolf (1956), Hendus and others (1959), Mikhailov (1960), Hikichi and Furuichi (1961), Ishida, Matsuo and Yamafuji (1962), O'Reilly (1962) and Saito (1963). The results of Ishida, Matsuo and Yamafuji shown in Figure 9.1 are typical of those reported by other workers. As the frequency increases the ε'' curves are seen to shift to higher temperatures but their shape is unchanged. Thus the method of reduced variables appears to be valid. This fact is further illustrated in Figure 9.2 which shows superposed plots of $\varepsilon''/\varepsilon''_{max}$ against $\log \omega/\omega_{max}$. The master curve is broader than that corresponding to a single relaxation time ($\bar{\beta} = 0.8$), but is narrower than that observed for the α relaxations of most other amorphous polymers, including PMA ($\bar{\beta} = 0.6$ to 0.7). However, as found for several other well-resolved dielectric α peaks (cf. polyacetaldehyde and polypropylene oxide, Chapter 14), it is slightly non-symmetrical, being broader on the high-frequency side. This shape is predicted by the theory of Yamafuji and Ishida (1962) which treats the dielectric α relaxation in terms of the cooperative rotational motions of dipolar groups within the main chain. This theory has been outlined in Section 5.3b, and results in Equation (5.48) for the complex dielectric constant. In Figure 9.2 the dielectric loss peak for PVAc is compared with theoretical curves (Equation 5.48) for y equal to 5 and 20 respectively. Ishida and coworkers concluded from the observed agreement that both the width and the asymmetry of the experimental loss curves arise from the intrachain interactions and are thus characteristic of the long chain nature of the molecules.

The magnitude of the dielectric α relaxation, $\varepsilon_R - \varepsilon_U$, was calculated by Ishida and coworkers from the area beneath plots of ε'' versus log frequency. From Figure 9.3a it is seen that $\varepsilon_R - \varepsilon_U$ decreases with increasing temperature, as found for most other amorphous polymers at temperatures above T_g.

The log frequency versus $1/T$ plot of the dielectric loss maxima for the α relaxation in PVAc is shown in Figure 9.4. This plot exhibits the curvature which is typical for glass–rubber relaxations of amorphous polymers and is consistent with the WLF equation (Saito, 1963). Over limited frequency regions, however, the plot is approximately linear, and Ishida and coworkers have quoted an apparent activation energy of 60 kcal/mole which agrees well with the value of 59.8 kcal/mole reported earlier by Mead and Fuoss (1941).

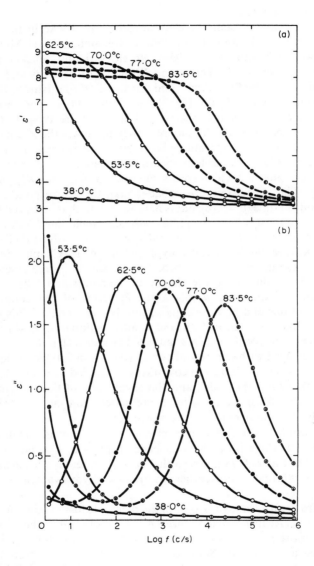

Figure 9.1. Frequency dependence of (a) ε' and (b) ε'' at various temperatures for polyvinyl acetate in the α relaxation region (Ishida, Matsuo and Yamafuji, 1962).

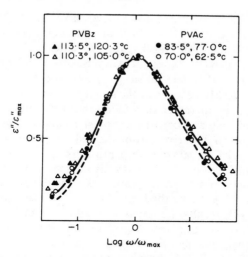

Figure 9.2. Superposed loss curves of polyvinyl acetate (○, ●) and polyvinyl benzoate (△, ▲). Temperatures are indicated in the accompanying key. The full and dashed lines are theoretical curves calculated from Equation (5.48) with y equal to 5 and 20 respectively (Ishida, Matsuo and Yamafuji, 1962).

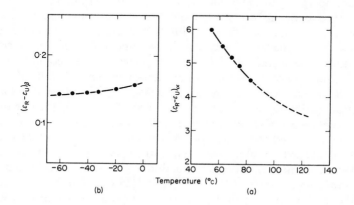

Figure 9.3. Plots of $\varepsilon_R - \varepsilon_U$ against temperature for PVAc in the region of (a) the α relaxation and (b) the β relaxation (Ishida, Matsuo and Yamafuji, 1962).

Dynamic mechanical investigations of the α relaxation in PVAc include those of Schmieder and Wolf (1953), Williams and Ferry (1954), Wolf and Schmieder (1955), Thurn and Wolf (1956), Hendus and others (1959), McKinney and Belcher (1963), and Kovacs, Stratton and Ferry (1963). Williams and Ferry (1954) determined J', J'', G' and G'' at frequencies ranging from 30 c/s to 5100 c/s and in the temperature range 50°C to 90°C. Within this range of temperature they found that the method of reduced variables was applicable and constructed master curves from their data. The shift factors (log a_T), derived from the master curve construction, conformed to the WLF equation with parameters $\alpha_f = 5.9 \times 10^{-4} \deg^{-1}$ and $f(T_g) = 0.028$ (Equation 5.61). They also agreed with the dielectric shift factors obtained from a reduction analysis using the dielectric data of Mead and Fuoss (1941). Using the log a_T factors we have estimated from Williams and Ferry's results the apparent frequencies of the G'' and J'' maxima as a function of temperature (Figure 9.4). It is seen that the dielectric ε'' (and tan δ_ε) maxima lie at frequencies or temperatures intermediate between those corresponding to the G'' and J'' maxima respectively. Similarly, as shown in Figure 9.5, the maximum of the dielectric relaxation spectrum is located at a relaxation time intermediate between the relaxation times corresponding to the maxima of the mechanical relaxation and retardation spectra respectively.

Recently, Kovacs, Stratton and Ferry (1963) have investigated the dynamic shear properties of PVAc in the glass-transition temperature region (20 to 39.2°C). Their measurements were made with a torsion pendulum operating in the frequency range 0.7 to 40 radians/sec. These workers were particularly interested in the effect of thermal history on the dynamic shear properties. Thus, prior to the dynamic measurements, samples were quenched from temperatures well above T_g to the appropriate temperature of measurement. The dynamic measurements could then be made as a function of the 'elapsed time' after quenching. Parallel dilatometric experiments were also made to measure the isothermal volume contraction during the elapsed time. From 31.25 to 39.2°C the dynamic measurements were made after sufficient time had elapsed such that the samples had reached their equilibrium volume. Above about 33°C the method of reduced variables was found to apply and the shift factors derived from the master curve construction followed the WLF equation with parameters $\alpha_f = 5.3 \times 10^{-4} \deg^{-1}$ and $f(25°C) = 0.0225$ (see Equation 5.61). As illustrated by the log a_T versus temperature plot (Figure 9.6), these parameters differ somewhat from those derived from the measurements of Williams and Ferry (1954) at temperatures well above T_g. Kovacs, Stratton and Ferry suggest, therefore, that the WLF equation

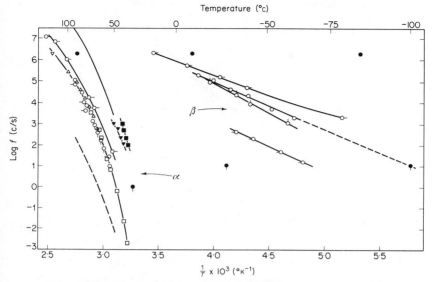

Figure 9.4. The frequency–temperature locations of the mechanical (filled points) and dielectric (open points) loss maxima for the α and β relaxations of polyvinyl acetate. In the following key (f) refers to the frequency of maximum loss in constant temperature experiments and (T) refers to the temperature of maximum loss in constant frequency experiments.

○ ε'' (f) (Ishida, Matsuo and Yama-fuji, 1962)
○ tan δ_ε (T) (Veselovskii and Slus-ker, 1955)
○ tan δ_ε (f) (Mikhailov, 1960)
□ ε'' (f) (Saito, 1963)
○ ε'' (f) (Mead and Fuoss, 1941)
△ tan δ_ε (f) (Broens and Müller, 1955)
▽ tan δ_ε (T) (Thurn and Wolf, 1956)
○ ε'' (f) (Hikichi and Furuichi, 1961)

● ultrasonic attenuation maxima (T) (Thurn and Wolf, 1956)
● tan δ_G (T) (Schmieder and Wolf, 1953)
■ bulk modulus (K'')(T)(McKinney and Belcher, 1963)
▼ bulk compliance (B'')(T)(McKin-ney and Belcher, 1963)
—— estimated G'' (f) (Williams and Ferry, 1954)
– – – estimated J'' (f) (Williams and Ferry, 1954)

requires some modification in the vicinity of T_g, and that the lower temperature limit of its validity may be at about $T_g + 10°C$. Below 33°C the method of reduced variables did not hold, as shown (Figure 9.7) by the unsuccessful attempt to form the G'' master curve using shift factors derived from the WLF equation. This behaviour was ascribed to a change in the shape of the relaxation spectrum with decreasing temperature.

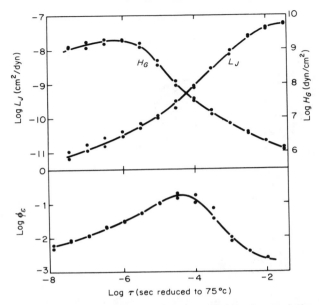

Figure 9.5. The mechanical relaxation (H_G) and retardation (L_J) spectra, and the normalized dielectric relaxation spectrum (ϕ_ε). Each spectrum plotted logarithmically against $\log \tau$, and reduced to 75°C (Williams and Ferry, 1954).

From 20 to 33·8°C, G' and $\tan \delta_G$ were determined (at about 0·2 c/s) at various time intervals during the isothermal volume contraction after quenching (Figure 9.8). From a detailed analysis of such data and in conjunction with the dilatometric data the authors concluded that the mechanical relaxation times depend on temperature largely through their dependence on free volume. However, about 20% of the total temperature dependence of relaxation times was attributed to an effect which was either independent of free volume or related to a small change of free volume with temperature which occurs instantaneously, the same above and below T_g.

The work of Kovacs, Stratton and Ferry illustrates that the dynamic shear properties of PVAc are greatly influenced by the fractional free volume. A direct investigation of volume (or free-volume) relaxation in PVAc has also been made by McKinney and Belcher (1963) (see also Marvin and McKinney 1965). These authors have determined the *bulk* dynamic storage (B') and loss (B'') compliance at frequencies from 50 to

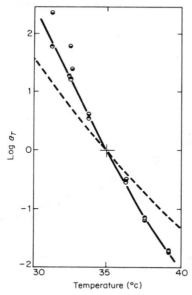

Figure 9.6. Log a_T plotted against temperature for PVAc in the α relaxation region (Kovacs, Stratton and Ferry, 1963).

⬒ empirical shifts of G' curves
◓ shifts of tan δ_G curves
—— calculated from WLF equation,
$\alpha_f = 5\cdot3 \times 10^{-4}\,\mathrm{deg}^{-1}, f_{35} = 0\cdot0225$

– – – WLF equation, parameters
from data of Williams and Ferry
(1954)

1000 c/s and in the temperature range 0 to 100°C. The effect of applying a static hydrostatic pressure P in the range 0 to 981 bars was also studied. At each value of P the volume relaxation region was observed to shift to higher temperatures with increasing frequency (Figure 9.9). Figure 9.4 includes a plot of the temperature of the B'' maximum as a function of frequency at $P = 0$. Also shown on this diagram are points corresponding to the frequency–temperature location of the maximum of the dynamic bulk *modulus* loss, K''. These points were obtained from K'' versus temperature loss peaks which were calculated from McKinney and Belcher's compliance results from the equation $K'' = B''/(B'^2 + B''^2)$. It will be observed from Figure 9.4 that the K'' maxima lie in a slightly higher frequency or lower temperature region than the corresponding G'' maxima. This slight deviation could be due to the presence of a small amount of plasticizer in the sample studied by McKinney and Belcher. Hence,

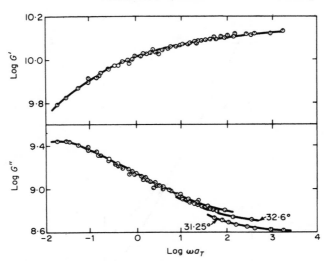

Figure 9.7. Master plots of G' and G'' for PVAc reduced to 35°C, using shift factors from the WLF equation with $\alpha_f = 5\cdot3 \times 10^{-4}\,\mathrm{deg}^{-1}$ and $f_{35} = 0\cdot0225$ (Kovacs, Stratton and Ferry, 1963).

it appears that there is a reasonable correlation between the locations of the G'' and K'' loss peaks, which suggests that similar molecular motions are responsible for the relaxation of both the bulk and shear moduli, (compare, however, Chapter 6.1). From Figure 9.9 it is also seen that, at constant frequency, the relaxation region shifts to higher temperatures with an increase in P. This effect is opposite to that observed by the addition of plasticizer and must be attributed to the increased hindrance to molecular motions caused by a decrease in fractional free volume. The magnitude of this shift is given approximately by the thermodynamic quantity $(\partial T/\partial P)_f$ for which McKinney and Belcher quote a value of about 0·020°C/bar. This value was in good agreement with the value of 0·022 reported by O'Reilly (1962) from a study of the effect of pressure on the dielectric relaxation of PVAc. McKinney and Belcher have also treated their data by the method of reduced variables in which they propose the use of a modified WLF equation to include the effect of static pressure.

Effect of Plasticizer

For amorphous polymers, the decrease in T_g and the α relaxation temperature after adding low molecular weight plasticizers is a well-known

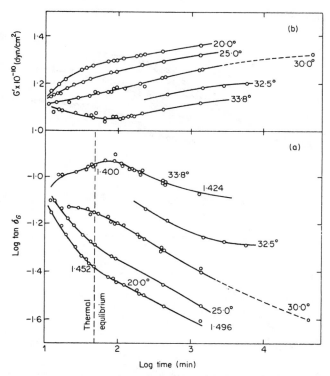

Figure 9.8. (a) Plot of tan δ_G against elapsed time during the isothermal contraction of PVAc specimens. Samples were initially quenched from 45°C to room temperature, installed in a torsion pendulum and heated to the temperature indicated on each curve. Vertical dashed line indicates time required for thermal equilibrium. Small numbers represent the frequency in radians/sec at constant moment of inertia; (b) Plot of G' against elapsed time as in (a) (Kovacs, Stratton and Ferry, 1963).

phenomenon which has already been noted in the case of PMMA. In constant temperature experiments the effect of plasticizer is likewise to shift the loss factor–frequency curves to higher frequencies. These effects are opposite to those produced by the application of an external hydrostatic pressure (see above), and are ascribed to the enhanced chain mobility arising from the increase in free volume (or decrease in intermolecular interactions). For PVAc plasticized with (non-polar) diphenylmethane, the effect is well demonstrated in Figure 9.10 by the dielectric

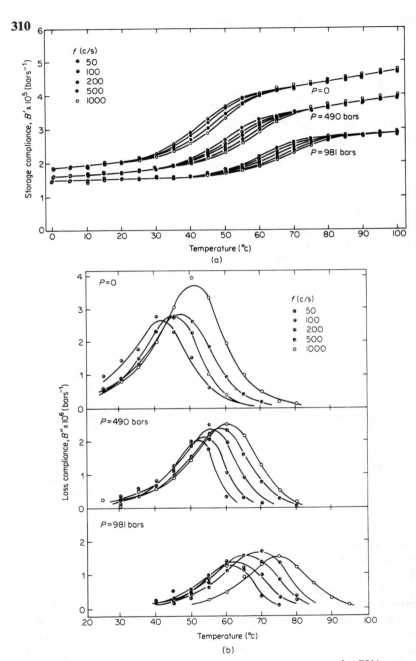

Figure 9.9. Plots of (a) B' and (b) B'' versus temperature for PVAc at several frequencies (f) and static pressures (P) (McKinney and Belcher, 1963).

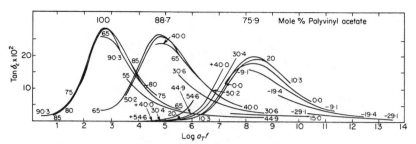

Figure 9.10. Illustration of an attempt to construct reduced master curves of the dielectric loss tangent for PVAc plasticized with diphenylmethane. Reference temperature for each curve is 65°C. The actual temperature of measurement and per cent PVAc are as indicated (Broens and Müller, 1955).

results of Broens and Müller (1955). In spite of the obvious failure of the reduced variables treatment for the mixed systems, it is clear that the addition of plasticizer causes the loss peaks to shift to higher frequencies and also to flatten and broaden somewhat.

Several studies have also been made of the effect of polar plasticizers over a wide range of composition. Figure 9.11 shows the dielectric results of Würstlin (1953) for various compositions of PVAc–benzyl benzoate mixtures. At low concentrations of plasticizer, a single peak is observed due to the relaxation of polymer segments modified by the presence of plasticizer. At very high plasticizer concentrations the loss peak is due predominantly to motions of the plasticizer molecules. Two peaks are observed at intermediate compositions due to overlapping relaxations involving the (mutually interacting) polymer and plasticizer respectively. According to Würstlin (1953), at low plasticizer concentrations the plasticizer molecules are bound to the polymer segments by dipole–dipole interactions. The 'critical' composition at which the plasticizer begins to show a separate peak then corresponds to the point at which the polymer is fully solvated. Furthermore, from the critical compositions for a number of PVAc–plasticizer systems, Hartmann (1957) has evaluated the number of PVAc repeat units per bound plasticizer molecule. However, as shown by Luther and Weisel (1957) and later by Thurn and Würstlin (1958), these results can also be accounted for on the basis of the simple superposition of two symmetrical loss peaks. The concept of bound plasticizer molecules, for which there is no direct physical evidence, must therefore be regarded with some uncertainty. An excellent review of this problem has been made by Curtis (1960).

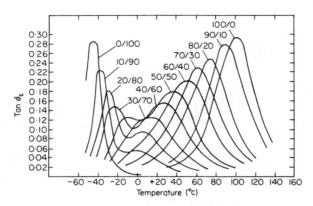

Figure 9.11. Tan δ_ε against temperature at 2×10^6 c/s for PVAc plasticized with benzyl benzoate. The ratio of polymer to plasticizer is shown for each curve (Würstlin, 1953).

Effect of Molecular Weight

An investigation of the dependence on molecular weight of both the dielectric and mechanical properties of PVAc (at 2×10^6 c/s) has been made by Thurn and Wolf (see Hendus and others, 1959). The results of this study are summarized in Figure 9.12. At low molecular weights the temperature of maximum loss increases rapidly with increasing molecular weight, but tends toward an asymptotic limit at a molecular weight of about 10,000. This behaviour has also been reported by Würstlin (1951, 1953), and Thurn has observed a similar effect in the case of polyiso-butylene (Hendus and others, 1959). Furthermore, it is generally observed for amorphous polymers that both T_g and the density also increase with molecular weight toward asymptotic limits which are reached at a molecular weight of about 10,000 (Fox, Gratch and Loshaek, 1956; Gibbs and Di Marzio, 1958). The density dependence suggests that the observed relaxation behaviour is closely related to free-volume changes. Thus, as the molecular weight is decreased below about 10,000, the concentration of chain ends becomes significant and, since the packing around chain ends is inevitably looser than along the main chain, the free volume increases. Below the critical molecular weight, therefore, the concentration of chain ends, the free volume, and hence the molecular mobility will increase with decreasing molecular weight. Above this critical weight the chain end concentration becomes negligible. As

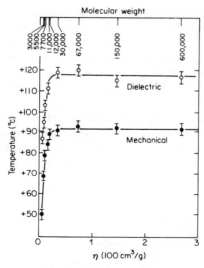

Figure 9.12. The temperature of the dielectric and mechanical loss maxima as a function of intrinsic viscosity or molecular weight (number average) for PVAc (Hendus and others, 1959).

noted by Würstlin (1951, 1953), the effect of chain ends is analogous to that produced by the addition of plasticizer. Fortunately, most studies of relaxation in amorphous polymers appear to have been made on samples having molecular weights above the critical value.

Comparison with Polymethyl acrylate

Three points are worth noting in comparing the α relaxation of PVAc with that of PMA. Firstly, the dielectric loss peak of PVAc is somewhat narrower than the PMA loss peak, as mentioned above. Secondly, the magnitude $(\varepsilon_R - \varepsilon_U)_\alpha$ is greater for PVAc than for PMA, and, thirdly, at a given frequency, both the dielectric and mechanical α relaxations occur at somewhat higher temperatures for PVAc than for PMA. The last fact shows that, with the side-chain oxygen atom adjacent to the main chain, the mobility of the main-chain segments is somewhat decreased.

9.2b The β Relaxation

The β relaxation in PVAc has been observed in the low-temperature dielectric studies of Veselovskii and Slusker (1955), Broens and Müller (1955), Mikhailov (1960), Hikichi and Furuichi (1961) and Ishida, Matsuo and Yamafuji (1962). From the results of Ishida and coworkers, shown in Figure 9.13, it can be seen that the β relaxation is of much smaller magnitude and broader than the α relaxation, as is also found for PMA. Owing to the very slow variation in the shape of the β absorption with temperature, Ishida and coworkers were able to construct a reduced master curve from plots of $\varepsilon''/\varepsilon''_{max}$ against $\log(\omega/\omega_{max})$ obtained at different temperatures. A satisfactory fit to this master curve (Figure 9.14) was obtained with a theoretical curve based on the local mode theory of Yamafuji and Ishida (1962) (Section 5.5b). This theory was originally devised to explain the β relaxations in polymers such as PVC (Chapter II, **1**) in which the polar side-groups are attached *rigidly* to the main chain. For such

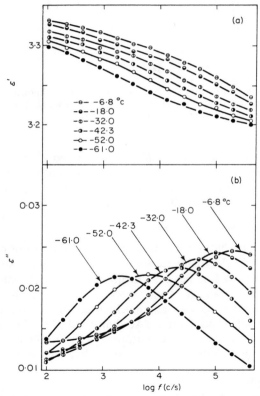

Figure 9.13. Frequency dependence of (a) ε' and (b) ε'' for PVAc at several temperatures in the β relaxation region (Ishida, Matsuo and Yamafuji, 1962).

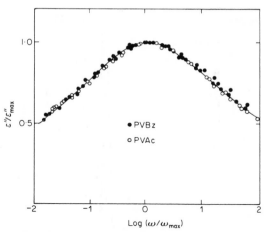

Figure 9.14. Master dielectric loss factor curves for PVAc (○) and PVBz (●) in the β relaxation region. A theoretical line is also drawn according to an equation derived by Yamafuji and Ishida (1962) (Ishida, Matsuo and Yamafuji, 1962).

polymers the dielectric β relaxation cannot arise from rotations of the side-groups about the bonds joining them to the main chain, since this mechanism does not involve a directional change of the side-group dipole. Some local movement *within* the main chain was thus postulated. In applying the theory to PVAc, for which independent rotations of the *flexible* side-groups could lead to the β relaxation, the authors suggest that the β mechanism arises from motions of the side-groups which lead to, or are accompanied by, local distortions of the main chain. In Figure 9.4 we have collected together the results of various authors on the log frequency versus $1/T$ plot. The discrepancy seen between Mikhailov's data and the results of the other workers cannot be explained. However, the activation energy of 10·5 kcal/mole calculated by Mikhailov agrees well with the values of 10, 10 and 8·6 kcal/mole obtained by Ishida and coworkers, Hikichi and Furuichi, and Veselovskii and Slusker respectively.

In mechanical shear measurements at about 10 c/s. Schmieder and Wolf (1953) observed two secondary maxima for PVAc at about -30 and $-100°$c respectively (see also Wolf and Schmieder, 1955). These data are illustrated in Figure 9.15(a). According to Figure 9.4 the peak at $-100°$c correlates well with Ishida's dielectric results. The $-30°$c peak seems less prominent than that at $-100°$c and has no dielectric

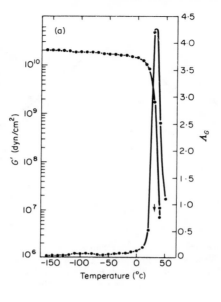

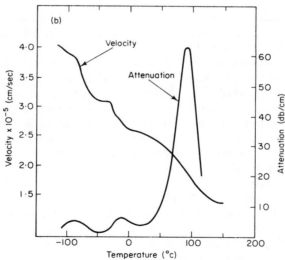

Figure 9.15. (a) Variation of torsion modulus and logarithmic decrement with temperature for PVAc (Wolf and Schmieder, 1955); (b) Dependence of bulk longitudinal wave velocity and attenuation (at 2×10^6 c/s) on temperature for PVAc (Thurn and Wolf, 1956).

equivalent. It may conceivably be due to impurities. Using the bulk longitudinal wave propagation method, Thurn and Wolf (1956) have also found two secondary attenuation maxima in PVAc at about -10 and $-85°$C (2×10^6 c/s) respectively (Figure 9.15b). The peak at $-10°$C appears to show some correlation with the dielectric β peak and consequently with the low-frequency mechanical peak at $-100°$C (Figure 9.4). However, the ultrasonic maximum at $-85°$C probably corresponds to some lower temperature mechanism which may possibly involve some more limited motion within the side-chain. As noted by Mikhailov (1960), the dielectric data also indicate that a third relaxation may exist for PVAc at temperatures below those of the β relaxation, since the limiting value of ε' for the β relaxation at high frequencies (Figure 9.13) is some 30% higher than the square of the refractive index.

The β mechanism in PVAc is generally thought to involve motions of the —OCOCH$_3$ side-groups. This process is presumably similar to that proposed for the β relaxation in PMA in which the —COOCH$_3$ side-groups are involved. In comparing the data for the β absorptions of these two polymers, we may first note from Figures 9.3(b) and 8.35 that the magnitude $(\varepsilon_R - \varepsilon_U)_\beta$ of the relaxation in PVAc is considerably less than that in PMA. This observation is opposite to that found for the α relaxations of the two polymers, so that the ratio $(\varepsilon_R - \varepsilon_U)_\alpha/(\varepsilon_R - \varepsilon_U)_\beta$ is greater for PVAc than for PMA. This ratio is greater than unity for both PVAc and PMA, unlike the case for conventional PMMA. Secondly, the activation energies of 8·6 to 10·5 kcal/mole quoted by various authors for the β relaxation in PVAc are similar to the values of 7·6 to 13 kcal/mole reported for the β process in PMA. However, a close inspection of Figures 9.4 and 8.37 reveals that at constant temperature the β loss maxima for PVAc occur at frequencies at least one decade below the frequencies of the corresponding PMA loss maxima. Hence, the side-group motions appear to be slightly more restricted when the side-chain oxygen atom exists adjacent to the main chain. A similar conclusion was stated above in connection with the motions of the chain backbone.

9.3 HIGHER VINYL ESTER POLYMERS

9.3a The α Relaxation

In addition to their study of PVAc, Schmieder and Wolf (1952, 1953) have made internal friction measurements at 1 c/s on polyvinyl propionate (PVPr; R = —CH$_2$—CH$_3$; **1**, page 300) and polyvinyl butyrate (PVBu; R = —CH$_2$—CH$_2$—CH$_3$). Thurn and Wolf (1956) have also investigated

PVPr both mechanically and dielectrically at 2×10^6 c/s, and have quoted a dielectric result at 50 c/s, for this polymer obtained by Würstlin. The data reported by the above authors are summarized in Figure 9.16 and Table 9.1. An increase in the length of the n-alkyl group in the side-chain is seen to decrease the temperature of the α relaxation, as noted also for the acrylate and methacrylate polymer series.

Table 9.1. The density, glass-transition temperature and temperatures of the mechanical and dielectric loss maxima for the α relaxation of polyvinyl acetate, polyvinyl n-propionate and polyvinyl n-butyrate (Wolf and Schmieder, 1955; Thurn and Wolf, 1956; Jenckel, 1942). Temperatures are in °C

Polymer	Structure of —R group in (1)	Density (g/cm³) (room temp.)	T_g	Dielectric tan δ_ε maximum 50 c/s	Dielectric tan δ_ε maximum 2×10^6 c/s	Mechanical 1 c/s (tan δ_G maximum)	Mechanical 2×10^6 c/s (attenuation maximum)
Acetate	—CH$_3$	1·17	28	54	122	33	93
Propionate	—CH$_2$CH$_3$	1·02		23	88	12	54
Butyrate	—(CH$_2$)$_2$CH$_3$					4	

Ishida, Matsuo and Yamafuji (1962) have made a detailed dielectric study of polyvinyl benzoate (PVBz; $R = C_6H_5$). Their data in the region of the α relaxation, when plotted in terms of $\varepsilon''/\varepsilon''_{max}$ against log (ω/ω_{max}), reduced to a single master curve. This master curve is somewhat broader than the curve for PVAc (see Figure 9.2) and its shape conforms to the shape of the theoretical curve drawn according to Equation (5.48). Ishida and coworkers observed also that $\varepsilon_R - \varepsilon_U$ for the α relaxation was smaller for PVBz than for PVAc, a fact which they attributed mainly to the larger dipole concentration in PVAc. As illustrated in Figure 9.16, the ε'' maxima for the α peak of PVBz are positioned at lower frequencies or higher temperatures than the corresponding maxima for PVAc.

Mead and Fuoss (1941) have presented dielectric results for polyvinyl chloroacetate (PVClAc; $R = $ —CH$_2$—Cl). As illustrated in Figure 9.16, the ε'' maxima for the α relaxation in this polymer have almost exactly the same frequency–temperature location as the corresponding maxima for PVAc. Also included in Figure 9.16 are some dielectric results of Mikhailov (1960) for polyvinyl chloropropionate

$$\text{(PVClPr}; R = \text{—CH}_2\text{—CH}_2\text{—Cl)}$$

in the α relaxation region. For this polymer the dielectric loss maxima lie at slightly higher temperatures or lower frequencies than the corresponding maxima for PVPr.

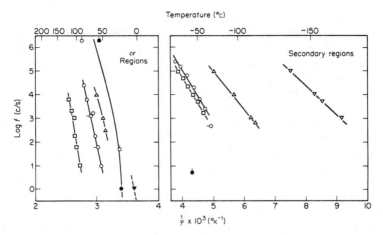

Figure 9.16. Frequency–temperature location of the mechanical (filled points) and dielectric (open points) loss maxima for several vinyl ester polymers. In the following key (f) refers to the frequency of maximum loss in constant temperature experiments and (T) refers to the temperature of maximum loss in constant experiments.

○ PVAc, ε'' (f) (Ishida, Matsuo and Yamafuji, 1962)

□ PVBz, ε'' (f) (Ishida, Matsuo and Yamafuji, 1962)

⊖ PVClAc, ε'' (f) (Mead and Fuoss, 1941)

△ PVClPr, tan δ_ε (f) (Mikhailov, 1960)

▽ PVBu, tan δ_ε (f) (Mikhailov, 1960)

◔ PVPr, tan δ_ε (T) (Thurn and Wolf, (1956)

● PVPr, tan δ_G (T) (Schmieder and Wolf, 1953) and ultrasonic attenuation maximum (T) (Thurn and Wolf, 1956)

▼ PVBu, tan δ_G (T) (Schmieder and Wolf, 1952)

In summarizing the above results, we note that, as in the case of the methacrylate, acrylate, and chloroacrylate polymer series, the main-chain mobility is increased when the flexible n-alkyl side-group is increased in size. This so-called 'internal plasticization' has been ascribed to the decreased interchain interaction resulting from an increased interchain spacing. However, the chain backbone motions are more restricted if the increase in size of the side-chain is effected by replacing the —CH₃ group by a bulky *rigid* group such as the benzyl group. It seems probable that the latter effect may arise from an increase in steric hindrance to rotations about main-chain bonds, which more than compensates for the effect of internal plasticization. Furthermore, if the size increase is effected by replacing one of the —CH₃ group hydrogen atoms by a chlorine atom, then the chain mobility is either unaffected or perhaps slightly reduced.

In this case the internal plasticization may be compensated for by increased dipole–dipole interactions between the polar C—Cl bonds. Recalling that the —CH$_3$ group and Cl atom have almost identical van der Waals' radii, the latter suggestion is supported by the observation that the respective α relaxation temperatures for PVClAc and PVClPr are higher than those for PVPr and PVBu respectively (Figure 9.16). The effects discussed above are similar to those found within the methacrylate, acrylate and chloroacrylate polymer series. For these polymers the cyclohexyl group provided the example of a bulky *rigid* side-chain substituent.

It is of interest to compare the α relaxation temperatures of polyvinyl propionate, polyethyl acrylate, and polymethyl methacrylate respectively, in view of the fact that their repeat units are isomeric. We may first note that the α relaxation temperature is somewhat higher for PVPr than for PEA (cf. Figures 9.16 and 8.40). Hence the main chain becomes less mobile when the side-chain oxygen atom is linked to the main chain. Although the reason for this is obscure, it is worth recalling that a comparison of PVAc and PMA (above) yielded a similar conclusion. Secondly, the α relaxation temperature is considerably higher for PMMA than for both PVPr and PEA. Clearly, therefore, the movement of a —CH$_3$ group from the side-chain to the main chain causes a large reduction in the main-chain mobility, and this effect is much larger than that produced by rearrangements within the side-chain itself. Recalling that T_g for both syndiotactic (115°C) and isotactic (45°C) PMMA is considerably higher than that for PEA (-23°C), the effect is seen to be larger for the syndiotactic form of PMMA.

9.3b Secondary Relaxations

Figure 9.16 summarizes the dielectric data reported for a number of vinyl ester polymers in the low-temperature region. The results shown for PVAc and PVBz are those of Ishida, Matsuo and Yamafuji (1962). Those for PVBu and PVClPr are from Mikhailov (1960), and the point shown for PVClAc was derived from the results of Mead and Fuoss (1941). The mechanical point shown for PVPr at -40°C and 5·4 c/s was taken from the paper of Schmieder and Wolf (1953).

Considering first the results of Ishida and coworkers for PVBz, it is seen from Figure 9.14 that the shape of the master dielectric loss curve is fairly symmetrical and almost identical with the curve for PVAc. Ishida and coworkers also found that $\varepsilon_R - \varepsilon_U$ for the β relaxation was only slightly lower for PVBz than for PVAc, a fact which they ascribed mainly to the lower dipole concentration in PVBz. Furthermore, at a given temperature the frequency of the secondary loss maximum is only slightly lower for PVBz than for PVAc (Figure 9.16). Hence this relaxation, which is thought to arise from motions of the —OCOR groups, is very little affected when the methyl group (R) is replaced by the benzyl group. This evidence supports the suggestion that the mechanism may involve local main-chain motions (Section 9.2b).

It is of interest to compare the low-temperature data for PVBu and poly-n-propyl acrylate (PnPA) since the alkyl group R (1, page 300, 3, page 284) is equal to —CH$_2$—CH$_2$—CH$_3$ for each of these polymers. From Figures 9.16 and 8.40 it is seen that the low-temperature peaks for these two polymers have a fairly similar temperature–frequency location, which suggests that the relaxation mechanisms may be related. This mechanism probably involves a movement initiated by the —CH$_2$—CH$_2$—CH$_3$ group, though since the motion is dielectrically active, the dipolar parts of the side-chains may also be involved.

Another interesting comparison concerns the low-temperature data for PVClPr (R = —CH$_2$—CH$_2$—Cl; 1). Figures 9.16 and 8.40 show that the secondary peaks for PVClPr and PClEA have almost identical locations. In fact Mikhailov (1960) has shown that the low-temperature peaks for these two polymers are also identical both in shape and magnitude. Furthermore, these peaks correlate very closely with the low-temperature peaks for poly-β-chloroethyl methacrylate (Section 8.3c) and poly-β-chloroethyl chloroacrylate (Mikhailov, 1965). The above comparisons provide strong evidence that the dipolar —CH$_2$—CH$_2$—Cl side-groups are mainly responsible for these secondary relaxations.

9.4 POLYVINYL ETHERS

$$[-\overset{\displaystyle H}{\underset{\displaystyle H}{C}}-\overset{\displaystyle H}{\underset{\displaystyle O-R}{C}}-]_n$$

(2)

The repeat unit of the polyvinyl ethers resembles that of the polyvinyl esters in having a polar side-group which is linked to the main chain via an oxygen atom. Several of the polyvinyl ethers have been prepared in stereoregular crystalline forms (Natta, 1960a, b; Iwasaki, 1962; Ketley, 1962; Kern and others, 1963), and evidence from the temperature dependence of the modulus (Kern and others, 1963) indicates that for polyvinyl methyl ether the glass-transition temperature is not significantly affected by changes in stereoregularity (in contrast to the methacrylate polymers). However, studies of the relaxation behaviour of the polyvinyl ethers have so far been confined to the conventional amorphous polymers. Since the polar side-chain may rotate independently of the main chain, we might expect these polymers to exhibit, in addition to the main-chain α relaxation, secondary relaxations if indeed these do arise from side-group rotations. Both primary and secondary relaxation regions have, in fact, been observed in both mechanical and dielectric investigations.

9.4a   The α Relaxation

Of the vinyl ether polymers, polyvinyl ethyl ether has probably been investigated most extensively in the region of the α relaxation (Schmieder and Wolf, 1953; Thurn and Wolf, 1956; Wetton, 1962, 1964). Some results of these investigations are conveniently summarized in Figure 9.17 which illustrates the frequency–temperature location of both the mechanical and dielectric loss maxima. Ishida (1960a) has also studied the dielectric properties of polyvinyl isobutyl ether in the α relaxation region. His measurements were made at frequencies ranging from 10^2 to 10^6 c/s and in the temperature range -13 to $+43°$c. The frequency–temperature location of the single broad ε'' peak which he observed is shown also in Figure 9.17.

As shown in Table 9.2 the α relaxation has been observed, in both dielectric and mechanical experiments, for several other polyvinyl ethers. The density, the glass-transition temperature, and the α relaxation temperature (at constant frequency) are each seen to decrease with an increase in length of the n-alkyl side-group. These trends provide yet another example of 'internal plasticization', an effect which has already been discussed for the methacrylate, chloroacrylate, acrylate and vinyl ester polymers. Furthermore, as illustrated by the results for the n-butyl, isobutyl and t-butyl vinyl ether polymers (Figure 9.18 and Table 9.2), the α loss peak shifts to higher temperatures as the alkyl side-group is shortened, whilst keeping its overall size approximately constant. These results serve to emphasize once again that the effect of internal plasticization is probably related to the length and flexibility of the side-chain alkyl group rather than to its overall size.

A comparison of Tables 9.2 and 9.1 and Figure 9.18 shows that the T_g values and the temperatures of the α loss peaks are lower for the polyvinyl ethers than for the corresponding polyvinyl esters. According to Würstlin (1950), these differences may partly be related to the lower dipole moments of the polyvinyl ethers. From group dipole moments, which were calculated from the dipole moments of simple low molecular weight compounds,

The densities, glass-transition temperatures and temperatures of the dielectric and mechanical loss peaks for the primary and secondary relaxations of the polyvinyl ethers

Polymer	Structure of —R group in (2)	Density (g/cm³) (room temp.)	T_g (°C)	α Relaxation — Mechanical: Attenuation maximum at 2×10^6 c/s	α Relaxation — Mechanical: Tan δ_G (freq. c/s in brackets)	α Relaxation — Dielectric (tan δ_ε maximum) 2×10^6 c/s	α Relaxation — Dielectric 50 c/s	Secondary — Mechanical: Attenuation maximum at 2×10^6 c/s	Secondary — Mechanical: Tan δ_G maximum (freq. c/s in brackets)	Secondary — Dielectric (tan δ_ε maximum) at 2×10^6 c/s
Polyvinyl methyl ether	$-CH_3$	1·03	< −20	29±2	−10(2·7)	37±2	−15	−49 −73		−50±10 −85±5
Polyvinyl ethyl ether	$-CH_2CH_3$	0·96	−43·6	26±2	−17(1·1)	45±2		−40±4 −88±5	−70(7·5)? −110(8)?	−45±10 −90? −125?
Polyvinyl n-propyl ether	$-(CH_2)_2CH_3$	0·94		17±2	−27(1·1)	42±2	−22	−30±5 −80±10	−70(4·2) −160(5·5)	−10? −50? −75±10 −125±10
Polyvinyl n-butyl ether	$-(CH_2)_3CH_3$	0·91		5±2	−32(0·8)	40±2	−10	−32±5 −57? −76±4	−70(8) −150(10·4)	−35±10 −72±7 −115±10 −165?
Polyvinyl isobutyl ether	$-CH_2-\overset{\displaystyle CH_3}{\underset{\displaystyle CH_3}{CH}}$			48±2	−1(1·2)	72±2	5	−52±5 −70?	−150(11·8)	−45±10 −70 −125±10
Polyvinyl t-butyl ether	$-\overset{\displaystyle CH_3}{\underset{\displaystyle CH_3}{C}}-CH_3$				83(1·7)		>100			

Densities and mechanical and dielectric data at 2×10^6 c/s from Thurn and Wolf (1956).
Mechanical data at low frequencies from Schmieder and Wolf (1953).
Dielectric data at 50 c/s and T_g for polyvinyl methyl ether from Würstlin (1950).
T_g for polyvinyl ethyl ether from Wetton (1962).

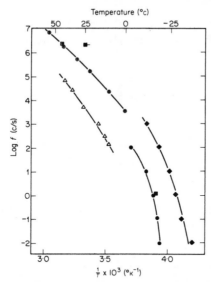

Figure 9.17. Plots of $\log f$ against $1/T$ for the dielectric (open points) and mechanical (filled points) loss maxima for the relaxation of polyvinyl ethyl ether. In the following key (f) refers to the frequency of maximum loss in constant temperature experiments and (T) denotes the temperature of maximum loss in constant frequency experiments.

⊙ ε'' (f) (Wetton, 1964)
⊡ $\tan \delta_\varepsilon$ (T) (Thurn and Wolf, 1956)
● $\tan \delta_G$ (T) (Wetton, 1962)
◆ G'' (T) (Wetton, 1962)

■ $\tan \delta_G$ (T) (Schmieder and Wolf, 1953)
▣ ultrasonic attenuation maximum (T) (Thurn and Wolf, 1956)

Also shown are the locations of the ε'' maxima (f) for polyvinyl isobutyl ether (△) (Ishida, 1960a).

Würstlin has computed dipole moments of 1·2 and 1·7 D per repeat unit for polyvinyl methyl ether and polyvinyl acetate respectively. Hence, the low T_g for polyvinyl methyl ether ($< -20°$c) compared with that for polyvinyl acetate (28°c) may be related to comparatively low interchain cohesive (dipole–dipole) forces. However, differences in intrachain flexibility might also be involved.

9.4b Secondary Relaxations

As illustrated in Table 9.2, the low-frequency mechanical results of Schmieder and Wolf (1953) and the dielectric results of Thurn and Wolf (1956) at 2×10^6 c/s indicate the existence of a number of secondary loss

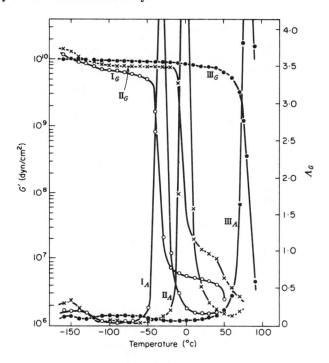

Figure 9.18. Temperature dependence of the logarithmic decrement and shear modulus for three polyvinyl butyl ethers at about 1 c/s. I (○) n-butyl ether; II (×) isobutyl ether; III (●) t-butyl ether (Schmieder and Wolf, 1953).

peaks for the polyvinyl ethers. Several of the observed loss peaks are of very small magnitude.

Recently, Saba, Sauer and Woodward (1963) have reported some low-frequency mechanical data for polyvinyl ethyl ether (Figure 9.19). Although there is no evidence of the peaks at -70 and $-110°C$ reported by Schmieder and Wolf for this polymer, a secondary loss peak clearly exists at about $-173°C$ ($100°K$) at 0.90 c/s. Similarly, the poly-n-propyl, -n-butyl and -isobutyl vinyl ethers almost certainly exhibit mechanical peaks in the region of $-150°C$ at 5 to 10 c/s (see Figure 9.18 and Table 9.2). The low-temperature loss maxima may be associated with local motions possibly involving the side-chains, and may correspond to peaks observed

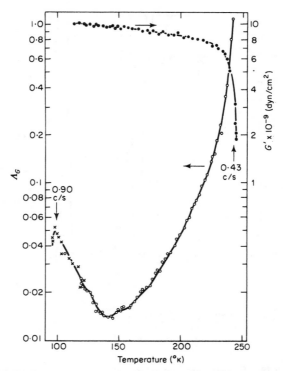

Figure 9.19. Temperature dependence of the shear modulus and logarithmic decrement at about 1 c/s for polyvinyl ethyl ether (Saba, Sauer and Woodward, 1963).

for poly-1-pentene

$$(-CH_2-CH-)_n$$
$$\quad\quad\quad |$$
$$\quad\quad CH_2-CH_2-CH_3$$

and poly-1-butene

$$(-CH_2-CH-)_n$$
$$\quad\quad\quad |$$
$$\quad\quad CH_2-CH_3$$

at about $-123°C$ (150°K) ($f \approx 10^3$ c/s) (Woodward, Sauer and Wall, 1961). They might also be related to the γ peaks observed for polyethylene and the n-propyl and n-butyl methacrylate and chloroacrylate polymers at about $-120°C$ ($10^2 \rightarrow 10^3$ c/s) (see Chapters 8 and 10).

9.5 POLYVINYL ALCOHOL

$$[-\overset{\overset{\displaystyle H}{|}}{\underset{\underset{\displaystyle H}{|}}{C}}-\overset{\overset{\displaystyle H}{|}}{\underset{\underset{\displaystyle O-H}{|}}{C}}-]_n$$

(3)

Polyvinyl alcohol, PVOH, (3) is normally prepared by the hydrolysis of polyvinyl acetate which itself is usually obtained by free-radical polymerization above room temperature. Samples of PVOH obtained by this method may typically have a degree of crystallinity of about 50%. This high crystallinity was earlier thought to indicate a high degree of isotactic stereoregularity. However, Bunn (1948) concluded from an x-ray analysis that the molecules of PVOH (prepared by the above method) are essentially atactic, and a similar conclusion is obtained from the high-resolution n.m.r. results of Satoh and others (1962). Thus, PVOH provides an exception to the apparently general idea that atactic polymers will not crystallize. Bunn (1959) attributed the high crystallinity of atactic PVOH to the fact that the hydrogen atoms and —OH groups are not so different in size as to seriously hinder the close regular packing of chains. He also suggested that strong hydrogen bonds between —OH groups on neighbouring chains would tend to hold the molecules in a regular structure in spite of the lack of stereoregularity.

PVOH has also been synthesized by methods thought to yield the polymer in both syndiotactic (Haas and others, 1956; Fordham and others, 1959; Cooper and others, 1963) and also isotactic (Murahashi and others, 1962) forms. The isotactic polymer is more soluble in water and is usually less crystalline than the syndiotactic polymer. From infrared studies (Murahashi and others, 1962) these differences would seem to result from the fact that the hydrogen bonds are predominantly *intra*molecular in the isotactic polymer and are largely *inter*molecular for the syndiotactic polymer. Before discussing the effect of stereoregularity on the relaxation behaviour of PVOH (Section 9.5c) it is convenient to consider the earlier studies on the atactic polymer.

Owing to the presence of the hydrophilic —OH substituents, PVOH is very hygroscopic and great care is necessary to ensure the absence of humidity during both dielectric and mechanical measurements.

9.5a Mechanical Studies

Investigations of the mechanical properties of PVOH include those of Fujino and others (1956, 1962), Yamamura and Kuramoto (1959),

Laible and Morgan (1961), Onogi and others (1962) and Takayanagi (1963). Takayanagi's dynamic data, shown in Figures 9.20, 9.21 and 9.22 provide a good overall illustration of the mechanical behaviour. The samples investigated by Takayanagi had a degree of polymerization of 1700 and contained 0·044 mole per cent residual acetyl group. Specimens were prepared by solvent evaporation from water solution at 90°C and the degree of crystallinity was varied by heat treatment as shown in Table 9.3. The per cent crystallinity was calculated using density values of 1·34 g/cm³ and 1·27 g/cm³ for the crystalline and amorphous components respectively. Figure 9.20 shows the temperature dependence of tan δ_E at 138 c/s for each sample listed in Table 9.3, and for a sample containing 30% water content. Corresponding E' and E'' versus temperature plots are shown in Figure 9.21. The data for the 46% crystalline sample illustrate most clearly the existence of five absorption regions centred at about 180, 135, 80, 35 and −60°C respectively. These regions may be labelled, respectively, α, β, γ, δ and ε.

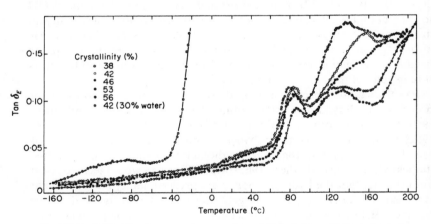

Figure 9.20. Tan δ_E versus temperature at 138 c/s for polyvinyl alcohol of crystallinities indicated (Takayanagi, 1963).

Considering first the temperature region above 100°C we note that the specimen of 38% crystallinity exhibits a single tan δ_E peak at about 140°C. As the degree of crystallinity increases the temperature of this loss peak increases, approaching about 200°C at 56% crystallinity. The shift of this (α) peak to higher temperatures reveals the existence of a second (β)

Table 9.3. Details of PVOH samples used for mechanical measurements (Takayanagi, 1963)

Sample	Temperature of heat treatment (°C)	Time of heat treatment (hr.)	Density (g/cm³)	Crystallinity (%)
1	90	9	1·2957	38
2	120	6	1·2989	42
3	150	1	1·3009	46
4	180	0·25	1·3061	53
5	210	5 min	1·3078	56

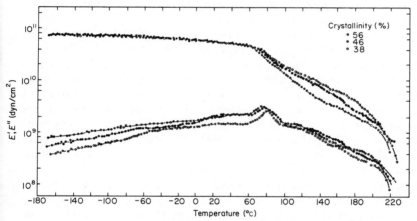

Figure 9.21. E' and E'' versus temperature at 138 c/s for PVOH of crystallinities indicated (Takayanagi, 1963).

peak which, for the samples of 46, 53 and 56 % crystallinity, is observed at 135°C. As suggested by Takayanagi, the β peak is probably submerged below the α peak for the samples of 38 and 42 % crystallinity. Both the α and β peaks, which are also apparent in the plots of E'' against temperature (Figure 9.21), are attributed by Takayanagi to motions within the *crystalline* phase of PVOH. More specifically, the β peak is ascribed to anharmonic twisting vibrations *around* the chain axis, a result reported to be consistent with the infrared observations of Nagai (1955). The α peak, on the other hand, is assigned to translational modes of chain segments *along the chain axis.* The latter process was thought to be associated with an increase in the long period, as observed by low-angle x-ray diffraction (Kakudo and others, 1955), during the annealing of PVOH in the temperature region of the α peak. It is of interest to note that the α and β mechanisms in PVOH are similar to the α and α' relaxations for polyethylene (Chapter 10). In the latter case, motions along the chain axes (the α process) are thought to be related to the increase in thickness of the crystalline lamellae as the annealing temperature is increased.

The sharp tan δ peak centred around 80°C increases in height with decreasing crystallinity and is accompanied by a sharp fall in modulus with increasing temperature. Since T_g for PVOH also lies in this temperature region, this γ peak evidently corresponds to the glass transition. It may thus be assigned to micro-Brownian motions of long chain segments in the *amorphous* regions of the polymer.

According to Takayanagi the broad (δ) peak in the region of 35°C results from local twisting motions about main-chain bonds. This local mode mechanism is discussed in Section 5.5b. The magnitude of this relaxation also decreases with increasing crystallinity. Thus, Takayanagi suggests that the local mode relaxation may occur both in the amorphous interlamellar regions and at the defects within the lamellae. Takayanagi also noted the close similarity between the shapes of the E'' versus temperature curves for PVOH and isotactic polypropylene. For both polymers the local mode relaxation was observed by a broad low-temperature shoulder to the primary amorphous peak. However, at about 100 c/s the primary and secondary peaks each occurred at considerably higher temperatures for PVOH than for isotactic polypropylene. Takayanagi has ascribed this difference to the fact that in PVOH the —OH groups form intermolecular hydrogen bonds which increase the resistance to both large-scale and local twisting vibrations of the main polymer chain.

The very weak and broad absorption in the region of $-60°$C appears to result from the presence of small amounts of absorbed moisture. Evidence for this suggestion was obtained from a study of the effects of added water content, the results of which are shown in Figures 9.20 and 9.22. The plasticizing effect of the added water is seen to shift the primary amorphous peak to lower temperatures. Hence the secondary peak, which occurs at 35°C for the 'dry' sample, is not observed for the wet specimens, being submerged beneath the primary peak. An additional peak also appears for the wet specimens and at 30% water content this peak is located at about $-90°$C. With decreasing water content this peak

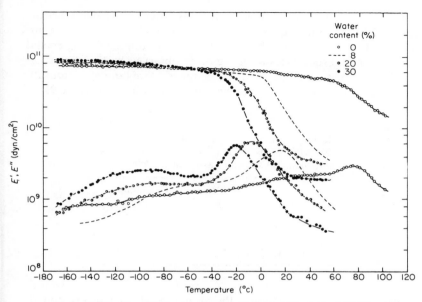

Figure 9.22. E' and E'' versus temperature at 138 c/s for PVOH with water contents indicated (Takayanagi, 1963).

decreases in magnitude and shifts to higher temperatures. It seems likely, therefore, that the peak at $-60°C$ for the 'dry' sample arises from the presence of a small quantity of absorbed moisture. Takayanagi proposed that the "water peak" is due to local twisting motions of those chain segments around which the intermolecular hydrogen bonds have been broken by the absorbed water molecules. Breaking of these hydrogen bonds would be expected to yield a loss peak at temperatures below 35°C, since the resistance to the local mode relaxation process should decrease.

9.5b Dielectric Studies

The dielectric properties of PVOH have been studied by Holzmüller (1940), Kurosaki and Furumaya (1960) and Ishida, Takada and Takayanagi (1960). Ishida and coworkers investigated PVOH having a degree of polymerization of 1460. Films of the polymer were prepared from water solution, and the density (or degree of crystallinity) was varied by heat treatment as shown in Table 9.4. The samples were dried over phosphorous pentoxide at 50°C and 10^{-3} mm Hg pressure for two weeks, and during the dielectric measurements care was taken to exclude humidity. Figure 9.23 shows the frequency dependence of ε' and ε'' at temperatures between 31 and $-36°C$ for the (unheat-treated) sample of 37% crystallinity. A single broad relaxation region is observed within the experimental frequency range at each temperature. Similar data were also obtained for the two samples of higher crystallinity listed in Table 9.4.

Table 9.4. Density and per cent crystallinity of PVOH samples used for dielectric measurements (Ishida, Takada and Takayanagi, 1960)

Sample	Density (g/cm^3)	Crystallinity (%)
Unheat-treated	1·296	37
Heat-treated at 140°C	1·301	49
Heat-treated at 180°C	1·312	58

From Cole–Cole arc diagrams the authors evaluated $\varepsilon_R - \varepsilon_U$ and the distribution parameter $\bar{\beta}$ as a function of temperature. As illustrated in Figures 9.24 and 9.25, both $\varepsilon_R - \varepsilon_U$ and $\bar{\beta}$ increase with temperature for each specimen investigated. Although the distribution parameter is little affected by the variations in crystallinity (Figure 9.25), $\varepsilon_R - \varepsilon_U$ decreases appreciably with increasing crystallinity (Figure 9.24) and, as shown in Figure 9.26, extrapolates to zero at 100% crystallinity. The latter result shows that the dielectric relaxation process is associated with the motions of dipoles in the *amorphous* regions of the polymer. From the plots of $\log f_{max}$ against $1/T$, shown in Figure 9.27, activation energies of 13·8, 13·2 and 12·5 kcal/mole were derived for the processes in the samples of 37, 49 and 58% crystallinity respectively. The activation energy is therefore little affected by the changes in degree of crystallinity. At any given temperature, however, the frequency of maximum loss increases slightly with increasing crystallinity. The latter result is somewhat surprising since motions within the disordered regions of a polymer are often found to be more restricted (i.e. f_{max} decreases) by an increase in the crystalline fraction (e.g. polyethylene terephthalate, Chapter 13).

The dielectric dispersion observed by Ishida and coworkers is probably *not* associated with the primary amorphous relaxation in PVOH, since the temperatures of measurement were below the glass-transition temperature. The authors also commented that the primary peak would be masked

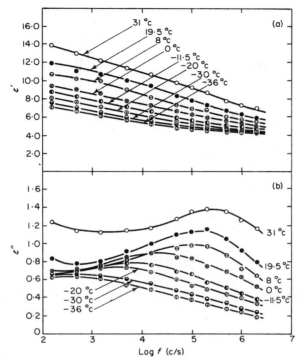

Figure 9.23. Frequency dependence of (a) ε' and (b) ε'' at various temperatures for unheat-treated PVOH (Ishida and others, 1960).

at very low frequencies by the low-frequency conduction. They thus attributed their data to the secondary amorphous relaxation, which presumably corresponds to the 'local mode' relaxation observed mechanically by Takayanagi (1963) at about 35°C (138 c/s) (Figure 9.20). An inspection of Figure 9.27, however, reveals a rather poor correlation between the locations of this mechanical peak and the dielectric loss maxima. The location of the dielectric absorption seems to correlate somewhat better with the small mechanical peak at about -60°C (Figure 9.22) which Takayanagi assigns to the presence of very small quantities of absorbed moisture.

Kurosaki and Furumaya (1960) have studied the dielectric behaviour of PVOH samples having degrees of polymerization of 2670 and 775 respectively. They have also investigated a specimen of deuterated PVOH.

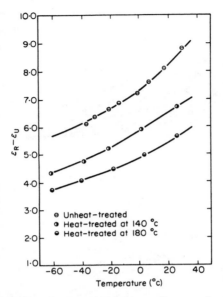

Figure 9.24. Temperature dependence of $\varepsilon_R - \varepsilon_U$ for PVOH (Ishida and others, 1960).

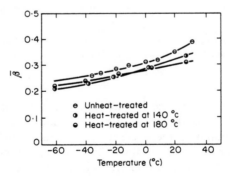

Figure 9.25. Temperature dependence of the Cole–Cole distribution parameter, $\bar{\beta}$, for PVOH (Ishida and others, 1960).

Their results were similar to those obtained by Ishida and coworkers. Kurosaki and Furumaya noted, in particular, that the activation energy for the dielectric relaxation (about 10 kcal/mole) was small despite the relatively long relaxation times (about 10^{-6} sec at room temperature).

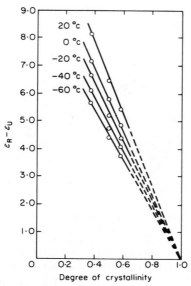

Figure 9.26. Dependence of $\varepsilon_R - \varepsilon_U$ on the degree of crystallinity for PVOH at various temperatures (Ishida and others, 1960).

In order to account for this result, and also for the high values of ε' and ε'', the authors suggested that the usual concept of the simultaneous hindered rotation of individual dipoles was not applicable. Instead they interpreted their data in terms of a chain mechanism, proposed by Sack (1952), involving successive dipole rotations within linear hydrogen-bond chains. In the case of PVOH it was assumed that the hydrogen-bond chains are formed by the linking together of —OH groups by either intra- or intermolecular hydrogen bonds. These hydrogen-bond chains are assumed to contain 'orientational fault sites' such as two protons intervening between two oxygen atoms (A),

$$\cdots\text{O—H}\cdots\text{O—H} \qquad \text{H—O}\cdots\text{H—O}\cdots \tag{A}$$

or no proton between two oxygen atoms (B).

$$\text{H—O}\cdots\text{H—O} \qquad \text{O—H}\cdots\text{O—H} \tag{B}$$

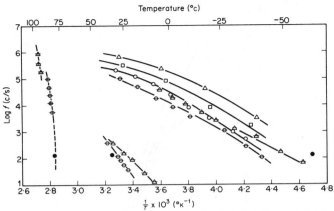

Figure 9.27. Plot of log f against $1/T$ for the dielectric (open points) and mechanical (filled points) *amorphous* peaks in polyvinyl alcohol. In the following key (f) refers to the frequency of maximum loss in constant temperature experiments and (T) denotes the temperature of maximum loss in experiments at constant frequency.

Atactic PVOH

● tan δ_E (T) (Takayanagi, 1963)

○ ε'' (f) for unheat-treated sample (Ishida and others, 1960)

□ ε'' (f) for sample heat treated at 140° (Ishida and others, 1960)

△ ε'' (f) for sample heat treated at 180° (Ishida and others, 1960)

Syndiotactic PVOH

⊿ ε'' (f) (Kajiyama and others, 1965)

⫛ ε'' (T) for heat-treated sample (Kajiyama and others, 1965)

Isotactic PVOH

⊖ ε'' (f) (Kajiyama and others, 1965)

⊕ ε'' (T) for heat-treated sample (Kajiyama and others, 1965)

The model adopted by Sack (1952) consists of N straight chains oriented in random directions in the material. Each chain comprises n elementary dipoles each having a dipole moment μ. The chains can each take up $n+1$ states depending on the position of the fault site, and in the absence of an electric field all states of a chain are assumed to be occupied with an equal probability ρ_0. It is further assumed that the energy barrier for rotation of an —OH dipole (about the C—O bond) is smaller at the fault site than at other sites along the chain. Hence, if a constant field acts along the chain axis the dipole at the fault site will turn. This elementary process will result in a transfer of the fault to an adjacent site followed by successive dipole rotations. Sack's theory yields the following expression for the complex dielectric constant in the frequency range around the maximum in ε'',

$$\varepsilon^* - \varepsilon_U = \frac{128}{3\pi^3} \frac{\mu^2 N \rho_0}{kT} \frac{(n+1)^2}{1 + i\omega\tau_n} \tag{9.1}$$

in which τ_n, the effective relaxation time, is related to τ_0, the relaxation time of an elementary transition, by the following equation,

$$\tau_n = \frac{2(n+1)^2}{\pi^2} \tau_0 \tag{9.2}$$

Hence, on account of the chain mechanism, τ_n is related to the length of the chain through the factor $(n+1)^2$. The activation energy, on the other hand, may be determined largely by the energy barrier for the elementary transition. If the hydrogen-bond chains are not of uniform length then Equation (9.1) can be generalized to include a distribution of relaxation times, rather than a unique value of τ_n.

Some limitations of Sack's theory deserve comment. Firstly, Equation (9.1) was derived on the assumption that the internal field acting on a dipole is about equal to the applied field. This assumption requires that $|\varepsilon - \varepsilon_U| \ll \varepsilon_U$. The latter condition is not fulfilled in the case of PVOH. However, Sack considers that even if the condition does not apply, deductions from Equation (9.1) will apply qualitatively. Secondly, only one type of fault site (either A or B above) must be present for the theory to hold generally. If, for example, there is a small but finite probability of sites (A) existing in addition to sites (B) then the theory applies for moderately long chains, but with increasing chain length τ_n will be independent of chain length. Thirdly, if the chains are not straight but coiled ε', ε'' and τ_n should each be independent of chain length. In fact Kurosaki and Furumaya suggest that in PVOH it is more natural to assume the hydrogen-bond chains to be coiled to some extent than to assume them to be linear. They propose, therefore, that ε', ε'' and τ_n should be proportional not to $(n+1)^2$ but to $(n+1)^\beta$ where $0 \leqslant \beta \leqslant 2$.

Like Ishida and coworkers, Kurosaki and Furumaya attribute the decrease in height of the ε'' peaks (or decrease in relaxation strength) following heat treatment to the fact that the relaxation process which they observe occurs in the amorphous regions of PVOH. The broad shape of the dispersion was ascribed to a wide range of hydrogen-bond chain lengths, resulting in a distribution of n or τ_n values. From the measured (average) values of τ_n, and in conjunction with τ_0 values calculated from dielectric data for (non-hydrogen-bonded) 2,4,6-tri-t-butylphenol, they estimated (Equation 9.2) a value of 1350 Å for the mean length of hydrogen-bond chains in the amorphous regions of PVOH. From the relative decrease of τ_n and τ_0 with increasing temperature, determined from the shifts of the appropriate ε'' peaks to higher frequencies, it was concluded from Equation (9.2) that n must decrease as the temperature increases. This effect could result from the increased probability of breaking hydrogen-bond chains as their kinetic energy is increased. However, according to Equation (9.1), a decrease in n is not consistent with the observed increase in $\varepsilon_R - \varepsilon_U$ and ε'' with increasing temperature. Kurosaki and Furumaya attempt to explain this inconsistency by suggesting that as the temperature is raised free —OH groups, which are not initially connected to hydrogen-bond chains, are able to form new chains owing to an increased probability of collision. Hence the increase in N, the number of chains per unit volume (Equation 9.1), may exceed the decrement in $(n+1)^\beta$. Their observation that the frequency of the ε'' maxima, and hence the average relaxation times, were independent of the degree of polymerization suggested to the authors that the hydrogen-bonds were intermolecular rather than intramolecular. This conclusion seems reasonable for an atactic chain in which the —OH groups within the chain are well separated. The fact that the ε'' peaks were located at similar frequencies for PVOH and deuterated PVOH suggested further that deuteration has little effect on the mean length of the 'hydrogen'-bond chains. The observation of Ishida and coworkers that the frequency of maximum ε'' increases somewhat after heat treatment might be interpreted as indicating a decrease in the length of hydrogen-bond chains in the amorphous regions of PVOH resulting from a reduction in the amorphous content. Unfortunately the above study of PVOH does not show conclusively whether or not the proposed structure and chain mechanism of relaxation are valid. However, the suggested mechanism is an interesting alternative to the local mode relaxation mechanism proposed by Takayanagi (1963). It also indicates that somewhat unusual types of relaxation processes may occur in hydrogen-bonded polymers such as PVOH.

9.5c Effect of Stereoregularity

Nagai and Takayanagi (1964) have determined the temperature dependence of tan δ_E, E' and E'' at 138 c/s for both syndiotactic and isotactic PVOH. Their data for the syndiotactic polymer were similar to those for the atactic polymer discussed in Section 9.5a. For the isotactic polymer the two high-temperature crystalline peaks were not observed, a result attributed to the low degree of crystallinity. However, a broad peak located in the 60 to 150°c temperature region, and assigned to the primary amorphous relaxation, could partially involve the crystalline phase. Evidence in favour of the latter suggestion comes from the observation that this peak shifted considerably to higher temperatures as the annealing temperature was increased. The height of the tan δ_E peak also increased with increasing annealing temperature. Nagai and Takayanagi also observed the secondary amorphous relaxation for isotactic PVOH at about room temperature.

Dielectric studies of both isotactic and syndiotactic PVOH have been reported by Kajiyama and others (1965). By extending their measurements to higher temperatures ($\approx 110°$c) these workers were also able to study the primary amorphous relaxation in addition to the secondary region studied earlier (Section 9.5b). Figure 9.28 illustrates their data in the region of the primary amorphous peak. For both the isotactic and syndiotactic specimens the ε'' peak decreases in height after heat treatment, a result responsible for the amorphous assignment. Also, the loss peaks are situated at somewhat higher temperatures for the syndiotactic than for the isotactic polymer, both for the unheat-treated and heat-treated samples. This result was ascribed to the fact that the syndiotactic configuration gives rise to a relatively high proportion of *inter*molecular hydrogen bonds, the latter causing relatively large restrictions to main-chain motions. Figure 9.27 shows that the dielectric results for the heat-treated samples correlate well with Takayanagi's mechanical data in the primary amorphous region.

In contrast to the earlier dielectric investigations of atactic PVOH (Figure 9.23), Kajiyama and coworkers observed *two* secondary peaks in plots of ε'' versus log f for both isotactic and syndiotactic PVOH. This result is illustrated in Figure 9.29 for the isotactic polymer. The location of the high-frequency component appears to correlate with that for the atactic polymer (Figures 9.23 and 9.27) and the low-frequency component seems to correlate with the secondary mechanical peak at 35°c (138 c/s) (see Figure 9.27). Each of the secondary relaxations was ascribed to local twisting motions of the main chain. For the isotactic polymer the authors suggested that the high-frequency loss component involved chain segments *intra*molecularly hydrogen bonded and that the low-frequency component involved chain segments which were linked by *inter*molecular hydrogen bonds. This interpretation was based on the observation that the high-frequency component was larger than the low-frequency component (Figure 9.29). As mentioned above (Section 9.5) infrared studies of isotactic PVOH suggest that it has a relatively high proportion of *intra*molecular hydrogen bonds.

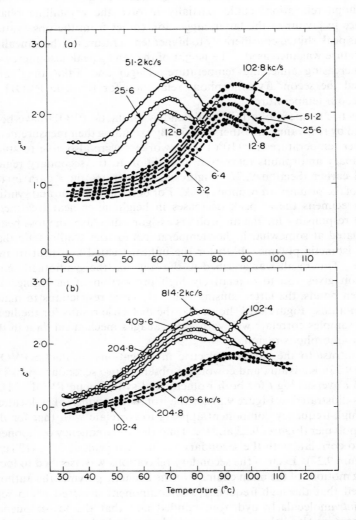

Figure 9.28. Plots of ε'' against temperature at the frequencies indicated for (a) isotactic and (b) syndiotactic PVOH. ○ unheat-treated samples; ● heat-treated samples (Kajiyama and others, 1965).

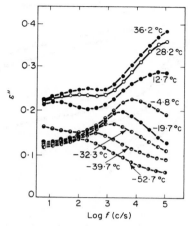

Figure 9.29. Frequency dependence of ε'' for isotactic PVOH at temperatures in the secondary amorphous relaxation regions (Kajiyama and others, 1965).

9.6 VINYL ALCOHOL–VINYL ACETATE COPOLYMERS

Fujino and others (1962) have studied the tensile stress relaxation behaviour of copolymers of vinyl alcohol and vinyl acetate. These polymers were prepared by the partial acetylation of PVOH. The samples investigated were labelled PVAc-3, PVAc-15 and PVAc-40, the numbers denoting the approximate mole per cent acetylation. Figure 9.30 shows the master stress relaxation curves reduced to 75°C, and Figure 9.31 gives the corresponding relaxation spectra obtained from the stress relaxation curves. It will be observed that the curves shift to longer times, or relaxation times, with increasing alcohol content. This result indicates that the main chains become less mobile with increasing vinyl alcohol content, a conclusion consistent with the fact that T_g for PVOH (about 80°C) is higher than that for PVAc (about 30°C). This effect may be due partly to an increase in interchain cohesion resulting from the existence of hydrogen bonds between vinyl alcohol units. For PVOH and PVAc-3, the only samples in which crystallinity could be detected by x-rays, the relaxation curves are relatively flat and broad, typical of the behaviour of partially crystalline polymers. Fujino and others suggested also that some local regions of crystallinity might exist in the PVAc-15 and PVAc-40 samples since the temperature dependence of the shift factors (a_T) derived from the master curve constructions did not conform to the WLF equation (5.57).

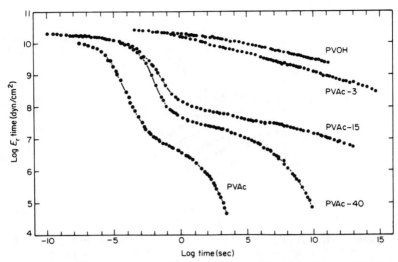

Figure 9.30. Stress relaxation master curves reduced to 75°C for PVOH, PVAc-3, PVAc-15, PVAc-40 and PVAc. Numbers refer to the mole per cent acetylation of the PVOH (Fujino and others, 1962).

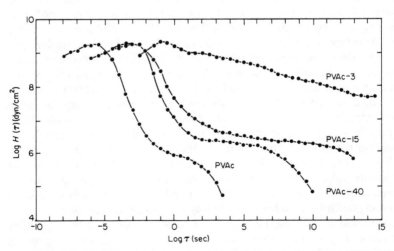

Figure 9.31. Relaxation spectra (reduced to 75°C) obtained from the master stress relaxation curves of Figure 9.30 (Fujino and others, 1962).

9.7 POLYVINYL ACETALS

$$[-\overset{\overset{\displaystyle H}{|}}{\underset{\underset{\displaystyle H}{|}}{C}}-\overset{\overset{\displaystyle H}{|}}{\underset{\underset{\displaystyle O}{|}}{C}}-\overset{\overset{\displaystyle H}{|}}{\underset{\underset{\displaystyle H}{|}}{C}}-\overset{\overset{\displaystyle H}{|}}{\underset{\underset{\displaystyle O}{|}}{C}}-]_n$$

$$\underset{\underset{\displaystyle R}{|}}{\overset{\diagdown}{\underset{\diagup}{CH}}}$$

(4)

Polyvinyl acetals have so far been obtained and studied in the amorphous state. Their repeat units (4) contain dipolar side-groups which are bridged across alternate chain carbon atoms. The permanent dipoles are thus unable to reorient independently of the main chain. Dielectric relaxation in these polymers must therefore involve simultaneous main-chain and side-chain motions, unlike most other amorphous polymers whose side-chain dipoles can rotate independently of the chain backbone.

The synthesis of the polyvinyl acetals involves, firstly, the preparation of polyvinyl alcohol (PVOH) by, for example, the hydrolysis of polyvinyl acetate. Subsequently, or concurrently, the PVOH is condensed with the appropriate aldehyde (R—CHO) in the presence of a strong acid. Owing to the fact that these reactions may not go to completion, poly-vinyl acetal samples may typically contain about 1 % and 10 %, respec-tively, of unreacted polyvinyl acetate and PVOH repeat units. In this sense, the polyvinyl acetals should strictly be regarded as random vinyl acetal–alcohol copolymers containing a relatively small proportion of PVOH units.

9.7a The α Relaxation

Fitzhugh and Crozier (1952) have made an extensive investigation of the physical properties of several polyvinyl acetals in relation to the structure of the side-chain —R group and the incompleteness of acetalization (i.e. the PVOH content). In particular, they determined the temperature dependence of the torsion modulus in the vicinity of the α relaxation or glass-transition temperature. From these measurements they evaluated the inflection temperature, T_i, at which an inflection point was observed in the log modulus–temperature plot, and the softening temperature, T_s, defined as the temperature at which the torsion modulus has the value 10,000 lb/sq in. As shown in Table 9.5, the values of T_i and T_s always agreed within 1°C. Both T_i and T_s may, in fact be regarded as estimates

Table 9.5. The inflection temperatures (T_i) and softening temperatures (T_s) of several polyvinyl acetals (Fitzhugh and Crozier, 1952; Fujino and others, 1962)

Polyvinyl acetal	Structure of –R group	PVOH (%)	T_i(°C)	T_s(°C)
Formal	—H	90·2	80	
		77·5	90	
		45·6	102	
		24·5	106	
Acetal	—CH$_3$	11·9		97
		9·1	81	82
Propional	—CH$_2$—CH$_3$	19·7		73
		16·7		71
		13·7	72	72
		10·3	72	72
n-Butyral	—(CH$_2$)$_2$CH$_3$	25·6		64
		18·0	61	62
		16·0		61
	CH$_3$	12·1	48	49
	$\mid$			
Isobutyral	—CH	25·6		75
	$\mid$	20·1		73
	CH$_3$	13·3		56
2-Ethyl butyral	—CH—CH$_2$—CH$_3$	21·8	55	55
	$\mid$	17·7		58
	CH$_2$—CH$_3$	15·5	56	57
n-Hexanal	—(CH$_2$)$_4$CH$_3$	24·6		48
		19·6	45	46
		16·0	44	44
		12·2	39	39
n-Heptanal	—(CH$_2$)$_5$CH$_3$	20·2		43
		19·5	41	42
		15·4	39	39
		13·0	29	29
2-Ethyl hexanal	—CH—(CH$_2$)$_3$—CH$_3$	19·7		43
	$\mid$	14·3	41	41
	CH$_2$—CH$_3$	12·8	35	35
		6·5	17	17

either of T_g or of the α relaxation temperature at low frequencies. It is probable that T_i and T_s each lie some 10 to 15°C above the dilatometric value of T_g. Although Fitzhugh and Crozier's investigation did not include measurements on polyvinyl formal, some stress relaxation and low-frequency dynamic data have been reported for partly to highly formalized polyvinyl alcohol by Fujino and others (1962). The temperatures of the primary dynamic modulus dispersions at 3×10^{-3} c/s for these polymers are included in Table 9.5 under the T_i column.

Summarizing the results shown in Table 9.5 we first note that for all of the polymers except polyvinyl formal T_i and T_s tend to increase with increasing PVOH content, whereas for polyvinyl formal the reverse effect is observed. From the relationship between T_g and composition for random copolymers, this result indicates that T_g for PVOH is lower than that for pure polyvinyl formal but somewhat higher than that for pure polyvinyl acetal and the higher polyvinyl acetals. Secondly, both T_i and T_s decrease with an increase in length of the side-chain —R group, and again this internal plasticization is larger for the linear n-alkyl than for the branched alkyl side-groups.

Takahashi (1961a) has also observed the mechanical α relaxation in polyvinyl butyral (36·6% vinyl alcohol content) using the vibrating cantilever experiment at a frequency of about 50 c/s. In Figure 9.32 the α relaxation is evident by a large peak in tan δ_E at about 75°C. The small peak at about -60°C arises from the β relaxation to be discussed in the following section.

Dielectric investigations of the α relaxation in the polyvinyl acetals include those of Funt (1952), Funt and Sutherland (1952), Sutherland and Funt (1953), Erlikh and Shcherbak (1955), Mikhailov (1955), Veselovskii (1956) and Takahashi (1961a, b). Takahashi's dielectric results for polyvinyl butyral at 60 c/s (Figure 9.32) correlate well with his mechanical data obtained on the same sample. The effect of internal plasticization is also evident from Mikhailov's data shown in Figure 9.33 for polyvinyl acetal, polyvinyl propional, polyvinyl butyral and polyvinyl octanal.

The dielectric measurements of Funt and Sutherland were made principally in the range 25°C to 135°C at frequencies from 50 to 100 kc/s. The polymers investigated were polyvinyl formal, polyvinyl acetal, polyvinyl butyral, polyvinyl isobutyral, polyvinyl hexanal, and polyvinyl-2-ethyl hexanal. Most of the samples investigated had approximately 12 to 13% PVOH content. For each polymer a single peak was observed in plots of ε'' against temperature, although at high temperatures and low frequencies an upswing in ε'' was found and attributed to d.c. conductance. With increasing size of the side-chain substituent, the loss peaks were

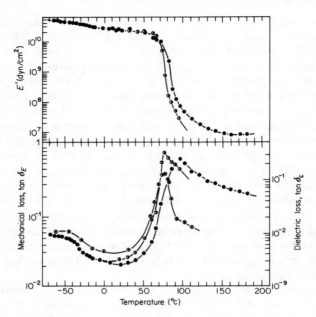

Figure 9.32. Temperature dependence of E' and the mechanical and dielectric tangents for polyvinyl butyral.

● Mechanical at 50 c/s, sample cured at 200°c for 30 minutes; ◑ Mechanical at 50 c/s, sample cured at 170°c for 30 minutes; ⊙ Dielectric at 60 c/s, sample cured at 200°c for 30 minutes (Takahashi, 1961a).

displaced to lower temperatures (at constant frequency) as shown by the plots of log f versus $1/T$ in Figure 9.34. This trend, attributed to internal plasticization, is in keeping with that observed by Mikhailov and by Fitzhugh and Crozier (see Table 9.5 for T_i and T_s). In fact, for polyvinyl isobutyral, polyvinyl hexanal and polyvinyl-2-ethyl hexanal, the temperature corresponding to a dielectric relaxation time of 1 sec, determined by extrapolation from a plot of log τ against $1/T$, was almost identical with the values of T_i and T_s for these polymers. Since *identical samples* were used both for the dielectric and mechanical tests for the three polymers, it was concluded that the dielectric and mechanical properties were in close correspondence. This correlation was ascribed to the fact that both the mechanical and dielectric relaxation in these polymers must result from motions *within* the chain backbone.

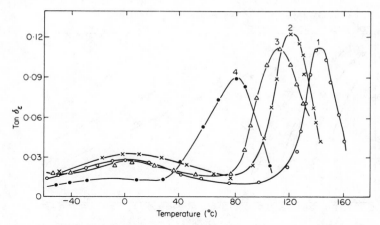

Figure 9.33. Temperature dependence of the dielectric loss tangent at 1000 c/s for: Curve 1 polyvinyl acetal; Curve 2 Polyvinyl propional; Curve 3 Polyvinyl butyral; Curve 4 Polyvinyl octanal (Mikhailov, 1955).

Figure 9.34 also summarizes the dielectric data obtained by Veselovskii (1956) for polyvinyl formal, polyvinyl acetal, polyvinyl propional, polyvinyl butyral and polyvinyl octanal in the α relaxation region. With the exception of the results for polyvinyl acetal a fairly good agreement is noted between the data of Veselovskii and Funt and Sutherland. The large discrepancy between the two sets of data for polyvinyl acetal is not understood. Veselovskii's data for this polymer seem to be anomalous when viewed in terms of the effect of internal plasticization.

As mentioned above, Fujino and others (1962) have studied the tensile stress relaxation behaviour of partly to highly formalized PVOH. Their master stress relaxation curves, reduced to 110°C, are shown in Figure 9.35 and the mechanical relaxation specta derived from these master curves are plotted in Figure 9.36. From an x-ray analysis, crystallinity was detected in the three samples (PVF-10, PVF-20 and PVF-50) containing more than 45·6% PVOH, whereas the sample containing 24·5% PVOH (PVF-80) was found to be amorphous. For the amorphous (copolymer) specimen the α relaxation is seen by the rapid fall in modulus at $\log t \approx 2\cdot5$ (Figure 9.35), and by the associated maximum in the relaxation spectrum at $\log \tau \approx 1$ (Figure 9.36). The fall-off in modulus at $\log t \approx 6$ to 10 and the corresponding peak in the relaxation spectrum at $\log \tau \approx 7$ probably result from the irreversible viscous flow process. The shift of the predominant α relaxation to shorter times (or relaxation times) as the PVOH

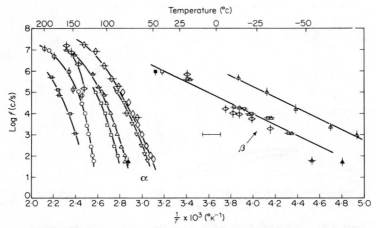

Figure 9.34. Plot of $\log f$ against $1/T$ for the dielectric (open points) and mechanical (filled points) loss maxima of the polyvinyl acetals. In the following key (f) refers to the frequency of maximum loss in constant temperature experiments and (T) denotes the temperature of maximum loss in experiments at constant frequency.

Polyvinyl formal

$\bigcirc$ $\varepsilon''(T)$ (Funt and Sutherland, 1952)
$\oplus$ tan δ_ε (f) (Veselovskii, 1956)

Polyvinyl acetal

$\square$ ε'' (T) (Funt and Sutherland, 1952)
$\boxminus$ tan δ_ε (f) (Veselovskii, 1956)
$\square$ tan δ_ε (T) (Kabin, 1960)
$\blacksquare$ tan δ_K (T) (Kabin, 1960)
$\longmapsto$ tan δ_ε (T) (Mikhailov, 1955)

Polyvinyl propional

$\oslash$ tan δ_ε (f) (Veselovskii, 1956)
$\longmapsto$ tan δ_ε (T) (Mikhailov, 1955)

Polyvinyl butyral

$\triangle$ ε'' (T) (Funt, 1952)
$\triangle$ tan δ_ε (f) (Veselovskii, 1956)

$\triangle$ ε'' (f) (Takahashi, 1961b)
$\triangle$ tan δ_ε (T) (Takahashi, 1961a)
$\blacktriangle$ tan δ_E (T) (Takahashi, 1961a)
$\longmapsto$ tan δ_ε (T) (Mikhailov, 1955)

Polyvinyl hexanal

∇ $\varepsilon''(T)$ (Sutherland and Funt, 1953)

Polyvinyl-2-ethyl hexanal

$\Diamond$ $\varepsilon''(T)$ (Sutherland and Funt, 1953)

Polyvinyl octanal

$\Diamond$ tan δ_ε (f) (Veselovskii, 1956)
$\longmapsto$ tan δ_ε (T) (Mikhailov, 1955)

content increases suggests that the chain backbone motions occur less rapidly in pure polyvinyl formal than in the *amorphous regions* of PVOH. This result is consistent with the suggestion from Table 9.5 that T_g is higher for pure polyvinyl formal than for PVOH. The lower mobility of the polyvinyl formal chains may result from a relatively rigid chain structure produced by the bridging of the side-chains across alternate chain carbon atoms.

Figures 9.35 and 9.36 also show that an increase in crystallinity (or PVOH content) leads to a broadening and flattening of the relaxation curves. Furthermore, the three partially crystalline samples show, in addition to the predominant relaxation, a small relaxation region at longer times, as evidenced by the small peak in the relaxation spectrum at $\log \tau \approx 4$. From dynamic mechanical experiments at 3×10^{-3} c/s this relaxation was also observed at about 120°C. The authors attributed this relaxation region to motions within the crystalline (PVOH) phase, but not to melting of these regions. This process probably corresponds to that responsible for the mechanical crystalline loss peak for pure PVOH observed by Takayanagi (1963) in the temperature range 120 to 180°C (see Section 9.5a). It should be added that the use of the shift procedure for the formation of the master relaxation curves in Figure 9.35 may not be strictly valid since the two overlapping relaxation processes may have different activation energies. This fact was apparent from the anomalous temperature dependence of the shift factors (a_T).

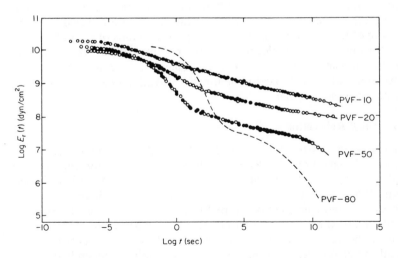

Figure 9.35. Stress relaxation master curves reduced to 110°C for partly to highly formalized polyvinyl alcohol. Per cent formalization as follows: PVF-10 (9·8 %), PVF-20 (22·5 %), PVF-50 (54·4 %) and PVF-80 (75·4 %) (Fujino and others, 1962). [Note PVF here should not be confused with polyvinyl fluoride, see 11.4.]

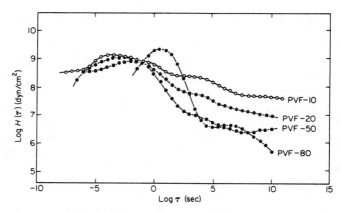

Figure 9.36. Mechanical relaxation spectra at 110°C determined from the master curves of Figures 9.35 (Fujino and others, 1962).

Yamamura and Kuramoto (1959) have studied the creep behaviour of PVOH and polyvinyl formals of varying degrees of formalization and at various temperatures and relative humidities. In the theoretical analysis of the creep curves it was assumed that the elongation resulted from the breaking of hydrogen bonds in the amorphous regions of the polymer by the diffusion of water molecules into the polymer films. Assuming also that the diffusion coefficient was time dependent, the authors deduced values of about 13 to 17 kcal/mole for the activation energy for diffusion of water into the polymer.

9.7b The β Relaxation

As already noted from Figures 9.32 and 9.33, the polyvinyl acetals also exhibit a β relaxation. Dielectric studies of this relaxation region have been made by Erlikh and Shcherbak (1955), Mikhailov (1955), Veselovskii (1956), Kabin (1960) and Takahashi (1961b). Figure 9.34 summarizes the frequency–temperature locations of the β loss maxima as determined by the different authors. As noted by Mikhailov (1955), these results indicate that the relaxation time and activation energy (10 → 12 kcal/mole) for the β mechanism are essentially independent of the length of the alkyl side-group. A similar result was obtained for the β process of the methacrylate polymers (see Section 8.3b).

Kabin (1960) has also detected the β relaxation in polyvinyl acetal from measurements of the velocity and attenuation of both transverse and longitudinal waves at 1 Mc/s. From these measurements he determined (Chapter 6) the components of G^* and M^* (the complex bulk longitudinal modulus) and evaluated also the components of the complex bulk modulus, K^*, where $M^* = K^* + 4G^*/3$. As illustrated in Figure 9.37 the β relaxation gives rise to a maximum in $\tan \delta_M (= M''/M')$ at about 50°C and in $\tan \delta_K (= K''/K')$ at about 45°C. However the *shear* loss tangent, $\tan \delta_G (= G''/G')$, does *not* show a peak in the temperature range 20°C to 80°C. A striking feature of these results (Figure 9.37c) is the close similarity between both the magnitude and location of the $\tan \delta_K$ and dielectric loss tangent peaks. This correlation would suggest that similar molecular mechanisms give rise to the dielectric and mechanical *volume* relaxations. Takahashi (1961a) has observed the mechanical β peak for polyvinyl butyral, containing 36·6% vinyl alcohol, from measurements of the temperature dependence of E' and $\tan \delta_E$ at 50 c/s. As illustrated in Figure 9.32 this result correlates well with dielectric data for the same sample.

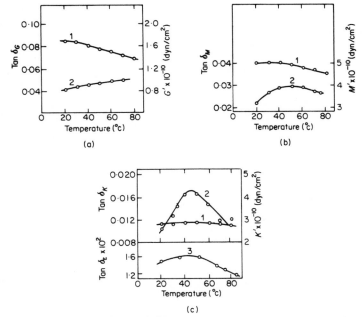

Figure 9.37. Temperature dependence of the mechanical and dielectric properties of polyvinyl acetal at 1 Mc/s. (a) (1) G' and (2) G''/G'; (b) (1) M' and (2) M''/M'; (c) (1) K', (2) K''/K' and (3) the dielectric loss tangent (Kabin, 1960).

The mechanism of the β relaxation in the polyvinyl acetals is unknown. It may possibly be related to the low-temperature mechanism in polyvinyl alcohol which occurs in approximately the same temperature–frequency region (compare Figures 9.27 and 9.34). Another possibility is that it involves local torsional oscillations of the main chain (i.e. the local mode mechanism, Section 5.5b), the side-chain dipoles being unable to rotate independently of the chain backbone.

10

Hydrocarbon Polymers

10.1 POLYETHYLENE AND COPOLYMERS

$$
\begin{array}{ccccccc}
\text{H} & \text{H} & \text{H} & \text{H} & \text{H} & \text{H} \\
| & | & | & | & | & | \\
-\text{C}-\text{C}-\text{C}-\text{C}-\text{C}-\text{C}- \\
| & | & | & | & | & | \\
\text{H} & \text{H} & \text{H} & \text{R} & \text{H} & \text{H}
\end{array}
$$

(1)

The polymerization of ethylene produces a molecule composed of long sequences of CH_2 groups with alkylidene side branches, R, of varied number and length. When polymerization is carried out at high pressure the molecule may possess as many as fifty or more side-branches, mainly ethyl and butyl, per 1000 carbon atoms (see Chapter 2). Heterogeneous catalysis, carried out at low pressure, using Zeigler or Phillips catalysts, yields a molecule with far fewer side-branches. Zeigler polyethylenes apparently contain only ethyl side-groups; approximately 5 per 1000 carbon atoms. No short branches have yet been positively detected in Phillips polyethylenes (Bunn, 1960). The polyethylene made at low pressure is generally more crystalline and therefore denser and stiffer than the polyethylene made at high pressure. The former is often known as low-pressure, high-density or linear polyethylene to distinguish it from the latter which is known as high-pressure, low-density, side-branched or free-radical polyethylene.

As indicated in Chapter 2, it is now quite clear that in solid polyethylene the crystals have the form of twisted lamellae. The breadth of the lamellae (ca. 1 micron) and the thickness of the lamellae (ca. 100 Å) both decrease with decreasing temperature of crystallization (Palmer and Cobbald, 1964; Keller and Sawada, 1964). A manifold increase in lamellar thickness may be produced by annealing at temperatures above 80°C (Fischer and Schmidt, 1962). Another effect of annealing is to produce a higher density. The way in which the lamellae thicken and the density increases is explained by the model of Fischer and Schmidt (1962), Figure 10.1.

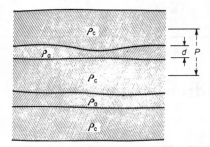

Figure 10.1. The model of Fischer and Schmidt for the structure of polyethylene.

If the periodicity is P and if the thickness of the amorphous regions between the crystalline lamellae is d, then clearly,

$$\rho P = \rho_c(P-d) + \rho_a d$$

ρ is the density of the solid and ρ_c and ρ_a the densities of the crystalline and amorphous fractions. This equation assumes that each fraction has a particular density which does not vary with P or d. The density of the solid is given by

$$\rho = \rho_c - \frac{d}{P}(\rho_c - \rho_a)$$

The effect of the annealing on the breadth of lamellae is not known. It is certain that neither the pitch of the lamellar helix nor the size of the spherulites (see Chapter 2) are changed by annealing below the melting point.

It is not at present clear to what extent the small side-branches can be accommodated within the crystal. Measurements of the unit cell dimensions at 25°c have shown that both a and b increase with increasing number of side-branches (Walter and Reding, 1956). For instance, a increases by 4·35% and b by 1·62% when the methyl content increases from 0 to 80 per 1000 carbon atoms. According to Bunn (1960) these results can mean either (1) methyl groups enter the lattice and so strain it internally or (2) methyl groups do not enter the lattice but strain it externally. The external strain is thought to arise because 'the very small crystalline regions composed of unbranched chain segments suffer unusually large strains originating at boundaries where branch points bring the regular structure to an end'. It is certain that bulky side-branches (e.g. acetate groups) cannot exist in the crystal and that the major effect of side-branches of any size is to increase the fraction of amorphous material.

The internal friction of LD polyethylene was studied first by Schmieder and Wolf (1953) at about 1 c/s. Later measurements in the same frequency range by Flocke (1962) on both LD and HD polyethylene are shown in Figure 10.2. The labelling scheme of Oakes and Robinson (1954) is used

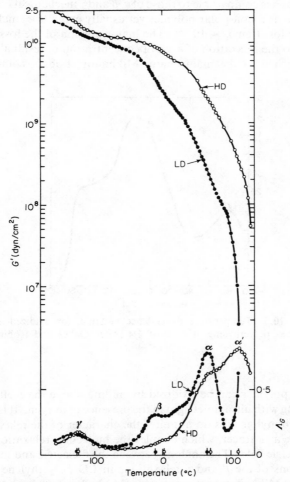

Figure 10.2. Temperature dependence of shear modulus and logarithmic decrement of high density (HD) and low density (LD) polyethylene at ca. 1 c/s (After Flocke, 1962).

in Figure 10.2 in preference to that of Flocke. Oakes and Robinson did not observe the relaxation at the highest temperature which we have labelled α'. The α' relaxation is observed only in mechanical relaxation at frequencies below those used by Oakes and Robinson.

Oakes and Robinson (1954), Figure 10.3, observed that oxidized LD polyethylene exhibits three dielectric relaxations in temperature and frequency ranges comparable to those of α, β and γ mechanical relaxations. Polyethylene is a non-polar polymer yet usually exhibits a small amount of dielectric loss (tan $\delta_\varepsilon \sim 10^{-4}$). The major portion of the loss has been attributed to the relaxation of a small concentration of residual carbonyl groups attached to the main chain (Mikhailov, Lobanov and Sazhin,

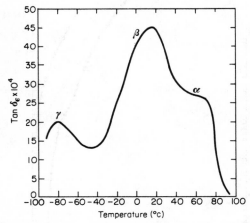

Figure 10.3. Temperature dependence of tan δ_ε for oxidized, side-branched polyethylene at $1 \cdot 1 \times 10^4$ c/s (After Oakes and Robinson, 1954).

1954). The polymer can be oxidized by milling above the melting point or irradiating with ultraviolet light in the presence of oxygen. It is assumed that the carbonyl groups do not alter the character of the relaxation but merely serve as a tracer, which couples the molecular relaxation process with the electric field. The analogy between the dielectric and mechanical manifestations of the β and γ relaxations in LD polyethylene was also observed by Mikhailov and coworkers (Mikhailov and Borisova, 1953; Mikhailov, Lobanov and Sazhin, 1954; Mikhailov, Kabin and Sazhin, 1955; Kabin, 1956; Mikhailov and Lobanov, 1958).

Values of the activation energies for the α and γ relaxations are given in Table 10.1 (Sandiford and Willbourn, 1960). The mechanical and electrical values are in good agreement, as would be expected if both types of relaxation had similar origins. Little at present is known of the activation energy of the α' relaxation. Sandiford and Willbourn (1960) give the activation energy of the β process determined both mechanically and dielectrically as 38 kcal/mole. This is not in agreement with Kabin (1956) who finds the value 16 kcal/mole to hold for both mechanical and dielectric relaxation. Other values observed for the β dielectric relaxation are 20 kcal/mole (Okamoto and Takeuchi, 1959) and 25 kcal/mole (Saito and Nakajima, 1959). The cause of this discrepancy is not known.

Table 10.1. Approximate activation energies (kcal/mole) (After Sandiford and Willbourn, 1960).

Process	Dielectric	Mechanical
α	28	25
γ	11	11–15

10.1a The α' Relaxation

The α' relaxation occurs at temperatures close to the melting point at 1 c/s. It cannot be observed, therefore, at high frequencies (above ca. 10^2 c/s) and is observed best in creep experiments. Nakayasu, Markovitz and Plazek (1961) were the first to study the α' relaxation in detail although its existence was noted by Schmieder and Wolf (1953).

Takayanagi (1963) has proposed that the α' relaxation is due to a translational motion of the molecule in the direction of the molecular axis. In the region above 80°C crystals of HD polyethylene recrystallize and thicken, a process which must presumably necessitate molecular motion parallel to the chain axis. This is the basis of Takayanagi's hypothesis. A more plausible hypothesis has been proposed by Iwayanagi (1962). According to Iwayanagi the α' relaxation is a boundary phenomenon analogous to grain boundary slip observed in metals. The Iwayanagi hypothesis has yet to be established.

The temperature of the α' relaxation was shown by Flocke (1962) to depend on the method of crystallization. If the specimen is quenched then the α' maximum occurs at a lower temperature than in a slowly cooled specimen. The α' peak shown in Figure 10.2 for HD polyethylene is for a quenched specimen. In a slowly cooled specimen the α' peak occurs ca. 20°C higher (Flocke, 1962). Another significant property of the α' relaxation is that it is not apparently observed in dielectric relaxation.

10.1b. The α Relaxation

The best hypothesis at present for the mechanism of the α relaxation is that it is due to vibrational or reorientational motion within the crystals (Rempel and others, 1957). These authors base their conclusion on the narrowing of the broad line observed in nuclear magnetic resonance experiments. In LD polyethylene the narrowing appears at about room temperature, and at higher temperatures in HD polyethylene. The hypothesis of Rempel and others has been supported by many authors studying bulk crystallized and also oriented specimens (Peterlin and Olf, 1964).

It is certain that the α relaxation is not observed in mechanical experiments unless polyethylene crystals are present. This was shown first by Schmieder and Wolf (1953) who, in a distinguished series of experiments, studied the properties of the α relaxation as a function of chlorine content in chlorinated polyethylene. Chlorinated polyethylene is prepared by passing chlorine gas through a solution of polyethylene in a suitable solvent. Initially the chlorine attacks the chain in a presumably random manner forming

$$
\begin{array}{cccc}
\text{H} & \text{H} & \text{H} & \text{H} \\
-\text{C} & -\text{C} & -\text{C} & -\text{C}- \\
\text{H} & \text{Cl} & \text{H} & \text{H}
\end{array}
$$

At higher chlorine contents the probability of two chlorine atoms per carbon atom increases. At low chlorine concentrations the principal effect of the chlorine atoms is to lower the crystallinity. It is therefore possible to chlorinate a given polyethylene to different degrees, thus producing different crystallinity levels, and so study the effect of varying crystallinity on internal friction.

The variation of logarithmic decrement with temperature for chlorinated LD polyethylene is given in Figure 10.4. The data are shown in two graphs: Figure 10.4(a) shows chlorine contents up to 28·2% and Figure 10.4(b) chlorine contents from 28·2 to 77·3%. At 0% chlorine the α peak is larger than the β peak and occurs at 67°C (1·7 c/s). The addition of 5·8% chlorine moves the α peak to 43°C (0·9 c/s) and lowers its magnitude so that it is comparable in size to the β peak. For 10·4% chlorine the α peak is further reduced in size and occurs at 24°C (1·4 c/s). At the highest

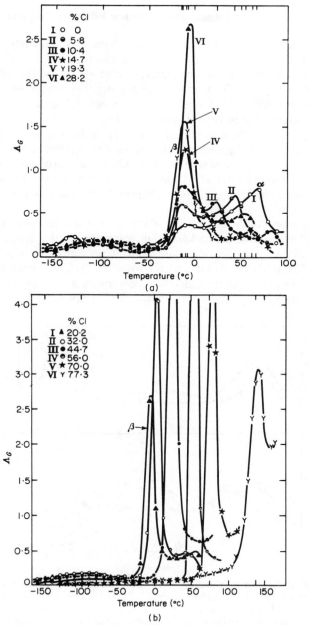

Figure 10.4. Temperature dependence of logarithmic decrement for chlorinated polyethylene. (a) Chlorine contents from 0 to 28·2%; (b) chlorine contents from 28·2 to 77·3% (After Schmieder and Wolf, 1953).

chlorine contents the α peak does not occur. Now crystalline reflections of x-rays are not observed at chlorine concentrations above ca. 40 %, which is approximately the chlorine concentration at which the last vestige of the α peak disappears. These results prove that the α peak occurs only when the specimens yield sharp lines in the x-ray diffraction pattern, i.e. when they possess long-range crystalline order.

The assignment of the α relaxation to the crystalline region is supported by the interesting mechanical experiments of Thurn (1960) and of Takay-anagi (1963) on single-crystal laminates. Other supporting work is due to Wada and Tsuge (1962) who observed the α relaxation in a suspension of polyethylene single crystals (see Chapter 2) in xylene (see also Tsuge and others, 1962).

The assignment of the dielectric α relaxation to the crystalline regions is due to Mikhailov and coworkers. The variation of the dielectric loss tangent with temperature for oxidized LD polyethylene is shown in Figure 10.5 (Mikhailov, Kabin and Krylova, 1957). The α peak does not appear in LD polyethylene which has been quenched from the melt to low temperatures; that is, in polyethylene with a low crystalline content.

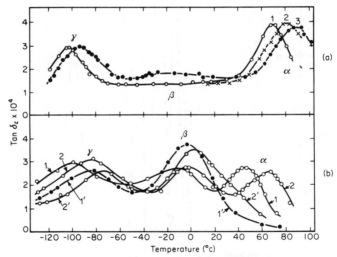

Figure 10.5. Temperature dependence of tan δ_ε for oxidized poly-ethylene. (a) HD polyethylene; curves 1, 2 and 3 at frequencies 1, 5 and 10 kc/s; (b) LD polyethylene; curves 1′ and 2′ at frequencies of 1 and 10 kc/s measured on a quenched specimen; curves 1 and 2 at frequencies of 1 and 10 kc/s measured on an annealed specimen (After Mikhailov and others, 1957).

If LD polyethylene is not quenched then the α peak is observed and is comparable in magnitude to the β peak.

The temperature of the α relaxation depends on both side-branch content (Schmieder and Wolf, 1953) comonomer content (vinyl acetate, Nielsen, 1960) and method of crystallization and post-crystallization annealing (Nielsen, 1962; Flocke, 1962). Any change in side-branch or comonomer content which lowers the melting point lowers the temperature of the α relaxation. Quenching appears to lower the temperature of the α relaxation but annealing moves it to higher temperatures (Nielsen, 1962; Tuijnman, 1963).

According to Takayanagi (1963) the α relaxation is due to the relaxation of CH_2 units in the crystals, and the molecular mechanism is the same as the γ mechanism. (It is generally agreed that the γ mechanism is due to the relaxation of CH_2 units in the amorphous region, (see below).) According to Takayanagi (1963) the same of molecular movements (local mode; Section 5.5b) occur at both γ and α relaxations and therefore Takayanagi labels these relaxations β_a and β_c, the subscripts standing for amorphous and crystalline. This hypothesis is plausible but has not yet proved fruitful.

Reddish and Barrie (1959) investigated the dielectric relaxation in a HD specimen which had been milled at 160°C for 1.5 hours in air. Figure 10.6 shows the plot of dielectric loss factor as a function of frequency at different temperatures. A striking feature of these curves is that the half-width is near 1.75 decades, independent of temperature, which is a value to be compared with 1.14 decades obtained for a single relaxation process. Most amorphous polymers exhibit a backbone relaxation (α or glass–rubber relaxation) whose half-width is between 1.7 and 2.5 decades of frequency. Crystalline polymers normally exhibit very broad relaxation curves, thus the α relaxation in oxidized polyethylene is unusually sharp, i.e. a relatively narrow distribution of relaxation times is present.

Reddish and Barrie evaluated the area A_α below the α relaxation curve of ε'' against log frequency at given temperatures. They found it had the form

$$A_\alpha = \frac{A_0}{T}(1 - CT) \qquad (10.1)$$

Here A_0 and C are constants. Reddish and Barrie combined the above relation with the equation of Sillars, (Sillars, 1939).

$$A_\alpha = (\varepsilon_R - \varepsilon_U)\pi/4\cdot606$$

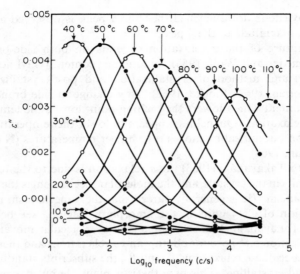

Figure 10.6. Frequency dependence of ε'' for oxidized HD polyethylene at temperatures between 0 and 110°C showing the α relaxation (After Reddish and Barrie, 1959).

The derived $(\varepsilon_R - \varepsilon_U)$ value was used in conjunction with the Fröhlich–Onsager relation and they evaluated an effective dipole moment $(g\mu^2)^{1/2}$, knowing the number of dipoles per cc. The latter was obtained from the infrared study of carbonyl concentration. The $(g\mu^2)^{1/2}$ values of 3·4 D at 0°K and 2·8 D at 293°K evaluated by Reddish and Barrie were compared with the group dipole moment of 2·3 D and the agreement was considered to be reasonable.

Tuijnman (1963a) found that the α peak moved to higher temperatures on annealing in accord with Mikhailov above. He also found that $(\varepsilon_R - \varepsilon_U)$ for the α relaxation decreased with increasing temperature in accord with Reddish and Barrie and proposed that the α process would disappear above the polymer melting point.

Tuijnman (1963b) proposed a model for the α relaxation process. Consider a planar zig-zag polyethylene chain containing the carbonyl group as follows:

This chain is assumed to pack into the normal polyethylene crystal lattice (for a drawing to scale, see Figure 2.7). The α mechanism is regarded as the rotation of the chain about its axis from one equilibrium position to another of equal energy via a potential barrier due to the crystalline field.

Tuijnman used the 'barrier diffusion' theory of Brinkman (1956) together with the theory of Fröhlich (1942) to evaluate the relaxation time for the α process. The average relaxation time is given by

$$\tau = \tau_0 \exp \frac{\Delta U}{kT} \qquad (10.2)$$

ΔU is the effective barrier to relaxation and is a function of the chain length and of the terms A^2 and B^2. A^2 is the order of the bond energy for one CH_2 group in the chain. B^2 is a force constant for the twisting of the chain. Tuijnman showed that

$$\Delta U = \frac{\pi^2}{4} AB \tanh \frac{B_m}{2A} \qquad (10.3)$$

where m is the number of CH_2 units in the relaxing chain. In order to test the theory, Tuijnman analysed experimental data for long chain ketones and esters of different chain length, in solid solution in paraffin wax. The slope $S_m = d \ln \tau/dm$ was evaluated at a given chain length, i.e. a given m, at a given temperature. From Equations (10.2) and (10.3)

$$AB \simeq 1.866 kTS_m \left(\frac{B}{A}\right)^{-1} \cosh^2 \left(\frac{B_m}{2A}\right) \qquad (10.4)$$

For a given value of S_m at temperature T, and a given m value, a plot of AB against B/A was constructed using Equation (10.4). Similar plots were made for a different value of m (hence a different value for S_m) at the same temperature. All these plots of AB against (B/A) intersected at a particular $[AB, B/A]$ coordinate, which gives the solution to Equation (10.3) for the experimental data considered, and also shows that the theory is applicable to solid solutions of long chain ketones in paraffin wax.

Tuijnman found $A^2 = 8.1 \times 10^{-12}$ erg/mon rad^2 and $B^2 = 4.3 \times 10^{-14}$ erg/mon rad^2. Müller (1936) had calculated B^2 for one CH_2 group in a solid paraffin to be either 14.6×10^{-14} erg/mon rad^2 or 7.7×10^{-14} erg/mon rad^2, i.e. Tuijnman's data is in reasonable agreement with Müller's calculation. A^2 has the order of magnitude of the bond energy of one CH_2 group which is about 5×10^{-12} erg/mon rad^2, again in agreement with Tuijnman's experimental value.

In this way, Tuijnman showed that the relaxation of long chain esters and ketones in solid paraffin wax was consistent with a model of rotation of the entire length of the 'visitor' chain about its C axis from one equilibrium position to another, when the chain in turning overcomes the intermolecular barrier B^2 and twists simultaneously (characterized by A^2).

Other experiments of Tuijnman (1963) support the view that the α relaxation occurs in the crystalline regions of the polymer. This author observed a large difference between the dielectric properties of HD polyethylenes (a) oxidized with ultraviolet light at room temperature and (b) oxidized by milling at 160°c in air. Figure 10.7 shows the results for the two cases. It is seen that the α region is prominent for the milled sample,

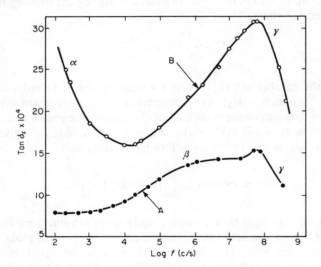

Figure 10.7. Frequency dependence of tan δ_ε for oxidized low-density polyethylene. Curve A artificially aged at 25°c by means of ultraviolet light; Curve B artificially aged at 160°c by milling for three hours (After Tuijnman, 1963a).

but is very much reduced in the ultraviolet oxidized sample. Ultraviolet oxidation at room temperature is a process controlled by the diffusion of oxygen into the partially crystalline polymer. The amorphous regions are preferentially oxidized because the diffusion coefficient for the amorphous phase is larger than that for crystalline phase. The milling method of oxidation is carried out on a molten sample, thus, on solidification, the crystalline regions of the resulting solid sample have a higher probability of containing carbonyl groups than was the case for the ultraviolet oxidized sample. Since the α peak is large in the milled sample and much smaller in the other sample, the α relaxation must result from the carbonyl rotation in the crystal lattice.

The assignment of the dielectric α relaxation to the relaxation of carbonyl groups in the crystals has been proved beyond all doubt by Booij (1965). According to this author the value of ε''_{max} for the α relaxation at room temperature should be proportional to the number of carbonyl groups per cc in the crystal lattice. If this is so then the value of ε'' at the α maximum divided by the oxygen content should be proportional to the weight of crystalline material per cc. Experimental results are given in Figure 10.8. The weight fraction of crystalline material is ϕ_k and ρ is the density. This quantitative result is the strongest evidence available concerning the crystalline origin of the α relaxation.

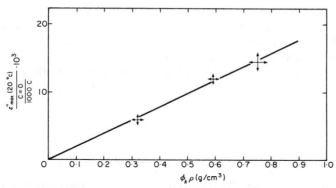

Figure 10.8. Demonstration of proportionality between the value of ε'' at the α relaxation peak divided by the carbonyl concentration and the weight fraction of crystalline material per cc, $\phi_k\rho$ (After Booij, 1965).

There is little doubt that the α relaxation as observed in dielectric relaxation and nuclear magnetic resonance is due to reorientation of molecules within the polyethylene crystals. It is a reasonable working hypothesis that the mechanical α relaxation is also due to reorientation of molecules within the crystals (Tsuge and others, 1962; Pechold, Blasenbrey and Woerner, 1963; Takayanagi, 1963). The way in which the reorientation produces mechanical relaxation has not been established.

It has been proposed that the α mechanism is due to the melting of crystals (sometimes termed premelting). However, Nakayasu, Markovitz and Plazek (1963) have observed the α relaxation in linear polyethylene in creep at temperatures as low as 150°C below the melting point. According to these authors this implies that the α relaxation cannot be due to melting. Takayanagi (1963) has also shown the melting explanation of the α mechanism to be very dubious.

10.1c   The β Relaxation

The relationship between the amount of side-branching in polyethylene and the magnitude of the β relaxation was demonstrated in the experiments of Kline, Sauer and Woodward (1956). Three specimens were prepared from three different polyethylenes with densities varying from 0·915 to 0·957 g/cc. The methyl content for each specimen, a measure of the number of side-branches since each side branch contains one methyl group, is given in Table 10.2 in units of methyl groups per 1000 carbon atoms. The results are shown in Figure 10.9. It is clear that the β peak (ca. 270°K) of specimen A is greater than that of specimen B which is greater than the β peak of C, if the very small dispersion in the region of 300°K in C can be assigned to the β mechanism. These results lead naturally to the generally held interpretation (due originally to Oakes and Robinson, 1954) that the β peak in LD polyethylene is due to the relaxation of branch points, specifically of the portion of the molecule containing the group,

$$
\begin{array}{ccc}
\text{H} & \text{H} & \text{H} \\
-\text{C}-\!\!-\text{C}-\!\!-\text{C}- \\
\text{H} & \text{R} & \text{H}
\end{array}
$$

in which R is a side-group. It is found that for low concentrations of R, the β relaxation always occurs at about the same temperature independent

Table 10.2. The methyl content and density of polyethylenes studied by Kline, Sauer and Woodward (1956)

Property	Polyethylene			Method of Measurement
	A	B	C	
Methyl groups per 1000 carbon atoms	32	16	1	Infrared
Density g/cc	0·915	0·922	0·957	Alcohol displacement

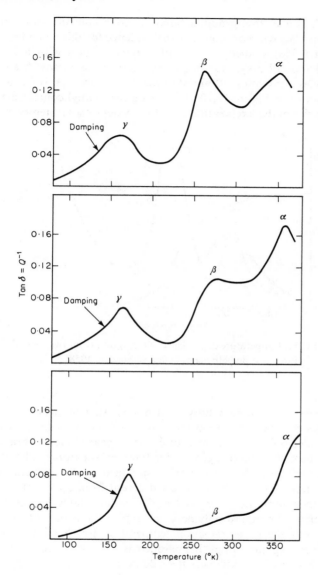

Figure 10.9. Showing the effect of side-branch content on the magnitude of the β relaxation in polyethylene. Upper, middle and lower curves for specimens containing 32, 16 and 1 methyl group per 1000 carbon atoms (After Kline, Sauer and Woodward, 1956).

of whether R is a methyl group, a butyl group, a chlorine atom, or an acetate group. We will now consider the evidence for this statement.

Willbourn (1958) studied a series of polymethylenes with a known number and type of alkyl side-branches. Results for methyl-branched polymethylenes containing 10 and 140 methyl groups per 1000 carbon atoms are shown in Figure 10.10. Increasing the methyl content greatly increases the size of the β relaxation and slightly depresses its temperature.

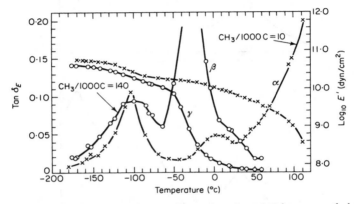

Figure 10.10. Temperature dependence of tan δ_E and E' for two methyl-branched polyethylenes (After Willbourn, 1958).

Polymethylenes containing n-butyl and n-amyl side branches (6 and 8 side-branches per 1000 carbon atoms respectively) exhibit the β peak at about 0°c, which is the same temperature as the β peak of methyl-branched polymethylene with 10 methyl groups per 1000 carbon atoms. That is, the length of the side branch does not affect significantly the temperature of the β relaxation. This result is entirely different from the effect of side-branch length in homo-polymers such as the methacrylates and the higher olefin polymers. For these polymers, increasing the length of the side-group decreases the glass-transition temperature (see Sections 8.3 and 10.3). The best explanation of this paradox is as follows. In the methacrylate and higher olefin polymers the side branch occurs on every other main-chain carbon atom. The temperature of the glass transition thus depends to some extent on interactions between the side-groups. For polyethylenes or polymethylenes with low side-branch contents the side-branches are sufficiently 'dilute' not to interact with each other. Increase in side-branch content therefore changes the magnitude of the relaxation but not the temperature.

Since chlorine atoms and alkyl side-branches have a comparable effect on the α peak in polyethylene, it might be anticipated that both would affect the β peak in the same way. At low concentrations of chlorine this is exactly what happens (Figure 10.4a). Thus, with increasing chlorine concentrations the β peak is rapidly enlarged so that at 28·2 % chlorine it is the dominant relaxation. Also the temperature of the β peak at low concentrations declines very slightly with increasing chlorine concentration.

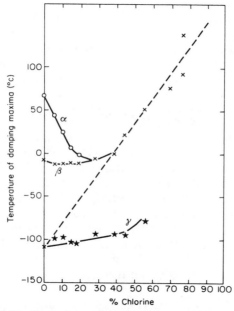

Figure 10.11. Dependence of the temperature of the α, β and γ peaks in chlorinated polyethylene on chlorine content (After Schmieder and Wolf, 1953).

In Figure 10.11 the temperature of the peaks in chlorinated LD poly-ethylene is plotted as a function of chlorine content. Above 40 % chlorine the temperature of the β peak increases rapidly with chlorine content (see Figure 10.4b). Schmieder and Wolf (1953) consider the effects of chlorine on the relaxation of the molecule to be twofold. Firstly, a steric pushing apart of the molecules such as would be almost equally performed by a methyl group. Secondly, a dipolar interaction between the C—Cl dipoles. The dipolar interaction predominates at high chlorine concentrations, thus raising the temperature of the β peak. At low concentrations the steric effect causes the β temperature to decline slightly before the dipolar interaction becomes dominant.

Results similar to those of Schmieder and Wolf have been obtained by Nielsen (1960) who studied the copolymers of ethylene and vinyl acetate. Between 0 and 60 volume per cent vinyl acetate the β peak remained at $-25°C$. From 60 to 100% vinyl acetate the β temperature increased linearly with volume per cent vinyl acetate towards the glassy relaxation peaks of pure polyvinyl acetate which occurs at 33°C(1·9 c/s), Figure 9.15a. If the line between 60 and 100% vinyl acetate was extended to 0% vinyl acetate an extrapolated β temperature of $-110°c$ was obtained. The same value was obtained by extrapolating the β temperature in the chlorinated polyethylenes to 0% chlorine, Figure 10.11. These results may be explained by means of the two-phase model in the following qualitative way (Nielsen, 1960).

When a small amount of ethylene is added to pure polyvinyl acetate there is a rapid reduction in the temperature of the glassy peak. It is to be expected that in the first approximation the glass relaxation temperature will follow the Gordon–Taylor equation (see Equation 2.2). Linear sequences of amorphous CH_2 units relax at $-110°C$ (see next section) and therefore the temperature of the glass relaxation extrapolates to $-110°C$. At approximately 40% ethylene, however, crystallization occurs. Thus for ethylene concentrations above 40% the ethylene concentration in the amorphous regions is lower than the gross concentration. Therefore the glass temperature decreases from 33°c for pure polyvinyl acetate to $-25°c$ for an ethylene content of 40% and then remains constant from 40 to 100% ethylene.

Reding, Faucher and Whitman (1962) reject the above explanation for the rapid increase in the temperature of the β relaxation. According to these authors a copolymer of ethylene and a vinyl monomer must be considered as a mixture of groups containing a few linear CH_2 units and structures containing a tertiary carbon atom, both of which relax independently. At high ethylene compositions the CH_2 groups relax at the γ relaxation ($-110°C$ at ~ 1 c/s), and the tertiary carbon structures at the β relaxation (-10 to $-50°C$ at ~ 1 c/s depending on the nature of the side group). When the ethylene composition decreases the magnitude of the γ relaxation decreases but the temperature of the relaxation does not change. This occurs because the CH_2 units relax independently of the rest of the chain. When the CH_2 content is less than about 40 weight per cent the tertiary carbon structures no longer relax independently of the rest of the chain due to the interaction of the side-groups. The β relaxation thus rises from the region of $-10°C$, finally reaching a value equal to T_g of the vinyl homopolymer.

Reding and others (1962) propose this hypothesis to explain the fact that in random ethylene–vinyl acetate copolymers the last vestige of crystallinity disappears at about 45 weight per cent vinyl acetate whereas the upward move of the β relaxation occurs at about 65 weight per cent vinyl acetate. According to this viewpoint, in the range 0 to 65 weight per cent vinyl acetate the temperature of the β relaxation changes very little because the pendent groups are isolated from each other whether or not the copolymers are crystalline.

Extending their hypothesis, Reding and others (1962) conclude that all vinyl homopolymers would have glass transitions in the -10 to $-50°C$ range if the steric hindrance between side-groups did not occur.

10.1d The γ Relaxation

The most significant property of the γ relaxation in polyethylene is its dependence on density. When the density is increased the internal friction peak is reduced in magnitude. For instance, in HD polyethylene of density 0·944 g/cc (23°C) the peak is of height $\Lambda = 0·20$ and for a density 0·982 g/cc, $\Lambda = 0·092$ (~ 1 c/s). The temperature of the internal friction peak and its shape do not vary with density. These characteristics are all shown by the γ peaks in polytetrafluoroethylene and polyoxymethylene (see Chapters 11 and 14).

Willbourn (1958) has drawn attention to the role of the $(CH_2)_n$ group in determining the magnitude of the γ peak in polyethylene and related polymers. When this group occurs in the main chain of a polymer it invariably leads to a relaxation which occurs in the same temperature and frequency region as the γ peak in polyethylene. Willbourn (1958) considers the γ relaxation to be the glass relaxation of the $(CH_2)_n$ group. The two-phase model leads to a simple interpretation of the density dependence of the magnitude of the γ peak. That is, the γ relaxation occurs in the amorphous regions and therefore its magnitude is a function of the volume fraction of amorphous polymer.

The soundest of the many molecular models proposed for the γ mechanism is due to Schatzki (1962, 1965). This so-called 'crankshaft mechanism' is discussed in Section 5.5a.

10.1e   The Glass Transition of Polyethylene

There are three arguments advanced which place the glass transition of polyethylene (a) in the β region just below $0°c$ (b) in the region of $-81°c$ (Tobolsky, 1961; Boyer, 1963), (c) in the γ region below $-100°c$.

According to Cole and Holmes (1960) T_g for polyethylene lies at $-20°c$ (see Chapter 2). These authors used the volumetric criterion for the glass transition.

Tobolsky (1961) also used the volumetric criterion but did not actually observe a knee in the V versus T curve of polyethylene. Tobolsky took the value of $-20°c$ for T_g for polypropylene obtained from $V-T$ measurements by Dannis (1959) together with his own measurement from a $V-T$ experiment on an ethylene–propylene copolymer (mole ratio 2/1) of $T_g = -59°c$ and, using the Gordon–Taylor equation (2.2), obtained an extrapolated T_g for polyethylene of $-81°c$. Boyer (1963) quotes several other arguments in favour of the temperature region of $-81°c$ for the T_g of polyethylene including (1) the brittle point of branched polyethylene is $-68°c$, (2) this value of T_g is approximately 0·5 times the melting point, T_m (in agreement with the empirical rule that $0·5 < T_g/T_m < 0·67$).

The properties of the amorphous regions of polyethylene have of course to be inferred from measurements on the composite, crystalline–amorphous, solid. It is therefore no longer possible to define T_g by a simple operational definition, such as, for example, 'T_g is the temperature of the knee in the volume–temperature curve (see Chapter 2) which occurs in the same temperature region as the glass–rubber relaxation for which $G_U/G_R \sim 10^3$ (see Chapter 3)'. This criterion can be used for completely amorphous polymers and thereby also for crystalline polymers which may be obtained in one way or another in the completely amorphous state (e.g., polypropylene, polychlorotrifluoroethylene, polyethylene terephthalate). The properties (specific volume, modulus) of the amorphous fraction of polymers which are unobtainable in the completely amorphous state cannot be measured directly. They may be inferred from measurements on the bulk, but this is not a simple matter. It is necessary to use other criteria to establish T_g for a partially crystalline polymer such as polyethylene.

According to Willbourn (1958) if it is accepted that T_g is the temperature at which the main polymer chain acquires large-scale mobility then crystallization will be possible above T_g and impossible below. To crystallize, molecules must be able to move from a tangled to an ordered array and they can only do this if the main polymer chain is mobile. For a copolymer in which only one type of unit crystallizes, according to this viewpoint, the idea of a single glass transition may become artificial (Willbourn, 1958). LD polyethylene can be considered as a copolymer consisting of methylene units and a few alkylidene units,

$$-CH_2- \qquad -\underset{\underset{C_nH_{2n+1}}{|}}{CH}-$$

If the proportion of alkylidene units is small then, since the crystals formed are almost completely composed of methylene units, T_g will be the temperature at which methylene units acquire sufficient mobility to crystallize. There is not much evidence on this point since polyethylene cannot be quenched to the amorphous state. Nevertheless, what evidence there is would suggest, according to Willbourn, that for polyethylene containing only a few alkylidene units, T_g is below $-100°C$, so that the γ relaxation is the glass relaxation. When the proportion of alkylidene units increases, however, this statement no longer holds. For instance, T_g for a methyl-branched polymethylene containing 1 methyl branch per 7 chain atoms is clearly $-30°C$, the relaxation temperature of the alkylidene units (Willbourn, 1958). For this polymer it is unnecessary to use the crystallization criterion since it is only 10% crystalline. The polymer, nevertheless, exhibits a γ relaxation due to CH_2 units (as indeed LD polyethylene exhibits a β relaxation due to alkylidene units).

Now, if it is accepted that at low alkylidene contents T_g is in the γ region and at high alkylidene contents T_g is in the β region, then clearly at some intermediate alkylidene content there must be a change over from one T_g to the other.

Consider the ethylene–propylene copolymer system. For linear polyethylene (Marlex) a knee in the high-frequency modulus ($\sim 10^4$ c/s) versus temperature curve occurs at $\sim -120°C$, Figure 10.12 (Baccaredda, Butta and Frosini, 1963). This knee coincides with the change in thermal expansion coefficient (see Figure 2.18) and we denote it by T_γ in Figure 10.12. At the frequencies used by Baccaredda and others, the γ loss peak occurs at a temperature almost 40°C higher than T_γ. (The temperature of the relaxation is frequency dependent whereas T_γ, defined by the knees in the high-frequency modulus (or the $V-T$ curve) is not frequency dependent.) In polypropylene the knee in the high-frequency modulus and in the $V-T$ curve occurs at about $-20°C$. This is the glass transition. T_g. (The associated relaxation in polypropylene is the β relaxation which, at ~ 1 c/s, occurs at 0°C, see Section 10.2.) In Figure 10.12 the transitions are labelled T_γ and T_β.

For an ethylene–propylene copolymer containing 78 mole per cent propylene two knees are observed, Figure 10.13. They are labelled T_γ and

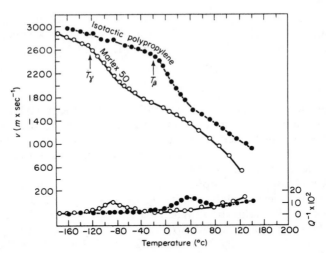

Figure 10.12. Temperature dependence of sound velocity and Q^{-1} in HD polyethylene and isotactic polypropylene (After Baccaredda, Butta and Frosini, 1963).

T_β. The knee at T_γ is poorly defined. The knee at T_γ becomes better and better defined as the propylene content is reduced from 78·0 through 48·7, 46·9 to 42·8 mole per cent, Figure 10.13. At these compositions the knee at T_β is also well defined. However, for linear polyethylene (Marlex, Figure 10.12) T_γ is well defined but if there is a knee which would correlate with T_β it is not easily discernable in the data of Baccaredda and others (1963). According to Flocke (1962), however, even in quenched HD polyethylene (Ziegler) there is a β relaxation at 2°C (3·1 c/s). At Flocke's frequencies the β relaxation occurs at a temperature too close to T_β for the knee to be resolved since it is somewhat overlapped by the modulus drop due to relaxation. According to Flocke (1962) the β relaxation in HD polyethylene is related to the β relaxation in polypropylene. The relationship between the two is demonstrated in Figure 10.14 which shows the locus of the β relaxation temperature in ethylene–propylene co-polymers. The data in Figure 10.14 are due to Tuijnman (1965) and the extrapolation to weight fraction of ethylene, $\phi_E = 1$, follows Flocke (1962). According to Flocke (1962) the incorporation of small amounts of ethylene in the polypropylene molecule plasticizes the β relaxation and so lowers it from ~ 0°C. The decrease in the β temperature with increasing ethylene continues until the ethylene crystals form. The crystals then

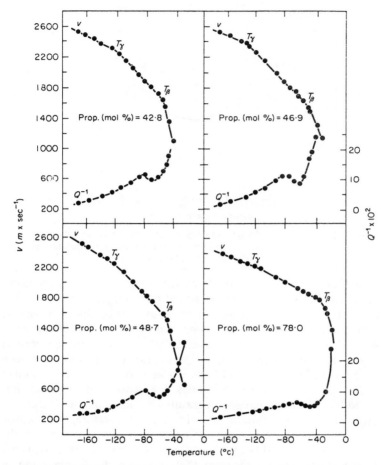

Figure 10.13. Temperature dependence of sound velocity and Q^{-1} in ethylene–propylene copolymers (After Baccaredda, Butta and Frosini, 1963).

exert a 'bracing' action on the amorphous regions causing a minimum in the β temperature locus, Figure 10.14. With increasing ethylene content the β temperature continues to rise, due to increasing 'bracing', reaching a value $\sim 0°c$ for 100% ethylene.

Clearly, the character of the β relaxation changes along the locus of the line shown in Figure 10.14. Between 0 and 60% ethylene it would be rash

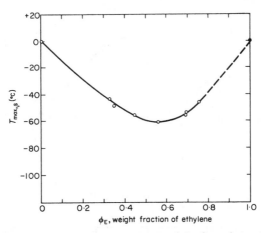

Figure 10.14. Locus of the temperature of the β maximum in ethylene–propylene copolymers (After Tuijnman, 1965; dashed extrapolation after Flocke, 1962).

to assume that the character of the β relaxation remains exactly the same. This cannot be so because between 0 and 60% ethylene an increasing portion of the molecule containing ethylene units is becoming mobile at the γ relaxation. Nevertheless, between 0 and 60% ethylene the character of the β relaxation, although changing, falls clearly into the general category, glass–rubber relaxation. But in the range 60 to 100% ethylene beyond the minimum, Figure 10.14, the character of the β relaxation must change drastically. Not only is the β relaxation reduced in magnitude but it is shifted by $\sim 50°c$ to higher temperatures by, according to Flocke, the bracing action of the crystals. It cannot be argued therefore, that because the β relaxations in polypropylene and polyethylene are connected by the locus shown in Figure 10.14 and because the β relaxation in polypropylene is the glass–rubber relaxation that therefore the β relaxation in polyethylene is the glass–rubber relaxation of polyethylene.

It is by no means certain that the assignment of a T_g in polyethylene is a meaningful activity. It cannot be denied that the disordered fractions of the polymer relax: that when the alkylidene content is low the major relaxation is the γ relaxation and when the alkylidene content is high the major relaxation is the β relaxation. Both the γ and β relaxations have some of the properties normally associated with the glass–rubber transition. But to call either of them 'the glass relaxation' may not be profitable. To use Boyer's (1962) aphorism there must be 'degrees of glassiness of the glassy state.'

10.2 POLYPROPYLENE

$$[-\overset{\overset{\displaystyle H}{|}}{\underset{\underset{\displaystyle H}{|}}{C}}-\overset{\overset{\displaystyle H}{|}}{\underset{\underset{\displaystyle CH_3}{|}}{C}}-]_n$$

(2)

Isotactic, syndiotactic and atactic forms of polypropylene can be obtained using different polymerization conditions. The isotactic and syndiotactic forms are crystalline (Natta, 1960). A partially crystalline polypropylene contains atactic material which can be extracted using suitable solvents. A typical method of describing the degree of isotacticity of a polypropylene sample is to quote the percentage weight which is insoluble in boiling heptane. The isotactic polymer is largely insoluble and the atactic polymer is soluble in boiling heptane. An example of this classification is given in Table 10.3 for five specimens of polypropylene (Flocke, 1962). Note that the specimen with the highest insoluble portion has the highest density.

Four types of spherulite (types I to IV) have been observed in polypropylene (Padden and Keith, 1959). There are no reports in the literature of the effects of various spherulitic forms on either mechanical or dielectric behaviour.

There is unanimous agreement in the literature that polypropylene is a two-phase solid. During growth the crystalline fibrils reject the atactic molecules which subsequently lie trapped between the crystalline fibrils (Keith and Padden, 1963).

10.2a Mechanical Studies

The first measurements were due to Baccaredda and Butta (1958), Sauer and others (1958), Willbourn (1958), McCrum (1959c), Newman and Cox (1960) and Van Schouten and others (1960). The most comprehensive studies are due to Flocke (1962) and Passaglia and Martin (1964). The specimens used by Flocke are described in Table 10.3, P1 being the most and P5 being the least isotactic. The specimens were cooled at the same rate from the crystallization temperature so that the decreasing density from specimen P1 to P5 reflects decreasing isotacticity.

Figure 10.15 shows the temperature dependence of the shear modulus and logarithmic decrement of specimens P1 to P5 measured at frequency ~ 1 c/s. There are three relaxations marked α, β, γ. The dominant relaxation is the β relaxation. This relaxation is clearly the glass–rubber relaxation of the amorphous portions of the solid since the β peak is greatest in specimen P5 and least in P1. This statement presupposes that for polypropylene the logarithmic decrement is a meaningful measure of relaxation magnitude. According to Passaglia and Martin (1964) this

Table 10.3. Description of the degree of isotacticity of five specimens of polypropylene studied in the experiments of Flocke (1962).

Specimen	Portion insoluble in boiling heptane (%)	Density at 22°c (g/cc)
P1	96	0·905
P2	90	0·894
P3	75	0·880
P4	20	0·875
P5	0	0·87

assumption of Flocke's (1962) is not true. Passaglia and Martin nevertheless concur with Flocke that the β relaxation is the glass–rubber relaxation of the amorphous portion of the solid. The dilatometric T_g of polypropylene is $-35°c$ (Natta and others, 1957), so clearly a large relaxation at $0°c$ (~ 1 c/s) must be assigned to the glass–rubber relaxation.

Passaglia and Martin (1964) studied five specimens prepared from the same polypropylene resin in different ways. One specimen was quenched from the melt and another crystallized isothermally at 135°C. The other three specimens were prepared from portions of quenched specimens by annealing at elevated temperatures. The temperature dependence of the shear modulus and logarithmic decrement of these five specimens is shown in Figures 10.16 and 10.17. From these values the temperature dependence of G'' and J'' was also computed by Passaglia and Martin and is shown in Figures 10.18 and 10.19. These data bring out the following points.

(1) The shear modulus G' at a particular temperature below $-20°C$ increases with increasing density. This is in agreement with the results of Flocke (1962) shown in Figure 10.15.

(2) The magnitude Λ_{max} of the γ peak decreases with increasing density. The values of J''_{max} for the γ peak also decrease with increasing density, Figure 10.19. G''_{max} for the γ peak, however, shows little dependence on density. This result is clearly due to the fact that in the γ region increasing density causes G' to increase but Λ to decrease. Thus the two effects annul each other when G'' is considered causing G''_{max} in the γ region to have no pronounced dependence on density. The two effects reinforce each other

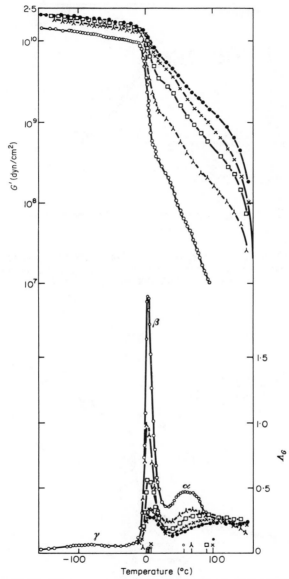

Figure 10.15. Temperature dependence of shear modulus and logarithmic decrement for the five polypropylene specimens described in Table 10.3. P1 (●), P2 (×), P3 (□), P4 (⅄), P5 (○) (After Flocke, 1962).

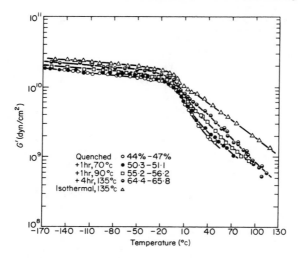

Figure 10.16. Temperature dependence of G' for five specimens of polypropylene crystallized from one resin with the thermal treatments indicated (After Passaglia and Martin, 1964).

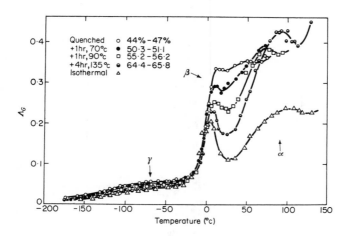

Figure 10.17. Temperature dependence of the logarithmic decrement for the five specimens of polypropylene described in Figure 10.16 (After Passaglia and Martin, 1964).

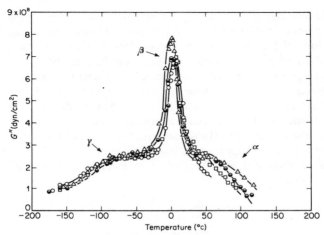

Figure 10.18. Temperature dependence of G'' for four specimens of polypropylene described in Figure 10.17. Symbols same as in Figure 10.16 (After Passaglia and Martin, 1964).

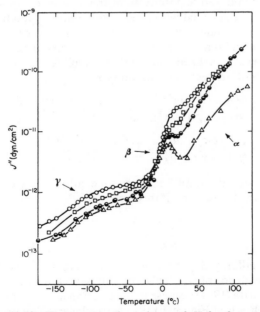

Figure 10.19. Temperature dependence of J'' for four specimens of polypropylene described in Figure 10.17. Symbols same as in Figure 10.16 (After Passaglia and Martin, 1964).

when J'' is considered causing J''_{max} to vary more with density than Λ_{max} (compare γ peaks in Figures 10.19 and 10.17). According to Flocke (1962) the behaviour of Λ_{max} with varying density implies that the γ mechanism occurs in the amorphous regions. Passaglia and Martin (1964) accept and extend this hypothesis and propose the relaxing unit to consist of a few chain segments and not to involve the motion of more complex morphological units. In this instance, Passaglia and Martin do not accept the density dependence of G''_{max} as meaningful evidence.

(3) Passaglia and Martin (1964) note that G'' varies little with density over the temperature range (Figure 10.18) whereas G' does (Figure 10.16). According to Passaglia and Martin these facts lead to the following deduction. Write,

$$G' = G_R + \sum_i \frac{G_i \omega^2 \tau_i^2}{1 + \omega^2 \tau_i^2} \tag{10.5}$$

$$G'' = \sum_i \frac{G_i \omega \tau_i}{1 + \omega^2 \tau_i^2} \tag{10.6}$$

in which G_R is the relaxed (equilibrium or low-frequency) modulus, τ_i is the relaxation time associated with a relaxing element of modulus G_i and ω is the frequency. Assume that because G'' varies little with density (or crystallinity χ) at different temperatures G'' will vary little with χ over various frequencies at a fixed temperature. Differentiating Equations (10.5) and (10.6) with respect to crystallinity χ

$$\frac{dG'}{d\chi} = \frac{dG_R}{d\chi} + \sum_i \frac{\omega^2 \tau_i^2}{1 + \omega^2 \tau_i^2} \frac{dG_i}{d\chi} + \sum_i \frac{2G_i \omega^2 \tau_i}{(1 + \omega^2 \tau_i^2)^2} \frac{d\tau_i}{d\chi}$$

$$\frac{dG''}{d\chi} = \sum_i \frac{\omega \tau_i}{1 + \omega^2 \tau_i^2} \frac{dG_i}{d\chi} + \sum_i \frac{G_i (\omega - \omega^3 \tau_i^2)}{(1 + \omega^2 \tau_i^2)^2} \frac{d\tau_i}{d\chi}$$

The vanishing of $dG''/d\chi$ at all frequencies (i.e. all temperatures) implies that $dG_i/d\chi$ and $d\tau_i/d\chi$ are zero. If this is so then

$$\frac{dG'}{d\chi} = \frac{dG_R}{d\chi} \tag{10.7}$$

That is, the crystallinity dependence of G' (Figure 10.16) is due to the crystallinity dependence of G_R. This interesting argument is qualified by Passaglia and Martin (1964) because in the α region (above 20°c) G'' varies considerably with χ (see Figure 10.19). Nevertheless, Equation (10.7) is certainly significant below 20°c, in which temperature region the principal effect of varying crystallinity on G'', Λ and J'' is caused by the effect of crystallinity on G_R. Note that this argument holds for crystallinity changes brought about by changes in thermal history and not for crystallinity changes brought about by changes in tacticity, such as in Flocke's experiments (see Figure 10.15).

McCrum (1964) studied resins of differing tacticity brought to essentially the same density by varying thermal treatments. Figure 10.20 shows the dependence of the β peak on isotacticity for resins labelled 'A' and 'B',

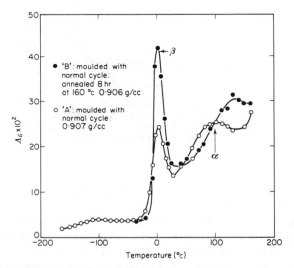

Figure 10.20. Temperature dependence of logarithmic decrement of two polypropylene resins 'A' and 'B' treated thermally as indicated to have essentially the same density (After McCrum, 1964).

B being more atactic than A. The thermal treatments necessary to obtain essentially equal densities are marked in Figure 10.20. Note that the β peak is much larger in resin B than in resin A, despite the fact they have essentially the same density. The interpretation of this is clear: the β relaxation is due to the relaxation of the atactic portion of the solid.

McCrum (1964) observed that the magnitude of the α peak and its relaxation time distribution both depend on thermal history. Resin B was brought to essentially the same density by two different routes, Figure 10.21. The α peak is much smaller in the specimen which was cooled slowly from the melt than in the quenched specimen. McCrum (1964) also observed that the α peak was moved to higher temperatures by

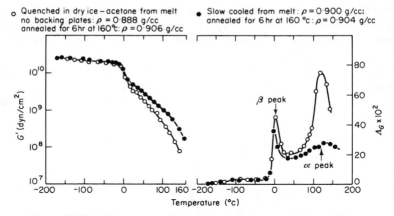

Figure 10.21. Temperature dependence of shear modulus and logarithmic decrement of two specimens of polypropylene 'B' prepared using the indicated thermal history so as to have essentially equal densities (After McCrum, 1964).

annealing. In polypropylene A the α peak maximum was moved by annealing from 100°C (1·36 c/s) to 138°C (1·06 c/s).

From the evidence of the last paragraph McCrum (1964) drew the hypothesis that the properties of the α relaxation depend on crystal morphology. This hypothesis leaves undecided the question whether the α relaxation in polypropylene is due to (I) a molecular relaxation within the crystal (for instance, the dielectric α relaxation in polyethylene is clearly due to a relaxation within the crystal, as described in Section 10.1), (II) a lamellar slip mechanism similar to that proposed by Iwayanagi (1962) for polyethylene. Passaglia and Martin (1964) consider the α mechanism in polypropylene to be of type II, following Scott and others (1962). Flocke (1962) considers the α relaxation in polypropylene to be an amorphous relaxation. The nature of the α relaxation is thus at present not known (unlike the β and γ relaxations, which are unanimously considered to be amorphous relaxations).

Takayanagi (1963) studied a crystalline polypropylene, and its hexane extract (atactic, amorphous) and its hexane residue (isotactic, highly crystalline). The β relaxation was very sharp in the hexane extract and very broad in the hexane residue, showing that the shape of the relaxation distribution function is a strong function of crystallinity, but the average activation energy is unaffected. Takayanagi used a series–parallel model (see Chapter 4) to describe the mechanical properties of the untreated partially crystalline sample in terms of the amorphous contribution (hexane extract) and the crystalline contribution (hexane residue). The calculated curves of E' and E'' against temperature were in excellent agreement with the experimental curves for the untreated polymer.

Sinnott (1959) has observed mechanical relaxations in polypropylene below liquid nitrogen temperatures at 53 to 19°K ($\sim$ 10 c/s) which are attributed to the onset of hindered rotation of CH_3 groups.

Sauer and others (1962) have observed the effects of pile irradiation on the mechanical relaxations in polypropylene. At large irradiation doses (3 to 4×10^{18} n.v.t.) the polymer becomes crosslinked and the crystallinity is destroyed.

10.2b Dielectric Studies

Polypropylene is essentially a non-polar polymer and thus should not exhibit dielectric relaxation. The dielectric properties found by Sazhin and others (1959), by Kishi and Uchida (1963) and by Kramer and Helf (1962) probably result from the oxidation of the polymer, as is the case for polyethylene. The dielectric constant is low and the dielectric loss factors are extremely small at all frequencies. Sazhin and others (1959) have given dielectric data for the main β relaxation region, in the frequency range 600 c/s to 5×10^9 c/s.

Kramer and Helf (1962) investigated a highly crystalline polypropylene between 150 c/s and 300 kc/s. Figure 10.22 shows their data. Two relaxation regions, α and β, are observed. The temperature variation of ε' is shown in Figure 10.22(b). Kramer and Helf analysed their data with the aid of Cole–Cole arcs and obtained ($\varepsilon_R - \varepsilon_U$) between 0·02 and 0·04 for the β relaxation, and a Cole–Cole parameter $\bar{\beta}$ for this process which ranged from about 0·01 at -40°c to about 0·3 at 20°c, i.e. the distribution of relaxation times narrows with increasing temperature. They obtain apparent activation energies of 150 kcal/mole and 28 kcal/mole for the α and β relaxations respectively. They suggest that the β relaxation is associated with the amorphous regions and the α relaxation is associated with the crystalline regions of the polymer.

Kishi and Uchida (1963) studied the dielectric properties of a polypropylene which had been oxidized using u.v. light for 100 hr at 40°c. It was observed that the oxidized samples gave losses nearly ten times larger than those for the non-oxidized sample. Measurements in the range 100 to 10^4 c/s and 0 to 150°c revealed two relaxation regions, very similar to the results of Kramer and Helf (1962) above. In a quenched specimen the

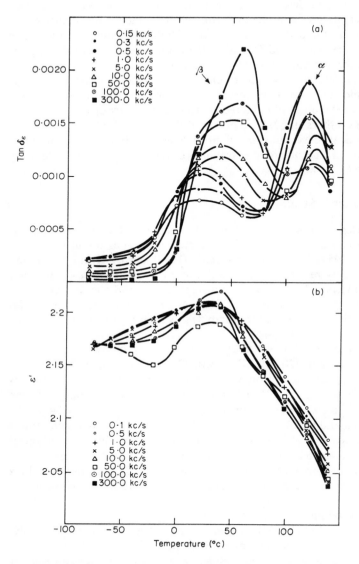

Figure 10.22. Temperature dependence of (a) the dielectric loss and (b) the dielectric constant at various frequencies for isotactic polypropylene (After Kramer and Helf, 1962).

α peak was smaller and the β peak larger than in a slow cooled specimen. This evidence supports the view that the α relaxation is of crystalline and the β relaxation of amorphous origin.

10.3 HIGHER POLYOLEFINS

$$[-\underset{\underset{H}{|}}{\overset{\overset{H}{|}}{C}}-\underset{\underset{R}{|}}{\overset{\overset{H}{|}}{C}}-]_n$$

(3)

Clark, Turner-Jones and Sandiford (1962) studied the mechanical properties of a number of higher polyolefins. The samples were made with a $TiCl_4$–LiAl $(Alk)_4$ catalyst where LiAl $(Alk)_4$ was made from the reaction of $LiAlH_4$ with monomer. Thus the resulting polymers did not contain spurious alkyl groups which would have been obtained from catalysts containing other alkyl groups such as ethyl, etc. The polymers were a mixture of stereoregular and atactic forms.

Table 10.4 shows the appearance, maximum melting point and T_α and T_β for the polymers studied. Here T_α and T_β refer to the temperatures of maximum tan δ_E for the two relaxation processes observed using a vibrating reed method, where the frequency varied from 50 to 300 c/s over the temperature range studied. It is seen from the table that polyhex-1-ene to polyhexadec-1-ene are essentially amorphous. They could, however, be crystallized by orientation into fibre form. The melting points given in the table refer to the maximum melting point obtained for a given polymer in fibre form. Polyethylene and polypropylene are included in the table for comparison with the higher polymers.

The melting point is seen to decrease on going from polyethylene to polyhept-1-ene, then increases continuously up to polyoctadec-1-ene. The initial decrease in melting point is due to the increase in alkyl side-group which makes the crystalline packing more open. At long alkyl side-chains, the side-chains themselves pack side by side. The melting point now corresponds to a normal paraffin, and thus increases with increasing alkyl length.

In Table 10.4 the T_α values correspond to the backbone relaxation temperatures at the given frequency. As the alkyl side-chain length increases, there is first a decrease to around $-40°C$, then from polydodec-1-ene onwards T_α increases to about 55°C. The initial decrease can be explained as an internal plasticization of the backbone relaxation by flexible alkyl side-chains, an effect which has been observed in the alkyl methacrylate and alkyl acrylates (see Chapter 8). The increase in T_α

Table 10.4. Some physical properties of the higher polyolefins

Polymers	Appearance at 20°C	Melting point (maximum) (°C)	T_α (°C/c/s)	T_β (°C/c/s)
Polyethylene		[a]138	(? -20 to -60)	[b]$-120/300$
Polypropylene		[a]180	$+15/100$	—
Polybut-1-ene		[a]132[1]	$+15/150$[2]	$-125/430$[2]
Polypent-1-ene		[a]80[3]	$+3/80$	$-130/300$
Polyhex-1-ene	rubber	—	$-23/170$	$-130/500$
Polyhept-1-ene	sticky rubber	17	$-31/180$	$-148/320$
Polyoct-1-ene	viscous gum	—	$-42/140$	$-160/300$
Polynon-1-ene	gum	19	$-47/170$	$-160/310$
Polydec-1-ene	rubber	34	$-35/180$	$-160/590$
Polydodec-1-ene	rubber	49	$-6/150$	$-145/520$
Polytetradec-1-ene	stiff rubber	57	$+10/60$	$-130/450$
Polyhexadec-1-ene	stiff rubber	67·5	$+40/30$	$-125/200$
Polyoctadec-1-ene	white fibrous powder	71	$+55/25$	$-110/170$

[a] Maximum m.p. obtained here by optical birefringence method.

[b] A main-chain process in polyethylene; in the α-olefins the side-chains are involved in subsidiary processes.

[1] Polybut-1-ene can exist in two crystal forms. The melting points are: rhombohedral 134·7°C, tetragonal 121·5°C.

[2] M. Takayanagi and others (1963) (*Rept. Progr. Polymer Phys., Japan*, **6**, 121) have found $T_\alpha = -10°C$ and $T_\beta = -145°C$ at 100 c/s.

[3] Polypent-1-ene may exist in two crystal forms, $T^I_{melting} \rightleftharpoons 130°C$ (stable modification), $T^{II}_{melting} = 80°C$.

above polydodec-1-ene onwards must in some way be associated with the crystallizability of the longer side-chains, as was the melting point described above.

The second process, which occurs between -100 and $-160°C$ is the same as that observed in polyethylene and its copolymers described above. This process is found in all the higher polyolefins and is further evidence for the conclusion that this process is due to $[CH_2]_n$ motion where $n > 3$.

Woodward, Sauer and Wall (1961) studied the mechanical properties of several poly-α-olefins as a function of temperature at frequencies between 200 and 3000 c/s. They found two relaxation regions, for polybut-1-ene, polypent-1-ene, poly-4-methylpent-1-ene ($R = (CH_3)_2CHCH_2CH_2$) and poly-3-methylbut-1-ene ($R = (CH_3)_2CHCH_2$). The data of Clark and others (above) and Woodward and others agree for polybut-1-ene and polypent-1-ene. It was found by Woodward and others that T_α moved about 30°c higher on going from polypent-1-ene to poly-4-methylpent-1-ene and also on going from polybut-1-ene to poly-3-methylbut-1-ene. This shift may be due to a hindering of the backbone relaxation due to the incorporation of a branch methyl group in the side-chain. A similar effect has been found for polyalkyl methacrylates and polyalkyl acrylates (see Chapter 8).

10.4 POLYISOBUTYLENE

$$[-\underset{\underset{H}{|}}{\overset{\overset{H}{|}}{C}}-\underset{\underset{CH_3}{|}}{\overset{\overset{CH_3}{|}}{C}}-]_n$$

(4)

Polyisobutylene (PIB) is normally prepared using a Friedel Crafts catalyst with isobutylene at low temperatures ($\simeq -80$°c). Samples prepared in this way are amorphous, but crystallize on stretching. Butyl rubber is a copolymer of isobutylene and isoprene where the isoprene content is as low as one or two per cent, and acts as a centre for cross-linking in vulcanization.

A feature of PIB is that the glass transition occurs near -71°c, which is far lower than that observed in highly branched polyethylenes, amorphous polypropylene, and higher polyolefins. This is unexpected since internal rotation in the PIB chain might be expected to be more hindered than, for example, in polypropylene, due to the additional methyl group. Turner and Bailey (1963) studied PIB ($T_g = -71$°c), poly-3,3-dimethylprop-1-ene ($T_g = -10$°c) and poly-4,4-dimethylbut-1-ene ($T_g = -20$°c) and again it is seen that PIB has an anomalous T_g. Turner and Bailey found that the n.m.r. spectrum of PIB differed from the spectrum of the two other compounds in that the methyl group proton lines were offset to lower fields, indicating greater steric hindrance in PIB. Thus the n.m.r. information would predict that the T_g for PIB would be greater than that for the other two compounds, which is the converse of the observed values. Turner and Bailey concluded that PIB was anomalous, and although the explanation is not clear, the n.m.r. indication of steric hindrance must be part of a future explanation.

A possible answer to the T_g anomaly comes from the thermodynamic theory of the glass transition due to Gibbs and DiMarzio (1958) (see Chapter 2) in which the *energy difference* between stable chain conformations determines T_g rather than the energy *barrier* between conformations. The energy difference for PIB may well be less than that for polypropylene, and the T_g would, on this basis, be lower for PIB. It is worth noting that a similar anomaly is found for polyvinyl chloride ($T_g \simeq 75°$c) and poly-vinylidene chloride ($T_g \simeq 18°$c), polyvinyl fluoride ($T_g \simeq 20$ to $40°$c) and polyvinylidene fluoride ($T_g \simeq -35°$c) (see Chapter 11). Würstlin (1950) has suggested that the T_g decrease is due to the partial compensation of dipoles in the repeat unit.

A standard unfractionated high molecular weight sample ($M_\eta = 1·35 \times 10^6$) of PIB was studied by mechanical experiments in several laboratories. This work has been extensively reviewed by Ferry (1961) and by Tobolsky (1960).

Fitzgerald, Grandine and Ferry (1953) studied the shear modulus and compliance in the frequency range 30 to 5×10^3 c/s over the temperature range $-45°$c to $+100°$c, for the standard sample. In this range the main rubber–glass relaxation was observed. The J' and J'' data were super-posed at the reference temperature 25·0°c. The shift factors log a_T were also obtained with respect to the reference temperature.

Ferry and others (1953) also calculated tan δ_G as a function of log ωa_T from their J', J'' master curves, and found that although J'' exhibited only one maximum, tan δ_G appeared to have an additional high-frequency shoulder with respect to the main relaxation peak. They compared their tan δ_G data for PIB with similar data for polystyrene and for plasticized PVC polymers. No anomalous tan δ_G structure was observed for these other polymers, and they suggested that the curious behaviour of PIB was associated in some way with steric hindrance present in the PIB chain.

Catsiff and Tobolsky (1955) reviewed the early work on the stress relaxation of PIB and presented new data for the standard sample des-cribed above. Measurement in the ranges 10^{-2} hr to several hours and -80 to $+25°$c gave Young's modulus, E, as a function of time and temperature. These data were used to form a master curve of $E(t)$ as a function of reduced time, at the reference temperature 25°c.

Catsiff and Tobolsky converted $E(t)$ into $E^*(\omega)$, and then converted $E^*(\omega)$ into the shear modulus $G^*(\omega)$, all at the reference temperature 25°c. In converting $E^*(\omega)$ into $G^*(\omega)$, the complex bulk modulus data of Marvin and others (1954) were taken into account. The result of the rather in-volved transformation of stress relaxation data into complex shear modulus data is seen in Figure 10.23. Here we see G' and G'' as functions of ωa_T. The continuous line corresponds to the converted stress relaxa-tion data, the circles are the master curve points of Ferry, Grandine and Fitzgerald (1953) and the crosses are the results of Philippoff on the same standard sample of PIB (Philippoff, 1953). The agreement between the

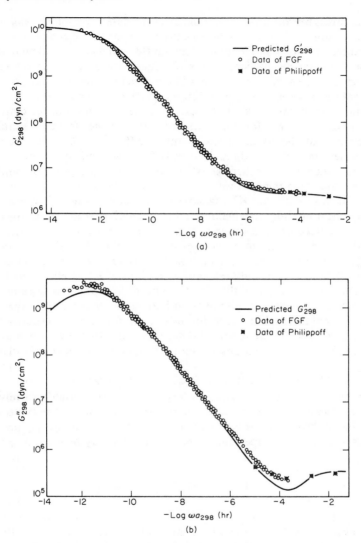

Figure 10.23. Master curves of (a) G' and (b) G'' as a function of reduced frequency at the reduction temperature 25°C for polyisobutylene. Continuous curve obtained by converting stress relaxation data into dynamic data (Catsiff and Tobolsky, 1955). Circles correspond to data of Ferry, Grandine and Fitzgerald (1953). Crosses correspond to data of Philippoff (1953).

converted stress relaxation data and the dynamic data is very good, and is a striking example of the time–temperature superposition principle.

According to Catsiff and Tobolsky the master $\tan \delta_G$ curve calculated from the dynamic mechanical points of Figure 10.23 is so subject to experimental scatter that it is not possible to conclude that there are two $\tan \delta_G$ peaks as suggested above by Fitzgerald, Grandine and Ferry.

Marvin, Aldrich and Sack (1954) described the evaluation of the complex bulk modulus K^* for the standard PIB sample, from accurate measurements of the longitudinal modulus $M^* = K^* + \frac{4}{3}G^*$, and the dynamic shear modulus G^* data. Evidently from M^* and G^*, K^* can be calculated. Marvin and others showed that there was significant bulk relaxation, and when the data were normalized to the reference temperature 25°C, it was found that K''_{max} occurred at the same reduced frequency ωa_T as the shear modulus loss maximum G''_{max}. This is of some significance since J''_{max} occurred at six decades of reduced frequency lower than G''_{max}. If one converted K^* data into compressibility B^* then B''_{max} differs little from K''_{max} in frequency since the limiting bulk moduli K_U, K_R differ by only a small factor. Thus bulk relaxation, in terms of K^* or B^* occurs in the same frequency range as that for G^* (and in fact E^*), and does not correspond to the compliance relaxation frequency. Marvin and others also showed that the a_T shift factors for bulk relaxation data corresponded to the a_T factors for shear modulus data, indicating that the activation energies for volume deformations are about the same as for shear deformations.

Figure 10.24 shows the collected frequency–temperature data for the glass–rubber relaxation in PIB. The large shift between modulus and compliance locations is seen. The plot curves towards lower temperatures in accord with the behaviour normally observed for amorphous polymers in the glass-transition temperature region. The plot may be fitted to the WLF equation

$$\log a_T = -\frac{C_1(T - T_g)}{C_2 + T - T_g}$$

with $C_1 = 16.5$, $C_2 = 104°$ for $T_g = -71.2°$C. Thurn and Wolf (see Hendus and others, 1959) studied the effect of molecular weight on the mechanical relaxation in PIB. Measurements at 5×10^5, 2×10^6 and 4.8×10^6 c/s showed that the temperature of $(\tan \delta_G)_{max}$ i.e. T_{max} rapidly increased with increasing molecular weight below a molecular weight of 2000. Between molecular weights of 2000 and 10^4, the T_{max} variation became gradual. Above 10^4, T_{max} was independent of molecular weight at each frequency (cf. polyvinyl acetate; Section 9.2a).

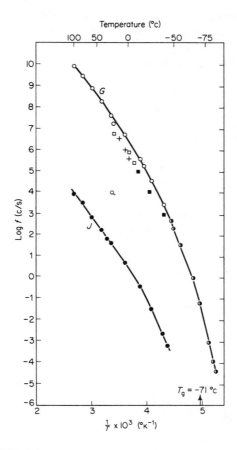

Figure 10.24. The frequency–temperature location of the maxima for dielectric and mechanical relaxation in polyisobutylene.

Mechanical data

⊙ G'' vs. $\log f$ (Ferry, Grandine and Fitzgerald, 1953)

◔ $\tan \delta_G$ vs. $\log f$ (Ferry, Grandine and Fitzgerald, 1953)

● J'' vs. $\log f$ (Ferry, Grandine and Fitzgerald, 1953)

◐ G'' vs. $\log f$. Points calculated from converted stress relaxation data of Catsiff and Tobolsky (1955)

+ $\tan \delta_M$ vs. T (Kabin and Mikhailov, 1956)

Dielectric

□ $\tan \delta_\varepsilon$ vs. T (Kabin and Mikhailov 1956)

■ $\tan \delta_\varepsilon$ vs. T (Herrman, unpubl. work)

The symmetry of the PIB chain renders the molecule essentially non-polar; thus one would not expect PIB to exhibit an intrinsic dielectric loss. Kabin and Mikhailov (1956) reported dielectric loss measurements on PIB. The losses were extremely small, as expected, and are probably due to oxidation of the polymer chain, the oxidized groups acting as a probe (as in polyethylene, etc.) for the study of the chain motions. One dielectric peak was observed and the frequency–temperature locations for $(\tan \delta)_{max}$ are shown in Figure 10.24.

From the frequency–temperature diagram (Figure 10.24) we see that the mechanical relaxation plot curves both in J and G as the temperature decreases. The G curve is more closely associated with T_g (dilatometric) than the J curve, if we assume that the time scale for T_g dilatometric is near 10^{-1} to 10^{-3} c/s in equivalent frequency. We also see that the dielectric points correspond to the G curve (mechanical relaxation) rather than the J curve (mechanical retardation). Although it has often been stated that ε^* and J^* are comparable via relaxation models, it is often found that ε^* and G^* (or E^*) compare better in frequency–temperature location, and polyisobutylene is a good example of such a comparison.

10.5 NATURAL RUBBER

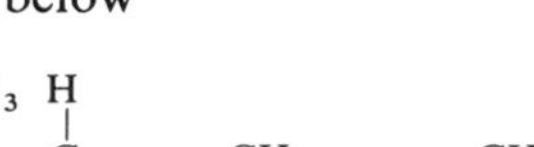

[—CH₂=C—CH=CH₂—]ₙ
CH₃

(5)

Hevea Rubber (or natural rubber) is obtained from the latex (or milk) of the tree *Hevea brasiliensis*. It has the chemical structure poly-*cis* 1,4-isoprene, which is shown below

CH₃
H
C
C
CH₂
CH₂
CH₂
CH₂
C
C
CH₂
CH₃
H

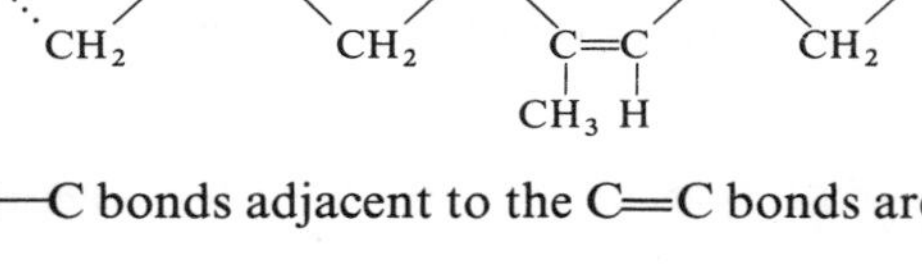

Note that the C—C bonds adjacent to the C=C bonds are fixed in the *cis* configuration.

The other naturally occurring polyisoprene is Gutta Percha (or balata) which has the structure poly-*trans*-1,4-isoprene.

CH₂
CH
CH₂
CH
CH₂
C
CH₂
C
CH₂
CH₃
CH₃

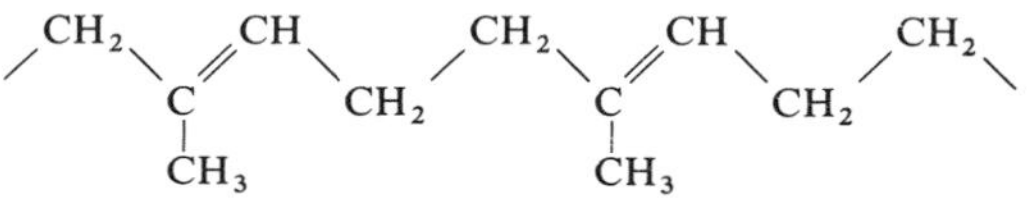

The physical properties of hevea and gutta percha are quite different. Hevea is amorphous and rubbery at room temperature, flows above 60°C and crystallizes on cooling below 0°C or on stretching. In the crystalline state the unit cell contains two right-hand and two left-hand cis-1,4-isoprene units. Gutta percha on the other hand is highly crystalline at ordinary temperatures. The difference between the two natural forms is clearly due to the greater symmetry of the chain in gutta percha.

The free-radical polymerization of isoprene leads to a mixture of cis-1,4, trans-1,4, -1,2 and -3,4 components in the resultant polymer chain. As a result the free-radical polymers are inferior to hevea in their physical properties. The synthesis of poly-cis-1,4-isoprene was not achieved until the Ziegler stereospecific catalysts were discovered. Shortly after this discovery the Goodrich Research Laboratories synthesized this isomer using a Ziegler catalyst, and the Firestone Laboratories also achieved the synthesis using metallic lithium, a stereospecific catalyst.

In commercial applications, hevea rubber is strengthened by vulcanization. This process involves the crosslinking of chains by sulphur bridges which are made at the C=C linkages. The product, known as a vulcanizate cannot flow due to crosslinks, and also has a higher modulus than the original material. Other crosslinking agents may be used, particularly peroxides. The mechanical properties of rubber are also improved by loading the polymer with carbon black.

The dielectric and dynamic mechanical properties of rubbers and their vulcanizates have been widely studied. We shall merely discuss the main features of the behaviour of natural rubber and the effect of vulcanization and loading on these properties.

10.5a Unvulcanized Natural Rubber

Mechanical Properties

Payne (1958) has described in detail the mechanical properties of natural rubber and its modifications. He studied the complex shear modulus of unvulcanized natural rubber in the ranges -75 to $+100°C$, and 10^{-2} to 50 c/s. These data were reduced by the Williams–Landel–Ferry shift procedure (Ferry, 1961), which yielded overall master curves for G' and G'' as a function of reduced frequency at the reduction temperature $273°K$. These plots are shown in Figure 10.25, and the excellence of superposition shows that the shape of the mechanical distribution of relaxation times $\phi(\ln \tau)$ is independent of temperature.

The shift factors a_T required for superposition yield the relaxation frequency as a function of temperature, since we have from Figure 10.25

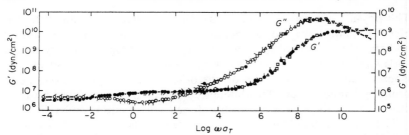

Figure 10.25. Master curves of G' and G'' as a function of reduced frequency at the reference temperature 273°K for unvulcanized natural rubber (Payne, 1958).

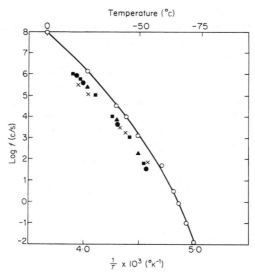

Figure 10.26. Frequency–temperature location of the relaxation maxima in unvulcanized rubber.

○ G'' vs. log f superposition procedure (Payne 1958)
■ tan δ_ε vs. T dielectric (Norman, 1953) dry purified rubber
▲ tan δ_ε vs. f dielectric (Norman, 1953) dry purified rubber
● tan δ_ε vs. T dielectric (Norman, 1953) dry crude rubber
× tan δ_ε vs. T dielectric (Scott and others, 1933) dry crude rubber

that log $2\pi f_{\max} = 8.8$ at 273°K. Figure 10.26 shows the frequency–temperature location of $G''_{\max}$.

We see that the plot curves strongly at lower frequencies and temperatures. This is the normal behaviour of a glass–rubber relaxation, and the mechanism for the relaxation is the reorientation of the backbone of the polymer (micro-Brownian motion).

Payne found that the shift factors could be expressed as a function of temperature according to the WLF equation

$$\log a_T = -\frac{8 \cdot 86(T - T_s)}{(101 \cdot 6 + T - T_s)}$$

Here $T_s = 248°\text{K}$, for unvulcanized natural rubber (note $T_g = 205°\text{K}$).

Dielectric Properties

Norman (1953) made a careful study of the dielectric properties of unvulcanized natural rubber. Care was taken during the preparation of samples to minimize oxidation effects and to ensure a high chemical purity of the rubber. Figure 10.27 shows the dielectric loss tangent as a function of temperature at given frequencies for dry, raw purified rubber. Although the losses are small, a relaxation region is clearly seen. Norman found that oxidation via exposure of the sample to dry air, increased the loss,

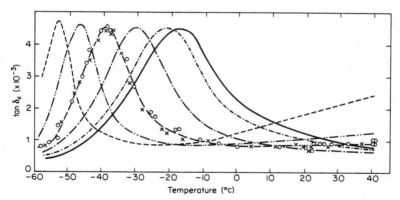

Figure 10.27. Dielectric loss tangent as a function of temperature at given frequencies for a sample of dry purified rubber (Norman, 1953).

- – – 50 c/s   — ·· — ·· — 800 c/s   — — — 10^4 c/s
- — · — · — 10^5 c/s   — · — · — · — 5×10^5 c/s   —— 10^5 c/s
- $\times$, $\bigcirc$ refer to samples at 10^4 c/s

particularly in the low-frequency–high-temperature region. It was found that the loss curves for dry, raw crude rubber were essentially identical with those for dry purified rubber shown in Figure 10.27. On exposure to moist nitrogen 0·5% water was absorbed compared with 0·1% for the purified rubber under the same humidity conditions. This quantity of water increased the loss of crude rubber very considerably, particularly in the low-frequency–high-temperature region.

The well-defined relaxation region, Figure 10.27, is due to micro-Brownian motions of the chain and is to be correlated with the mechanical relaxation of Figure 10.25. The frequency–temperature location of the dielectric process is shown in Figure 10.26, where we see that the dielectric curve lies at slightly lower frequencies than the mechanical G''_{max} curve.

In Figure 10.27 at 50 c/s, the loss increases with increasing temperature above $-10°$C. This loss region is associated with d.c. conductivity or Maxwell–Wagner polarization, due in both cases to impurities, particularly absorbed water.

Scott and others (1933) measured the dielectric properties of purified rubber with essentially the same results as given above. These data are included in Figure 10.26.

Since the polyisoprene chain is essentially non-polar, a dipole relaxation process should not occur in the polymer. It is likely that the observed loss is due to oxidation of the chain, as was the case for polyethylenes.

10.5b Crosslinked Natural Rubber

Mechanical Properties

If natural rubber is mixed with sulphur and heated, vulcanization occurs. This is partially a crosslinking of polyisoprene chains with sulphur bridges which occur at the ethylenic links. The resultant polymer cannot flow, and has an increased rubbery modulus. Crosslinks may also be produced by reacting natural rubber with organic peroxides.

Payne (1958) has shown that vulcanization with sulphur decreases the mechanical relaxation frequency. Ebonite, which is a very highly cross-linked rubber exhibits its glass–rubber relaxation some six decades lower in frequency than the unvulcanized polymer, and also has a rubbery modulus at least five times larger than the natural product.

Schmieder and Wolf (1953) have investigated the effect of sulphur vulcanization on the relaxation region in natural rubber. They find, using a torsion pendulum, that the temperature of tan δ_{max} changes from $-40°$c for 5% sulphur to $+63°$c for 30% sulphur. Figure 10.28 shows the plot of the temperature location of the maximum mechanical loss tangent for given sulphur contents at given frequencies, obtained from Schmieder and Wolf's data.

Mason (1964) has investigated the mechanical and thermal expansion properties of natural rubber crosslinked to different degrees using dicumyl

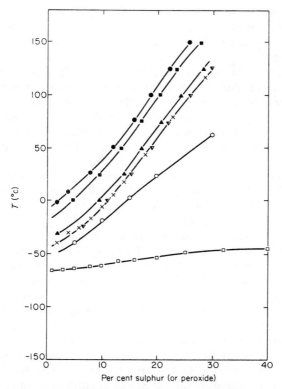

Figure 10.28. The variation of loss maximum location with sulphur content for rubber vulcanizates at given frequencies.

Mechanical
○ $(\tan \delta_G)_{max}$ vs. T (1 c/s) (Schmieder and Wolf, 1953)
□ G'' vs. T (1 c/s) (Mason, 1964)

Dielectric (Scott and others, 1933)
● $\tan \delta_\varepsilon$ vs. T (3×10^5 c/s)   ▲ $\tan \delta_\varepsilon$ vs. T (1×10^3 c/s)
■ $\tan \delta_\varepsilon$ vs. T (1×10^5 c/s)   × $\tan \delta_\varepsilon$ vs. T (60 c/s)
▽ 'Short time' conductivity results

peroxide as the crosslinking agent. The mechanical measurements were performed using a torsion pendulum around 1 c/s, and in Figure 10.28 we show the temperature location of G''_{max} as a function of combined dicumyl peroxide. The feature of these data is the very slow variation in the location of the relaxation region compared with that observed for sulphur vulcanizates. Thus a peroxide-cured sample containing 30 % crosslinking agent has a glass transition near −49°C, whereas a 30% sulphur vulcanizate has a glass transition near $+62°C$, at a frequency around 1 c/s. It is quite clear from this contrast in behaviour that the reaction of sulphur with poly-*cis*-1,4-isoprene is chemically quite different from that of dicumyl peroxide with the polymer.

It has been shown (Moore and Watson, 1956) that the reaction between natural rubber and di-*t*-butyl peroxide produces a network crosslinked by C—C bonds. In sulphur vulcanization, only part of the combined sulphur goes to form polysulphide crosslinks between chains. The remainder of the combined sulphur is combined as heterocyclic groups arranged randomly along the main chain (Farmer and Shipley, 1947; Bloomfield, 1947, 1949).

With this knowledge of the chemical structure of the sulphur- and peroxide-cured rubbers, the difference in mechanical properties is qualitatively understood. The mechanical relaxation process requires a cooperative motion along the chain and between chains. For 20% combined dicumyl peroxide the molecular weight between crosslinks is around 10^3, or twelve monomer units. This is sufficient to impair micro-Brownian motions and is reflected in a raised temperature for relaxation, as is observed in Figure 10.28. The actual motional units along the chain are not seriously affected by the crosslinks. The reaction with sulphur, on the other hand, mainly affects the relaxation properties via the heterocycles formed along the chain, which slow down the chain motions. The crosslinking with sulphur would only affect the relaxation to the same degree as that observed for the peroxide-cured rubbers. Heinze and others (1962) are in agreement with this result. It is also possible that some of the sulphur in the sulphur vulcanizates acts as a filler in a similar manner to carbon black. The action of carbon black on the mechanical properties will be described below (Section 10.5c).

Mason (1964) observed that the mechanical relaxation curves broaden with increased degree of crosslinking. If one considers that the micro-Brownian motions are controlled in part by available free volume, a broadened relaxation curve could correspond to a broadened distribution of free volume in the system. Mason applied the concept of a distribution of free volumes to explain the decreased sharpness of the dilatometric glass temperature as the degree of crosslinking is increased. Taking a Gaussian distribution of free volume, with a standard deviation parameter σ, Mason showed that $\Delta\alpha \cdot \Delta T = 3 \cdot 29\sigma$. Here ΔT is the range over which the volume transition from rubber to glass occurs, and $\Delta\alpha$ is the change in thermal expansion coefficients through the transition. In this way Mason supported the hypothesis that the distribution of free volumes increases with increasing crosslink density.

Stratton and Ferry (1963) investigated six natural rubber vulcanizates, each crosslinked in a different way. One sample (385) was prepared by γ irradiation, another (73) was prepared using tetramethyl thiuram disulphide (TMT) as the crosslinking agent. Sample 77 was dicumyl peroxide crosslinked rubber. Samples 74, 75 and 76 were sulphur vulcanizates containing 3 to 4% sulphur, and the curing times were 60, 40 and 15 min respectively at 140°C. All samples were lightly crosslinked with a molecular weight between crosslinks around 4×10^3 for samples 74, 75 and 76, and 7×10^3 for samples 385, 73 and 75. Figure 10.29 summarizes the retardation spectra of the six vulcanizates calculated by the method of reduced variables from J' and J'' in the ranges 0·1 to 1600 c/s, -10 to 55°C. We see from Figure 10.29 that all samples show the glass–rubber relaxation in approximately the same log τ region at 25°C. The difference between samples clearly lies in the long relaxation time region. Sample 385 has a long tail to the glass–rubber relaxation, whereas sample 76 exhibits no tail. Stratton and Ferry point out that the irradiated sample (385) and dicumyl peroxide-cured samples (77) have crosslinks that are covalent and bulky. The TMT sample is believed to crosslink via a single interstitial sulphur atom (Studebaker and Nabors, 1959). In conventional sulphur vulcanizates (74, 75 and 76) the crosslinks are complex.

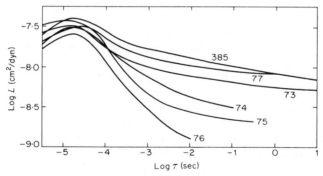

Figure 10.29. Retardation spectra for six vulcanizates, reduced to 25°C (Stratton and Ferry, 1963).

Although the chemical difference between the six different vulcanizates is fairly well established, Stratton and Ferry conclude that the long relaxation time processes (low frequency) are puzzling and no adequate theory is available to explain them.

Ferry and others (1964) followed up the earlier work of Stratton and Ferry (1963) by studying sulphur and dicumyl peroxide vulcanizates in which the crosslink spacings were known. M_c varied from 5×10^3 to 2×10^4 for five samples of dicumyl peroxide vulcanizate. The vulcanizates were examined in the frequency range 0·1 to 10^3 c/s, and the temperature range -18 to $+55°$c. The data were reduced to form master plots. As in the data of Stratton and Ferry, the low-frequency losses were much larger for lowly crosslinked samples (M_c large). As M_c decreased so the low-frequency 'tail' of the main relaxation decreased. The same behaviour was observed for the sulphur vulcanizates. Thus M_c determines the low-frequency (long relaxation time) losses. Although no clear explanation of this effect was given by Ferry and others it seems that the overall effect of the crosslinks is to abstract the low-frequency modes from the Rouse spectrum (Section 5.3a) so that the highly crosslinked material will not have the long low-frequency tail observed for the lightly crosslinked material.

Dielectric Properties

Scott and others (1933) made a comprehensive study of the effect of sulphur vulcanization on the dielectric constant and loss tangent of natural rubber. Measurements were performed in the ranges 60 to 3×10^5 c/s, -75 to $+235°$c, 0 to 32% combined sulphur. It was found that the static dielectric constant increased from about 2·3 for 0% sulphur to 4·5 for 30% sulphur. The unvulcanized polymer exhibited a very small relaxation region; the intensity of the relaxation increased and its location was shifted to higher temperatures continuously with increased sulphur content. The latter effect is in accord with the behaviour observed in the mechanical experiments, and is in the opposite direction to plasticization.

Scott and others were among the first to present dielectric data in the form of a contour diagram. Figure 10.30 shows the dielectric loss tangent (tan δ_ε) as a function of temperature and combined sulphur at 60 c/s. The ridge represents the locus of the loss maximum, and we see that as the sulphur content increases so the temperature of the loss maximum increases. We have evaluated the temperature of tan $\delta_{\varepsilon,\text{max}}$ for given sulphur contents and given frequencies, and Figure 10.28 shows these data in the range 60 to 3×10^5 c/s. The constant frequency plots are parallel but curved. The effect on the dielectric relaxation is very similar to that observed for the mechanical relaxation.

It is of interest to note that Scott and others measured the apparent d.c. conductivity of rubber–sulphur vulcanizates at prescribed times after the application of a step voltage. It was found that the apparent conductivity

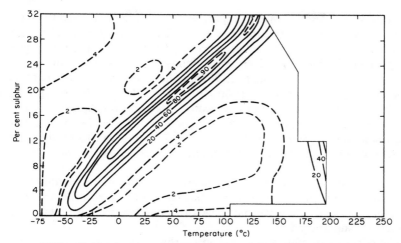

Figure 10.30. Contour diagram of dielectric loss tangent as a function of temperature and combined sulphur at 60 c/s. Tan δ values given in units of 10^{-3} (Scott and others, 1933).

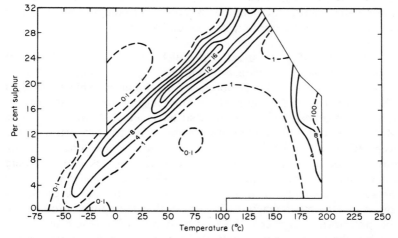

Figure 10.31. Contour diagram of 'short time' conductivity (0·002 sec) as a function of temperature and combined sulphur. Conductivity values given in units of 10^{-12} mho/cm (Scott and others, 1933).

was a strong function of time, the value decreasing with increased elapsed time. Figure 10.31 shows the contour diagram for the apparent conductivity at 0·002 sec after the application of the step voltage. Figure 10.31

shows a strong resemblance to Figure 10.30 and this was noted by Scott
and others but no explanation was given. The explanation is now clear
in view of the transform between a.c. and transient experiments (see
Chapter 7). A time of 0·002 sec corresponds to a transform frequency of
$f = 0·1/0·002 = 50$ c/s (Williams, 1962b). The dielectric loss factor at this
frequency is given by $\varepsilon''_f = (1·80 \times 10^{12}\sigma_i)/f$. Here σ_i is the specific
conductivity (ohm cm). Thus the dielectric loss factor is proportional to σ_i
at the prescribed frequency, and the maximum in σ_i in Figure 10.31 is
equivalent to the maximum in loss at 50 c/s. Thus Figure 10.30 and Figure
10.31 have similar forms since the difference between 60 c/s and 50 c/s is
trivial.

Included in Figure 10.28 is the temperature of maximum conductivity
as a function of sulphur content at 0·002 sec obtained from Figure 10.9.
As expected the 60 c/s curve and the short time ($f = 50$ c/s) conductivity
curves are superposed.

The effect of sulphur vulcanization on the dielectric properties arises
due to two processes. The sulphur crosslinks hinder the motion of the
chains to the same degree as that observed in the mechanical behaviour of
peroxide-cured rubber. This effect is small, and the dominant effect of
sulphur is the heterocyclic polar chains attached to the backbone. These
stiffen the chain and contribute a large dipole intensity to the observed
motions.

Schallamach (1951) investigated the dielectric properties of natural
rubber vulcanizates in the frequency range 10^2 to 10^8 c/s. The same results
as Scott and others (1933) were obtained for sulphur vulcanizates studied
in the range 0 to 6% combined sulphur. Using accelerators for the
vulcanization process, it was found that tetramethyl thiuramdisulphide
(TMT) accelerator with sulphur gave the same electric properties for the
vulcanizate as the vulcanizate with sulphur and no accelerator. Using
diethyl dithiocarbamate (ZDC) as an accelerator the loss maxima were
smaller than the TMT-Sulphur and ZDC-Sulphur vulcanizates of the
same sulphur content. Since ZDC favours the formation of crosslinks
in preference to other reactions it is clear that the contribution to the
dipole polarization from crosslinks is less than that from the heterocycles
along the chain. This agrees with the explanation given above for the
difference between sulphur- and peroxide-cured natural rubber.

Schallamach (1951) also shows that the ZDC accelerated vulcanizate
has a shorter relaxation time than a TMT accelerated vulcanizate or an
unaccelerated vulcanizate for a given sulphur content. Again this is
consistent with preferred crosslinks in the ZDC compound with fewer
complexes along the chain compared with the other vulcanizates. The

smaller the fraction of combined sulphur along the chain compared with that forming crosslinks, the smaller is the hindrance to molecular motion. Waring (1951) has given dielectric data in the range 0·1 to 100 Mc/s for a number of natural rubber–sulphur vulcanizates containing different additives. It was shown that the dielectric technique is a convenient way of following the progress of curing reactions. Dipole moments were calculated assuming a dilute solution equation for dipoles in a non-polar medium. The dipoles were assumed to be disulphide or monosulphide linkages. The dipole moment data were consistent with either linkage.

10.5c Carbon black filled Rubber Vulcanizates

Mechanical Properties

It is well known that natural rubber vulcanizates may be considerably reinforced by compounding with colloidal carbon black. The modulus of the carbon black vulcanizates is higher in both the rubbery range and the glassy range than that of rubber. It has been found that the carbon particles interact with the rubber matrix ; the act of reinforcement is not therefore a simple one of a mixture of high modulus (carbon) and low modulus (rubber). The exact nature of the carbon–rubber link has not been determined, but strong bonding does exist, and these are formed from the reaction of rubber free radicals with active sites on the carbon particle (see Billmeyer, 1962).

The dynamic modulus data for carbon black filled vulcanizates has been presented by Payne (1958, 1960, 1962a, 1962b). He noted that many studies on the mechanical properties of this system have found non linearity in modulus as a function of strain amplitude. The effect was absent for pure gum vulcanizates and for inert filler vulcanizates. Payne worked at 0·1 c/s, and found Young's modulus and shear modulus to be independent of amplitude at low amplitudes. Figure 10.32(a) shows the Young's modulus data for samples containing various proportions of MAF black. The curves are reminiscent in shape of modulus–temperature plots where the modulus drop has an absorption region associated with it. In fact, Payne found that tan δ_E showed a relaxation-type peak and these data are shown in Figure 10.32(b), and the peak positions correspond to the inflection points of the modulus–strain amplitude plots.

Payne normalized the shear modulus–strain amplitude data, and showed that G_0, the limiting low-amplitude modulus increased continuously with carbon loading. In particular, the loaded vulcanizate containing 38·4 volume per cent carbon had a modulus 125 times larger than the pure gum vulcanizate. Payne showed that G_0 decreased rapidly

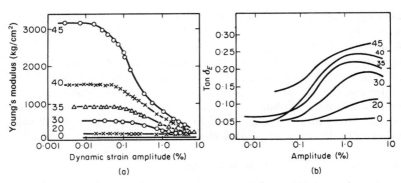

Figure 10.32. (a) Young's modulus as a function of dynamic strain amplitude for natural rubber vulcanizates containing various proportions of medium abrasion carbon black. Figures 0, 20, 30, 35, 40 and 45 give the different amounts of carbon loading (Payne, 1962a); (b) $\tan \delta_E$ as a function of dynamic strain amplitude for samples of Figure 10.32(a) (Payne, 1962a).

with increasing temperature; this is in the opposite direction to that observed for a pure gum rubber. Payne studied G_0 values for 'furnace' carbon blacks and for 'thermal' carbon blacks. It was found that G_0 increased with the increase of surface area per unit mass of carbon particle. This correlates with an increased bonding of the rubber to the carbon particle as its surface area is increased via mode of preparation.

In a later paper (Payne, 1962b) the limiting high strain shear modulus G_∞ was investigated. It had been found earlier that G_∞ was far larger than the modulus of the pure gum rubber, and the increase was due to two factors (a) an inert filler effect (b) an enhancement factor associated with the bonding of the carbon particles to the rubber.

The limiting modulus G_∞ was written as $G_\infty = G_{\text{pure gum}} \times F(f, c) \times F(A)$. Here $F(f, c)$ is a function of the shape factor f of the filler and c its volume concentration. $F(f, c)$ is the 'inert' filler (contribution (a) of the previous paragraph) and was determined from model vulcanizates using glass spheres as the filler particles. Payne has reviewed the theoretical relations which have been derived by several workers to express $F(f, c)$. He showed that none of these expressions are adequate to express G_∞ up to high inert filler concentrations (i.e. above about 30 % filler). $F(A)$ is the enhancement factor due to bonding between the carbon particles and the rubber matrix. Payne showed that $F(A)$ could be estimated in a relative way by comparing the swelling properties of carbon loaded vulcanizate with pure gum vulcanizate. Since the modulus G was approximately inversely proportional to the molecular weight between crosslinks, M_c, the measurement of M_c from toluene swelling measurements on samples yielded (G_{carbon}/G_{gum}) ratios taking into account only the carbon particle–rubber interaction. Thus, $F(A)$ was evaluated.

Payne showed that $F(A)$ increased from 1·0 to 3·0 on going from zero to 38 volume per cent carbon black. He then calculated (G_∞/G_{gum}) using the $F(f, c)$ and $F(A)$ values derived in the above way and found that the calculated values agreed with experiment with good accuracy up to about 36 % carbon black.

Thus Payne has clearly shown that the 'reinforcing' effect of carbon black is due to (a) $F(f, c)$ (the 'hydrodynamic' effect) and (b) $F(A)$ (the direct linking between carbon and rubber molecules).

The hydrodynamic effect $F(f, c)$ is 5·42 at 40 % carbon black, while $F(A)$ is 3·0 at 38·4 % carbon black. The factors are therefore of the same order of magnitude, and their product gives the overall increase in the modulus of the reinforced rubber.

Dielectric Properties

Carbon particles have a large intrinsic conductivity. The inclusion of carbon black into a rubber vulcanizate gives rise to a complex dielectric behaviour. The dielectric properties arise from the following components (a) normal dipole loss from the sulphur–rubber chains, (b) conductivity losses due to direct conduction paths of linked carbon particles in the matrix ; (c) Maxwell–Wagner losses due to the conducting carbon particles in a non-conducting rubber medium and (d) 'structural losses' characterized by a loss which is independent of temperature and frequency at a given carbon composition. The losses (d) are a function of the actual structure formed by the carbon black in the mixture and give rise to a high-frequency dielectric constant ε'_∞, which is larger than the high-frequency dielectric constant of the normal unfilled sulphur vulcanizate.

The dielectric properties of carbon black–rubber sulphur vulcanizates have been studied by a number of workers, notably Thirion and Chasset (1951), Desanges, Chasset and Thirion (1957, 1958), Chasset, Thirion and Ngyuen (1959), Dogadkin and Lukomskaya (1953) and Lukomskaya and Dogadkin (1960). The results obtained from the above studies together with many other studies have been reviewed by Lukomskaya and Dogadkin (1960).

The first analysis of the ε' and ε'' plots as a function of frequency for various tread vulcanizates was made by Schmeider and others (1945) and Carter and others (1946). The loss curves were found to have two maxima, one at $7 \cdot 10^3$ c/s and the other at 5×10^6 c/s. At low frequencies the loss increased rapidly with decreasing frequency. The higher-frequency loss maximum was due to dipole orientation, the lower-frequency maximum was shown to be due to Maxwell–Wagner polarization and the low-frequency rising loss was due to d.c. conduction in the sample. It was noted that the higher-frequency dielectric constant for a neoprene tread (polychloroprene) was $6 \cdot 3$, whereas ε_∞ for the pure gum rubber was $3 \cdot 03$. No explanation for this difference was given, but we shall see below that the 'structural' factor is responsible for this effect.

Lukomskaya and Dogadkin (1953, 1960) made a more careful study of the frequency dependence of ε' and ε'' and also investigated the effect of temperature. It was found that a further loss region could be resolved which was independent of frequency and temperature. As the carbon black content in natural rubber vulcanizates was increased, so this loss region increased. This loss region is known as the 'structural' loss since its magnitude depends not only on the amount of carbon black present, but also on the type of carbon black used. The limiting high-frequency dielectric constant ε'_∞ was found to be far larger than expected in agreement with the work of Carter and others above. Lukomskaya and Dogadkin expressed ε'_∞ as

$$\varepsilon'_\infty = \varepsilon'_1(1 + 3\phi p)$$

Here ε'_1 is the dielectric constant of the unloaded rubber, p is the weight concentration of carbon black and ϕ is a 'shape factor'. $\phi = 1$ for spherical particles, and $\phi > 1$ if the particles join to form structures (chains or a network). They found that ϕ was 1 for thermal black–rubber vulcanizates, but ϕ was near 3 for 10% channel gas black–rubber vulcanizates.

Lukomskaya and Dogadkin (1960) have outlined the method of separating the dielectric properties into the conductivity, Maxwell–Wagner, 'structural' and dipolar contributions, and have illustrated the method for experimental systems.

Chasset, Thirion and others (1957, 1958, 1959) observed the background 'structural' loss, and noted that the magnitude was associated with the structure or state of aggregation of the carbon black in the matrix. The carbon black structures are broken down when the rubber is swollen or stretched, and they found that the 'structural loss' decreased significantly in both of these cases. For example, no 'structural loss' was observed on swelling in 80% n-heptane solution. They suggest that the 'structural loss' arises from the series arrangement of carbon black particles, i.e. the formation of series microcapacitors. Lukomskaya and Dogadkin (1960) have pointed out that this model would in fact lead to frequency- and temperature-dependent losses.

10.6 POLYSTYRENE

$$[-\overset{\overset{\displaystyle H}{|}}{\underset{\underset{\displaystyle H}{|}}{C}}-\overset{\overset{\displaystyle H}{|}}{\underset{\underset{\displaystyle \bigcirc}{|}}{C}}-]_n$$

(6)

Various forms of polystyrene may be obtained using different polymerization conditions. Free-radical initiators give atactic (amorphous) polystyrene. The polymerization may be carried out in solution, suspension or using emulsion polymerization techniques (see Sorensen and Campbell, 1961). Using sodium naphthalene as catalyst, styrene may be polymerized via an anionic mechanism to give a monodisperse 'living' polymer. Further addition of monomer to the solution results in polymerization to a higher (monodisperse) molecular weight, hence the name 'living polymer'. The polymer is atactic and amorphous (Szwarc, 1960). If an organometallic (stereospecific) catalyst is used, isotactic polystyrene results, which is crystalline, melting at 240–250°C (see Sorensen and Campbell, 1961).

The atactic polymer is of the greatest commercial significance. Due to its commercial importance, the mechanical properties have been extensively studied. The non-polar nature of the polymer chain leads to excellent electrical properties in the field of insulation (e.g. capacitors). The dielectric constant of pure polystyrene is 2·55 at 25°C, and the square of refractive index is 2·53. Dielectric losses are very low with tan $\delta < 2 \times 10^{-4}$ over the range 10^2 to 10^{10} c/s at 25°C (see Von Hippel, 1954).

10.6a Mechanical Properties

The early studies were confined to the atactic amorphous polymer. Schmieder and Wolf (1953) use a torsion pendulum to observe the α peak (in tan δ_G) at 116°C for 0·9 c/s. Inspection of their data suggests a small shoulder to the α peak in the 50°C region, and a small peak near -140°C. They also showed that the location of the α peak depended on molecular weight with T_{max} (tan δ) at 116°C for $\overline{M} = 1\cdot2 \times 10^5$ and 100°C for $\overline{M} = 8 \times 10^4$. The α peak is clearly due to the glass–rubber relaxation since T_g occurs at 100°C in the atactic polymer. Sauer and Kline (1955), Buchdahl and Nielsen (1955), Becker (1955), Merz, Nielsen and Buchdahl (1951) and Jenckel (1954) have given details of the mechanical properties of atactic polystyrene, with particular reference to the α region.

In later studies, Illers and Jenckel (1958, 1959), Illers (1961) and Sinnott (1962) showed that, in addition to the α relaxation, β, γ and δ relaxations could be observed in atactic polystyrene. Figure 10.33 shows tan δ_G as a function of temperature at 0·5 c/s (Illers and Jenckel, 1958). The β and γ relaxations are seen to occur in the same positions as observed in the data of Schmieder and Wolf (1953). A δ relaxation at 38°K (5·59 c/s) was observed by Sinnott (1962) and is shown in Figure 10.34. It is noted that the strengths of the β, γ and δ peaks in tan δ are very small compared with the α peak.

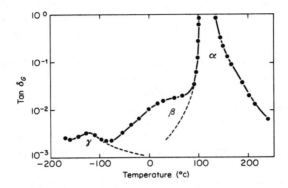

Figure 10.33. Tan δ_G as a function of temperature for lightly cross-linked (1 % triacrylformal) atactic polystyrene $f = 0·5$ c/s (Illers and Jenckel, 1958).

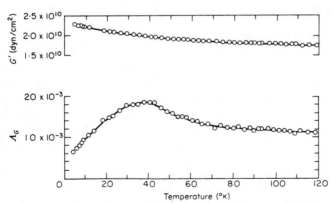

Figure 10.34. Temperature dependence of the shear modulus G' and logarithmic decrement Λ_G for atactic polystyrene at low temperatures (Sinnott, 1962).

Illers and Jenckel (1959) and Illers (1961) showed that the β relaxation is only resolved at low frequencies. At frequencies higher than 40 c/s, the peak is merged into the α region. They evaluated an apparent activation energy of 35–40 kcal/mole, and suggested that this process was due to the motion in parts of the polymer chain in which the phenyl groups have less steric hindrance associated with them compared with the bulk of the chain. Buchdahl and Nielsen (1955), however, noted that the β peak was observed if the polymer chain ends had bulky substituents or if very low molecular weight material was present. Takayanagi (1963) observed the β peak (E'') at 100 c/s as a low-temperature shoulder to the α peak near 50°C, and suggested that it was due to the local relaxation mode of the twisting of main chains (Section 5.5b).

The γ peak of Figure 10.33 was studied by Illers and Jenckel (1958, 1959) and by Illers (1961). Since it was first observed by Illers and Jenckel (1958) for a lightly crosslinked polystyrene, they studied pure uncrosslinked polystyrene (Illers and Jenckel, 1959), and again it was found. Illers and Jenckel (1959) showed that the γ peak was unaffected by stretching and was not due to water or monomer contamination. They tentatively associated the γ peak with the irregularities in the polymer chain of the following form. Normally head-to-tail polymerization occurs, but there is a possibility of head-to-head units occurring. Illers and Jenckel (1958, 1959) and Illers (1961) associated the γ relaxation with the head-to-head units of the chain. Takayanagi (1963), however, suggests that the broad γ relaxation he observed at 100 c/s was associated with the restricted rotation of the phenyl groups.

The δ relaxation observed by Sinnott (1962) was attributed to the motion of the phenyl groups. This result is in general accord with the specific heat measurements of Sochava and Trapeznikova (1958), who reported a specific heat anomaly near 70°K, and concluded that it was due to the rotational vibrations of the phenyl group.

It is quite clear from the preceding paragraphs that the assignment of relaxations in polystyrene to particular mechanisms is extremely tentative.

The temperature variation of the internal friction of isotactic crystalline polystyrene ($\sim 40\%$ crystalline) was studied by Natta, Baccaredda and Butta (1959). Both crystalline and amorphous polymers gave the α peak at approximately the same temperature, but the peak in the crystalline material was comparatively small. A similar result was obtained by Wall, Sauer and Woodward (1959) and Newman and Cox (1960). Newman and Cox studied the anisotropy of the dynamic complex modulus in stretched (uniaxially oriented) crystalline polystyrene, using a vibrating reed apparatus. Takayanagi (1963) studied atactic and isotactic and blends of atactic/isotactic samples. The α relaxation was discussed in terms of a series–parallel model for isotactic and atactic contributions to E''. The α relaxation was sharp in the atactic sample and broad in the isotactic (crystalline) sample, but the actual location of the peak was unchanged. The broadening effect was attributed to restrictions imposed by the crystallites on the amorphous phase. It is found for polyethylene tere- phthalate (see Chapter 13) that increased crystallinity broadens the α relaxation curves and also removes them to higher temperatures. This is clearly also attributable to restriction imposed by the crystallites. In the case of polystyrene, however, the temperature of maximum loss is independent of the degree of crystallinity.

Recently Baccaredda, Butta and Frosini (1965) studied atactic and isotactic samples at low temperatures (50 to 300°K) and high frequencies (10^4 c/s). The γ peak was seen at 195°K (tan δ) in normal and γ irradiated atactic samples. In addition, a rising loss was seen below 100°K in the samples, which corresponds to the δ relaxation observed by Sinnott (1961). A quenched isotactic sample gave a large peak at 160°K and an annealed (crystallinity 40%) sample gave a peak in the same position (160°K) but the peak height was smaller than that observed in the quenched (amorphous) sample. No indication was seen of the δ relaxation observed in the atactic sample. Baccaredda, Butta and Frosini point out that if the γ relaxation in atactic polystyrene is due to head-to-head couplings in the chain, the γ relaxation will be missing in isotactic polymers since the stereospecific polymerization should not allow such defects to occur along the chain. They suggest that the 160°K relaxation peak corresponds to the δ relaxation observed in the atactic polymer, only it occurs some 110°K higher. Baccaredda, Butta and Frosini suggest that the 160°K relaxation is due to the oscillation of the benzene rings.

Kono (1960) determined the temperature dependence of the imaginary parts of the shear and bulk moduli G'' and K'', in the α relaxation region for atactic polystyrene. The measurements were made at 0·5, 1·0 and $2·25 \times 10^6$ c/s. It was found that the K'' maxima occurred about 20°C higher than the maxima for G'' and that the activation energy for shear deformation was smaller than that for volume deformation. Thus, the molecular mechanism for the two types of deformation may be different.

An interesting empirical method of determining molecular weights for polystyrenes has been given by Cox, Isaksen and Merz (1960). Using the temperature of *minimum* tan δ_G observed above the α relaxation near 150°C, at 1 c/s, they obtained the relation $M_n = T_{min}^{-1·9} \times 10^{-5}$. Here M_n is the number average molecular weight.

10.6b Dielectric Properties

The dielectric properties of atactic polystyrene were first measured by Broens and Müller (1955) and by Baker, Auty and Ritenour (1953). Broens and Müller (1955) found the α relaxation peak in the ranges 120 to 145°C over the frequency range 10^2 to 10^5 c/s. The peaks were small with tan $\delta_{max} \simeq 4 \times 10^{-3}$ in the frequency plots. The dipole moment was expected to be near that of toluene ($\mu = 0.30$ D), and was calculated via the Onsager relationship to be 0.2 D/monomer unit.

Baker and others (1953) studied atactic polystyrene (I), crosslinked atactic polystyrene (6% divinylbenzene) (II), and atactic polystyrene plasticized with diphenyl (III). It was found that (II) gave the α relaxation at higher temperatures than (I), and (III) gave the α relaxation at lower temperatures than (I). Thus, hindrance to main-chain motions occurred on crosslinking, whereas plasticizer had the usual effect of increasing chain mobility. These data were fitted to a Davidson–Cole empirical equation (Davidson and Cole, 1951), and apparent activation energies of 86, 91 and 45 kcal/mole were evaluated for (I), (II) and (III) respectively. Broens and Müller obtained 79 kcal/mole for (I) which compares favourably with Baker and others.

Saito and Nakajima (1959b) studied the dielectric α relaxation in polystyrene in the frequency range 10^{-4} to 10^6 c/s, with results similar to those given above.

Figure 10.35 shows the frequency–temperature locations of the mechanical α and β relaxations and the dielectric α relaxation. A good correlation exists between the α relaxations observed by the two techniques. The δ and γ relaxations are not given in Figure 10.35 since the amount of information available is small.

10.6c Plasticized Polystyrenes

Illers and Jenckel (1958) have studied atactic polystyrene plasticized with diethyl phthalate, with dibutyl phthalate, and with dioctyl phthalate. The mechanical measurements of tan δ_G at 1 c/s showed the α relaxation moving rapidly to lower temperatures with increased plasticizer content. The β relaxation was submerged below the α relaxation in the plasticized systems, and the γ relaxation moved to *higher* temperatures at the given frequency. The γ relaxation was also studied as a function of frequency (10^{-1} to 10^4 c/s) over a temperature range. Figure 10.36 shows a plot of $\log f_{max}(\tan \delta)$ at given temperatures for diethyl phthalate/polystyrene and dibutyl phthalate/polystyrene systems. Also shown in the figure are $T_{max}(\tan \delta)$ for given frequencies. For the diethyl phthalate/polystyrene systems, the γ relaxation moves to higher temperatures, with

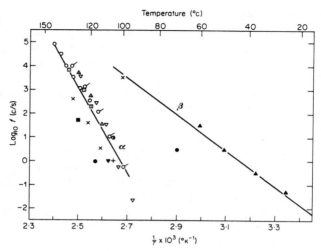

Figure 10.35. The frequency–temperature locations of the α and β relaxations in polystyrene.

Mechanical

● tan δ_G vs. T Atactic (Schmieder and Wolf, 1953)

▲ G'' vs. T Atactic, β process (Illers, 1961)

■ Q^{-1} vs. T Atactic (Wall, Sauer and Woodward, 1959)

▼ tan δ_G vs. T Atactic (Illers and Jenckel, 1958)

× tan δ_E vs. T (Becker, 1955)

+ tan δ_G vs. T Isotactic, amorphous and crystalline (Newman and Cox, 1960)

◑ E'' vs. T Isotactic, atactic and blends (Takayanagi, 1963)

Dielectric

○ tan δ vs. log f Atactic (Broens and Müller, 1955)

⊡ tan δ vs. T Atactic (Broens and Müller, 1955)

△ ε'' vs. T Atactic (Baker and others, 1953)

◔ ε'' data Atactic (Kästner, and Schlosser and Pohl, 1963)

▽ ε'' vs. f Atactic (Saito and Nakajima, 1959b)

increasing activation energy in a continuous fashion as the plasticizer content is increased. However, little change is observed in the dibutyl phthalate/polystyrene systems. It would appear from Figure 10.36 that as little as 10% dibutyl phthalate gives the maximum effect of shifting the γ relaxation, and corresponds to about 40% diethyl phthalate.

Broens and Müller (1955) studied benzyl benzoate/polystyrene mixtures using the dielectric technique, and Baker and others (1953) studied polystyrene/diphenyl mixtures.

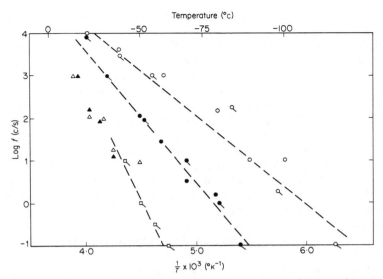

Figure 10.36. The frequency–temperature location of the maximum of the mechanical γ relaxation in plasticized polystyrenes (Illers and Jenckel, 1958).

○ $\log f_{max}$ (tan δ) 9·7% diethyl phthalate
◔ T_{max}(tan δ) 9·7% diethyl phthalate
● $\log f_{max}$(tan δ) 20% diethyl phthalate
◕ T_{max}(tan δ) 20% diethyl phthalate
◻ T_{max}(tan δ) 41·7% diethyl phthalate
△ $\log f_{max}$ 10% dibutyl phthalate
▲ $\log f_{max}$ 21% dibutyl phthalate

10.7 SUBSTITUTED POLYSTYRENES

10.7a Mechanical Properties

$$[-CH_2CH(\langle\!\!\!\bigcirc\!\!\!\rangle\!\!-X)-]_n$$

(7)

Illers (1961) studied the poly-para-halogenated polystyrene series (7) with X = H, F, Cl, Br and I. Using a torsion pendulum at 1 c/s the maximum loss temperature (T_{max}) for G'' was found to move to higher temperatures throughout the series. T_{max} = 106, 109, 125, 132 and 156°C for X = H, F, Cl, Br and I respectively. This compared favourably with the variation of the dilatometric glass temperature T_g, which was 92, 95, 110 and 118°C for X = H to Br. Also it was shown that T_g varied continuously with the van der Waals radius of X. Thus, as the mass and volume of the *para* substituent increases, so the chain motions become more difficult, and T_{max} and T_g move to higher temperatures throughout the series.

The low-temperature region was studied, and the γ peak (G'') was observed for X = H at $-138°$C together with an increasing loss at the very lowest temperatures indicative of the δ relaxation. For X = F, the γ peak was enlarged and also the tail of the δ process. For X = Cl, a double peak was seen, with peaks at -130 and $-160°$C. A single γ peak was observed for X = Br and X = I at -117 and $-110°$C respectively. The data strongly suggest that the δ peak moves to higher temperatures as the mass of X is increased; it is seen in the p-chloro compound, and is merged with the γ peak in X; Br and X = I. Thus as the mass of the substituent is increased, so the phenyl group motions become more difficult and T_{max} moves to higher temperatures. This behaviour is to be compared with that observed by Baccaredda, Butta and Frosini (1965) for atactic and isotactic polystyrenes. In an earlier note, Illers and Jenckel (1959) gave data for X = Cl and X = Br in accord with the later work described above.

In view of the correlation between the temperature location of the mechanical α process at low frequencies with the glass temperature (T_g) it is of some interest to note that Barb (1959) and Overberger, Frazier, Mandelman and Smith (1953) have tabulated T_g for a large number of substituted polystyrenes. It is found that poly-α-methyl styrene has $T_g = 180°$C, a large increase over that of polystyrene. Also for the *para*-n-alkyl substitution, T_g *decreases* rapidly with alkyl chain length up to *para*-n-dodecyl ($-52°$C), then actually increases again at longer chain lengths of substituent.

10.7b Dielectric Studies

Dielectric studies of polystyrene and its halogen derivatives have been made by Fattakov (1952) who found one relaxation attributed to the micro-Brownian motions of the chain. Mikhailov and Borisova (1962) studied poly-2-fluoro-5-methyl styrene and found one dielectric peak in the neighbourhood of 115°C at 20 c/s. On the basis of the low-temperature non-symmetry of the peaks, they analysed them into two components (α and β). The higher-temperature (α) component was assigned to the micro-Brownian motions of the chain. The resolved β component was correlated with the mechanical β relaxation observed by Illers and Jenckel (1958). They attribute this process to the independent motion of the benzene groups. Measurements were made down to $-160°$C and no evidence was seen of the γ relaxation observed mechanically at very low temperatures. The root mean square dipole moment per monomer unit calculated from the dielectric constant data was found to increase with increasing temperature reaching a plateau level above 110°C of 1·44 D. Curtis (1962) studied the dielectric behaviour of poly-*para*-chlorostyrene and poly-*meta*-chlorostyrene. For the *para* compound the α relaxation was observed near 5 c/s at 145°C in the ε″ versus frequency plot. Lower-temperature relaxations were not observed.

The *meta* compound gave two relaxation regions, α and β. Below the dilatometric T_g, the small β peak was observed in the loss factor–frequency plot (10^{-2} to 10^7 c/s). In the softening region, a mixed α and β peak was seen, and above the softening region the merged (α, β) peak was seen. Curtis associated the α relaxation with the micro-Brownian motions of the chain, and the β peak with the hindered rotation of the phenyl side-group. This β peak correlates with the β peak in the mechanical data of Illers and Jenckel (1958). The absence of the dielectric β peak in the *para* compound is understood since the hindered rotation of the *para*-chlorophenyl group does not involve an orientation of dipole moment.

10.8 POLYSTYRENE COPOLYMERS AND PHYSICAL MIXTURES

The glass transition of a random copolymer is found to depend on (a) the weight concentration of the two components, (b) the T_g values of the homopolymers of the two components (see Section 2.3a, particularly Equation 2.2). It is therefore to be expected that a copolymer should exhibit one α relaxation whose temperature should depend also on these variables. On the other hand, physical mixtures of two polymers exhibit two α relaxation regions corresponding to the α relaxations of the two polymers. Between these two extremes lie block copolymers which sometimes yield two α relaxations (associated with the α relaxations of the two homopolymers) and sometimes a single α relaxation (as would a random copolymer). The reader may well imagine that the deciding factor is quite simply whether or not the solid, however prepared, consists of discrete regions of pure homopolymer or whether the solid is in a single phase. Some of the experimental results on which this rationalization is based are described in the following paragraphs.

Schmieder and Wolf (1953) studied styrene–isobutylene copolymers using a torsion pendulum, and observed that the α relaxation temperature decreased linearly with increasing isobutylene content. Catsiff and Tobolsky (1954) studied random copolymers of styrene and butadiene. These copolymers are well known as SBR or GR-S rubbers and are perhaps the widest used synthetic rubbers employed at the present time. Catsiff and Tobolsky studied the stress relaxation behaviour of 40, 50 and 70 % (weight) styrene copolymers, in the glass–rubber relaxation region. These data were reduced to master curves, showing that the mechanical distribution function $\phi(\ln \tau)$ had a shape independent of temperature. Also, the relaxation curves move to shorter times as the butadiene content increases, i.e. the molecular motions occur more rapidly at a given temperature as the butadiene content increases. This is in agreement with the behaviour observed by Schmieder and Wolf (1953) described above. Other studies of styrene–butadiene copolymers have been made by Nielsen, Claver and Buchdahl (1951).

Block copolymers of styrene–butadiene and styrene–isoprene have been made using anionic catalysts and studied using a torsion pendulum by Angelo, Ikeda and Wallach (1965). Figure 10.37 shows the shear modulus and damping as a function of temperature for styrene (S) and butadiene (B) block copolymers from the pure polystyrene to pure polybutadiene. The immediate observation is that the relaxations of the two components are observed at 0°C and 110°C and are seen to be independent of the composition. The block copolymers therefore differ entirely from the random copolymers of styrene–butadiene. The two-phase character exhibited in Figure 10.37 is not observable visually since all the samples were transparent. (A physical mixture of polystyrene and polybutadiene is opaque.) A physical blend of polystyrene and polybutadiene was compared with a block copolymer of corresponding composition. It was found that the G'' versus $1/T$ curves were very similar for the two systems, but the G' versus $1/T$ curves gave a rubbery plateau modulus (above 0°C) of 10^9 dyn/cm^2 for the blend and less than 10^8 dyn/cm^2 for the block copolymer. The copolymers of Figure 10.37 correspond to a central block of polystyrene to which is attached blocks of polybutadiene on each end, i.e. $(B)_m–(S)_n–(B)_m$. Data were obtained for the block copolymers $(S)_{m'}–(B)_n–(S)_{m'}$, and the mechanical behaviour was found to be identical with that obtained for the copolymers having central styrene blocks.

Baer (1964) studied the mechanical properties of block copolymers of styrene–methyl methacrylate, styrene–α-methylstyrene and styrene–ethylene oxide. $(MMA)_n–(S)_m–(MMA)_n$ block copolymers gave two regions of mechanical loss maxima (Λ_G) corresponding to polymethyl methacrylate (133°C) and polystyrene (110°C) components. The $(EO)_n–(S)_m–(EO)_n$ block copolymer also showed two peaks, at 110°C and -49°C, corresponding to the two components. Both the styrene–MMA and the styrene–EO block copolymers were optically clear, yet the mechanical behaviour was similar to that of a two-phase blend. Presumably the

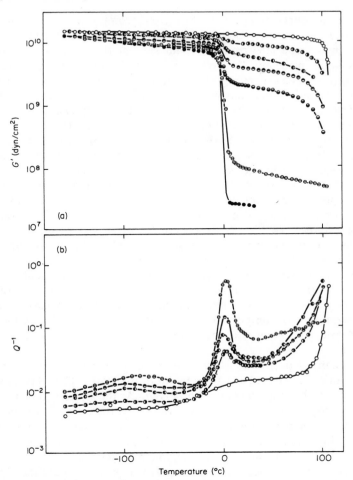

Figure 10.37. Styrene–butadiene block copolymers. Shear modulus and mechanical loss Q^{-1} as a function of temperature (Angelo and others, 1965).

○ Polystyrene, ◑ S/B$_1$, ◐ S/B$_2$, ◕ S/B$_3$, ◔ S/B$_4$, ⊖ S/B$_5$, ● Polybutadiene. Here S/B$_1$, S/B$_2$, S/B$_3$, S/B$_4$, S/B$_5$ correspond to block copolymers $(B)_n$–$(S)_m$–$(B)_n$. The compositions 74, 59, 49, 39 and 19% styrene for S/B$_1$ to S/B$_5$ respectively.

discrete regions were of such a size as to be optically homogeneous whilst being mechanically inhomogeneous. Styrene–α-methylstyrene block copolymers gave an interesting but curious result. Both $(\alpha MeS)_n$–$(S)_m$–$(\alpha MeS)_n$ and $(S)_n$–$(\alpha MeS)_n$–$(S)_n$ copolymers gave a single relaxation peak near 150°C for 1:1 styrene–α-methylstyrene composition. This was intermediate between the relaxation peaks for polystyrene (110°C) and poly-α-methylstyrene (183°C), and would suggest that these block copolymers were mechanically homogeneous. However, the moulded samples were opaque suggesting a heterogeneous two-phase system. This particular system is anomalous when compared with the styrene–MMA, styrene–EO blocks studied by Baer and also with the styrene–butadiene and styrene–isoprene blocks studied by Angelo and others (1965).

Tanaka and Matsumoto (1963) prepared graft copolymers of vinyl acetate on polystyrene. The graft copolymer gave the polystyrene relaxation peak (near 120°C) but no polyvinyl acetate peak near 60°C. A physical mixture gave the two peaks. They explained the absence of two peaks in the graft copolymer as being due to the shortness of the branch lengths of polyvinyl acetate. They also studied polyvinyl acetate to which was grafted polystyrene. In this case the polyvinyl acetate peak was seen, with little evidence of the polystyrene peak, and was explained as being due to shortness of the polystyrene branches.

Fukada and Takamatsu (1962) studied polystyrene–polyvinyl chloride graft copolymers. The relaxation in the copolymer lay between relaxations of the homopolymers. They explained the variation in maximum loss temperature with composition in terms of an extension of the Gibbs and DiMarzio (1959) theory for random copolymers to the case of graft copolymers.

The mechanical properties of styrene–butadiene random copolymers blended with styrene have been given by McIntire (1961). The relaxation peaks due to both components were observed. As the styrene content of the random copolymer component was raised, so the relaxation region due to the copolymer component moved to higher temperatures as expected. Hughes and Brown (1961) have given mechanical data for a number of heterogeneous styrene–polymethyl acrylate copolymers. Other work on heterogeneous styrene copolymers has been published by Buchdahl and Nielsen (1955) and by Nielsen (1962).

Takayanagi (1963) has given a detailed description of the mechanical behaviour of polymer blends which include various styrene copolymers. In all samples the component polymer relaxations were observed, and these data were interpreted in terms of a series–parallel model for the heterogeneous systems.

11

Halogen Polymers

11.1 POLYVINYL CHLORIDE

$$[-\overset{\displaystyle H}{\underset{\displaystyle H}{C}}-\overset{\displaystyle H}{\underset{\displaystyle Cl}{C}}-]_n$$

(1)

On account of the chlorine substituents on alternate chain carbon atoms, polyvinyl chloride (PVC) is a strongly polar polymer. Evidence from x-ray (Natta and Corradini, 1956), infrared (Krimm, 1960; Takeda and Iimura, 1962; Germar, 1963) and high-resolution n.m.r. (Johnsen, 1961; Bovey and others, 1963; Satoh, 1964) studies suggests that conventional PVC, prepared by free-radical polymerization above room temperature, has a fairly high degree ($\sim 65\%$) of syndiotactic stereoregularity. Also, conventional PVC has a predominantly head-to-tail structure (Marvel and others, 1939) and is somewhat branched (George and others, 1958). As ordinarily prepared, PVC is only slightly crystalline and according to Alfrey and others (1949) the crystallite dimensions are small. Consequently, in many of its properties, PVC resembles an amorphous rather than a partially crystalline polymer.

Other studies (Fordham and others, 1959; George and others, 1958) have shown that if the temperature of the free-radical polymerization is lowered, then the amount of branching decreases and the degree of syndiotacticity increases, resulting in a more highly crystalline polymer. Furthermore, from modulus–temperature plots for pure and plasticized PVC samples, Reding and others (1962) have estimated that both the glass-transition temperature and the melting temperature increase as the polymerization temperature is lowered (see Table 11.1). In this respect PVC resembles polymethyl methacrylate (see Chapter 8) for which both the degree of syndiotacticity and T_g increase as the (free-radical) polymerization temperature is lowered.

Table 11.1. Effect of the (free-radical) polymerization temperature on the glass-transition temperature (T_g) and melting temperature (T_m) of polyvinyl chloride. Data estimated from modulus–temperature plots on pure and plasticized PVC samples (Reding and others, 1962)

Polymerization temp. (°C)	T_g (°C)	T_m (°C)
125	68	155
90	75	—
40	80	220
−10	90	265
−80	100	> 300

A further point of interest concerns the fact that T_g for conventional PVC ($\approx 74°$C) is considerably higher than the T_g value for polypropylene ($\approx -35°$C). Recalling that the chlorine atom and the —CH_3 group have almost identical van der Waals' radii (1·8 Å), the steric hindrance to rotations about main-chain bonds should be similar for these two polymers. Hence the above result illustrates the increase in T_g, or reduction of chain mobility, produced by strengthening the interchain cohesive forces by an increase in polarity of the main chain. However, this result could be determined partly by differences in stereoregularity.

Most mechanical and dielectric studies of PVC have been made on the conventional polymer. These studies have shown that at least two relaxation processes, α and β, occur. An additional mechanical loss peak of very small amplitude has been reported at 18°K and 7225 c/s (Crissman, Sauer and Woodward, 1964). In the following sections we will discuss the α and β relaxations separately, and will consider also the effect of plasticizer which has been investigated by several authors.

11.1a   The α Relaxation

Numerous workers have studied the dielectric behaviour of PVC in the primary or α relaxation region. These workers include Davies, Miller and Busse (1941), Fuoss (1941a, b), Fuoss and Kirkwood (1941), Würstlin (1943, 1949, 1951), Nielsen, Buchdahl and Lavreault (1950), Dyson (1951), Wolf (1951), Thurn (1955), Thurn and Würstlin (1958), Reddish (1959), Ishida (1960a), Kiessling (1961), Koppelmann and Gielessen (1961a, b),

Saito (1962, 1963), Koppelmann (1963), Kästner, Schlosser and Pohl (1963) and Reddish (1965).

The dielectric data of Ishida (1960a) are illustrated in Figure 11.1. From the plots of ε' and ε'' against frequency Ishida has constructed Cole–Cole plots and evaluated $\varepsilon_R - \varepsilon_U$ and $\bar{\beta}$ at each temperature. As

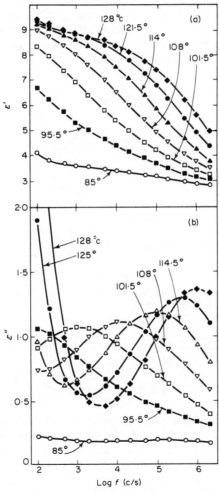

Figure 11.1. Frequency dependence of (a) ε' and (b) ε'' for PVC at various temperatures in the α relaxation region (Ishida, 1960a).

shown in Figure 11.2 $\varepsilon_R - \varepsilon_U$ decreases with increasing temperature for the α relaxation. It is also seen (Figure 11.3) that $\bar{\beta}$ decreases from a limiting value at high temperatures to a small value in the vicinity of T_g ($\approx 74°$C). The latter result, which was also found by Fuoss and Kirkwood (1941), shows that the dielectric α relaxation is broadening rapidly as the temperature decreases towards T_g. This effect appears to be characteristic of partially crystalline polymers such as PVC. It might result from the heterogeneity in structure of these systems which could give rise to a range of barrier heights or distribution of activation energies. Hence the construction of master curves by the reduced variables method is clearly not valid for the dielectric α relaxation of PVC.

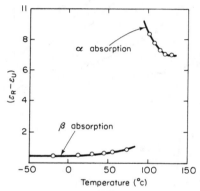

Figure 11.2. Temperature dependence of $\varepsilon_R - \varepsilon_U$ for PVC in both the α and β relaxation regions (Ishida, 1960a).

The dielectric results of several authors are collected together on the log frequency versus $1/T$ plot shown in Figure 11.4. The curvature exhibited by the plot for the α relaxation has been shown to be consistent with an equation of the WLF form (Payne, 1958; Saito, 1962, 1963; Ferry, 1961).

Dynamic mechanical studies of the α relaxation in PVC include those of Nielsen, Buchdahl and Lavrealt (1950), Wolf (1951), Schmieder and Wolf (1953), Becker (1955), Buchdahl and Nielsen (1955), Thurn (1955), Koppelmann (1955, 1963), Thurn and Würstlin (1958), Sommer (1959), Reding and others (1962), Heydemann (1963), Takayanagi (1963) and Kästner and others (1963). The investigations of Becker (1955) and Sommer (1959) are particularly noteworthy since they each covered an

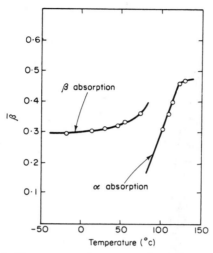

Figure 11.3. Temperature dependence of the Cole–Cole distribution parameter, $\bar{\beta}$, for PVC in the α and β relaxation regions (Ishida, 1960a).

unusually wide frequency range. Becker determined E' and $\tan \delta_E$ in the frequency range 2 to 10^4c/s and temperature range 4·5 to 153°c. Sommer measured E', E'' and $\tan \delta_E$ from 10^{-5} to 10^4 c/s and at temperatures ranging from 23 to 120·5°c. He also measured the stress relaxation modulus, $E_r(t)$, in the time range 10^{-1} to 10^5 sec. Sommer's results are shown in Figures 11.5 and 11.6 respectively. Using second-order approximation methods (see Section 4.1d), Sommer evaluated the mechanical relaxation spectrum, H_E, from both the dynamic mechanical ($H_{E'}$) and stress relaxation (H_{E_r}) data at each temperature. Figure 11.7 shows plots of $H_{E'}$ and H_{E_r} against $\log \tau$ at 79°C. The discrepancy observed between $H_{E'}$ and H_{E_r} is of interest since at first sight it would appear to indicate that the Boltzmann superposition principle (see Section 4.1a) is invalid. According to Sommer, however, both the dynamic mechanical and stress relaxation results were obtained at sufficiently small strains such that the stress–strain relationship was linear. Furthermore, the discrepancy was outside the limits of error both of the original measurements and in the evaluation of the spectra. However, it is difficult to understand Sommer's interpretation that different molecular mechanisms, involving different fractional free volumes, can be operative in the static and dynamic experiments respectively. The differences could be related in part to differences involving the thermal history of the specimens. In particular, results obtained at temperatures around and below T_g (74°C) are likely to be time

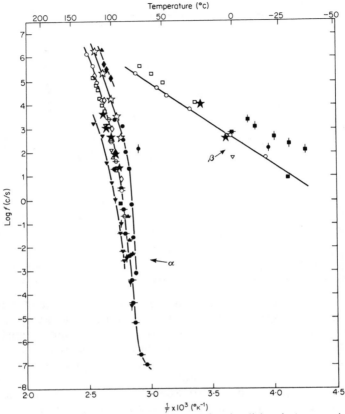

Figure 11.4. Plot of log f against $1/T$ for the dielectric (open points) and mechanical (filled points) loss maxima for PVC. In the following key (f) refers to the frequency of maximum loss in constant temperature experiments and (T) denotes the temperature of maximum loss in experiments at constant frequency.

○ ε'' (f) (Ishida, 1960a)

□ ε'' (f) (Dyson, 1951)

▽ ε'' (T) (Fuoss, 1941a)

△ ε'' (T) (Davies, Miller and Busse, 1941)

⊖ ε'' (f) (Kästner and others, 1963)

◇ ε'' (T) (Koppelmann, 1963)

☆ $(1/\varepsilon)''$ (T) (Koppelmann, 1963)

● E'' (f) (Sommer, 1959)

⬤ tan δ_E (f) (Sommer, 1959)

★ tan δ_E (f) (Becker, 1955)

▼ D'' (f) (Koppelmann, 1963, from data of Becker, 1955)

▼ D'' (f) (Koppelmann, 1963, from data of Sommer, 1959)

▼ D'' (f) (Kästner and others, 1963)

▲ E'' (f) (Kästner and others, 1963)

■ Λ (T) (Schmieder and Wolf, 1953)

● E'' (T) (Takayanagi, 1963)

■ E'' (T) (Tanaka 1962)

■ Q^{-1} (T) (Sauer and Kline, 1956)

▲ tan δ_M (T) (Thurn and Würstlin, 1958)

◆ tan δ_M (T) (Koppelmann, 1963)

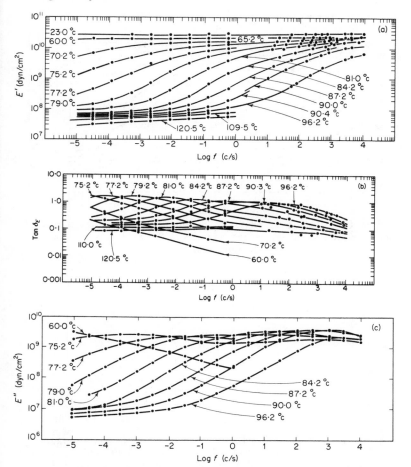

Figure 11.5. Frequency dependence of (a) E', (b) tan δ_E and (c) E'' for PVC at various temperatures in the α relaxation region (Sommer, 1959).

dependent owing to the very slow establishment of volume equilibrium subsequent to the sample having attained the temperature of measurement. Such effects have been observed by Kovacs, Stratton and Ferry (1963) in the case of polyvinyl acetate (see Chapter 9). From the temperature dependence of the frequency of the tan δ_E maximum (Figure 11.4), Sommer has estimated that the apparent activation energy (H_{app}) for the α relaxation in PVC passes through a maximum at temperatures some 5°c above

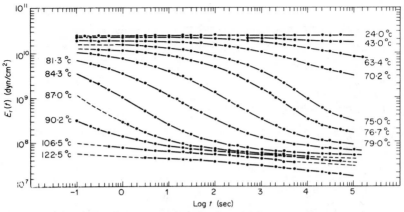

Figure 11.6. The stress relaxation modulus, $E_r(t)$, plotted against log time for PVC at several temperatures in the α relaxation region (Sommer, 1959).

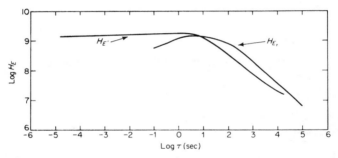

Figure 11.7. Mechanical relaxation spectra plotted against $\log \tau$ for PVC at 79°C. $H_{E'}$ was calculated from the data (at 79°C) of Figure 11.5, and H_{E_r} was calculated from the data (at 79°C) of Figure 11.6 (Sommer, 1959).

T_g. The significance of this result, which is illustrated in Figure 11.8, is not understood. However, similar effects have been observed for PMMA from both creep (Bueche, 1955) and stress relaxation (McLoughlin and Tobolsky, 1952) data. Furthermore, from very low frequency dielectric data, Saito and Nakajima (1959b) have observed a similar effect for poly-n-butyl methacrylate, polyethylene terephthalate, and copolymers of vinyl chloride and vinylidene chloride. It is probable that the precise shape

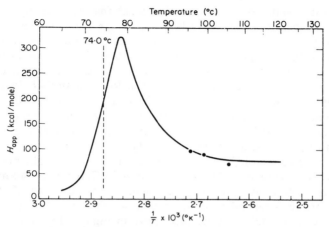

Figure 11.8. Temperature dependence of the apparent activation energy, H_{app}, for the α relaxation of PVC. The glass-transition temperature of about 74°C is indicated by the vertical dotted line (Sommer, 1959).

and position of the curve in Figure 11.8 will again be largely dependent on the thermal history of the sample, particularly on the rate at which the temperature is varied within the region of T_g.

The dynamic mechanical data of several authors are summarized in Figure 11.4 which illustrates the frequency–temperature locations of various loss maxima. It is seen from this diagram that, at each temperature, the E'' maxima lie at frequencies some 3 or 4 decades higher than the frequencies of the corresponding compliance (D'') maxima, and that the $\tan \delta_E$ maxima occur at intermediate frequencies. A similar result is generally obtained for the α relaxations of amorphous polymers.

It will also be observed in Figure 11.4 that the D'' maxima lie at lower frequencies than the analogous dielectric ε'' maxima, a result generally observed for the α relaxations of amorphous polymers. It is also interesting to note (Figure 11.4) that the locations of the E'' maxima correlate well with the positions of the analogous $(1/\varepsilon)''$ maxima (see Equations 4.16b and 4.33). This correlation has been noted and discussed by Koppelmann (1963).

Effect of Molecular Weight

An apparent discrepancy has appeared in the literature with regard to the molecular-weight dependence of T_g and the α relaxation temperature of PVC. For several amorphous polymers it has been found (see, for example, the results for PVAc, Section 9.2a) that the density, glass-transition temperature, and α relaxation temperature increase with increasing molecular weight at low molecular weights, but tend to asymptotic values at a molecular weight of about 10^4 (degree of polymerization about 200). This result is usually ascribed to the fact that the molecular packing around chain ends must be looser than that along the main chains. Since the concentration of chain ends decreases with increasing degree of polymerization (DP), an increase in DP results in a decrease in fractional free volume (increase in intermolecular cohesion), and therefore an increase in both the density and T_g. However, at a DP above about 200, the chain end concentration becomes negligible and T_g becomes independent of DP. In the case of PVC, therefore, we might also expect both T_g and the frequency–temperature location of the α relaxation to be independent of molecular weight, providing that the DP is above about 200. However, as shown in Figure 11.9, Saito (1962,

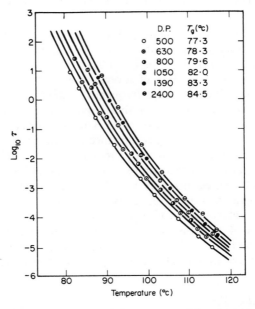

D.P.	T_g(°C)
○ 500	77·3
◑ 630	78·3
◔ 800	79·6
◉ 1050	82·0
● 1390	83·3
◕ 2400	84·5

Figure 11.9. Dependence of the average dielectric relaxation time, $\tau (= 1/2\pi f_{max})$, on temperature and degree of polymerization (DP) for PVC in the α relaxation region. Values of DP and T_g are shown in the key (Saito, 1962, 1963).

1963) has observed that T_g and the temperature of the dielectric α peak increase with increasing DP in the DP range 500 to 2400. On the other hand, Reding and others (1962) have found the same value of T_g for two PVC specimens, each prepared at the same polymerization temperature (40°C), but varying widely in molecular weight. However, a marked increase in T_g was observed on lowering the polymerization temperature (see Table 11.1), an effect attributed to an increase in syndiotactic stereoregularity. Thus, a possible explanation for Saito's anomalous results is that the samples of higher DP had also a higher degree of syndiotacticity. This could be the case if the increase in DP was achieved by lowering the polymerization temperature (see Flory, 1953). It is also of interest to note that Saito observed a 15°C increase in T_g for polymethyl methacrylate with an increase in DP from 850 to 11,000, since T_g for this polymer is also known to increase with increasing syndiotacticity. For polyvinyl acetate, on the other hand, the α relaxation temperature was found to be independent of DP in the range 460 to 6090. The above suggestion is somewhat speculative, however, since the methods of preparation of Saito's samples were not disclosed, and the samples were apparently uncharacterized with respect to stereoregularity. For PVC and PMMA, particularly, there is clearly a need for relaxation studies on samples characterized with respect to both molecular weight and stereoregularity.

The effect of molecular-weight distribution on the mechanical α relaxation of PVC has been discussed by Buchdahl and Nielsen (1955). Using the torsion pendulum method they found that, at constant frequency, the temperature and shape of the α peak was essentially the same for a fractionated sample as for the unfractionated polymer. Chain branching also had little effect on the α peak.

Effect of Plasticizer

The plasticization of PVC is of considerable significance since *at room temperature* the polymer is transformed from a brittle to a tough, flexible material having many commerical applications. Consequently numerous investigations have been made of the mechanical and dielectric properties of plasticized PVC. Table 11.2 shows some of the plasticizers which have been studied most extensively. In addition to those listed, several others have also been studied, notably by Würstlin (1943) and Hartmann (1956).

Some dielectric results of Fuoss (1941) are illustrated in Figure 11.10 for PVC containing from 0 to 20% of the (non-polar) plasticizer diphenyl. An increase in plasticizer content is seen to shift the ε' and ε'' curves (at 60 c/s) to lower temperature. Also, for amounts of plasticizer above about 3%, the height of the ε'' peak and the increment in ε' (i.e. the relaxation

Table 11.2. Dielectric and mechanical studies of plasticized PVC

Plasticizer	Reference	Type of Measurement
Diphenyl	Fuoss (1941a, b)	D[a]
	Fuoss and Kirkwood (1941)	
Dimethylthianthrene	Davies, Miller and Busse (1941)	D
(DMT)	Fitzgerald and Miller (1953)	D
	Fitzgerald and Ferry (1953)	M[b]
Tricresyl phosphate	Davies, Miller and Busse (1941)	D
(TCP)	Würstlin (1943, 1949)	D
	Dyson (1951)	D
	Hartmann (1956)	D
	Alfrey and others (1949)	M
	Sabia and Eirich (1963)	M
Di-n-butyl phthalate	Hartmann (1956)	D
	Thurn and Würstlin (1958)	D
Dioctyl phthalate	Davies, Miller and Busse (1941)	D
(DOP)	Dyson (1951)	D
	Hartmann (1956)	D
	Reddish (1959)	D
	Saito (1963)	D
	Reding and others (1962)	M
	Heydemann (1963)	M
Palatinol	Würstlin (1943)	D
	Wolf (1951)	M

[a] D = Dielectric.
[b] M = Mechanical.

magnitude) decrease with increasing plasticizer content, probably as a result of the decreased dipole concentration. Fuoss has also shown that the PVC dispersion region is narrowed, both with respect to frequency and temperature, by the addition of diphenyl. The latter result is somewhat unusual, since the addition of non-polar plasticizer to an amorphous polymer more generally leads to a broadening of the dispersion. This effect is exemplified by the results for the polyvinyl acetate/diphenyl-methane system shown in Chapter 9. In the case of PVC, the narrowing of the dispersion region by the addition of non-polar plasticizer might arise from a reduction in the interaction between the amorphous and microcrystalline regions.

Several workers have also studied the effects of *polar* plasticizers on the relaxation behaviour of PVC. Dielectric studies have shown that such systems usually exhibit two loss peaks at certain polymer/plasticizer

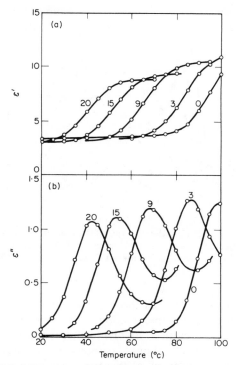

Figure 11.10. Temperature dependence of (a) ε' and (b) ε'' at 60 c/s for PVC plasticized with diphenyl. Numbers on each curve denote the per cent diphenyl (Fuoss, 1941b).

compositions, as was observed for the system polyvinyl acetate/benzyl benzoate (see Chapter 9). Figure 11.11 illustrates the dielectric data of Würstlin (1949) for the system PVC/tricresyl phosphate. At lower concentrations of plasticizer, the loss peak arises predominantly from motions of polymer chain segments as modified by the plasticizer molecules. At higher plasticizer concentrations the loss peak results largely from the relaxation of plasticizer molecules. According to Würstlin the composition at which the plasticizer first appears to exhibit a separate peak corresponds to the point at which the polymer molecules are fully solvated. Further addition of plasticizer is assumed to have simply a diluting effect. On the basis of this interpretation Hartmann (1956) has calculated that, for several PVC/plasticizer systems, there are between about four and six polymer repeat units per 'bound' plasticizer molecule. However, Luther and Weisel (1957), and also Thurn and Würstlin (1958),

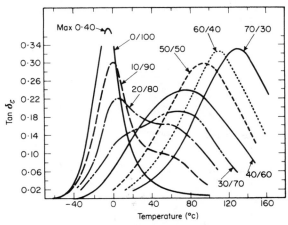

Figure 11.11. Temperature dependence of the dielectric loss at 10 Mc/s for PVC plasticized with tricresyl phosphate (TCP). Numbers on each curve indicate the composition PVC/TCP (Würstlin, 1949).

have shown that results such as those shown in Figure 11.11 may be accounted for by the simple superposition of two symmetrical loss peaks. Hence the above interpretation, involving the concept of 'bound' and 'free' plasticizer molecules, should perhaps be regarded somewhat tentatively. A very good review of this topic has been made by Curtis (1960).

11.1b   The β Relaxation

The β relaxation in PVC was apparently first observed in the dielectric work of Fuoss (1941a, b). Subsequent dielectric studies of the β relaxation include those of Dyson (1951), Reddish (1959), Ishida (1960a) and Koppelmann and Gielessen (1961). Figure 11.12 illustrates Ishida's data which were obtained in the temperature range 73·5 to −61°C. From Figure 11.2 it is seen that $\varepsilon_R - \varepsilon_U$ for the β relaxation is much smaller than that for the α relaxation and decreases slightly as the temperature decreases. The broadening of the loss peak with decreasing temperature is represented in Figure 11.3 by a decrease in the Cole–Cole distribution parameter. From the linear dependence of the frequency of the ε'' maximum on reciprocal temperature (Figure 11.4), Ishida has calculated an activation energy of 15 kcal/mole for the β process. This value is in good agreement with the value of 14·3 kcal/mole obtained by Dyson (1951). Fuoss (1941b)

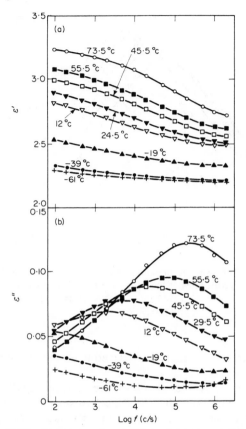

Figure 11.12. Frequency dependence of (a) ε' and (b) ε'' at various temperatures for PVC in the β relaxation region (Ishida, 1960a).

has shown that the addition of diphenyl to PVC depresses and finally eliminates the dielectric β peak.

The β relaxation has also been found from dynamic mechanical studies by Schmieder and Wolf (1953), Becker (1955), Sauer and Kline (1956) (see also Woodward and Sauer, 1958), Tanaka (1962), Okuyama and Yanagida (1963) and Takayanagi (1963). Some results of Tanaka (1962), obtained in the frequency range 116 to 2320 c/s, are presented in Figure 11.13. From these data (see also Figure 11.4) Tanaka has evaluated an activation energy of 13 kcal/mole for the mechanical β mechanism.

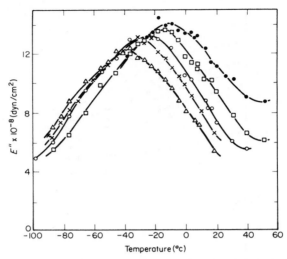

Figure 11.13. Temperature dependence of E'' for PVC in the β relaxation region. Frequencies of measurement as follows (Tanaka, 1962).

△ 116 c/s	○ 440 c/s	● 2320 c/s
× 228 c/s	□ 1180 c/s	

This value compares well with those derived from the dielectric measurements.

Since the chlorine substituents in PVC are attached rigidly to the main polymer chain, it seems likely that the β mechanism, like the α mechanism, must involve some main-chain movement. Tanaka (1962) has, in fact, attributed the β peak to the local mode relaxation mechanism outlined in Chapter 5.

11.2 POLYVINYLIDENE CHLORIDE

$$[-\underset{\underset{\text{H}}{|}}{\overset{\overset{\text{H}}{|}}{\text{C}}}-\underset{\underset{\text{Cl}}{|}}{\overset{\overset{\text{Cl}}{|}}{\text{C}}}-]_n$$

(2)

The molecules of polyvinylidene chloride (PVDC) contain two chlorine substituents on alternate chain carbon atoms. Like PVC, therefore, PVDC is a strongly polar polymer. Due to the symmetrical substitution, the main-chain carbon atoms are not asymmetric and the question of different stereoregular forms does not arise. The regularity of the PVDC structure is probably responsible for the fact that PVDC is far more crystallizable than PVC which has a fairly irregular stereochemical configuration. The crystalline structure of PVDC has been investigated by x-ray (Fuller, 1940) and infrared (Krimm, 1960) methods. PVDC melts at about 200°c.

Boyer and Spencer (1944) obtained a value of $-18°c$ for the dilatometric glass-transition temperature of PVDC. It is of interest to compare this value with the T_g value of about $-71°c$ obtained for polyisobutylene (PIB). This comparison illustrates the increase in T_g produced by an increased polarity of the main chain. Steric factors would be expected to be similar for PVDC and PIB since the —Cl atom and —CH$_3$ group are almost identical in size. It is also interesting to note that T_g for PVDC is about 90°c lower than the value of 74°c for PVC. According to Würstlin (1951) this difference arises from lower dipole–dipole interactions in the case of PVDC owing to a partial compensation of the two C—Cl dipoles. The evidence cited in support of this interpretation was that within the series of compounds CH_3Cl, CH_2Cl_2, $CHCl_3$ and CCl_4 the dipole moment shows a progressive decrease. However, a comparison with the case of the non-polar PIB suggests that steric factors may be largely involved. PIB has a T_g value ($-71°c$) considerably lower than the values of about $-25°c$ reported for many non-polar vinyl polymers such as polypropylene. As noted by Turner and Bailey (1963) it is also lower than the values of $-10°c$ and $-20°c$ found for poly-3,3,-dimethylpropane

$$(-CH_2-CH_2-\overset{\overset{\displaystyle CH_3}{|}}{\underset{\underset{\displaystyle CH_3}{|}}{C}}-)_n$$

and poly-4,4-dimethylbutane

$$(-CH_2-CH_2-CH_2-\overset{\overset{\displaystyle CH_3}{|}}{\underset{\underset{\displaystyle CH_3}{|}}{C}}-)_n$$

respectively. Hence the symmetrical substitution of two —CH$_3$ groups on *alternate* chain carbon atoms gives rise to a surprisingly low value of T_g. This 'anomaly' may in some way be related to the steric repulsion between —CH$_3$ groups on alternate chain atoms, which is thought to lead to a distortion of main-chain valence angles (Bunn and Holmes, 1958). In view of its structural similarity with PIB, PVDC might be anomalous for similar reasons. One possible explanation for these anomalies has been suggested by Gibbs and DiMarzio (1958) on the basis of their theory of the glass transition. According to this theory, T_g is a manifestation of a true thermodynamic second-order transition temperature, which is determined by the *difference*, u, (Figure 2.3b) between potential energy minima corresponding to the more favourable rotational conformations of main-chain bonds. Gibbs and DiMarzio have thus proposed that u is lower for PVDC than for PVC even though, for steric reasons, the absolute value of the potential energy is larger for PVDC for all rotational conformations.

Mechanical relaxation in PVDC has been investigated by Schmieder and Wolf (1953). As shown in Figure 11.14, two mechanical loss peaks were observed at temperatures below the melting point. The lower temperature (β) relaxation, which occurs at 15°C (11 c/s), gives rise to the largest drop in modulus and is certainly associated with the glass transition in PVDC. This conclusion is supported by the fact that the β peak lies in the vicinity of T_g ($-18°C$) determined dilatometrically. According to Schmieder and Wolf the α peak (80°C, 5·5 c/s) may arise from motions within strained amorphous regions.

Dielectric studies of PVDC have been described by Saito and Nakajima (1959a) and also by Ishida, Yamamoto and Takayanagi (1960). Saito and Nakajima investigated an unplasticized sample in powder form

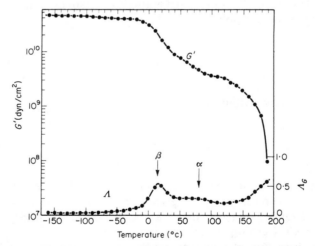

Figure 11.14. Temperature dependence of the shear modulus and logarithmic decrement at low frequencies for PVDC (Schmieder and Wolf, 1953).

whereas Ishida and coworkers studied a moulded sample containing 10% phenyl glycidyl ether plasticizer. Results of Ishida and coworkers are given in Figure 11.15. From Cole–Cole plots, values of $\varepsilon_R - \varepsilon_U$ and the distribution parameter, $\bar{\beta}$, were evaluated, and these are shown as a function of temperature in Figure 11.16. Above 0°C $\varepsilon_R - \varepsilon_U$ shows little variation with temperature. However, below 0°C, Saito and Nakajima (1959a) have found that $\varepsilon_R - \varepsilon_U$ decreases rapidly with decreasing temperature and approaches a very small value at temperatures below the glass-transition temperature. At -18°C (the T_g value measured dilatometrically) $\varepsilon_R - \varepsilon_U$ was found to have decreased to about a half its limiting value above 0°C. Figure 11.16 also shows that $\bar{\beta}$ is decreasing rapidly (i.e. the loss peaks are broadening) as T_g is approached. As noted by Saito and Nakajima (1959b), similar results have been obtained for several *crystallizable* polymers such as PVC, polyethylene terephthalate, polyacrylonitrile and polychlorotrifluoroethylene. For *non-crystallizable* polymers (e.g. polyvinyl acetate) $\varepsilon_R - \varepsilon_U$ and $\bar{\beta}$ do not exhibit this sudden decrease in the vicinity of T_g determined dilatometrically.

A comparison of Figure 11.16 and 11.2 shows that $\varepsilon_R - \varepsilon_U$ is considerably smaller for PVDC than for the α relaxation in PVC. Ishida and coworkers have attributed this observation to two factors. Firstly, as noted by Würstlin (1951), the dipole moment of the repeat unit is probably somewhat smaller for PVDC than for PVC. Secondly, PVDC is more highly crystalline than PVC and therefore contains a relatively small number of dipoles within the amorphous phase. The latter factor depended on the (very reasonable) assumption that the two relaxations being compared were both associated with the amorphous glass transition. However, a strict comparison of the $\varepsilon_R - \varepsilon_U$ values should also consider probable differences in the equilibrium chain conformation between PVDC and PVC in the rubbery state.

The frequency–temperature locations of the dielectric and mechanical loss maxima are illustrated in Figure 11.17. The points obtained from Saito and Nakajima's data lie at somewhat higher temperatures or lower frequencies than the corresponding dielectric points obtained from Ishida's results. This difference may be due largely to the fact that the sample employed by Ishida and coworkers contained 10% of plasticizer. The log f versus $1/T$ plot obtained from Saito and Nakajima's data exhibits an inflection point at about 5°C. The apparent activation energy must therefore pass through a maximum at about 20°C above the dilatometric value of T_g. It will be recalled that a similar effect has been observed from mechanical results for the primary relaxations of PMMA (Section 8.2a) and PVC (Section 11.1a). From Figure 11.17 a good correlation is

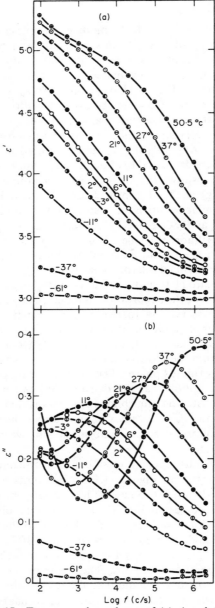

Figure 11.15. Frequency dependence of (a) ε' and (b) ε'' at various temperatures for PVDC containing 10% phenyl glycidyl ether (Ishida, Yamamoto and Takayanagi, 1960).

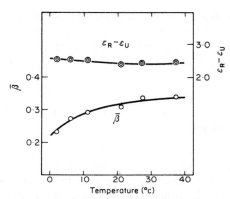

Figure 11.16. Temperature dependence of $\varepsilon_R - \varepsilon_U$ and the Cole–Cole distribution parameter, $\bar{\beta}$, for the β relaxation in PVDC containing 10% phenyl glycidyl ether (Ishida, Yamamoto and Takayanagi, 1960).

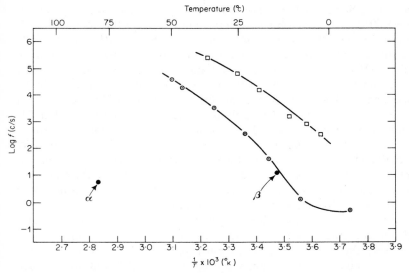

Figure 11.17. Plot of $\log f$ versus $1/T$ for the dielectric (open points) and mechanical (filled points) loss maxima for PVDC. In the following key (f) refers to the frequency of maximum loss in constant temperature experiments and (T) denotes the temperature of maximum loss in experiments at constant frequency.

□ ε'' (sample contained 10% phenyl glycidyl ether) (f) (Ishida, Yamamoto and Takayanagi, 1960)

⊙ ε'' (f) (Saito and Nakajima, 1959a)

● Λ (T) (Schmieder and Wolf, 1953)

seen between the location of the mechanical (Λ) β *peak* and the dielectric (ε'') peak determined by Saito and Nakajima (1959a). This correlation provides additional evidence that the observed dielectric peak and the mechanical β *peak* are each associated with similar mechanisms related to the glass transition.

11.3 VINYLIDENE CHLORIDE–VINYL CHLORIDE COPOLYMERS

The mechanical relaxation behaviour of vinylidene chloride (VDC)–vinyl chloride (VC) copolymers has been studied by Schmieder and Wolf (1953). Figure 11.18 illustrates some results for a copolymer containing 30% by weight vinyl chloride and 70% vinylidene chloride. Sample I was an amorphous sample prepared by quenching from the liquid state to about $-70°$C. Measurements on this sample were made only up to about 60°C to prevent crystallization. Sample II was prepared in a similar manner to Sample I but measurements were extended to temperatures above 60°C. The development of crystallinity above this temperature is reflected by a marked increase in the modulus. Sample III was a partially crystalline sample which was obtained by slow cooling from the melt. Sample IV was also a partially crystalline sample prepared by initially quenching from above the melt to $-70°$C (as Sample I) followed by annealing for 3 hours at 70°C. For this copolymer the β peak occurs at about 27°C. The very large magnitude of the β peak for Samples I and II confirms the conclusion that the β relaxation is associated with the amorphous glass transition. The α peak is observed at about 60 to 70°C. Since the onset of crystallization for Sample II occurs within this temperature region, the α peak might be due to motions related to the annealing process and thus involving, in part, the crystalline phase. It is also seen that the α peak is larger for Sample IV than for Sample III whereas the reverse situation exists for the amorphous β peak. This result would also indicate that the crystalline regions are involved in the α mechanism. However, the annealing process must presumably involve both the crystalline and amorphous phase so that the α peak could also involve 'strained' amorphous regions as proposed by Schmieder and Wolf (1953). The dependence on copolymer composition of the temperatures of the α and β peaks and the temperature of the melting region (corresponding to the fall in G at the highest temperatures) is shown in Figure 11.19. An increase in vinyl chloride content is seen to depress the melting point of PVDC as predicted by the theory of melting for copolymers (Flory, 1953). Also, the temperature of the β peak (or T_g) increases continuously from

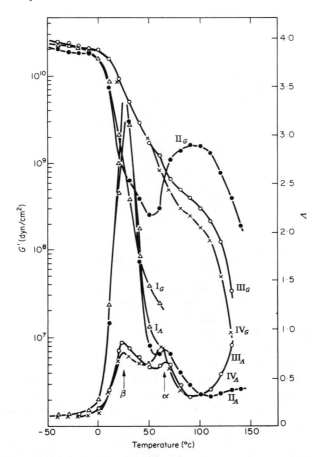

Figure 11.18. Temperature dependence of the shear modulus and logarithmic decrement for a VDC–VC copolymer containing 70% by weight VDC. Details of Samples I ($\triangle$), II ($\bullet$), III ($\bigcirc$) and IV ($\times$) are given in text (Schmieder and Wolf, 1953).

about 10°C for PVDC to about 90°C for PVC, a result which is consistent with the Gordon–Taylor equation (Equation 2.2). Hence, at intermediate compositions, the melting temperature and T_g are separated by only about 50°C. The temperature of the α peak, at which annealing takes place, is relatively insensitive to copolymer composition but appears to exhibit a shallow minimum at intermediate compositions.

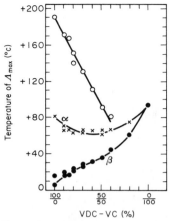

Figure 11.19. Dependence of the temperature of the melting region ($\bigcirc$) and the temperatures of the α ($\times$) and β ($\bullet$) mechanical Λ peaks on composition for copolymers (Schmieder and Wolf, 1953).

Saito and Nakajima (1959a, b) have studied the dielectric behaviour of unplasticized VDC–VC copolymers containing, respectively, 90, 80 and 70% VDC. Each of these unplasticized samples was studied in powder form and the determined values of ε' and ε'' were thus subject to large errors. More accurate results (shown in Figure 11.20) were obtained for a moulded film containing 80% VDC and 5% plasticizer. The low-frequency ε'' peak clearly corresponds to the α peak observed mechanically, since at 83·5°C it is located in the region of 1 c/s. The high-frequency loss peak corresponds to the mechanical β peak and also to the single dielectric peak observed for pure PVDC (see above). For each copolymer investigated, $T \times (\varepsilon_R - \varepsilon_U)$ was observed to decrease from an approximately constant value at temperatures some 30°C above T_g to a very small value below T_g. This result, which was also noted for PVDC (see above), is illustrated in Figure 11.21 for the plasticized and unplasticized copolymers containing 80% VDC. Values of T_g, estimated from the temperature at which $T \times (\varepsilon_R - \varepsilon_U)$ had decreased to a half its limiting value, were found to increase with increasing VC content for the unplasticized samples. Also, in the absence of plasticizer, the loss peak shifted to higher temperatures (or lower frequencies) with increasing VC content, consistent with the observations of Schmieder and Wolf (1953). Figure 11.22 shows the $\log f_{max}$ versus $1/T$ plots for the plasticized and unplasticized copolymers containing 80% VDC content. Each of these plots is seen to have a sigmoidal shape corresponding to a maximum apparent activation energy

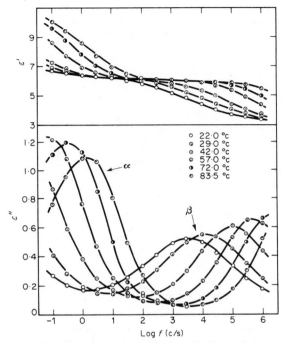

Figure 11.20. Frequency dependence of ε' and ε'' at various temperatures for a VDC–VC copolymer containing 80% VDC and 5% plasticizer (Saito and Nakajima, 1959a).

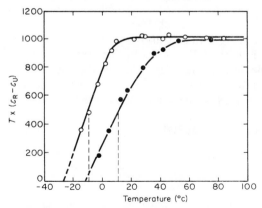

Figure 11.21. Temperature dependence of $T \times (\varepsilon_R - \varepsilon_U)$ for VDC–VC copolymers containing 80% VDC. ● unplasticized sample; ○ sample containing 5% plasticizer (Saito and Nakajima, 1959a).

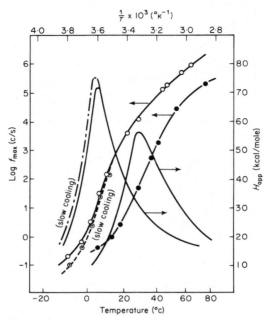

Figure 11.22. Plots of $\log f_{max}$ versus $1/T$ for the dielectric loss maxima of VDC–VC copolymers containing 80% VDC. ● unplasticized; ○ sample containing 5% plasticizer. The apparent activation energies, H_{app}, are also shown as a function of temperature. For the plasticized sample, the effect of the cooling rate prior to taking measurements is illustrated by the curves marked 'slow cooling', Saito and Nakajima, 1959a).

at temperatures some 10°c to 20°c above the estimated T_g value. However, as illustrated in Figure 11.22, results in this temperature region for the plasticized sample were strongly dependent on the rate at which the sample was cooled, probably as a. result of the slow establishment of volume equilibrium (see Kovacs, Stratton and Ferry, 1963).

11.4 POLYVINYL FLUORIDE

$$[-\underset{\underset{H}{|}}{\overset{\overset{H}{|}}{C}}-\underset{\underset{F}{|}}{\overset{\overset{H}{|}}{C}}-]_n$$

(3)

Mechanical relaxation in polyvinyl fluoride (PVF) has been investigated by Schmieder and Wolf (1953) whose results are shown in Figure 11.23. At temperatures below the melting region (>190°C) two loss peaks are observed as in the case of PVC. The peak at 41°C (1·7 c/s), labelled α, is probably associated with the glass transition in the amorphous regions, since this relaxation gives rise to a relatively large modulus drop. The temperature of the α peak, and hence T_g, is about 50°C lower for PVF than for PVC. This difference probably arises from lower steric hindrances to main-chain rotations in the case of PVF since the atomic radius of fluorine is less than that of chlorine. Differences in polarity seem unlikely to be involved to a large extent, as evidenced by the fact that the dipole moment of CH_3Cl (1·87 D) is only slightly larger than that of CH_3F (1·81 D) (Maryott and Buckley, 1953). The relatively small β peak (−20°C, 5·4 c/s) must presumably involve some limited main-chain motions, similar to those proposed for the β relaxation in PVC.

Reddish (1962) has compared some dielectric results for PVF with mechanical data obtained at about the same frequency (70 to 300 c/s). As illustrated in Figure 11.24, a rather close correlation exists between the shapes of the dielectric and mechanical β peaks if the data are plotted in

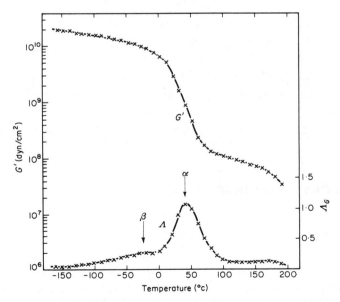

Figure 11.23. Temperature dependence of the shear modulus and logarithmic decrement at low frequencies for PVF (Schmieder and Wolf, 1953).

terms of ε'' and the mechanical compliance ($D'' \approx \tan \delta_{E'}/E'$) respectively. This correlation suggests that the dielectric and mechanical β relaxations have a closely related molecular origin. The data shown in Figure 11.24 seem also to be consistent with Schmieder and Wolf's data (Figure 11.23) if allowance is made for the difference in frequency. However, no attempt has been made to calculate an activation energy for the β process from a direct comparison of Figures 11.23 and 11.24, since Λ and D'' plots are not comparable for this purpose. Dielectric data recently reported by Ishida and Yamafuji (1964) yield an activation energy of about 10 kcal/ mole for the β relaxation in PVF. This value is consistent with a mechanism involving local chain motions.

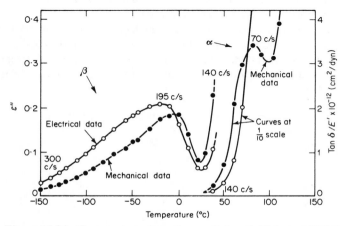

Figure 11.24. Temperature dependence of ε'' and D'' ($\approx \tan \delta/E'$) for PVF at the indicated frequencies (Reddish, 1962).

11.5 POLYVINYLIDENE FLUORIDE

$$[-\underset{\underset{\text{H}}{|}}{\overset{\overset{\text{H}}{|}}{\text{C}}}-\underset{\underset{\text{F}}{|}}{\overset{\overset{\text{F}}{|}}{\text{C}}}-]_n$$

(4)

Polyvinylidene fluoride (PVDF) is a partially crystalline polymer having a glass-transition temperature of about $-35°$C (Mandelkern, Martin and Quinn, 1957). This T_g value is considerably lower than the value of 20 → 40°C for PVF. Since the dipole moment of CH_2F_2 (1·93D) is slightly larger than that of CH_3F ($1\cdot81\text{D}$) (Maryott and Buckley, 1953), the repeat unit of PVDF should have a somewhat larger dipole moment than the repeat unit of PVF. This difference in polarity might be expected to yield a somewhat higher T_g value for PVDF than for PVF. A similar result might also be expected for steric reasons, since the atomic radius of fluorine is larger than that of hydrogen. However, as discussed in the case of PIB (Chapter 10) and PVDC (Section 11.2), vinylidene-type polymers, containing symmetrically substituted groups on *alternate* chain atoms, appear to have anomalously low T_g values. According to the theory of Gibbs and DiMarzio (1958), this effect is attributed to a relatively low energy *difference* between the potential minima of the stable rotational conformations of main-chain bonds.

Some dielectric results of Kabin and others (1961) for PVDF are shown in Figure 11.25. Two dielectric relaxation regions are observed. The low-temperature loss peak is probably associated with the glass transition in PVDF since at low frequencies it is located in the region of T_g. The latter point is illustrated by the curved plot of $\log f_{max}$ against $1/T$ (Figure 11.26). The high-temperature peak is irreversible with respect to temperature changes, suggesting that it may be connected with the annealing process and possibly involves the crystalline phase.

The dielectric behaviour of PVDF has also been studied by Ishida, Watanabe and Yamafuji (1964) in the frequency range 10 to 10^6 c/s and at temperatures from $-10\cdot9$ to $-73\cdot1°\text{C}$. In the range $-10\cdot9$ to $-35\cdot5°\text{C}$ they observed a single symmetrical loss peak which correlates fairly well with the low-temperature peak of Kabin and others (see Figure 11.26). However, below $-39\cdot5°\text{C}$ this peak tended to separate into two components. The high-frequency component (Figure 11.26) corresponds to a mechanism occurring below T_g and presumably involves some limited main-chain motion.

11.6 POLYVINYL BROMIDE

$$[-\overset{\displaystyle H}{\underset{\displaystyle H}{C}}-\overset{\displaystyle H}{\underset{\displaystyle Br}{C}}-]_n$$

(5)

Reference has been made by Zutty and Whitworth (1964) to some unpublished mechanical data of Faucher on polyvinyl bromide (PVBr). According to these authors PVBr exhibits a low-frequency loss peak at $100°\text{C}$ which is attributed to the glass transition. Hence T_g for PVBr is

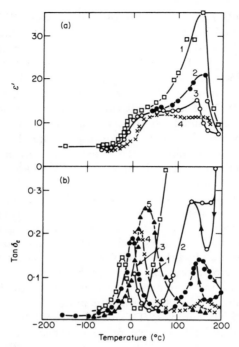

Figure 11.25. Temperature dependence of (a) ε' and (b) $\tan \delta_\varepsilon$ for PVDF at various frequencies. (a) Curves 1, 2, 3 and 4 correspond to frequencies of 500, 5×10^3, 2×10^4 and 8×10^5 c/s; (b) Curves 1, 2, 3, 4 and 5 correspond to frequencies of 500, 2×10^4, $1 \cdot 2 \times 10^5$, 8×10^5 and 8×10^6 c/s (Kabin and others, 1961).

higher than that for PVC (74°C) and PVF (20 → 40°C). For steric reasons this comparison is consistent with the fact that the atomic radius increases in the series F → Cl → Br.

11.7   POLYTETRAFLUOROETHYLENE

$$[-\underset{\underset{\text{F}}{|}}{\overset{\overset{\text{F}}{|}}{\text{C}}}-\underset{\underset{\text{F}}{|}}{\overset{\overset{\text{F}}{|}}{\text{C}}}-]_n$$

(6)

Polytetrafluoroethylene (PTFE), is a non-polar polymer containing two fluorine substituents on each main-chain carbon atom. There is

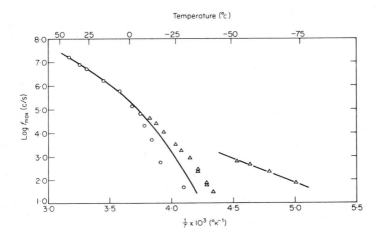

Figure 11.26. Plot of log f_{max} versus $1/T$ for the dielectric loss maxima of the low-temperature peaks in PVDF.

○ tan δ_ε (Kabin and others, 1961)
△ ε'' (Ishida, Watanabe and Yamafuji, 1964)

considerable evidence to suggest that the molecules of PTFE are essentially linear or unbranched (see, for example, Sperati and Starkweather, 1961). PTFE obtained from the polymerization reaction usually has a very high degree of crystallinity (93 → 98 %), and has a crystalline melting point of 327°C.

Within the crystal, the PTFE molecule exhibits a helical conformation (Bunn and Howells, 1954; Pierce and others, 1956). The crystal structure shows two sharp changes at about 19°C and 30°C (Pierce and others, 1956). Below 19°C the conformation can be described as a zig-zag which has been given a twist of 180° for each 13 CF_2 groups. The unit cell is then triclinic, and there is essentially perfect three-dimensional order (Clark and Muus, 1957, 1958). At temperatures above 19°C the helix contains 15 CF_2 groups per twist of 180°, and the unit cell becomes hexagonal. Between 19 and 30°C the packing of molecules on the hexagonal lattice is somewhat disordered owing to small angular displacements of chain segments about the chain axis. These displacements, which may arise from small-angle oscillations of segments about their long axes, are symmetrically distributed about a preferred crystallographic direction. Above the 30°C transition the preferred crystallographic direction is lost and the molecular segments oscillate about their long axes with a random angular orientation in the lattice (Clark, 1959). The 'crystal-disordering' transitions at 19 and 30°C were first observed by discontinuous changes in density (Rigby and Bunn, 1949), and later by specific heat maxima (Furukawa, McCoskey and King, 1952). Consequently they can be considered first order in the thermo-dynamic sense.

Unlike most other crystalline polymers PTFE, as ordinarily prepared, is not spherulitic. Bunn, Cobbald and Palmer (1958) examined fracture surfaces using electron microscopy and observed the solid polymer to consist largely of very long bands of width between 1 and 0·2 microns. Perpendicular to these bands were narrow (~ 300 Å) striations, the polymer chains being parallel to the striations. According to Sperati and Starkweather (1961), this well-defined band structure suggests that the chains are folded as in other structures of crystalline polymers.

The density of solid PTFE is not a reliable guide to the crystalline con-tent since the sample, after moulding, may contain a small amount of voids. The voids, which are caused by an insufficient sintering technique, are about 1 micron in size and may occupy as much as several per cent of the total volume (Thomas and others, 1956). The crystalline content may, however, be determined reliably from x-ray diffraction, infrared absorp-tion (Moynihan, 1959) or shear modulus (McCrum, 1959a).

Many workers have studied the relaxation processes occurring in PTFE using mechanical techniques. These workers include Schmieder and Wolf (1953, 1955), Sauer and Kline (1955, 1956), Robinson (see Smith, 1955), Kabin (1956a, b), Schulz (1956), Maeda (1957), Baccaredda and Butta (1958), Hellwege, Kaiser and Kuphal (1958), Illers and Jenckel (1958a), Nagamatsu, Yoshitomi and Takemoto (1958), McCrum (1958, 1959b), Ohzawa and Wada (1963) and Takayanagi (1963). The studies of Kabin (1956a, b), Illers and Jenckel (1958a), McCrum (1958, 1959a), and Takayan-agi (1963) are particularly instructive since they included an investiga-tion of the effect of variations in density or degree of crystallinity. McCrum's low-frequency measurements are shown in Figure 11.27 from 4·2°K ($-268\cdot8$°C) to the crystal melting point (600°K) for specimens with crystalline contents (determined by an infrared method) of 48, 64, 76 and 92% respectively. The only well-resolved peak, labelled γ in Figure 11.27, occurs at 176°K (-97°C) at 1 c/s. Also indicated in Figure 11.27 are an α peak at about 400°K (127°C) and a β relaxation region at around room temperature. Although the α and β relaxations appear to be fairly well resolved in the modulus–temperature plots, the loss peaks due to the α and β relaxations are partially merged and the β region is complicated by

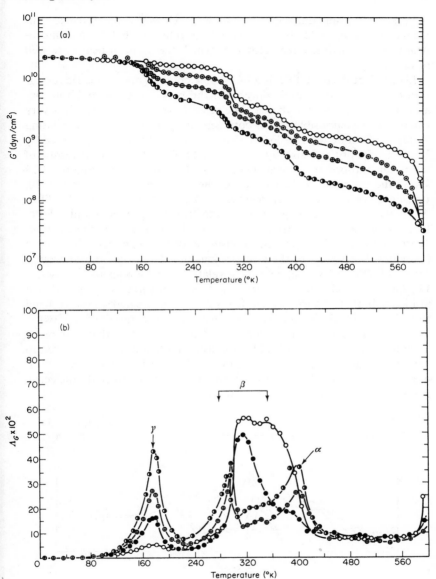

Figure 11.27. Temperature dependence of (a) the shear modulus and (b) the logarithmic decrement at ∼1 c/s for PTFE samples of 92% (○), 76% (●), 64% (⊕) and 48% (◑) crystallinity (McCrum, 1959b).

the first-order transitions at 19°C and 30°C. However, the above assignments are supported by results for tetrafluoroethylene-hexafluoropropylene copolymers (see below) for which the α and β peaks are well resolved.

Like polyethylene, PTFE is a non-polar polymer and dielectric relaxation can presumably result only from the presence of polar impurities attached to the polymer chains. Dielectric losses in PTFE are thus very small, a fact responsible for its commercial application as an insulating material. Despite the low losses, dielectric peaks have been observed by Kabin (1956a, b) and also by Krum and Müller (1959). Some results of the latter authors are shown in Figures 11.28 and 11.29. Figure 11.28 shows the temperature dependence of the dielectric loss tangent at 1, 10, 100, and 316 kc/s for a commercial, unheat-treated, sample of PTFE. At 1 kc/s the γ relaxation gives rise to the dielectric peak at about $-85°C$. By comparison with the mechanical data the large peak occurring at about 165°C (1 kc/s), which decreases considerably in height with increasing frequency, is assigned to the α relaxation, and the peaks centred around 15°C and 80°C at 1 kc/s appear to be related to the β relaxation. Figures 11.29 and 11.30 compare the dielectric results for three samples of PTFE at 1 kc/s. In these diagrams Sample I is the unheat-treated sample referred to in Figure 11.28. Sample II, which was quenched rapidly from 340°C, was more transparent, and probably more amorphous, than Sample I. Sample III was slowly cooled (2°/min) from 340°C and was considered to have the highest degree of crystallinity. The log frequency versus $1/T$ plot shown in Figure 11.31 illustrates the correlation between the dielectric and mechanical results of several authors.

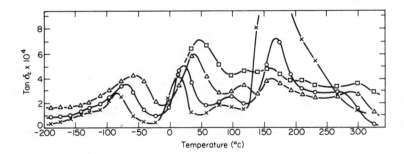

Figure 11.28. Temperature dependence of the dielectric loss tangent for a commercial sample of PTFE at 1 ($\times$), 10 (○), 100 (△) and 316 (□) kc/s (Krum and Müller, 1959).

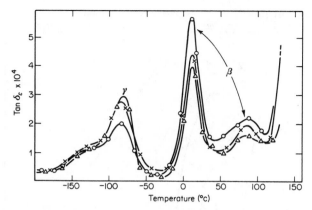

Figure 11.29. Variation of the dielectric loss tangent with temperature (below 120°C) at 1 kc/s for three samples of PTFE. The β and γ relaxation regions are indicated. Sample I unheat-treated commercial sample ($\times$); Sample II most amorphous ($\triangle$); Sample III most crystalline ($\bigcirc$) (Krum and Müller, 1959).

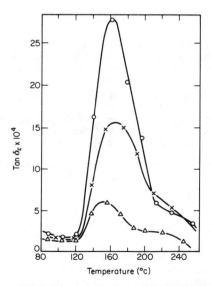

Figure 11.30. Temperature dependence of the dielectric loss tangent at 1 kc/s in the α relaxation region. Point symbols as in Figure 11.29 (Krum and Müller, 1959).

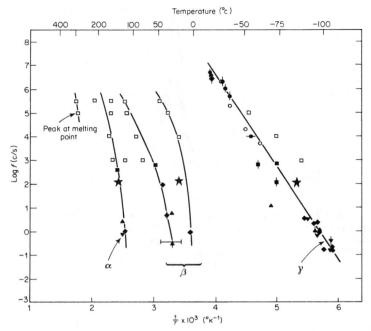

Figure 11.31. Plot of $\log f$ versus $1/T$ for the dielectric (open points) and mechanical (filled points) loss maxima in PTFE. All points refer to the temperature of maximum loss from experiments at constant frequency.

▼ and ⊢⊣ Λ (McCrum, 1959b, c)	◆ attenuation maxima (Maeda, 1957)
▲ Λ (Schmieder and Wolf, 1953)	
◆ tan δ_G (Illers and Jenckel, 1958a)	★ tan δ_E (Takayanagi, 1963)
■ Q^{-1} (Sauer and Kline, 1955, 1956)	▼ Λ (Schulz, 1956)
▉ Q^{-1} (Robinson, as quoted by Smith, 1955)	●— Q^{-1} (Baccaredda and Butta, 1958)
▲ Λ (Hellwege, Kaiser and Kuphal, 1958)	□ tan δ_ε (Krum and Müller, 1959)
● tan δ_M (Kabin, 1956b)	○ tan δ (Kabin, 1956b)

11.7a The α Relaxation

Some uncertainty and difference of opinion exists as to whether the α relaxation in PTFE originates in the amorphous or crystalline phase. Illers and Jenckel (1958a) and McCrum (1958, 1959b) inferred from their mechanical data (see, for example, Figure 11.27) that the α peak decreased in magnitude with increasing crystallinity. They suggested, therefore, that the α relaxation occurred in the amorphous regions, assuming the validity of the two-phase model. However, this assignment has been questioned. Firstly, the α peak is not well resolved, being superposed on the high-temperature wing of the β peak. Secondly, if the magnitude of the α relaxation is estimated from the height of the G'' peak rather than the tan δ_G peak then, according to Ohzawa and Wada (1963), it appears to *increase* slightly with increasing crystallinity. The latter observation raises the unsolved problem as to whether the G'' or tan δ_G peaks should be used in comparing the relative magnitudes of observed relaxations (see Section, 4.3). Thirdly, a dielectric peak in the α region as illustrated in Figure 11.30, *increases* in height with increasing crystallinity, which would suggest a crystalline phase mechanism.

Although the relaxation studies leave some doubt as to the origin of the α relaxation, the dilatometric and x-ray investigation of Satokawa and Koizumi (1962) (see also Takayanagi, 1963; Rydel, 1965) supports the view that the α relaxation is associated with the amorphous or disordered regions. These workers observed a break in the specific volume–temperature curve for PTFE at 130°C. This break was more marked for the samples of lower crystallinity. It was also found that the expansion coefficient determined from the spacing of an *amorphous* x-ray peak increased at 130°C whereas no such change was found from the spacing of the crystalline (100) peak.

11.7b The β Relaxation Region

As shown in Figures 11.27 and 11.29, both the mechanical and the dielectric loss peaks in the β relaxation region increase in height with increasing degree of crystallinity. These results show conclusively that the β relaxation is associated with the crystalline regions of PTFE. It seems probable that the β mechanism involves torsional oscillations of chain segments around the chain axes within the crystals since, as noted above, such motions are thought to be responsible for the crystal disordering transitions at 19 and 30°C. A similar mechanism has been proposed by Takayanagi (1963).

An ingenious argument has been given by Illers and Jenckel (1958a) to explain the complexity of the mechanical loss peaks in the β region. Since the transition at 19°C is first order, it is likely that it will induce an abrupt change in both the average relaxation time and the shape of the relaxation spectrum. In a plot of Λ versus T, such as in Figure 11.27(b), the internal friction will thus follow a smooth characteristic curve at temperatures up to 19°C. At the transition temperature the internal friction will change discontinuously and then follow the relaxation curve characteristic of the modified crystal structure stable at higher temperatures. Such an argument might explain the presence of two mechanical loss peaks for the 48 and 64 % crystalline materials (Figure 11.27) and of two dielectric peaks (Figure 11.28) in the β region (see also Figure 11.31).

Stress relaxation results of Nagamatsu, Yoshitomi and Takemoto (1958) support the conclusion of Illers and Jenckel (1958a) that the 19°c transition induces an abrupt discontinuity in the distribution of relaxation times. Superposed relaxation curves, reduced to 20°C, are shown in Figure 11.32. The curve marked 'below 19°c', formed by the superposition of the stress relaxation curves obtained below 19°c, has a slope clearly greater than the curve marked 'above 20°c', which was constructed from the data above 20°c. The method of reduced variables must therefore break down for PTFE in the region of 20°c.

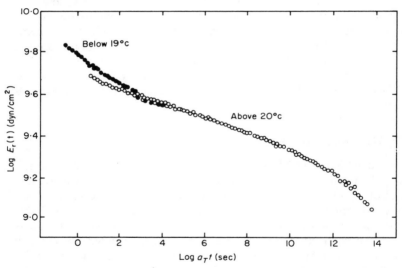

Figure 11.32. Stress relaxation master curve reduced to 20°c for PTFE (Nagamatsu, Yoshitomi, and Takemoto, 1958).

11.7c   The γ Relaxation

The activation energy for the γ relaxation of PTFE, which was evaluated from the linear plot of $\log f$ versus $1/T$ (Figure 11.31), is 18 kcal/mole. Using this value for the activation energy (H) the shape of the γ peak may be analyzed by plotting Λ/Λ_{max} against $H/RT + \ln \omega \tau_0$, (see Chapter 1.4, particularly Equation 1.51). Such plots are shown in Figure 11.33 for the three specimens of 48, 62 and 76 % crystallinity. The continuous line is the theoretical curve for a single relaxation time. The experimental curve is considerably broader than that for a single relaxation time, showing that a broad distribution of relaxation times would be required to describe the γ peak. Furthermore, the asymmetry of the experimental peak suggests that the relaxation spectrum is broadening as the temperature decreases, although the shoulder to the dielectric γ peak at about $-140°C$ might indicate the existence of an additional lower-temperature mechanism. It is interesting to note also that the shape of the observed peak is independent of the degree of crystallinity.

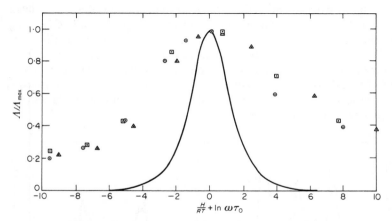

Figure 11.33. Plots of Λ/Λ_{max} against $H/RT + \ln \omega\tau_0$ for three samples of PTFE in the γ relaxation region. The samples have degrees of crystallinity of 48 % (⊙), 62 % (□) and 76 % (△). The continuous line corresponds to the result expected for a single relaxation time.

Both the mechanical (Figure 11.27) and dielectric (Figure 11.29) results show that the γ peak in PTFE decreases in height with increasing density or crystallinity. In terms of the two-phase model this result shows that the γ relaxation occurs in the amorphous regions of the polymer.

Since both the α and γ relaxations may involve motions in the disordered regions of the polymer, the question arises as to whether either of these relaxations is associated with a glass transition in PTFE. The relatively low (and temperature-independent) activation energy of 18 kcal/mole would suggest that the γ relaxation is not associated with a glass transition.

It seems more likely that the γ relaxation results from limited motions of short chain segments as involved in the crankshaft or local mode relaxations outlined in Section 5.5. The dilatometric evidence (see above) and high activation energy (see Figure 11.31) might also suggest that relatively large-scale backbone motions, characteristic of a glass transition, are involved in the α process. It should, however, be added that the concept of a glass transition may not be valuable for a highly crystalline polymer such as PTFE.

11.8 TETRAFLUOROETHYLENE–HEXAFLUOROPROPYLENE COPOLYMERS

It is of interest to consider the relaxation data for copolymers of tetrafluoroethylene (TFE) and hexafluoropropylene (HFP; $CF_3CF=CF_2$) since they aid in the understanding of the three loss regions exhibited by PTFE. Figure 11.34 shows the temperature dependence of the logarithmic decrement at 1 c/s for TFE–HFP copolymers containing, respectively, 0, 5·5, 7·5 and 14 mole per cent HFP (McCrum, 1959c). From an x-ray study these specimens were estimated to have degrees of crystallinity of 48, 52, 50 and 33 % respectively.

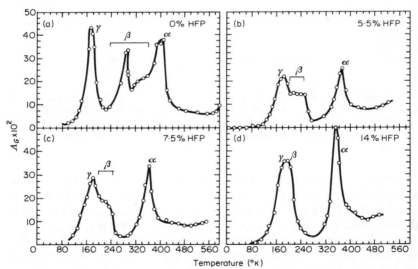

Figure 11.34. Temperature dependence of the logarithmic decrement at 1 c/s for copolymers of tetrafluoroethylene and hexafluoropropylene (HFP) (McCrum, 1959c).

The most striking feature of Figure 11.34 is the large shift of the β relaxation region to lower temperatures with increasing HFP content. As a consequence of this shift the α peak is well resolved for the copolymers, whereas for PTFE it is partially merged with the β peak. However, the γ peak, which is resolved for PTFE, is merged with the β peak for the copolymers.

11.8a The α Relaxation

From Figure 11.34 we observe that the α peak increases in magnitude when the HFP content is increased from 5·5 to 14 mole per cent. Although it appears from Figure 11.34(a) that the magnitude of the α peak is greater for PTFE than for the copolymers containing 5·5 and 7·5 mole per cent HFP, this conclusion is probably invalid since the apparent magnitude of the α peak for PTFE contains a contribution from the superposed β peak. The above observations would suggest that both linear $(CF_2)_n$ sequences and segments containing —CF_3 side-groups were involved in the α mechanism. Since the degree of crystallinity decreases with increasing HFP content, these results would also seem to favour the view that the α relaxation occurs in the amorphous phase.

Figure 11.35 shows a plot of the temperature of the α peak against mole per cent HFP. The point shown at 100% HFP corresponds to the temperature (162°C at 1 c/s) of the main glass–rubber relaxation of polyhexafluoropropylene. If the α peak in PTFE, is, in fact, assigned to the glass transition then a minimum must exist in the curve of Figure 11.35 between 14 and 100% HFP. This result would not conform to the Gordon–Taylor equation (Equation 2.2) which theoretically relates T_g to copolymer composition. However, the lack of agreement with the predictions of the Gordon–Taylor equation does not preclude the possibility that the α peak in PTFE is related to T_g, since similar discrepancies have been observed for the copolymer systems methyl methacrylate–acrylonitrile, styrene–methyl methacrylate, and vinylidene chloride–methyl acrylate (Beevers and White, 1960; Illers, 1963).

The dependence of the magnitude and temperature of the α peak on HFP content is reminiscent of the observations of Willbourn (1958) on the effect of methyl branching on the β peak in polyethylene (see Chapter 10). Although in the homopolymer, polymethylene, the β peak is not well, if at all, resolved, it clearly occurs at 0°C (100 c/s) for 1 methyl side-group per 100 carbon atoms. For 14 methyl side-groups per 100 carbon atoms the β peak occurs at −60°C and in addition is extremely large. However, the addition of further methyl groups ultimately raises the temperature

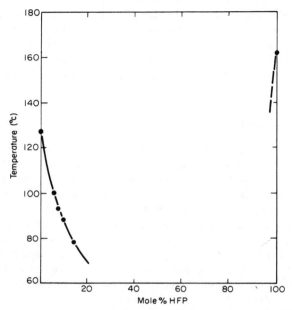

Figure 11.35. Dependence of the temperature of the mechanical α peak (∼1 c/s) on mole per cent HFP for TFE–HFP copolymers containing between 0 and 14 mole per cent HFP. The point plotted at 100 mole per cent HFP is for the α peak of the amorphous polyhexafluoropropylene.

of the β peak again, since in polypropylene it occurs at 0°C (see Chapter 10). These observations are in agreement with results obtained for the ethylene–propylene copolymer system (Natta and Crespi, 1960; Manaresi and Giannella, 1960; Flocke, 1962) illustrated in Figure 10.14.

11.8b   The β Relaxation Region

We have mentioned the hypothesis earlier that the β relaxation in PTFE involves oscillations of the chain segments around the chain axes in the crystals, and, further, that it is associated with the crystal disordering transitions at 19 and 30°C. Figures 11.27, 11.34 and 11.36 show that the β peak shifts to lower temperatures both with an increase in HFP content and with a decrease in crystallinity. According to McCrum (1959c) these observations appear to result from a decrease in the average 'crystallite size'. The 'crystallite size' is considered here to mean the average length

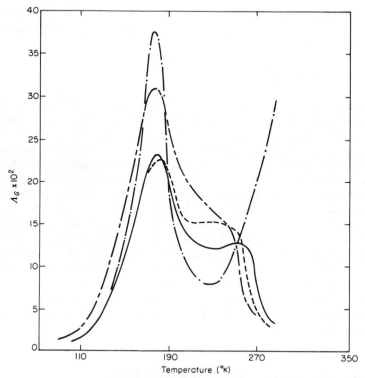

Figure 11.36. Temperature dependence of the logarithmic decrement at 1 c/s for TFE–HFP copolymers containing 5·5 mole per cent HFP in the mixed β and γ region. Degrees of crystallinity as follows: Sample A (————————) 43%; Sample B (————), 52%; Sample C (———), 52%. Although Samples B and C have the same crystallinity, C was annealed for 40 hours at 250°C whereas B was annealed for 15 hours at 250°C. Data also shown for PTFE (———·——) of 48% crystallinity (McCrum 1959c).

of the chain segments arranged side by side in the crystal. If the chain-folded crystal structure applies, it may presumably be regarded as the average length of chain segments between the folds. The above interpretation is consistent with the observation that perfluorohexadecane $(CF_3(CF_2)_{14}CF_3)$, which has a crystallite size of about 25 Å, undergoes a crystal disordering transition at $-170°C$ (Bunn and Howells, 1954), whereas for PTFE (crystallite size about 300 Å) the transition occurs at 19°C. It is to be expected that the average crystallite size would decrease with decreasing crystallinity (less annealing) and also with an increase in HFP content, since the —CF$_3$ side-groups cannot be easily accommodated within the crystals.

11.8c The γ Relaxation

It is clear from Figure 11.36 that the γ peak is considerably smaller for the copolymers than for PTFE itself. The probable depression of the γ peak with increasing HFP content is obscured for the copolymers, however, by the overlap of the β relaxation. Despite this overlap it is also clear that the temperature of the γ peak is essentially independent of HFP content (up to 14 mole per cent). These observations, which contrast with those made for the α peak, suggest that the —CF$_3$ side-groups do not participate in the γ relaxation, and that the introduction of these side-groups merely causes a reduction in the number of short $(CF_2)_n$ segments that are able to relax. Figure 11.36 also illustrates that, at constant HFP content, the γ peak is depressed in magnitude with increasing crystallinity or density. This result agrees with the observed depression of the γ peak in PTFE itself with increasing crystallinity, and confirms the suggestion that the γ mechanism occurs within the disordered regions of the polymer. The γ relaxation could thus arise from a crankshaft- or local mode-type mechanism (Section 5.5).

11.9 POLYCHLOROTRIFLUOROETHYLENE

$$[-\overset{\overset{\displaystyle F}{|}}{\underset{\underset{\displaystyle F}{|}}{C}}-\overset{\overset{\displaystyle F}{|}}{\underset{\underset{\displaystyle Cl}{|}}{C}}-]_n$$

(7)

Relaxation processes occurring in polychlorotrifluoroethylene (PCTFE) are readily detected by the dielectric method owing to the dipolar C—Cl bonds attached to alternate chain carbon atoms. Very low molecular weight forms of this polymer are either liquids or low-melting waxes. Higher molecular weight forms are partially crystalline polymers which have melting points of about 220°C and dilatometric glass transitions at about 52°C (Hoffman and Weeks, 1958). By appropriate thermal treatments, degrees of crystallinity between about 0 and 80%, as determined from density measurements, have been achieved.

According to Kaufman (1953) the x-ray fibre diagram of crystalline PCTFE indicates a hexagonal unit cell with $a_0 = 6.5$ Å and $c_0 = 35$ Å. Also, Liang and Krimm (1956) have found a fibre repeat distance of 43 Å. This very long fibre repeat distance suggests that the chains in the crystals are helical, twisting through 360° every 14–16 monomer units. Despite the high melting point and high attainable degree of crystallinity, high-resolution n.m.r. results (Tiers and Bovey, 1963) indicate that the stereochemical configuration of the PCTFE chains is very irregular, though not strictly atactic. For steric reasons syndiotactic placements appear to be slightly favoured over isotactic placements during the propagation reaction (Tiers and Bovey, 1963).

11.9a Mechanical Studies

Dynamic mechanical investigations of PCTFE have been made by Schmieder and Wolf (1953), Schulz (1956), Maeda (1957), Illers and Jenckel (1959a), Baccaredda and Butta (1960), McCrum (1962) and Passaglia, Crissman and Stromberg (1965). McCrum measured the torsion modulus and logarithmic decrement at 1 c/s for three samples of PCTFE having degrees of crystallinity of 27, 42 and 80% respectively. As shown in Figure 11.37, the sample of 80% crystallinity exhibits a small α peak at 140°C. From dielectric evidence it appears that this α relaxation is connected with the occurrence of chain-folded lamellar spherulites. Two prominent loss peaks are also observed at about 100°C (β peak) and between −10 and −40°C (γ peak). The β peak decreases in magnitude as the per cent crystallinity increases (Figure 11.37), showing that this relaxation originates in the amorphous regions of the polymer. The relatively large modulus drop for the specimen of 27% crystallinity suggests further that the β relaxation is associated with the glass transition which is found dilatometrically at 52°C (Hoffman and Weeks, 1958). The γ peak sharpens on the high-temperature side and its maximum shifts to lower temperatures with increasing crystallinity. On the basis of this result McCrum proposed that the γ relaxation comprised, in fact, two overlapping relaxations centred at −10 and −40°C respectively. The relaxation at −10°C, which increases in magnitude with decreasing crystallinity, was attributed to the amorphous regions. The −40°C relaxation was assigned to the crystalline regions, since it is apparently dominant in the highest crystallinity sample. The mechanical data of the various authors are generally in good agreement. This fact is partially illustrated by the log f versus $1/T$ plot shown in Figure 11.42 below.

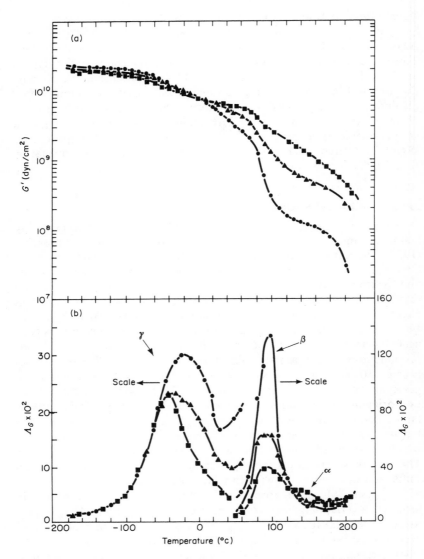

Figure 11.37. Temperature dependence of (a) the shear modulus and (b) the logarithmic decrement at about 1 c/s for three samples of PCTFE having degrees of crystallinity of 27% (●), 42% (▲) and 80% (■) (McCrum, 1962).

11.9b Dielectric Studies

Dielectric investigations of PCTFE include those of Reynolds and others (1951), Hartshorn and others (1953), Mikhailov and others (1957, 1958, 1959), Nakajima and Saito (1958), Krum and Müller (1959) and Scott and others (1962). The work of Scott and coworkers is particularly noteworthy since it represents probably the most extensive single investigation of a partially crystalline polymer. Table 11.3 outlines the methods of crystallization and gives the degrees of crystallinity (χ) of the various specimens studied by these authors. Each sample is numbered according to its degree of crystallinity which was determined from its specific volume ($\bar{v}$) at 23°C (Hoffman and Weeks, 1958). The dielectric measurements were made within the frequency range 0·1 c/s to 8·6 kMc/s and temperature range $-50°$c to $+250°$c.

Some results of Scott and coworkers are presented in Figure 11.38 which shows plots of ε'' against temperature at 1 c/s for samples 0·80, 0·73, 0·44 and 0·12. For sample 0·80 three loss peaks are clearly observed,

Table 11.3. Samples of PCTFE used for dielectric measurements (Scott and others, 1962)

Method of crystallization and sample characteristics	Specific volume ($\bar{v}$) at 23°C (cm³/g)	Degree of crystallinity (χ) and sample number
Crystallized slowly from melt. Highly spherulitic sample. White cloudy appearance	0·4625	0·80
Quenched in ice water from melt. No visible spherulites. Rather transparent	0·4711	0·44
Prepared by annealing sample 0·44 at 190°C for two weeks (i.e. quench-anneal method). No visible spherulites. Fairly transparent	0·4642	0·73
Film 0·15 mm thick quenched into ice water from melt. Transparent	0·4788	0·12
Sample studied in amorphous state above melting point (up to 250°C) and down to about 200°C by supercooling		0·0

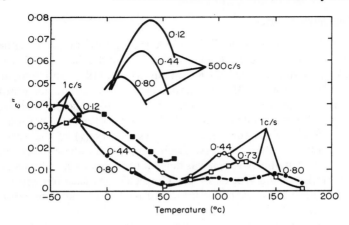

Figure 11.38. Variation of ε'' with temperature at 1 c/s and 500 c/s for PCTFE samples of various crystallinities (Scott and others, 1962).

and a comparison of Figure 11.38 with Figure 11.37 shows that the α, β and γ peaks each lie at temperatures close to those found, respectively, from the mechanical measurements. For samples 0·73 and 0·44 the α and β peaks have merged into a single broad peak. The predominant γ relaxation is illustrated in greater detail by plots of ε' and ε'' against frequency for sample 0·80 (see Figure 11.39). The α and β peaks are also detectable in the larger-scale plots of ε'' against frequency shown in Figure 11.40 and 11.41 respectively. Figure 11.42 summarizes the frequency–temperature location of the various dielectric and mechanical loss maxima. In addition to the α, β and γ relaxations a fourth relaxation region is also apparent from Figure 11.39 by the rise in ε'' with increasing frequency in the region of 1000 to 10,000 Mc/s (i.e. microwave region). This very high frequency of relaxation, which was also observed by Hartshorn and others (1953), is not observed in the ε'' versus temperature plots (Figure 11.38). According to the authors it would give rise, at 1 c/s, to an additional small (δ) peak at about $-100°$c which is below the temperature range investigated. For each specimen Scott and coworkers constructed Cole–Cole plots at each temperature investigated. Three examples of such plots are presented in Figure 11.43. From procedures involving an extrapolation of the Cole–Cole plots, the *total* magnitudes of all dipole relaxations, $\varepsilon_R - \varepsilon_U$, were evaluated. The quantity $\bar{v}(\varepsilon_R - \varepsilon_U)$, which is proportional to the effective contribution of the dipole orientation polarization to the

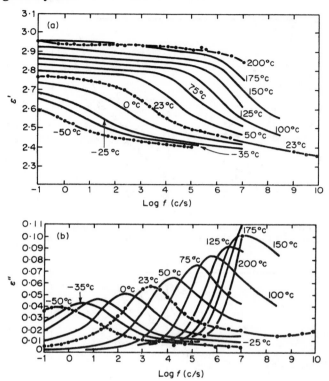

Figure 11.39. Frequency dependence of (a) ε' and (b) ε'' at various temperatures for PCTFE (sample 0·80) showing the predominant γ peak (Scott and others, 1962).

static dielectric constant, is shown as a function of temperature in Figure 11.44. In the following sections the origin and mechanisms of the four different relaxation regions will be discussed.

11.9c The α Relaxation

Since no break was observed in the $\bar{v}-T$ curve at about 120°C for sample 0·80 the authors concluded that the α relaxation was not associated with an amorphous 'glass-transition'. This conclusion was supported by the fact that a WLF-type equation was not necessary to describe the temperature dependence of the frequency of the resolved α peak in sample 0·80 (Figure 11.42). Instead a temperature-independent activation energy of

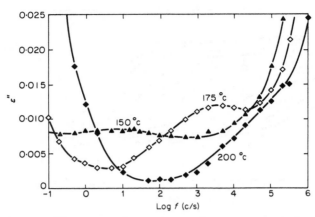

Figure 11.40. ε'' versus frequency at various temperatures showing the α loss peak for PCTFE sample 0·80 (Scott and others, 1962).

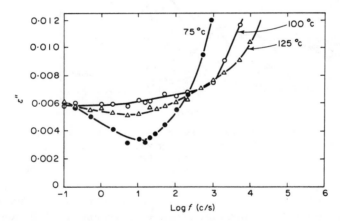

Figure 11.41. ε'' versus frequency showing evidence for the β relaxation in PCTFE sample 0·80 (Scott and others, 1962).

80 kcal/mole was found to describe the results within experimental error. The authors proposed, therefore, that the α mechanism in PCTFE involved the chain-folded crystal lamellae.

The dielectric α peak is resolved for sample 0·80 but not for sample 0·73 (Figure 11.38). This result is particularly interesting since, although these samples had very similar degrees of crystallinity at the temperature of the α peak (115°C, 1 c/s), lamellar spherulites were visible in the polarizing

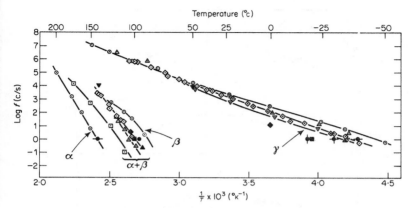

Figure 11.42. Plot of log f against $1/T$ for the dielectric (open points) and mechanical (filled points) loss maxima in PCTFE. In the following key (f) refers to the frequency of maximum loss in constant temperature experiments and (T) denotes the temperature of maximum loss in experiments at constant frequency.

⊙ $\varepsilon''(f)(\chi = 0.80)$ (Scott and others, 1962)

⊡ $\varepsilon''(f)(\chi = 0.73)$ (Scott and others, 1962)

△ $\varepsilon''(f)(\chi = 0.44)$ (Scott and others, 1962)

▽ $\varepsilon''(f)(\chi = 0.12)$ (Scott and others, 1962)

◇ $\varepsilon''(f)$ ($\chi = 0.43 \rightarrow 0.48$, as estimated from given densities by Scott and others, 1962) (Nakajima and Saito, 1958)

◆ $\Lambda(T)$ (Schmieder and Wolf, 1953)

■ $\Lambda(T)(\chi = 0)$ (Illers and Jenckel, 1959a)

▲ $\Lambda(T)$ (Schulz, 1956)

▼ $Q^{-1}(T)$ (Baccaredda and Butta, 1960)

● $\Lambda(T)(\chi = 0.27, 0.42$ and $0.80)$ (McCrum, 1962)

◕ $\Lambda(T)(\chi = 0.80)$ (McCrum, 1962)

◑ $\Lambda(T)(\chi = 0.42)$ (McCrum, 1962)

◔ $\Lambda(T)(\chi = 0.27)$ (McCrum, 1962)

microscope for sample 0·80 but not for sample 0·73. Scott and others thus suggested that the α process was not a property of the bulk crystalline phase, but that it involved the surfaces of the chain-folded lamellar spherulites. A decrease in crystallite or spherulite size could be involved in the shift of the α peak to lower temperatures (to merge with the β peak) for sample 0·73. It is of interest to recall that the α mechanisms in other crystalline polymers, such as polyethylene (Chapter 10), polyoxymethylene (Chapter 14) and polyvinyl alcohol (Chapter 9) have also been assigned to the chain-folded lamellar regions within the polymer.

11.9d The β Relaxation

Comparing the data for specimens 0·80 and 0·44 in Figure 11.38, it is clear that the height of the β peak increases with decreasing crystallinity,

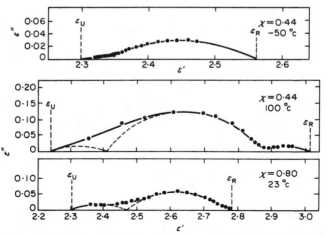

Figure 11.43. Typical Cole–Cole plots obtained from the dielectric data for PCTFE (Scott and others, 1962).

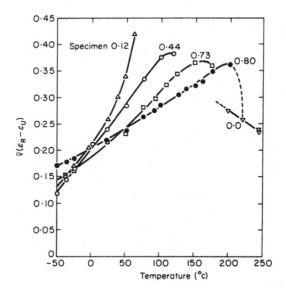

Figure 11.44. Temperature dependence of $\bar{v}(\varepsilon_R - \varepsilon_U)$ for the various samples of PCTFE (Scott and others, 1962).

despite the merging of the α and β peaks in specimen 0·44. This comparison supports the mechanical evidence which attributes the β relaxation to the amorphous phase. Furthermore, Scott and coworkers have shown that the curved plot of $\log f_m$ versus $1/T$ for the resolved β process in sample 0·80 (Figure 11.42) is consistent with the WLF equation with $T_g = 52°$C. The apparent activation energy, determined from this plot, varies from 34 kcal/mole at 120°C to 67 kcal/mole at 90°C. The above evidence confirms the conclusion obtained from the mechanical data that the β mechanism involves motions of large chain segments and is associated with the glass transition at 52°C.

11.9e   Mixed α and β Relaxations

From Figure 11.38 it is seen that the single high-temperature dielectric peak, which has been attributed to merged α and β peaks, shifts to higher temperatures as the degree of crystallinity increases from 0·44 to 0·73. It will be recalled that specimens 0·44 and 0·73, unlike specimen 0·80, do not contain large, visible spherulites, having been crystallized during or subsequent to quenching. The above result might possibly indicate that the β peak shifts to higher temperatures (i.e. T_g increases) with increasing crystallinity. Such an effect could arise if the presence of crystallites caused some restriction to motions in the amorphous regions, an effect commonly expected if one chain molecule can participate in several crystallites (as in the fringed-micelle model). This effect might not be expected to be very large for the chain-folded crystal structure, since the number of connections between chain-folded lamellae should be far less than in the fringed-micelle structure.

From $\bar{v}–T$ plots, however, specimens 0·44 and 0·73 were found to have the same T_g value (within 5°C) as specimen 0·80. Also, the mechanical results for samples 0·27 and 0·42 (Figure 11.37) suggest that there is no upward shift in the temperature of the 'mixed' α and β peak with increasing crystallinity. The authors gave the following explanation for the apparent disparity between the dielectric and mechanical data. As shown by the data for specimens 0·80 (Figures 11.37 and 11.38) the β peak is of much greater magnitude than the α peak in the mechanical results, whereas dielectrically the β peak is slightly less prominent than the α peak. Owing to this difference in the relative intensity of the two peaks, if the α peak shifts down in temperature and merges with the β peak to form a single broad peak, the mixed dielectric peak would appear at a higher temperature than the mixed mechanical peak. It thus appears that the apparent shift of the mixed dielectric peak to higher temperatures with increasing crystallinity is a result of a change in relative magnitude of the α and β relaxations and *not* a result of an increase in T_g.

11.9f The Very High Frequency (δ) Relaxation

The relaxation process occurring at very high frequencies, apparent from the 23°C results of Figure 11.39, gives rise to the asymmetry of the Cole–Cole plots (Figure 11.43) in the regions of low ε'. From a resolution of the Cole–Cole plots Scott and coworkers estimated the magnitude of the very high frequency relaxation $(\varepsilon_R - \varepsilon_U)_\delta$ at each temperature studied. Values of $\bar{v}(\varepsilon_R - \varepsilon_U)_\delta$ are plotted in Figure 11.45. It is observed from this diagram that $\bar{v}(\varepsilon_R - \varepsilon_U)_\delta$ increases with temperature for each sample and, within experimental error, is directly proportional to the degree of crystallinity.

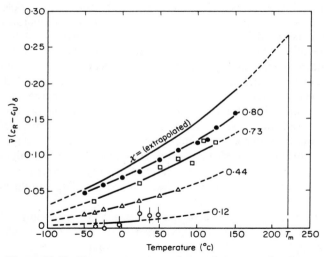

Figure 11.45. Temperature dependence of $\bar{v}(\varepsilon_R - \varepsilon_U)_\delta$ for the very high frequency relaxation in PCTFE (Scott and others, 1962).

The authors concluded from these data that the very high frequency process involved the *bulk crystalline phase* which is becoming more disordered with increasing temperature. This high-frequency crystal phase relaxation is also responsible for the observation (Figure 11.44) that at 200°C $\bar{v}(\varepsilon_R - \varepsilon_U)$ is considerably higher for sample 0·80 than for the supercooled amorphous sample 0·0. Reynolds and coworkers (1951) also found a marked increase in the static dielectric constant following the crystallization of low molecular weight PCTFE samples.

In discussing the mechanism of this crystal phase relaxation the authors first noted that the average relaxation time ($\tau \approx 10^{-11}$ sec at 23°C) associated with it was extremely short, indicating a low activation energy, and also that its magnitude $(\varepsilon_R - \varepsilon_U)_\delta$ was rather large. They also pointed out that, owing to the helical chain conformation, the resultant dipole moment of long chain segments would have only a very small component perpendicular to the chain axis due to vectorial cancellation. Hence it is unlikely that the mechanism involves the rotations of long rigid chain segments about the chain axis, since this process would not be expected to lead to a high value of $(\varepsilon_R - \varepsilon_U)_\delta$. Also, such a mechanism would be expected to have a longer relaxation time than that found experimentally, particularly if chain folds are located at the ends of the rotating segments. The authors suggested therefore that the electric field might distort short sections of the chains within the crystal, such that the local uncoiling of the helix produces a large local dipole moment. A relaxation mechanism involving the local uncoiling of the polymer chain is consistent with the observation that $d[\bar{v}(\varepsilon_R - \varepsilon_U)_\delta]/dT$ is positive since this result shows that the application of the electric field must cause an increase in the entropy or disorder of the system (Fröhlich, 1949).

11.9g The γ Relaxation

Figures 11.37, 11.38 and 11.39 illustrate that the γ relaxation is by far the most prominent dielectric, though not the largest mechanical, relaxation. From Figure 11.38 it will also be observed that the dielectric γ peak, both at 1 and 500 c/s, shifts to lower temperatures with increasing crystallinity. At 1 c/s the temperature of the dielectric ε'' maximum decreases from a value of -13 ± 5°C for sample 0·12 to a value of -41 ± 5°C for sample 0·80. Correspondingly, Figures 11.42 and 11.46 show that at constant temperature (below about 50°C) the γ peak shifts somewhat to higher frequencies with increasing crystallinity. Furthermore, the activation energy obtained from Figure 11.42 decreases from a value of 16·7 kcal/mole for sample 0·12 to a value of 14·6 kcal/mole for sample 0·80.

Figure 11.47 illustrates the temperature dependence of $\bar{v}(\varepsilon_R - \varepsilon_U)_\gamma$, which represents the contribution of the γ relaxation to $\bar{v}(\varepsilon_R - \varepsilon_U)$. It is clear from the curves marked $\chi = 0.44$, $\chi = 0.12$ and $\chi = 0$ (extrapolated) that the *amorphous* phase is largely involved in the γ mechanism for the moderate and weakly crystalline specimens. However, it also appears from the curve marked $\chi = 1$ (extrapolated) that an appreciable contribution to $\bar{v}(\varepsilon_R - \varepsilon_U)_\gamma$ is *not* derived from the normal amorphous phase. According to Scott and coworkers this so-called anomalous contribution

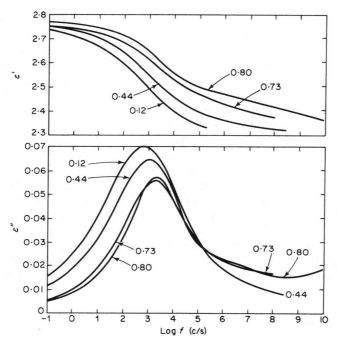

Figure 11.46. Frequency dependence of ε' and ε'' at 23°C for PCTFE samples of different crystallinities (Scott and others, 1962).

is not due to the very high frequency crystal relaxation, since at 23°C this relaxation is widely separated from the main γ relaxation region (see Figure 11.46). In fact, the anomalous contribution to $\bar{v}(\varepsilon_R - \varepsilon_U)_\gamma$ is obtained from the areas under the resolved γ peaks of Figure 11.46 after extrapolation to $\chi = 1$. The authors proposed therefore that the anomalous contribution may arise either from disordered regions within the crystalline phase, or from an abnormal amorphous phase produced at high crystallinities by the large fraction of lamellar crystals. The results and conclusions outlined above are not inconsistent with McCrum's suggestion, based on the mechanical results of Figure 11.37, that the γ relaxation comprises two overlapping relaxations centred at about $-10°$C (amorphous relaxation) and $-40°$C (crystalline relaxation) respectively at 1 c/s.

The activation energy estimated by Scott and coworkers for the γ relaxation lies between 17·4 kcal/mole ($\chi \to 0$) and 13·5 kcal/mole ($\chi \to 1$). These values, which compare well with those obtained by other workers,

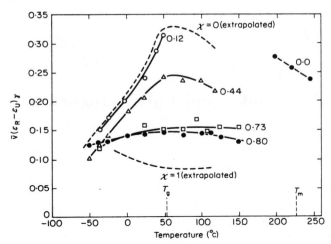

Figure 11.47. Temperature dependence of $\bar{v}(\varepsilon_R - \varepsilon_U)_\gamma$ for the main γ relaxation in PCTFE samples of different crystallinities (Scott and others, 1962).

are essentially independent of temperature. Since the molecules of PCTFE do not contain polar side-groups capable of independent rotations, the γ relaxation must involve motions *within* the chain backbone. The relatively low activation energy further suggests that only a small number of chain segments can be involved. Scott and coworkers have proposed a kinking or 'jump rope' type of motion which seems similar to the crankshaft mechanism (Section 5.5a). The so-called local mode type of motion (see Section 5.5b) might also be suggested as an alternative mechanism. PCTFE thus provides another example of a polymer in which limited chain backbone motions are indicated at temperatures below the proposed glass-transition temperature (52°C).

12

Polyamides and Polyurethanes

12.1 POLYAMIDES

$$[-\underset{\underset{H}{|}}{N}-(CH_2)_x-\underset{\underset{H}{|}}{N}-\overset{\overset{O}{\|}}{C}-(CH_2)_{y-2}-\overset{\overset{O}{\|}}{C}-]_n$$

(1)

The structural formula (1) represents one class of polyamides which are prepared by the condensation of diamines and dicarboxylic acids. The best-known polyamide of this type is polyhexamethylene adipamide, which is synthesized from hexamethylene diamine $H_2N(CH_2)_6NH_2$ and adipic acid $HOOC(CH_2)_4COOH$. This polymer is also known as nylon 66 since it is represented by the above formula with $x = 6$ and $y = 6$. Similarly, the polyamide made from hexamethylene diamine and sebacic acid $HOOC(CH_2)_8COOH$, is known as nylon 610.

In addition to the large series of polymers which may be prepared in the above manner, other polyamides can be obtained from the self-condensation of ω-amino acids. These have the general formula (2).

$$[-\underset{\underset{H}{|}}{N}-(CH_2)_{x-1}-\overset{\overset{O}{\|}}{C}-]_n$$

(2)

A well-known example in this series is poly-ε-caprolactam for which $x = 6$. According to the accepted nomenclature, this polymer is termed nylon 6.

The physical properties of the polyamides are dominated by the inter-molecular $>N-H\cdots O=C<$ hydrogen bond. In nylons 66 and 6 the crystals are formed of hydrogen-bonded sheets (Holmes, Bunn and Smith, 1955), and in both polymers the molecules are arranged in the crystal so that the hydrogen bonds are perfectly formed (Figure 2.1). Despite early reports to the contrary it now seems probable from infrared evidence that the amount of unformed hydrogen bonds in the polyamides and their copolymers is generally less than 1 % at room temperature (Trifan and Terenzi, 1958). However, an important exception is nylon 6. If nylon 6 is rapidly quenched from the melt the resulting solid is amorphous and contains approximately 40 % free $>$NH groups (Hendus and others, 1960). This form of nylon 6 is known as the δ-modification. Above 20°C the δ-modification transforms into the γ-modification in which there is complete formation of hydrogen bonds. The γ-modification is pseudocrystalline and converts to the crystalline, monoclinic α-modification above 100°C. A triclinic β-modification also exists in stretched specimens (Ziabicki, 1959; Kinoshita, 1959; Hendus and others, 1960). Investigations of the crystal structures of other polyamides include those of Slichter (1959) and Vogelsong and Pearce (1960).

On crystallizing from solution, polyamides may form spherulitic structures composed of lath-like lamellae (Keller, 1959) or single crystals (Geil, 1960a, 1963). The single crystals contain lamellae which are 50–100 Å thick and in which the molecules are folded. The bulk polymers are usually spherulitic unless specimens of thin cross-section are rapidly quenched from the melt into ice water (Starkweather and Brooks, 1959). In the spherulite the molecules are oriented in the tangential plane with the hydrogen bonds along the radii (Barriault and Gronholz, 1955). It is not clear at present whether the molecules are folded in the bulk polymer. The crystalline content of polyamides is usually in the range 20 to 55 % but in copolymers it may be considerably lower (Sandeman and Keller, 1956; Starkweather and Moynihan, 1956).

Polyamides absorb water in quantities inversely proportional to the density. On the basis of the two-phase model this result leads naturally to the assumption that water molecules do not enter the crystalline phase. The equilibrium uptake at 100 % relative humidity is greater for nylon 66 than for nylon 610; at the same density nylon 66 absorbs 8·5 % water at 23°C and nylon 610, 3 % (Starkweather and others, 1956). It is quite reasonable to assume on these grounds, and others too, that the water molecules are hydrogen bonded at the carbonyl or $>$N—H groups. There are numerous possible ways in which a water molecule may be bonded to or between carbonyl and amide groups but unfortunately there is at present no experimental evidence heavily in favour of any one bonding mechanism.

Mechanical relaxation in the polyamides has been studied by many workers including Schmieder and Wolf (1953), Tokita (1956), Woodward and others (1957), Becker and Oberst (1957), Thomas (1957), Deeley, Woodward and Sauer (1957), Yamamoto and Wada (1957), Willbourn

(1958), Kawaguchi (1959), Quistwater and Dunell (1959), Illers and Jenckel (1959b), Illers (1960), Illers and Jacobs (1960), Woodward, Crissman and Sauer (1960), Onogi and others (1962), Meredith and Hsu (1962), Howard and Williams (1963), Takayanagi (1963), Lawson, Sauer and Woodward (1963) and Kolarik and Janacek (1965). Some results of Takayanagi (1963) on nylon 6 are given in Figure 12.1 for three samples of different crystallinity. The least crystalline specimen was obtained by quenching from the melt to $-78°$C. The sample of intermediate crystallinity was prepared by crystallizing at $200°$C, and the preparation of the sample of highest crystallinity involved the removal of residual monomer (ε-caprolactam). Three loss peaks, labelled α, β and γ respectively, are clearly observed in Figure 12.1. In addition to these three peaks a further relaxation region is indicated by the shoulder on the high-temperature side of the α peak. Since this shoulder is most prominent for the specimen

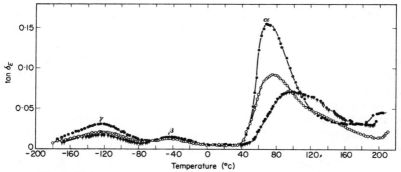

Figure 12.1. Temperature dependence of the mechanical loss tangent, tan δ_E, at 100 c/s for nylon 6 samples of different crystallinity. (●) sample quenched from the melt to $-78°$C (least crystalline); (○) sample crystallized at $200°$C (intermediate degree of crystallinity); (◖) ε-caprolactam removed (highest crystallinity) (Takayanagi, 1963).

of highest crystallinity, it has been attributed by Takayanagi to a relaxation process involving the *crystalline* regions. Takayanagi has also indicated that this relaxation might be related to a crystal structure change from monoclinic to hexagonal, apparent from x-ray and specific volume–temperature data. According to Takayanagi an additional crystalline relaxation mechanism might also be operative in the region of $190°$C, although the detection of this relaxation might be hindered by the relatively low ($< 50\%$) crystalline contents of the polyamides. Although the evidence for these high-temperature crystal relaxations is perhaps not very convincing, it should be recalled that crystalline relaxations have been clearly observed for polyethylene (Chapter 10), polypropylene (Chapter 10), polyoxymethylene (Chapter 14), polyvinyl alcohol (Chapter 9), polychlorotrifluoroethylene (Chapter 11) and polyethylene terephthalate (Chapter 13). However, since the existence of these high-temperature relaxations has not been conclusively established for the polyamide series in general, we will adhere to the well established α, β, γ labelling indicated in Figure 12.1.

Dielectric studies of the polyamides have been described by Baker and Yager (1942), Rushton and Russell (1956), Boyd (1959), McCall and Anderson (1960), Ishida (1960a), Curtis (1961), Dahl and Müller (1961) and Bares, Janacek and Cefelin (1965). These studies have also revealed the existence of the α, β, and γ relaxations together with a very low frequency–high temperature polarization. Figure 12.2 illustrates the temperature dependence of ε' and $\tan \delta_\varepsilon$ at various frequencies for nylon 66 (Rushton and Russell, 1956). The α peak is observed in the region of 90°C (10 kc/s) to 100°C (1 Mc/s). At frequencies between 65 c/s and 10 kc/s the β peak appears in the temperature range -20 to $+10$°C, whereas at 1 Mc/s the β peak is merged with the low-temperature tail of the α peak. The γ relaxation is also seen at 1 Mc/s by the peak centred at about -20°C.

The large rise in $\tan \delta_\varepsilon$ at the highest temperatures (e.g. 130°C at 10 kc/s) shown in Figure 12.2 can be accounted for only partially by a d.c. ionic conduction mechanism. This fact has been established with the aid of d.c. resistivity measurements, from which computed values of ε'' are much smaller than the measured values in this high-temperature region (Curtis, 1961). Also, Curtis has found that the ε'' values in this region can only be partially reduced by drying the samples, so that an ionic conduction mechanism involving water molecules could only account for part of the loss magnitude. That some additional relaxation mechanism must be involved is also evidenced by the large increase in ε' (Figure 12.2) at the highest temperatures. Baker and Yager (1942) proposed that both the d.c. conduction and the low-frequency polarization involved the inter-molecular hydrogen bond,

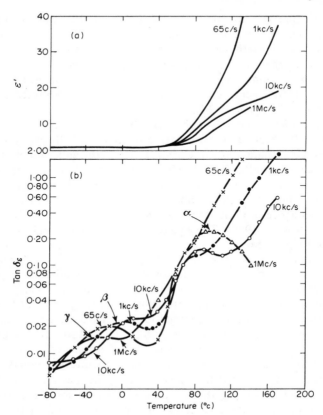

Figure 12.2. Temperature dependence of (a) ε' and (b) tan δ_ε for nylon 66 at frequencies indicated (Rushton and Russell, 1956).

such that the virtually ionic hydrogen oscillated (over a barrier) between potential minima, one provided by the oxygen and the other by its 'own' nitrogen. McCall and Anderson (1960) observed that the d.c. conductivity was greatly reduced if the $>$N—H group hydrogens were replaced by methyl groups in nylon 1010. They forwarded this evidence in favour of the suggestion that the amide protons acted as current carriers. McCall and Anderson (1960) also proposed that the low-frequency polarization was a Maxwell–Wagner-type loss process, closely associated with the d.c. conduction mechanism. Curtis (1961) has suggested an alternative mechanism for the low-frequency polarization, based on the relaxation mechanisms thought to be operative in long chain alcohols and other systems in which the hydrogen-bonding groups are associated throughout the structure. In these systems long-range molecular orientation can occur by the breaking and reforming of hydrogen bonds. As discussed in connection with polyvinyl alcohol (Chapter 9), such chain processes should be characterized by high dielectric constants and long relaxation times.

A summary of the mechanical and dielectric loss data for nylons 6, 66, 610 and 612 is presented on the frequency–temperature diagrams shown in Figures 12.3(a), 12.3(b), 12.3(c) and 12.3(d). Mechanical results for several other polyamides are listed in Table 12.1. It is clear from these results that the α, β and γ relaxations occur for all polyamides so far investigated at similar temperatures and frequencies. Hence, in discussing the proposed mechanisms of these three relaxations, it is convenient and instructive to consider collectively the data for all polyamides.

12.1a The α Relaxation

Figure 12.1 shows that the mechanical α peak for nylon 6 decreases in height with increasing crystallinity suggesting that the α process involves motions within the amorphous phase. As illustrated in Figure 12.4 Schmieder and Wolf (1953) have obtained a similar result for poly-hexamethylene methyl pimelamide,

$$[-N-(CH_2)_6-\overset{\overset{\displaystyle H}{|}}{N}-\overset{\overset{\displaystyle}{\underset{\underset{\displaystyle O}{\|}}{C}}}-(CH_2)_u-\overset{\overset{\displaystyle H}{|}}{\underset{\underset{\displaystyle CH_3}{|}}{C}}-(CH_2)_v-\overset{\overset{\displaystyle O}{\|}}{C}-]_n$$

where $u + v = 4$. From this diagram the height of the α peak is seen to be considerably larger for the sample which had been rapidly quenched from the melt than for the sample that had been annealed for two hours at 90°C. For the quenched sample, the sudden rise in modulus at temperatures just above the α peak temperature indicates that crystallization occurs more readily in this region than at temperatures below the α peak temperature. From dielectric measurements on nylon 610, Curtis (1961) has also observed that the intensity of the α peak decreases with increasing crystallinity, as illustrated in Figure 12.5. A comparison of x-ray intensity diagrams with ε'' values for the samples referred to in Figure 12.5 showed qualitatively that the α peak was associated with the disordered regions.

Other evidence which supports the amorphous assignment includes the effects produced by copolymerization. For example, a terpolymer prepared from the monomers of nylons 6, 66 and 610, which is considered

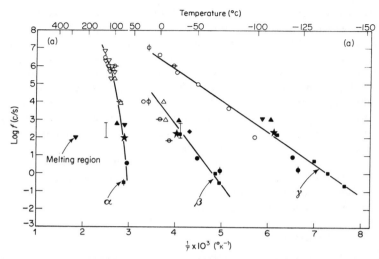

Figure 12.3. Plots of log f against $1/T$ for the dielectric (open points) and mechanical (filled points) loss maxima of (a) nylon 66, (b) nylon 6, (c) nylon 610 and (d) nylon 612. In the following key (f) refers to the frequency of maximum loss in constant temperature experiments and (T) denotes the temperature of maximum loss from constant frequency experiments.

(a) Nylon 66

○ ε'' (f) (Curtis, 1961)
△ ε'' (f) (Boyd, 1959)
⦶ ε'' (f) (Rushton and Russell, 1956)
⊖   tan δ_ε (T) (Rushton and Russell, 1956)
▽ ε'' (f) (McCall and Anderson, 1960)
● Λ (T) (Schmieder and Wolf, 1953)
■   tan δ_G (T) (Illers and Jenckel, 1959b)
▲ Q^{-1} (T) (Woodward and others, 1957)
▼ Q^{-1} (T) (Deeley, Woodward and Sauer, 1957)
◆   tan δ_E (T) (Kawaguchi, 1959)
$\rrbracket$ tan δ_E (T) (Meredith and Hsu, 1962)
★   tan δ_E (T) (Willbourn, 1958)
▧ Λ (T) (Thomas, 1957)

(b) Nylon 6

□ ε'' (f) (Bares, Janacek and Cefelin, 1965)

○ ε'' (f) (Ishida, 1960a)
● Λ (T) (Schmieder and Wolf, 1953)
■   tan δ_E (T) (Kawaguchi, 1959b)
▲   tan δ_E (T) (Becker and Oberst, 1957)
▼   attenuation maxima (T) (Yamamoto and Wada, 1957)
◆   attenuation maxima (T) (Wada and Yamamoto, 1956)
⬖   tan δ_E (f) (Tokita, 1956)
▨   tan δ_E (T) (Takayanagi, 1963)
◕ Λ (T) (Thomas, 1957)

(c) Nylon 610

○ ε'' (f) (Curtis, 1961)
□ ε'' (f) (Baker and Yager, 1942)
● Λ (T) (Schmieder and Wolf, 1953)
■ Q^{-1} (T) (Woodward and others, 1957)
▲ Λ (T) (Thomas, 1957)

(d) Nylon 612

●   tan δ_G (T) (Illers, 1960)

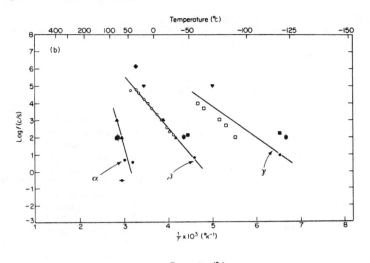

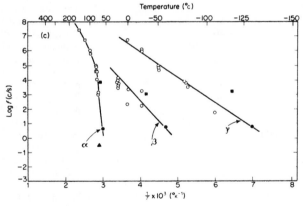

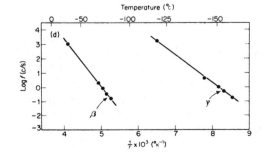

Table 12.1. Temperatures of mechanical loss peaks for several polyamides (nylons)

	Location of tan δ peaks						
	α		β		γ		
Nylon	$T(^\circ C)$	$f(c/s)$	$T(^\circ C)$	$f(c/s)$	$T(^\circ C)$	$f(c/s)$	Reference
4			-40	160			Kawaguchi (1959)
			-60	0·295	-146	0·31	Lawson, Sauer and Woodward (1963)
7	83	71	-50	134	-120	166	Kawaguchi (1959)
8	90	71	-50	133			Kawaguchi (1959)
	49	5·2	-50	8·1	-130	10	Schmieder and Wolf (1953)
9	90	61	-47	134			Kawaguchi (1959)
10	94	70	-39	129	-120	179	Kawaguchi (1959)
11	83	70	-45	125	-117	173	Kawaguchi (1959)
	56	3·8	-55	6·8	-115	8·3	Schmieder and Wolf (1953)
210	65	5·5	-55	8	-120	9·8	Schmieder and Wolf (1953)
49			-41	200			Kawaguchi (1959)
67	58	3·7	-50	6	-115	7·3	Schmieder and Wolf (1953)
68	64	3·1	-50	5·4	-120	6·8	Schmieder and Wolf (1953)
69			-40	199			Kawaguchi (1959)
77			-39	191			Kawaguchi (1959)
79			-38	203			Kawaguchi (1959)
88			-39	198			Kawaguchi (1959)
89			-37	207			Kawaguchi (1959)
99			-38	192			Kawaguchi (1959)
109			-36	198			Kawaguchi (1959)
1010			-42	194			Kawaguchi (1959)
	67	600			-100	1200	Woodward, Crissman and Sauer (1960)
1010 (17% N methylated)	42	500			-105	1000	Woodward, Crissman and Sauer (1960)
1010 (58% N methylated)	12	500			-110	1500	Woodward, Crissman and Sauer (1960)

to be largely amorphous, has a larger α relaxation magnitude than each of the more crystalline homopolymers in both mechanical (Figure 12.6) (Woodward and others, 1957) and dielectric (McCall and Anderson, 1960) studies. Methylation of the $>$N—H groups in nylon 1010, a procedure also considered to lower the crystalline content, has also been found to increase the height of the mechanical α peak (Woodward, Crissman and Sauer, 1960) although the resolved dielectric α relaxation does not appear to be similarly affected (McCall and Anderson, 1960).

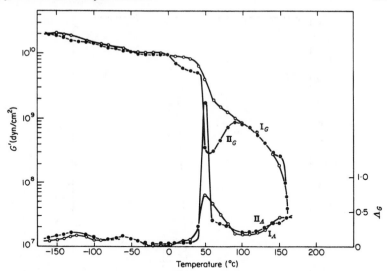

Figure 12.4. Temperature dependence of the shear modulus (G') and logarithmic decrement (Λ) at about 5 to 10 c/s for polyhexamethylene methyl pimelamide. Sample I ($\bigcirc$) annealed for two hours at 90°C; Sample II ($\bullet$) cooled rapidly from the melt prior to measurements (Schmieder and Wolf, 1953).

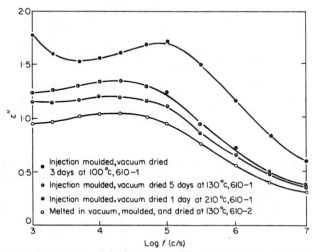

Figure 12.5. Dependence of ε'' on frequency at 80°C for nylon 610. Pretreatment of samples indicated in the key to point symbols (Curtis, 1961).

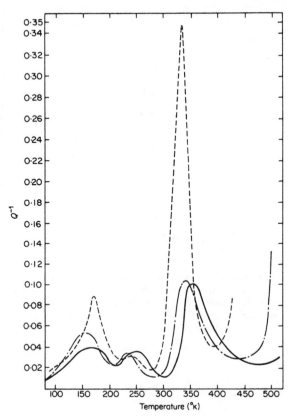

Figure 12.6. Mechanical damping (Q^{-1}) as a function of temperature for nylon 66 (———), nylon 610 (— —) and a copolymer prepared from the monomers of nylons 6, 66 and 610 (– – –) (Woodward and others, 1957).

The effect of irradiation on the α relaxation has been studied both by the mechanical and the dielectric method. Deeley, Woodward and Sauer (1957) found, in the case of nylon 66, that the magnitude of the mechanical α peak was considerably increased when the polymer was irradiated in a reactor at dosages up to 5.5×10^{18} nvt (Figure 12.7). This result was ascribed to the destruction of crystallinity produced by the radiation induced crosslinking. Boyd (1959) found that the dielectric α peak of nylon 66 was diminished and finally almost eliminated by irradiation with 2-Mev electrons for various exposure times (Figure 12.8).

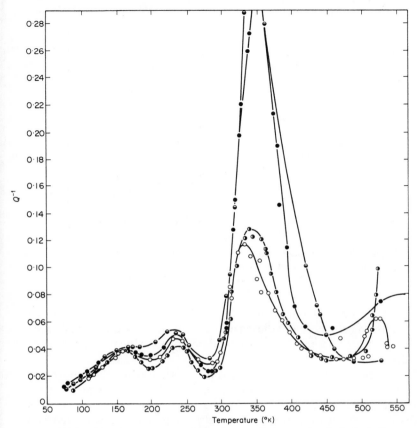

Figure 12.7. Temperature dependence of the mechanical damping
(Q^{-1}) for nylon 66 irradiated in a reactor to various dosages. (◑) 0 nvt;
(○) 0.3×10^{18} nvt; (●) 2.8×10^{18} nvt; (◓) 5.5×10^{18} nvt (Deeley,
Woodward and Sauer, 1957, see also Woodward and Sauer 1958).

Boyd estimated the degrees of crosslinking produced by the irradiation
as the ratio of the exposure times to the exposure time required to reach
the gel point (one crosslink per weight average molecule). The α peak was
found (Figure 12.8) to have essentially disappeared for a sample containing
about 10 crosslinks per molecule. Since the nylon 66 had a number
average molecular weight of 18,000, and an estimated weight average of
about 36,000, Boyd concluded that the moving segment contained about

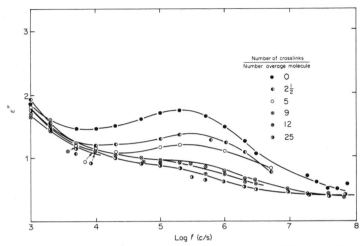

Figure 12.8. Dependence of ε'' on frequency at 100°C for nylon 66 crosslinked by irradiation with 2 Mev electrons. The crosslink densities are indicated in the key (Boyd, 1959).

15 amide groups. The difference observed between the effects of irradiation on the mechanical and dielectric α peak is possibly due to differences in dosage or to differences in the thermal treatment of the specimens during irradiation.

For a series of polyamides Thomas (1957) has found that the temperature of the mechanical α peak, T_α (at 0·3 c/s), decreases as the temperature of the melting region, T_m, decreases. This result is shown by the data listed in Table 12.2. The temperatures of the melting regions were estimated from the data of Schmieder and Wolf (1953). Table 12.2 also gives the α peak temperatures obtained by Schmieder and Wolf. Despite the higher frequencies (about 5 c/s) employed by Schmieder and Wolf the α peak temperatures observed by them are somewhat lower than those reported by Thomas. According to Thomas this difference could be due to residual monomer or moisture in the samples studied by Schmieder and Wolf. However, another reason for the difference might be related to the fact that Thomas studied (oriented) nylon *fibres* whereas Schmieder and Wolf investigated moulded samples. The latter suggestion is of some interest since Thomas also observed that the α peak temperature was increased when the samples were further oriented by increasing the draw ratio. A similar effect was reported by Thompson and Woods (1956) for polyethylene terephthalate. If the α peak in the polyamides is related to the

Table 12.2. Temperatures of the α peak (T_α) and the melting region (T_m) for various polyamides

	$T_\alpha(°C)$		$T_m(°C)$
Nylon	Thomas (1957) $(f = 0.3 \text{ c/s})$	Schmieder and Wolf (1953) $(f \approx 5 \text{ c/s})$	Schmieder and Wolf (1953)
66	78	65	250
68	75	64	230
610	70	60	220
67	73	58	220
6	65	40	210
8	63	49	195
11	53	56	190

amorphous glass transition, some relationship between T_m and T_α is perhaps to be expected. Current theories of melting (Starkweather and Boyd, 1960) and the glass transition (e.g. Gibbs and DiMarzio, 1958) suggest that both T_m and T_g are related to factors such as chain stiffness and intermolecular cohesion. One other interesting, and apparently unexplained, result from Table 12.2 is that both T_m and T_α are lower for nylon 6 than for nylon 66, despite the fact that these two polymers have equal densities of amide groups. This problem has been commented on by Holmes, Bunn and Smith (1955).

The α relaxation in the polyamides has properties similar to the α relaxation in polyethylene terephthalate and related polymers. The principal similarity is that these relaxations are the most dominant in both polymers. Also, for both systems the activation energies are high and T_α depends both on T_m and the degree of orientation. A major difference, however, is that crystallization cannot occur in polyethylene terephthalate at temperatures below the α region, whereas in the polyamides crystallization can occur at temperatures below the temperature of the α peak. Now according to Willbourn (1958), since the glass-transition temperature is commonly regarded as the temperature at which the main chain acquires large-scale mobility, then, for crystalline polymers which can be obtained in the quenched amorphous state, T_g is the temperature at which the crystallization rate becomes measurable in the time scale of the experiment (see Section 10.1). If we accept this statement as one criterion for T_g, then we are led to the conclusion that the α mechanism in the terephthalate polymers is related to the glass transition. This view is supported by evidence based on other definitions of T_g such as those involving volumetric and specific heat discontinuities (see Chapter 2). However, the crystallization criterion would suggest perhaps that T_g for the polyamides is located below $-36°$C (Willbourn, 1958) and not in the region of the α peak. On the other hand, as noted above in connection with Figure 12.4, the crystallization *rate* in the polyamides would seem to be much lower at temperatures below than just above T_α, a fact which favours some relationship between T_α and T_g.

Several authors (e.g. Woodward and others, 1957) have suggested that the main-chain motions responsible for the α mechanism must involve the rupture of hydrogen bonds in the amorphous regions of the polyamides. This suggestion seems reasonable particularly if fairly long chain segments participate in the motion. Also, the implication that the rupture of hydrogen bonds cannot occur at temperatures *below* that of the α peak is in keeping with the infrared study of Trifan and Terenzi (1958) which suggests that *at room temperature* only a very small fraction of hydrogen bonds are unformed in the polyamides. However, the question again arises as to why crystallization, which would seem to require the breaking and reforming of hydrogen bonds, occurs at temperatures below the α region. The infrared results might be explained by assuming that the reformation of hydrogen bonds, subsequent to rupture, occurs very rapidly.

Several investigations of the polyamides have shown that the α peak is shifted to lower temperatures by the absorption of water and other liquids. This effect is well illustrated by the results of Illers (1960) for 612 nylon shown in Figure 12.9. In this study samples were cooled slowly from above the melting temperature and dried for four days at 150°C *in vacuo* over calcium chloride. After this treatment the specimens were considered to be 'dry'. Required water contents were then achieved by allowing specimens to absorb a predetermined amount of water, and then ensuring that the water molecules were uniformly distributed by heating in a sealed tube at 100°C for 14 days. The crystalline content (undefined determination by x-rays) was found to be 28 % in both wet and dry specimens. The water contents of 0, 14 and 40 mole per cent (Figure 12.9) were evaluated as the concentration in the *amorphous regions*. The plasticizing effect of water on the mechanical α peak has also been reported by Woodward, Crissman and Sauer (1960) for nylon 66, and Boyd (1959) has noted a similar effect for nylon 66 from dielectric measurements. Since water is believed to enter only the amorphous regions of the polyamides the above results are consistent with the view that the α mechanism is related to the amorphous regions and possibly to the glass transition.

The initial extent of the depression of the temperature of the α peak

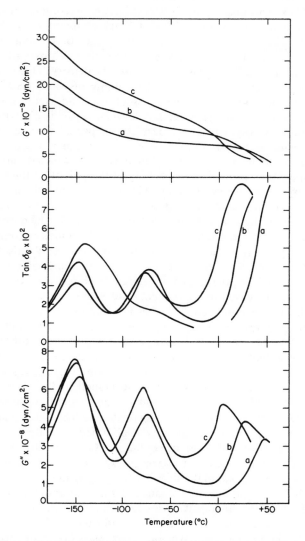

Figure 12.9. Temperature dependence of G', tan δ_G, and G'' at 1 c/s for nylon 612 of 28 % crystallinity. Water contents as follows: Curve a 0 %; Curve b 14 mole %; Curve c 40 mole % (Illers, 1960).

by absorbed water is quite extraordinary; 30°C for the first weight per cent (Illers, 1960). This observation is particularly interesting when viewed in connection with the sorption data of Starkweather (1959). Starkweather's results have shown that the partial specific volume of absorbed water at low concentrations is 0·46 cc/g. Hence, although nylon is more dense than water, the addition of water increases the density of the mixture. Similar low values of the partial specific volume of water (≈ 0.5 cc/g) occur in the water–cellulose and water–sulphuric acid systems at low water concentrations. On the other hand, when water is absorbed by polyvinyl acetate or polymethyl methacrylate the partial specific volume of the water remains essentially at 1 cc/g. Starkweather has proposed that in polyvinyl acetate and polymethyl methacrylate, which do not have hydrogen bonds, the mechanism of packing of the polymer molecules is not affected by the addition of water. However, for anhydrous nylon, cellulose and sulphuric acid the addition of water affects the mode of packing of the molecules so as to increase the density. In the absence of water, the strict steric requirements of the hydrogen bonds prevents the molecules from packing as closely as they would if van der Waals' forces were dominant. Bonded water molecules lower the number of interchain hydrogen bonds enabling the chains to pack more closely and so increase the density. The decreased interchain cohesive forces resulting from the breaking of hydrogen bonds could be responsible for the shift of the α peak to lower temperatures. Clearly the free-volume concept of the glass transition must break down in this case, since it would suggest that the α peak should be shifted to *higher* temperatures with increasing density.

For the α relaxation in nylon 66, a 'time-humidity superposition' procedure has been proposed by Quistwater and Dunell (1959). By analogy with the time–temperature superposition principle, these authors constructed master curves of E' and E'' against a reduced frequency by horizontally shifting experimental curves obtained at different water contents onto a curve at some reference humidity. Onogi and others (1962) have applied a similar procedure to stress relaxation data on nylon 6. In the latter investigation a plot of the shift factor log a_H against water content at 50°C and 60°C showed a sudden change in shape at about 1 % sorbed water. On the basis of this result it was suggested that up to about 1 % water content the chain mobility was largely affected by the breaking of hydrogen bonds in the amorphous regions. The relatively small effect produced by larger water contents was ascribed to an increase of free volume. Both mechanical (Illers and Jacobs, 1960) and dielectric (Boyd, 1959) studies have shown that the α peak in the polyamides is also plasticized by alcohols.

12.1b The β Relaxation

For the mechanical β relaxation in nylon 612 Illers (1960) has calculated an activation energy of 13 kcal/mole. This value compares well with the value of 14·5 kcal/mole reported by Ishida (1960a) for the dielectric β process in nylon 6.

The mechanical results of Illers (1960), shown in Figure 12.9, illustrate that the β peak increases in height with increasing water content. At room temperature the dielectric β peak occurs at a frequency of about 10^4 c/s. As illustrated by the dielectric results of Curtis (1961) for nylon 66 (Figure 12.10), the dielectric β peak also decreases in height, and is almost eliminated, by 'drying'. Similar effects of drying on the dielectric β peak have been noted by Baker and Yager (1942) for nylon 610 and by Rushton and Russell (1956) for nylon 66. The latter authors, who could not completely eliminate the β peak by drying, suggested that the dielectric β relaxation could result from motions of the dipolar $-NH_2$ and $-OH$ chain end-groups. This interpretation is interesting since a condensation reaction

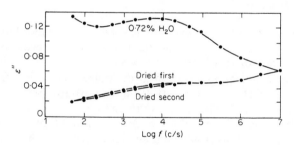

Figure 12.10. Variation of ε'' with frequency at 23°c for nylon 66. Data shown for a sample containing 0·72% water and after drying (Curtis, 1961).

can occur between $-NH_2$ and $-OH$ end-groups, resulting in a reduction in the number of end-groups (increase in molecular weight), when samples are dried by heating to or above 100°c (Curtis, 1961). On these grounds it might be argued that the decrease in size of the β peak after 'drying' arises from a reduction in the end-group concentration. However the end-group hypothesis is inconsistent with the observation of Curtis (1961) that the β peak increases in size when water is *added* to the polymer under conditions unlikely to regenerate end-groups by hydrolysis.

An early interpretation of the β mechanism (Woodward and others, 1957) assigned it to the motions of chain segments including amide groups which were not hydrogen bonded to neighbouring chains. This interpretation seems to imply that water molecules are not directly involved in the motion. However, the magnitude of the dielectric β relaxation strongly suggests that water molecules do, in fact, participate directly in the β process. For example, Curtis (1961) has found that ε' increases by about 0·54 at 50 c/s when 0·72 % of water is added to nylon 66. Also, this quantity of water caused an increase of 0·506 in the dispersion magnitude of ε' between 50 c/s and 10 Mc/s. These increases are in fairly good agreement with the value of about 0·56 to be expected on the grounds that pure (100 %) water has a dielectric constant of 78 at room temperature. On the basis of this result Curtis has proposed that the β relaxation involves the motion of a water–polymer complex. This interpretation is similar to that given by Illers (1960), who suggested that the β relaxation was due to the motions of carbonyl groups to which water molecules are attached by hydrogen bonds.

12.1c   The γ Relaxation

The mechanical γ peak for dry nylon 66 is somewhat broader but occurs at almost the same temperature and frequency as the γ peak for polyethylene. Evidence of this kind is responsible for the widely accepted view that the mechanical γ relaxation in the polyamides arises from the motions of the $(-CH_2-)_n$ units between amide groups (Schmieder and Wolf, 1953; Willbourn, 1958). As shown in Table 12.1 Lawson, Sauer and Woodward (1963) have observed a γ peak for nylon 4 (polypyrrolidone) at $-146°$C (0·31 c/s). This result is in agreement with the hypothesis of Willbourn (1958) that the γ mechanism requires at least three $-CH_2-$ groups in the methylene sequences.

The dielectric γ peak is illustrated in Figure 12.11 by the low-temperature data of Curtis (1961) for nylon 66. The existence of a dielectric peak indicates that the dipolar amide groups, in addition to the non-polar $(-CH_2-)_n$ sequences, are partially responsible for the γ relaxation. A similar conclusion has recently been obtained by Kolarik and Janacek (1965) from observations that the magnitude and temperature of the mechanical $(G'')\gamma$ peak of polycaprolactam decrease after the absorption of small quantities of water. Hence the above interpretation of the mechanical γ peak probably requires some modification.

Illers' (1960) data for the mechanical γ relaxation in nylon 612 yields an activation energy of 9 kcal/mole. From his dielectric data for nylon

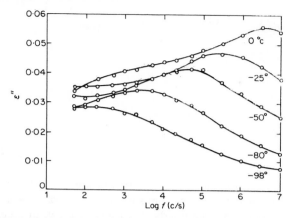

Figure 12.11. ε'' versus frequency for nylon 66 at the indicated low temperatures (Curtis, 1961).

66 Curtis (1961) has reported an activation energy of 13 kcal/mole. However, the latter value seems to be in error. A re-evaluation of the activation energy from Curtis's data (Figure 12.3a) also gives a value of 9 kcal/mole (McCall, 1965). For polycaprolactam an activation energy of about 10 kcal/mole has been obtained both from dielectric (Bares, Janacek and Cefelin, 1965) and mechanical (Kolarik and Janacek, 1965) data. On the basis of their dielectric results Bares, Janacek and Cefelin (1965) have proposed that the γ mechanism is of the local mode type (Section 5.5b).

12.2 POLYURETHANES

$$[-\underset{\underset{H}{|}}{N}-(CH_2)_x-\underset{\underset{H}{|}}{N}-\overset{\overset{O}{\|}}{C}-O-(CH_2)_y-O-\overset{\overset{O}{\|}}{C}-]_n$$

(3)

The addition reaction between a diisocyanate, $OCN-(CH_2)_x-NCO$, and a diol, $HO-(CH_2)_y-OH$, yields a polymer having the general structure **(3)** and known as a polyurethane. The polyurethane structure closely resembles the polyamide structure **(1)**, the essential difference involving the replacement of the

$$-N-\overset{\overset{\displaystyle O}{\|}}{C}-$$
$$\underset{H}{|}$$

group by the

$$-N-\overset{\overset{\displaystyle O}{\|}}{C}-O-$$
$$\underset{H}{|}$$

group. Like the polyamides, the polyurethanes are partially crystalline polymers whose crystalline regions comprise planar zig-zag chains having strong intermolecular hydrogen bonds. Figure 12.12 illustrates the crystal structure reported by Jacobs and Jenckel (1961a).

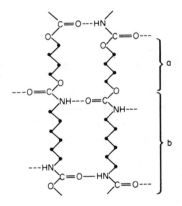

Figure 12.12. Crystal structure of polyurethane prepared from 1,4-butanediol and 1,6-hexamethylene diisocyanate (Jacobs and Jenckel, 1961a).

Polyurethanes of the above structural type have been studied mechanically by Becker and Oberst (1957), Illers and Jacobs (1960) and Jacobs and Jenckel (1961a, b). Flocke (1963) has also reported mechanical data for polyurethane elastomers prepared from an adipic acid–ethylene glycol polyester and 1,5-naphthalene diisocyanate. Figure 12.13 illustrates

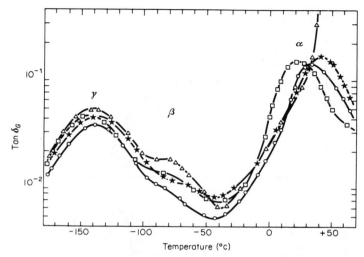

Figure 12.13. Temperature dependence of tan δ_G at 1 c/s for poly- urethanes prepared from hexamethylene diisocyanate and, respectively, 1,4-butanediol (★), 2,5-hexanediol (△), 1,6-hexanediol (○) and 1,10- decanediol (□) (Jacobs and Jenckel, 1961a).

some mechanical results of Jacobs and Jenckel (1961a) for four poly- urethanes prepared from hexamethylene diisocyanate ($x = 6$) and, respectively, 1,4-butanediol, 2,5-hexanediol, 1,6-hexanediol and 1,10- decanediol. As in the case of the polyamides, three relaxation regions, labelled α, β and γ respectively, are observed. Each of these relaxation regions is centred at a temperature close to the temperature of the corres- ponding relaxation in the polyamides. The α peak is seen to shift to lower temperatures as the $(CH_2)_n$ sequences increase in length, a result also found for the polyamides (Table 12.2).

The influence of water on the three loss regions is also similar to that found for the polyamides. In particular (Figure 12.14) with increasing water content the α peak is moved to lower temperatures, and the magni- tude of the β peak is increased. Hence there is clearly a very close correla- tion between the mechanisms of the three relaxations found, respectively, in the polyamides and polyurethanes. As discussed in Section 12.1 the α relaxation is probably related to the amorphous glass transition, the β relaxation involves the absorbed water molecules, and the γ relaxation is probably associated largely with motions of the $(CH_2)_n$ sequences.

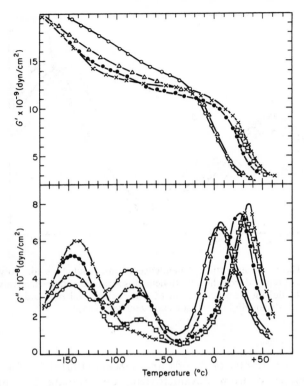

Figure 12.14. Temperature dependence of G' and G'' at 1 c/s for polyurethane prepared from 1,4-butanediol and 1,6-hexamethylene di-
isocyanate. Water contents as follows: ($\times$) 0, ($\square$) 3·7, ($\bullet$) 5·6, ($\triangle$) 24·5
and ($\bigcirc$) 28·8 mole % (Jacobs and Jenckel, 1961b).

13

Polyesters and Polycarbonates

13.1 POLYMETHYLENE TEREPHTHALATES

(1)

The relaxation behaviour of a number of polymethylene terephthalates has been investigated where x in (1) above lies between 2 and 10.

The best known member of the series is polyethylene terephthalate (which we abbreviate as PET), for which $x = 2$. This polyester, which is known commercially as Terylene, Dacron, Mylar or Melinex, may be prepared by the condensation polymerization of terephthalic acid with ethylene glycol (see Chapter 2). PET is a crystallizable polymer which melts at 260°C. The dilatometric glass-transition temperature T_g is 67°C for the amorphous material. Crystalline PET exhibits a higher T_g value which depends upon the degree of crystallinity, but occurs in the region of 80°C (Kolb and Izard, 1949; Uematsu and Uematsu, 1959, 1960). Since T_g is well above room temperature, the polymer is easily obtained in the amorphous state by quenching rapidly to below room temperature from the melt. If an amorphous sample is annealed at different temperatures above room temperature, crystallization does not occur until 90°C. At this temperature PET exhibits a 'stepwise' crystallization, and about 30% crystallinity (x-ray) is achieved. Kilian, Halboth and Jenckel (1960) have studied the relationship between annealing temperature and degree of crystallization. It was shown that no further crystallization takes place in the temperature range 90°C to about 170°C. Above 170°C, the degree of crystallization begins to rise and reaches a maximum value of about 50% at 230°C. Above 230°C, no further increase is obtained with increasing annealing temperature.

If an amorphous sample of PET is heated at a temperature exceeding 90°C, the rate of crystallization increases with increased temperature.

501

The crystallization process can be stopped at a given time by cooling rapidly to temperatures below T_g. In this way samples of different degrees of crystallinity can be obtained at one annealing temperature. It is also possible to obtain samples of the same nominal degree of crystallinity (as judged by density measurements), prepared at different annealing temperatures.

Owing to its commercial importance, and also because its degree of crystallinity can be varied between wide limits, PET has been studied extensively in relation to its relaxation behaviour. Both the dielectric and mechanical studies have revealed the presence of α and β relaxation regions in PET. The α and β relaxations occur mechanically in the higher polymethylene terephthalates, for $x = 3$ to 10. Before discussing these relaxations in detail, it is convenient to outline the general behaviour of PET, and to list some references to both dielectric and mechanical studies.

13.1a   Dielectric Studies of PET

The first detailed study of PET was made by Reddish (1950). Subsequent studies include those of Huff and Müller (1957), Krum and Müller (1959), Saito and Nakajima (1959b, c, d), Yamafuji (1960), Hellwege and Langebin (1960), Berestneva and others (1960), Mikhailov and others (1961), Ishida, Yamafuji, Ito and Takayanagi (1962), Saito (1963, 1964).

Figure 13.1 shows a solid model of the dielectric constant and loss tangent as a function of frequency and temperature for a partially crystalline sample of PET, as obtained by Reddish (1950). Figure 13.1a shows the α and β relaxation regions: at a frequency of 10^2 c/s, the α peak occurs near $100°$C, and the β peak occurs near $-50°$C. The rising loss above $140°$C at low frequencies in Figure 13.1a was shown by Reddish to be due to a superposition of a d.c. conductivity process and a Maxwell–Wagner process. Both α and β dispersions are shown in the ε' plot, Figure 13.1b.

Reddish showed that the α process was due to the micro-Brownian motions of the chain, in the amorphous regions, and was associated with the glass-transition temperature T_g. This result has been confirmed by many authors. The principal evidence is that the magnitude $(\varepsilon_R - \varepsilon_U)$ of the α relaxation decreases with increased degree of crystallinity.

From a study of dry amorphous and crystalline PET, and water-containing amorphous and crystalline PET, Reddish concluded that the β relaxation was due to the motion of the terminal OH groups of the polymer. The magnitude of the β process increased with increased water content, but the frequency location of the peak was unaffected. Using the dipole moment of the OH group, and the observed magnitudes of the β

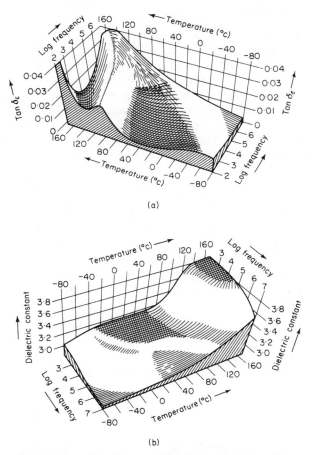

Figure 13.1. Solid models of (a) tan δ_ε and (b) ε' as a function of frequency and temperature for partially crystalline PET (Reddish, 1950).

process, Reddish calculated experimental values for the concentration of OH groups in dry and wet PET. These values for dry and wet PET agreed reasonably well with direct infrared determinations of the OH content. The interpretation of Reddish for the mechanism of the β process has been queried in later work (see Ishida and others, 1962) and it is now considered that the β region is due to limited motions of the polymer chain.

13.1b Mechanical Studies of PET

Many studies of the mechanical relaxation of PET have been made and these include Wolf and Schmieder (1955), Thompson and Woods (1956), Kawaguchi (1958), Yoshino and Takayanagi (1959), Farrow, McIntosh and Ward (1959, 1960), Takayanagi (1961, 1963), and Illers and Breuer (1963). An excellent review of the dielectric and mechanical relaxation of PET is included in the paper by Illers and Breuer (1963).

Figure 13.2 illustrates the mechanical behaviour of PET at different degrees of crystallinity (Takayanagi, 1963). The low crystallinity sample exhibits the α relaxation peak at 80°C, and the β peak at -55°C. For the higher crystalline samples, the α peak occurs near 105°C, while the β process is essentially unchanged in position and shape from that observed in the amorphous sample. It is evident from Figure 13.2 that the α process is a function of degree of crystallinity, both in position and shape. The mechanical α process is due to the micro-Brownian (glass–rubber) motions in the amorphous phase of the polymer, is related to T_g, and correlates in location with the dielectric α process. The origin of the mechanical β region is uncertain, and there is some evidence that it results from the superposition of several processes (Illers and Breuer 1963).

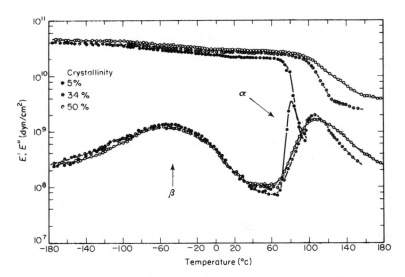

Figure 13.2. E' and E'' as a function of temperature at 138 c/s for PET samples of differing degrees of crystallinity (Takayanagi, 1963).

Figure 13.3(a) and (b) show the frequency–temperature locations of the mechanical and dielectric relaxations for PET samples of various degrees of crystallinity and for oriented samples. In Figure 13.3(a) the locations of the dielectric and mechanical α relaxations for amorphous PET are essentially identical. The data for crystalline specimens are somewhat scattered, due to the dependence of the location of the α process on the degree of crystallinity of the polymer. All the points, both mechanical and dielectric, for the crystalline samples lie at higher temperatures at a given frequency than those obtained for the amorphous polymer. The location of the mechanical α relaxation for the crystalline samples agrees fairly well with that obtained for the dielectric α relaxation. The α process in oriented crystalline PET occurs at higher temperatures at a given frequency than that obtained for the unoriented crystalline material. Due to the scatter in the experimental data shown in Figures 13.3(a) and 13.3(b), it is difficult to quote an activation energy for the α and β processes in PET. Also, as seen below, Saito (1963, 1964) has found that the α relaxation follows the WLF equation. In order to give an order of magnitude to the α and β activation energies, we quote Illers and Breuer (1963), who give $H_\alpha = 184$ kcal/mole for both the mechanical and dielectric α relaxation in amorphous PET. For crystalline PET, $H_\alpha = 87$ kcal/mole for the dielectric α relaxation, and no figure can be obtained for the mechanical relaxation. They give $H_\beta = 17$ kcal/mole for the mechanical β relaxation in dry amorphous and crystalline PET and $H_\beta = 12.9$ kcal/mole for the dielectric β relaxation in dry and wet, amorphous and crystalline PET.

Having given a general outline of the dielectric and mechanical relaxation in PET, the detailed nature of the results and their interpretation will now be given for this class of aromatic polyesters.

13.1c The α Relaxation

The magnitude, location and breadth of the α relaxation in PET are a function of the degree of crystallinity and also the manner in which a particular degree of crystallinity is attained. These aspects have been thoroughly studied for PET, but the experimental data on the higher poly-n-alkyl terephthalates are far less extensive. As a result, the following discussion is mainly concerned with the behaviour of PET.

The Effect of Crystallinity on the Magnitude of the α Relaxation

Ishida and others (1962) found that the magnitude ($\varepsilon_R - \varepsilon_U$) of the dielectric α relaxation decreased with increased degree of crystallinity for PET. Crystallinities between 5 and 51% were obtained by heating amorphous

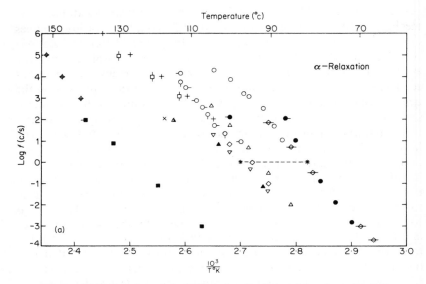

Figure 13.3. Log f against $(1/T°\text{K})$ for the dielectric and mechanical relaxations in PET. In the following key (f) refers to the frequency of maximum loss in constant temperature experiments and (T) refers to the temperature of maximum loss in constant frequency experiments. Figures 13.3(a) and 13.3(b) refer to the α and β regions, respectively.

Dielectric

○ ε'' (f) 5 % crystalline (Ishida and others, 1962)
○- ε'' (f) 32 % crystalline (Ishida and others, 1962)
◔ ε'' (f) 41 % crystalline (Ishida and others, 1962)
-○ ε'' (f) 51 % crystalline (Ishida and others, 1962)
□ ε'' (f) amorphous (Saito, 1964)
◈ ε'' (f) amorphous (Saito and Nakajima, 1959b)
□ tan δ (T) crystalline (Huff and Müller, 1957)
◇ tan δ (T) crystalline oriented (Huff and Müller, 1957)
+ tan δ (f) 40 % crystalline (Reddish, 1950)
△ ε'' (f) 27 % crystalline (Saito, 1963)
▽ ε'' (f) 28 % crystalline (Saito, 1963)
◇ ε'' (f) 31 % crystalline (Saito, 1963)

Mechanical

● tan δ_E (T) 2 % crystalline (Thompson and Woods, 1956)
▲ tan δ_E (T), α region, 48 % crystalline (Thompson and Woods, 1956)

■ tan δ_E (T) 65 % crystalline oriented (Thompson and Woods, 1956)
● E'' (T) 4 % crystalline (Takayanagi, 1961)
◖ E'' (T) 37 % crystalline (Takayanagi, 1961)
■ E'' (T) crystalline oriented (Yoshino and Takayanagi, 1959)
▲ tan δ_E (T) crystalline (Farrow, McIntosh and Ward, 1960)
-- G'' (T) 33 to 0 % crystalline respectively. --- indicates that intermediate crystallinities gave maxima lying between the two extremes (Illers and Breuer, 1963)
▲ G'' (T) 0 to 26 % crystalline, β region (Illers and Breuer, 1963)
▲ G'' (T) 0 % crystalline, β region (Illers and Breuer, 1963)
▲ G'' (T) 46 % crystalline, β region (Illers and Breuer, 1963)
◆ tan δ_G (T) crystalline (Wolf and Schmieder, 1955)
+ tan δ (T) amorphous (Kawaguchi, 1958) β region
× tan δ (T) 41 % crystalline (Kawaguchi, 1958)

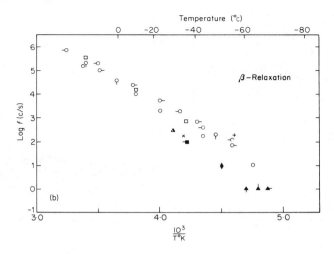

PET above its softening point for different times. Figure 13.4 shows a portion of the data and it is seen that the α process in the 51 % crystalline PET is far smaller than that obtained for the 5 % crystalline sample. This result clearly indicates that the α relaxation occurs in the non-crystalline phase of the polymer, since $(\varepsilon_R - \varepsilon_U)$ is proportional to the number of relaxing species per unit volume. A similar result was obtained by Saito (1964) who studied specimens of PET with different crystallinities obtained by crystallizing at a low and a high temperature. The L series specimens were crystallized at 118°C and the H series specimens at 202°C. Both series of samples contained spherulites, and if a sample of the H series was compared with an L series sample of the same degree of crystallinity as judged by density, the H series sample had a smaller number of spherulites per cc than the L series sample, but the average size of the spherulites was greater for the H series sample than for the L series sample. Inspection of Table 13.1 shows that $(\varepsilon_R - \varepsilon_U) = \Delta\varepsilon$ for the α process decreases with increasing degree of crystallinity for both the L and H series, in agreement with the data of Ishida and others (1962).

Saito and Nakajima (1959b) found that the magnitude $(\varepsilon_R - \varepsilon_U)$ of the dielectric α relaxation in amorphous PET increased in the usual manner with decreasing temperature until 75°C was reached. Below this temperature $(\varepsilon_R - \varepsilon_U)$ decreased rapidly and reached half of its maximum value near 68°C. This effect is known as the 'dielectric transition' and has been

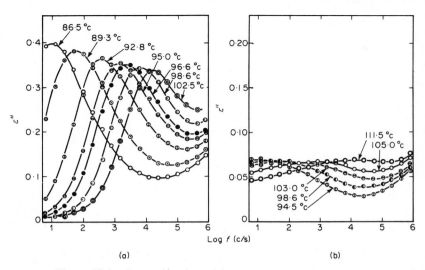

Log *f* (c/s)

(a)   (b)

Figure 13.4. ε'' as a function of frequency and temperature in the α relaxation region for (a) 5% and (b) 51% crystalline PET (Ishida and others, 1962).

Table 13.1. Properties of specimens of polyethylene terephthalate studied by Saito (1964) in which χ is the volume fraction crystallinity and $\Delta\varepsilon = (\varepsilon_R - \varepsilon_U)$, f_{max} the frequency of maximum loss and $\bar{\beta}$ the Cole–Cole parameter for the α relaxation at 85°C

Sample		χ	$\Delta\varepsilon$	f_{max} (c/s)	$\bar{\beta}$
Amorphous		0	1·80	7	0·54
	1	0·04	1·73	5	0·54
	2	0·17	1·31	4	0·48
L Series	3	0·24	0·96	2	0·35
	4	0·31	0·84	0·08	0·32
	5	0·35	0·81	0·02	0·30
	1	0·08	1·63	5	0·56
	2	0·18	1·45	4	0·51
H Series	3	0·39	0·90	2	0·49
	4	0·44	0·82	1·7	0·43
	5	0·48	0·75	1·1	0·39

observed by Saito (1964) in polybutyl methacrylate, amorphous and crystalline polycyclooxabutane, polyacrylonitrile and polyvinylidene chloride. The 'dielectric transition' temperature was defined as the temperature at which $(\varepsilon_R - \varepsilon_U)$ reached half its maximum value. It was found that this critical temperature was near to the dilatometric glass-transition temperature in all of the polymers given above (Saito and Nakajima, 1959b; Saito, 1963, 1964). Saito (1964) considers that the dielectric transition is an equilibrium phenomenon and is due either to a 'freezing in' of dipoles in the polymer or is due to a sudden change in the relative arrangement of the polar groups in the polymer, resulting in a net cancellation and hence a sudden reduction in $(\varepsilon_R - \varepsilon_U)$ as the temperature is reduced. Ishida and Yamafuji (1961) commented on the 'dielectric transition' described by Saito and Nakajima (1959b), and proposed an order–disorder theory to explain this effect. Williams (1963a) applied the Gibbs and DiMarzio theory of the apparent second-order transition to the case of amorphous polyacetaldehyde in an attempt to explain the apparent dielectric transition observed in this polymer. Hence the 'dielectric transition' which is found in amorphous PET is not an isolated anomaly but is a more general phenomenon.

The magnitude of the mechanical α relaxation to crystallinity for PET was examined by Takayanagi (1963) and by Illers and Breuer (1963). Takayanagi found that E''_{max} decreased with increased crystallinities, and yielded an extrapolated value of $E''_{max} = 0$ at 100% crystallinity. This result is not to be easily interpreted (see Chapter 4.3), as was shown by Illers and Breuer (1963). Illers and Breuer studied the mechanical properties in shear at 1 c/s for samples ranging from 0 to 48% crystalline. Figure 13.5 shows these data as a function of temperature in the α relaxation range, and they are seen to be similar to those of Takayanagi (1963) shown in Figure 13.2. Illers and Breuer found that although G''_{max} decreased with increasing degree of crystallinity, the area $\int G'' \, d(1/T)$ below the α relaxation curve actually *increased* with increased crystallinity. This behaviour was not explained by Illers and Breuer, but it is probable that the apparent anomaly is resolved as follows. The area $\int G'' \, d(1/T)$ is given by (Read and Williams, 1961b, see Chapter 4)

$$\int G'' \, d(1/T) = (G_U - G_R)\frac{\pi}{2}R(1/H)_{av}$$

G_U and G_R are the limiting moduli with respect to the α relaxation, and $(1/H)_{av}^{-1}$ is a statistical average activation energy for the process. For amorphous PET, the effective activation energy H_{eff} for the mechanical α process is near 180 kcal/mole while H_{eff} for crystalline PET is near

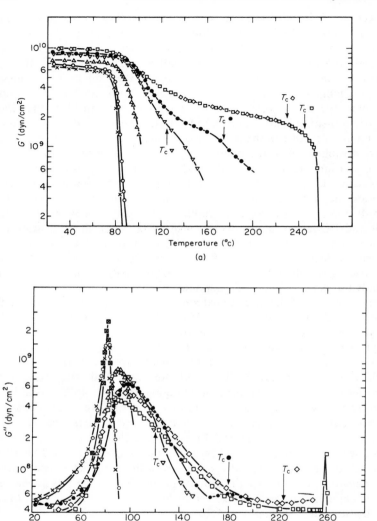

Figure 13.5. (a) G' and (b) G'' as a function of temperature for the mechanical α relaxation region in samples of PET (Illers and Breuer, 1963).

⊠, × 0 % crystalline; ○ 2·5 %; △ 16 %; ▽ 26 %; ● 33 %; ◇ 40 %; □ 46 % crystalline samples. T_c corresponds to the crystallization temperature of a given sample.

100 kcal/mole. Identifying H_{eff} with $(1/H)^{-1}$, it is predicted that $\int G''\, d(1/T)$ will increase on going from amorphous to crystalline PET if $(G_U - G_R)$ remains constant. Inspection of Illers and Breuer's data (Figure 13.5) shows that $(G_U - G_R)$ increases as the degree of crystallinity is raised. G_U is far greater than G_R up to about 33 % crystallinity, and G_U increases with increasing crystallinity; hence $(G_U - G_R)$ also increases. Thus $\int G''\, d(1/T)$ is expected to increase with increasing degree of crystallinity (despite the fact that the α relaxation is clearly an amorphous relaxation), and this is observed by Illers and Breuer. Inspection of Figure 13.5 shows that there is a considerable broadening of the loss curve as the crystallinity is raised, which is consistent with an increase in the breadth of the distribution of mechanical relaxation times. As a result, G''_{max} falls with increasing degree of crystallinity. In this way the apparent anomaly between the G''_{max} and $\int G''\, d(1/T)$ methods of the presentation of the data is resolved. It is noted that G_R increases rapidly as the degree of crystallinity is raised (see Figure 13.5). It is expected that if samples of extremely high degrees of crystallinity were studied, the area $\int G''\, d(1/T)$ would decrease with increasing crystallinity via the decrease in $(G_R - G_U)$ in this range. There is some evidence for this, since Illers and Breuer found that the 46 % crystalline sample gave a lower area than that obtained for the 40 % crystalline sample.

Effect of Crystallinity on the shape of the α Relaxation

The shape of the dielectric α process is a strong function of the degree of crystallinity. Figure 13.6 shows the normalized loss curves as a function of normalized frequency taken from the data of Ishida and others (1962). The 5 % crystalline sample gives a curve of asymmetric shape, being broader on the high-frequency side. This is in qualitative accord with the theory of Yamafuji and Ishida (1962) for the model of micro-Brownian motions of the main chain (Section 5.3b). The half-width of the curve, however, is broader than the maximum half-width obtainable from this molecular theory. The higher crystalline samples give loss curves which are broader on the low-frequency side than the high-frequency side, the asymmetry increasing with increasing degree of crystallinity. This strongly suggests that the crystallites exert an effect on the non-crystalline phase of the polymer. A similar effect is observed in the mechanical data of Takayanagi (1963) and Illers and Breuer (1963). In Figures 13.2 and 13.5(b) it is seen that the amorphous specimen gives a loss curve which is very sharp with a broadening on the low-temperature side of the peak, which is analogous to the dielectric high-frequency broadening in the 5 %

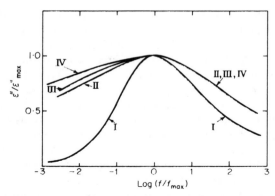

Figure 13.6. Normalized plots for the dielectric α relaxation in PET samples of different degrees of crystallinity. Here f_{max} is the frequency of maximum loss (ε''_{max}). Samples I, II, III and IV are 5, 32, 41 and 51% crystalline (Ishida and others, 1962).

crystalline sample in Figure 13.6. As the degree of crystallinity is raised, the loss curves in Figure 13.5(b), for example, are broader on the high-temperature side of the loss peak, which is analogous to the low-frequency broadening observed in Figure 13.6. It is evident that an increase in crystallinity broadens both the dielectric and mechanical relaxation curves in a similar manner. This broadening is consistent with the condition that the crystalline polymer contains non-crystalline regions under a variety of constraints imposed by the crystallites. The low-frequency (or high-temperature) broadening of the loss curves means that the broad distribution is enhanced in the long relaxation time region of the spectrum (with respect to the average relaxation time). The relaxation processes in this region arise from motions in the non-crystalline phase which are most influenced by the crystallites. Saito (1964) has also shown that the distribution broadens with increasing degree of crystallinity for the dielectric α relaxation. Table 13.1 includes the Cole–Cole parameter $\bar{\beta}$ (see Chapter 4) for the L and H series of samples. $\bar{\beta}$ decreases with increasing degree of crystallinity.

Effect of Crystallinity and Solvents on the Location of the α Relaxation

The frequency–temperature location of both the dielectric and mechanical α relaxations is a strong function of the degree of crystallinity. Figure 13.3(a) shows the collection of experimental data from various authors.

The α process in the crystalline specimens occurs at higher temperatures at a given frequency than those obtained for the amorphous specimen. This indicates that the crystallites restrict the motions of the non-crystalline phase, which is in agreement with the interpretation given above for the change of shape observed for the dielectric and mechanical α process as the degree of crystallinity is changed. Saito (1963, 1964) showed that the location of the dielectric α process in PET depended upon the thermal treatment used to prepare a given degree of crystallinity. In Table 13.1 it is seen that the frequency of maximum loss decreases continuously with increasing crystallinity for the H series. The same is true for the low crystallinity range in the L series, but an abrupt change occurs on going from 24 to 31 % crystallinity. Comparison of a sample of the L series with one of the H series at high crystallinities shows that the L series sample has a very much lower f_{max} value than that obtained for the H series sample of the same degree of crystallinity (as judged by density). The spherulites of the L series are smaller and more densely distributed than those of the H series at the same degree of crystallinity. According to Saito, at low crystallinities the crystallites exert little effect on the non-crystalline phase, and the L and H series give similar results. At high crystallinities the non-crystalline phase in the L series is under a more severe constraint than in the H series and this leads to the unusual decrease in log f_{max} as indicated in Table 13.1.

Illers and Breuer (1963) found that the temperature of maximum loss (G'') increased with increasing degree of crystallinity up to about 30 % crystallinity, but at higher degrees of crystallinity the temperature of maximum loss, T_{max}, actually decreased continuously with increased crystallinity. This can be seen in Figure 13.5(b). Their samples were prepared by annealing amorphous PET at a given crystallization temperature. The crystallization process was continued until no further changes in crystallinity with time occurred. It is therefore not possible to directly compare the dielectric behaviour of log f_{max} with crystallinity observed by Saito (1964) with the behaviour of T_{max} observed by Illers and Breuer (1963). The latter authors interpreted the unusual behaviour of T_{max} in terms of the x-ray study of the crystallization of PET made by Kilian, Halboth and Jenckel (1960). These authors showed that crystalline PET, prepared by annealing amorphous PET at a crystallization temperature between 90 and 170°C, consisted of small crystallites whose density distribution increased as the crystallization temperature was raised. Samples prepared in a similar way at crystallization temperatures exceeding 170°C consist of larger crystallites in which order is present in the longitudinal direction of the chains. Illers and Breuer found that the maximum value of T_{max} occurred for the sample crystallized at 150°C, corresponding to a degree of crystallinity of 30%. Thus, below 150°C crystallization temperature, T_{max} increases with increased crystallinity, and above 150°C crystallization temperature, T_{max} decreases with increased crystallinity. Illers and Breuer conclude, in view of the x-ray study of Kilian and coworkers that below 150°C the small crystallites interfere with the amorphous regions and hinder the molecular motions of the amorphous phase. However in specimens crystallized above 170°C the larger crystals do not interact so much with the amorphous regions (in these specimens both amorphous and crystalline regions are taken to be coarser than in specimens crystallized at lower temperatures). Consequently, according to Illers and Breuer, T_{max} decreases with increasing crystallinity for such specimens.

Saito (1963, 1964) showed that the dielectric α relaxation in amorphous and crystalline PET obeyed the WLF equation (see Section 5.3c). The parameters C_1 and C_2 were evaluated and it was found that both C_1 and C_2 increased as the crystallinity was increased. For example, $C_1 = 17.1$ and $C_2 = 31.7$ deg. for amorphous PET and $C_1 = 30.4$ and $C_2 = 106.6$ deg. for 31% crystalline PET. Using these data, Saito (1963, 1964) calculated $f(T_g)$, the fractional free volume at the dilatometric glass-transition temperature, and found that $f(T_g)$ decreased with increasing degree of crystallinity. Saito (1964) also found from measurements on amorphous PET that the apparent activation energy for the dielectric α process went through a maximum value in the neighbourhood of T_g. This is inconsistent with the WLF behaviour described above. The dielectric transition in PET described above also occurs near T_g, and a similar result is obtained for other polymers, such as polyvinylidene chloride (Sections 11.2 and 11.3). Therefore, the maximum in the apparent activation energy and the dielectric transition may be related to each other. As discussed in Section 11.3 these phenomena may be determined by the very slow establishment of volume equilibrium at temperatures close to T_g. If so, they should depend on the initial cooling rate, and they may not indicate an equilibrium transition in the polymer.

It has been indicated above that the dielectric and mechanical α relaxation in PET is due to the chain motions in the amorphous phase of the polymer. Further evidence for this proposal is obtained from the study of PET containing small amounts of solvent. Illers and Breuer (1963) examined a 28% crystalline PET sample containing up to 17.2 weight per cent dioxane. The temperature of maximum loss T_{max} was found to decrease continuously with increasing amount of absorbed dioxane. Similar results were obtained for a sample containing 3.2% benzene and a sample containing 19 % nitrobenzene. The lowering of T_{max} with solvent is explained as a plasticization of the amorphous phase of the polymer which eases the backbone motions of the chain. It was found that 0·2, 0·5 and 1·2 weight per cent of absorbed water also gave a progressive decrease in T_{max} for the mechanical α process. At 1 c/s, $T_{max} = 82°C$ for amorphous dry PET, while $T_{max} = 62°C$ for amorphous PET containing 1·2 % water.

Effect of Methylene Sequence Length on the α Relaxation

Farrow, McIntosh and Ward (1960) studied the mechanical properties of a series of crystalline polymethylene terephthalates. The methylene sequence between the benzene rings was varied from $x = 2$ to $x = 10$ (1). All samples were crystalline and were stabilized by annealing for one hour at a temperature 20 to 50°C above their glass transitions. Measurements were performed near 100 c/s using a cantilever technique. The results are shown in Table 13.2, where T_α is the temperature of maximum loss tangent for the α process, and f_α is the frequency at which the maximum loss occurs. T_α is seen to decrease with increasing x in a manner roughly parallel to the decrease in the crystal melting points of the polymers (compare polyamides, Section 12.1a). Farrow and others associated the α process with the micro-Brownian motions of the chains in the amorphous part of the partially crystalline polymers. This conclusion was substantiated by an n.m.r. study of the poly-n-alkyl terephthalates (see also Ward, 1960). The n.m.r. derivative spectrum of PET at 22°C was a broad line. Above 110°C the derivative spectrum exhibited structure

Table 13.2. Temperature of maximum loss (tan δ_E) for α and β relaxations for eight polymethylene terephthalates[a] (Farrow, McIntosh and Ward, 1960)

No. of CH$_2$ groups per glycol residue	$T_{melting}$ (°C)	T_α (°C)	f_α (c/s)	T_β (°C)	f_β (c/s)
2	265	115	165	−30	257
3	233	95	275	−30	530
4	232	80	215	−60	412
5	134	45	215	−95	405
6	154	45	160	−110	380
8	132	45	352	−100	334
9	85	35	366	−115	724
10	129	25	102	−125	225

[a] Note that the α and β relaxations in Table 13.2 were labelled as β and γ relaxations respectively by Farrow, McIntosh and Ward (1960).

corresponding to a broad and a narrow component. It was shown with the aid of a study of the spectra of deuterated derivatives of PET that the narrow component was due to the motion of methylene group protons and benzene ring protons. The broad component was shown to arise from rigid benzene ring protons. The temperature at which the narrow line appeared may be regarded as that temperature at which the main chain begins to undergo motion with respect to the resonance frequency ($\sim 10^5$ c/s). The temperature of maximum mechanical loss for the α process in PET at 165 c/s occurred at 115°C, thus this process is correlated with the n.m.r. non-motion–motion 'transition' at 110°C. As a result, Farrow, McIntosh and Ward conclude that the mechanical α process involves both the motions of the glycol residue and the motions of the benzene rings in the non-crystalline regions of PET. For polytetramethylene terephthalate, structure appeared in the n.m.r. spectrum near 90°C. For polyhexamethylene terephthalate and polydecamethylene terephthalate, no composite structure was observed, but the line width narrowed to the value observed in PET at 50 and 40°C, respectively. This decrease in the n.m.r. temperature for the line narrowing process is clearly analogous to the decrease in T_α shown in Table 13.2. Thus the mechanical α process in the higher poly-n-alkyl terephthalates is also due to the micro-Brownian motions of the main chain. Illers and Breuer (1963) studied polytetramethylene terephthalate and found that the G'' plot showed the α peak at 50°C, which is considerably lower than that observed in PET, in agreement with Farrow and others (1960).

Effect of Orientation

Figure 13.3(a) shows that the mechanical and dielectric α relaxations for oriented PET lie at higher temperatures at a given frequency than those for the unoriented polymer. Huff and Müller (1957) and Reddish (1962) studied the dielectric properties of oriented PET, and Thompson and Woods (1956), Yoshino and Takayanagi (1959) and Illers and Breuer (1963) studied the mechanical properties of oriented PET. The molecular motions in the non-crystalline phase are impeded by orientation of the polymer. A similar effect has been observed for the polyamides (see Section 12.1a). It is interesting to note that Illers and Breuer (1963) found a small additional loss peak at temperatures a little below the α relaxation regions for freshly drawn material. This loss region was observed by drawing dry amorphous PET at 80°C, then quenching the sample to $-50°$C. Measurements were made at 1 c/s with rising temperature and the small peak was found near 50°C. The sample was then annealed at 62°C, and it was found that the loss at this temperature gradually disappears with time. Subsequent measurements on cooling the polymer through 50°C gave results essentially the same as for the undrawn polymer. This behaviour was explained by Illers and Breuer in the following way. The orientation process followed by quenching freezes in a non-equilibrium state in the polymer. It is this state which gives rise to the loss peak. If the temperature is raised to 62°C, this non-equilibrium state relaxes out slowly with time, and the loss process disappears.

13.1d The β Relaxation

Reddish (1950) studied the dielectric properties of amorphous and crystalline PET and found that the frequency–temperature location and the shape of the β process were independent of crystallinity, but the magnitude was found to decrease with increased degree of crystallinity. Also, the magnitude increased as the absorbed water content of the polymer was raised. It was suggested that the β process was due to the reorientation of hydroxyl groups in the non-crystalline phase of the polymer. This interpretation was substantiated by Reddish who calculated the hydroxyl content from the magnitude $(\varepsilon_R - \varepsilon_U)$ of the β process. It was assumed that $(\varepsilon_R - \varepsilon_U)$ was related to the number of hydroxyl groups in the non-crystalline phase according to the Debye equation (Equation 3.62). The dielectric estimates of the hydroxyl content for dry and wet PET agreed reasonably with the values for hydroxyl content determined by an infrared technique.

Ishida and others (1962) and Saito (1964) found that the magnitude of the dielectric β process decreased with increased degree of crystallinity, but the frequency location and the shape of the loss process were unaffected by changes in crystallinity. These authors consider that the β process occurs in the non-crystalline phase, but, unlike Reddish, suggest that the mechanism is one of local motions of the chain, involving the COO groups. The magnitude of the mechanical β process decreases only slowly with increased crystallinity in PET (Takayanagi, 1963; Illers and Breuer, 1963). In view of this behaviour it might be concluded that the mechanical β process arises both from the non-crystalline and crystalline regions of the polymers. The n.m.r. study of Ward (1960) revealed only a gradual decrease in the second moment with increasing temperature in the temperature range in which the mechanical β process is observed. The decrease was interpreted as being due to the hindered motion of the methylene sequences of the chain, in both the non-crystalline and crystalline regions of the polymer.

However, a detailed examination by Illers and Breuer (1963) revealed a degree of structure within the overall β process for PET. This can be seen to some extent in Figure 13.7 in the asymmetry of the loss curve. The features were found by Illers and Breuer to be better resolved in a plot of $(\partial G'/\partial \log f)_T$ at 1 c/s against temperature. The quantity $(\partial G'/\partial \log f)_T$ was obtained from measurements of G' in the range 0·1 to 10 c/s at given temperatures. The plot gave three peaks—at -165, -105 and $-70°C$. The $-165°C$ peak was attributed to the hindered rotations of CH_2 groups by analogy with the mechanical results for polymers containing CH_2 sequences (see Chapter 10). The peak at $-105°C$ was attributed to the motions of COO groups associated with the *gauche* conformation of the polymer chain, and the peak at $-70°C$ was attributed to motions of COO groups associated with the *trans* conformation of the polymer chain. These interpretations were based on the study by Schmidt and Gay (1962) of the conformational structure of PET.

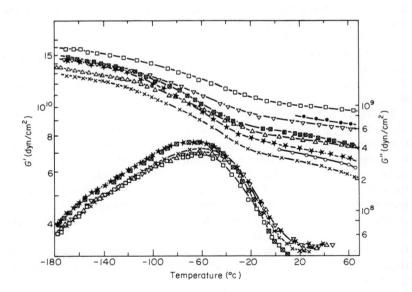

Figure 13.7. G' and G'' against temperature in the β relaxation region for samples of PET of different degrees of crystallinity. Key as in Figure 13.5 and ★ corresponds also to 0% crystallinity (Illers and Breuer, 1963).

Thus, according to the hypothesis of Illers and Breuer the mechanical β region appears to contain three distinct processes. The lowest-temperature process is to be correlated with that observed in the study by Ward (1960), and the two higher-temperature processes (which dominate the overall β process) are to be correlated with the dielectric β process. The latter process is considered to involve main-chain motions, which, in order to be active in the dielectric experiment, must involve the motions of the dipolar COO units.

Farrow, McIntosh and Ward (1960) studied the mechanical β process in a number of polymethylene terephthalates. These data are given in Table 13.2. The temperature of maximum loss tangent T_β decreases with increasing sequence length of (CH_2) units. These authors noted that the loss curve was asymmetric, being steeper on the low-temperature side than on the high-temperature side of the peak. As the methylene sequence was increased, the 'sense of the asymmetry reversed through the series'. In polynonyl terephthalate, the main loss peak occurred at $-115°C$, and a definite shoulder was observed at $-70°C$. Farrow, McIntosh and Ward suggested that the β region contained two superposed loss processes, and correlated their results with the n.m.r. study of Ward (1960) described above. It is noted that the asymmetry observed by Farrow and others in tan δ_E plots is the reverse of that observed by Illers and Breuer (see Figure 13.7) using G'' as the loss quantity. Illers and Breuer (1963) also studied polytetramethylene terephthalate and observed that the G'' plot in the β region indicated at least two overlapping processes.

It was found (Illers and Breuer, 1963) that dioxane, nitrobenzene and benzene absorbed in crystalline PET caused a decrease in the height (G''_{max}) of the mechanical β process, and a smearing of the peak over a very broad temperature range. A similar result was obtained by Kawaguchi (1958) for samples containing alcohols and carbon tetrachloride. The reason for this effect is not clear. Small amounts of absorbed water did not affect the mechanical loss (G'') in the β region at temperatures exceeding $-60°C$, but an increased loss was observed between -80 and $-110°C$. The low-temperature shoulder seen near $-150°C$ in Figure 13.7 for the dry sample is not present in samples containing 1.2% water. Illers and Breuer (1963) suggested that water influences only the COO groups responsible for the $-105°C$ process within the β region described above, but the action of small molecules on the mechanism of the overall β relaxation is poorly understood. Reddish (1950) found (as was noted above) that the magnitude ($\varepsilon_R - \varepsilon_U$) of the dielectric β relaxation was increased by addition of water to the polymer. This effect is consistent with the observation of Illers and

Breuer in the -80 to $-110°c$ region of mechanical loss at 1 c/s. It may be taken as evidence for the suggestion that the dielectric β relaxation is due to small motions of the chain which involve the COO groups. However, the specific role of water on these motions is not understood.

13.2 RELATED AROMATIC POLYESTERS AND POLYCARBONATES

The behaviour of the aromatic polycarbonates will be considered in this section together with the behaviour of the aromatic polyesters. A carbonate

$$R_1-O-\underset{\underset{O}{\|}}{C}-O-R_2$$

may be regarded as an ester of the alcohols R_1OH and R_2OH and carbonic acid H_2CO_3.

Commercially, the most common class of polycarbonates has the struc-ture (2).

$$[-\underset{\underset{O}{\|}}{C}-O-\langle\bigcirc\rangle-\underset{\underset{R_2}{|}}{\overset{R_1}{C}}-\langle\bigcirc\rangle-O--]_n$$

(2)

R_1 and R_2 may be alkyl or aryl groups. These polymers may be obtained from readily available starting materials. For example, acetone reacts with phenol to give bisphenol A (3). This compound is also known by the useful abbreviation 'dian'. Dian combines with phosgene, $COCl_2$, to give the polycarbonate (4). The chemical name of this polymer is

$$HO-\langle\bigcirc\rangle-\underset{\underset{CH_3}{|}}{\overset{CH_3}{C}}-\langle\bigcirc\rangle-OH$$

(3)

$$[-\underset{\underset{O}{\|}}{C}-O-\langle\bigcirc\rangle-\underset{\underset{CH_3}{|}}{\overset{CH_3}{C}}-\langle\bigcirc\rangle-O-]_n$$

(4)

poly-4,4'-dioxydiphenyl-2,2-propane carbonate. Other names which are used are poly(bisphenol A carbonate) and polydian carbonate. For information on the properties of this and other polycarbonates the reader is referred to Christopher and Fox (1962) and Schnell (1956).

Table 13.3 lists the aromatic polyesters and polycarbonates which have been studied by dielectric (D) and mechanical (M) methods. Polymers containing the dian group are denoted by the abbreviated name rather than the full chemical name for convenience in the discussion of these polymers to be given below. Table 13.3 also contains T_g values for several polycarbonates taken from the paper by Schnell (1956).

Table 13.3. Aromatic polyesters and polycarbonates which have been studied with regard to mechanical (M) and dielectric (D) relaxation

Polymer	T_g(°C)	Type of measurement	Ref.
Polydian sebacate		D	Mikhailov and Eidelnant (1960a)
Polydian terephthalate		D	Mikhailov and Eidelnant (1960a)
Polydian isophthalate		D	Mikhailov and Eidelnant (1960a)
Polydian carbonate (or poly-4,4'-dioxydiphenyl-2,2-propane carbonate) commonly known as 'polycarbonate'	149 (142)	D	Mikhailov and Eidelnant (1960a), Schnell (1956), Müller and Huff (1959), Krum and Müller (1959), Saito (1964), Ishida and Matsuoka (1965)
		M	Reding and others (1961), Illers and Breuer (1961, 196? Nielsen (1962)

Table 13.3—*continued*

Polymer	T_g (°C)	Type of measure-ment	Ref.
Poly-4,4'-dioxydiphenyl-2,2-butane carbonate	138	D	Schnell (1956)
Poly-4,4'-dioxydiphenyl-1,1-ethane carbonate	130	D	Schnell (1956)
Poly-4,4'-dioxydiphenyl-1,1-butane carbonate	123	D	Schnell (1956)
Poly-4,4'-dioxydiphenyl-1,1-isobutane carbonate	149	D	Schnell (1956)
Poly-4,4'-dioxydiphenyl-1,1-cyclohexane carbonate	171	D	Schnell (1956)

Table 13.3—*continued*

Polymer	T_g(°C)	Type of measurement	Ref.
 Poly(bis-*ortho*-cresol A carbonate)		M	Reding and others (1961)
 Poly(bisphenol of acetophenone carbonate)		M	Reding and others (1961)
 Poly-(*ortho*-tetrachlorobisphenol A carbonate)		M	Reding and others (1961)
 Polycarbonate of 1,5-naphthalenedi-(β-oxyethyl ether)		M	Illers and Breuer (1961)
 Poly-*p*-ethylene oxybenzoate		D	Ishida, Yamafuji, Ito and Takayanagi (1962)

Table 13.3—*continued*

Polymer	T_g (°C)	Type of measurement	Ref.
$-OC$ ⬡ $COOCH_2$ ⬡$-H-$ $CH_2-O-]_n$ Poly-1,4-bishydroxymethylcyclohexane tere-phthalate		D	Reddish (1962)
$[-OC$ ⬡ $-COOCH_2CH_2-O-]_n$ Polyethylene isophthalate		M	Kawaguchi (1958)
$[-OC(CH_2)_4COO$ ⬡ $-O-]_n$ Polyhydroquinone adipate		M	Kawaguchi (1958)
$[-OC(CH_2)_8COO$ ⬡ $-O-]_n$ Polyhydroquinone sebacate		M	Kawaguchi (1958)
$[-OC(CH_2)_4COO$ ⬡$-H-$ $O-]_n$ Poly-*trans*-quinitol adipate		M	Kawaguchi (1958)
$[-OC(CH_2)_8COO$ ⬡$-H-$ $O-]_n$ Poly-*trans*-quinitol sebacate		M	Kawaguchi (1958)

The most comprehensive study of the aromatic polyesters and polydian carbonate was made by Mikhailov and Eidelnant (1960a). Figure 13.8 shows the frequency–temperature locations for the dielectric α and β relaxations observed by these authors. The α and β relaxations were observed in both crystalline and amorphous samples. There is no comparable mechanical study of these polymers. Most of the mechanical data have been obtained at a single frequency, over a wide range of temperature.

Polydian carbonate has been extensively studied by both mechanical and dielectric techniques. Special emphasis will be given to this polymer in the following sections.

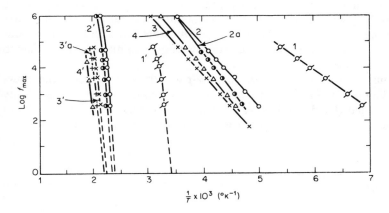

Figure 13.8. Log f_{max} against $(1/T°\text{K})$ for the dielectric α and β relaxations in several polyesters and polydian carbonate (Mikhailov and Eidelnant, 1960a). Here, f_{max} is the frequency of tan δ_{max} at a given temperature.

1	β relaxation, polydian sebacate	2	α relaxation, amorphous polydian carbonate
2	β relaxation, amorphous polydian carbonate	2'	α relaxation, partially crystalline polydian carbonate
2a	β relaxation, partially crystalline polydian carbonate	3'	α relaxation, amorphous polydian isophthalate
3	β relaxation, polydian isophthalate	3'a	α relaxation, partially crystalline polydian isophthalate
4	β relaxation, polydian terephthalate		
1'	α relaxation, polydian sebacate	4'	α relaxation, polydian terephthalate

13.2a   The α Relaxation

In view of the variation in the mechanical and dielectric α relaxations with the degree of crystallinity in the polymethylene terephthalates, a similar behaviour might be expected for the polyesters and polycarbonates listed in Table 13.3. The only extensive studies which have been made are those by Mikhailov and Eidelnant (1960a) for several polyesters and polydian carbonate, and by Ishida, Yamafuji, Ito and Takayanagi (1962) for poly-para-ethylene oxybenzoate.

Effect of Crystallinity

Mikhailov and Eidelnant (1960a) found that the dielectric α relaxation for polydian terephthalate, polydian isophthalate and polydian carbonate was decreased in magnitude (tan δ_{max}) as the degree of crystallinity was raised. In all three cases the loss peak moved to higher temperatures at a given frequency as the degree of crystallinity was raised. This indicates that the α relaxation in these polymers is associated with the amorphous phase. It is noted that this variation with crystallinity is very similar to that observed in PET.

The dielectric α relaxation in poly-*para*-ethylene oxybenzoate shows a behaviour similar to the α relaxation in PET as the degree of crystallinity is changed (Ishida and others, 1962). Figure 13.9 shows the results of these authors for (a) an amorphous and (b) a 38 % crystalline sample.

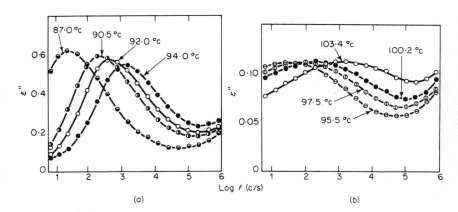

Figure 13.9. ε'' against log f (c/s) (α relaxation) at different temperatures for two samples of poly-*para*-ethylene oxybenzoate. (a) Amorphous and (b) 38 % crystalline (Ishida and others, 1962).

The decrease in magnitude as crystallinity is increased is very marked, and means that the dielectric α relaxation occurs in the non-crystalline phase of the polymer. Figure 13.10 shows the normalized plots of loss factor against frequency for this polymer, for different crystallinities, and this figure is essentially identical with Figure 13.6 given above for PET. The frequency–temperature location for the α process in this polymer is the same as that observed for PET if the comparison is made at the same degree of crystallinity. As the crystallinity is raised for poly-*para*-ethylene oxybenzoate, the α relaxation moves to higher temperatures at a given frequency. This indicates that, as in PET and the polymers studied by Mikhailov and Eidelnant, the crystallites exert a considerable influence on the motions in the non-crystalline phase.

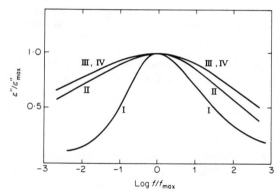

Figure 13.10. $(\varepsilon''/\varepsilon''_{max})$ against $\log f/f_{max}$ for samples of poly-*para*-ethylene oxybenzoate. I, II, III and IV correspond to 0, 27, 34 and 38% crystalline samples, respectively (Ishida and others, 1962).

Illers and Breuer (1961) studied the mechanical behaviour of two samples of the polycarbonate of 1,5-naphthalenedi-(β-oxyethyl ether). These data are shown in Figure 13.11. The α process is narrow in the lower crystalline sample, but broadens and moves to higher temperatures for the annealed (higher crystallinity) sample. This behaviour is again similar to that observed in the polymethylene terephthalates (e.g. Figure 13.2). Nielsen (1962) has given mechanical data for polydian carbonate and has shown that the α relaxation moves to higher temperatures at a given frequency as the degree of crystallinity is raised.

Thus the available dielectric and mechanical data for the α relaxation in this class of polyesters and polycarbonates shows a strong correlation

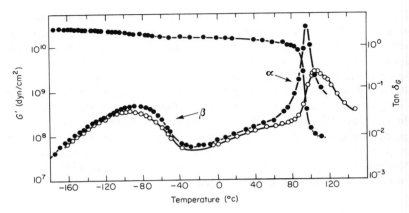

Figure 13.11. G' and $\tan \delta_G$ against temperature at 1 c/s for the poly-carbonate of 1,5-naphthalenedi-(β-oxyethyl ether) ● unannealed; ○ annealed (Illers and Breuer, 1961).

with the behaviour of the α relaxation in PET and the polymethylene terephthalates with regard to crystallinity effects. The α relaxation is due to the micro-Brownian motions of the chain in the non-crystalline phase of the polymer, and crystallites exert a large effect on these motions.

Effect of Chemical Structure

Although the degree of crystallinity is an important factor in determining the frequency–temperature location of the α relaxation in these polymers, chemical structure also has a large effect. Several studies have been concerned with the variation in the 'dynamic T_g' as the structure of the repeat unit is varied (Schnell, 1956; Mikhailov and Eidelnant, 1960a; Reding, Faucher and Whitman, 1961).

Figure 13.8 shows $\log f_{max}$ against $(1/T)$ for the polymers studied by Mikhailov and Eidelnant (1960a), and Table 13.4 gives $\tan \delta_{max}$ and the temperature of $\tan \delta_{max}$ at 20 c/s for the dielectric α process in these polymers. Their data for PET is included in the table for comparison. Polydian sebacate gives $(T_{max})_\alpha$ at a lower value than that observed in PET. This is due to the fact that although polydian sebacate contains the dian group which stiffens the chain, the sebacic acid residue between the aromatic components of the chain leads to a net increase in chain flexibility compared with PET. The other dian polymers in Table 13.4 exhibit higher $(T_{max})_\alpha$ values than that observed in PET. As the 'aromatic' content of the repeat unit is increased, $(T_{max})_\alpha$ also increases, and is at its greatest in polydian terephthalate.

Table 13.4. Dielectric relaxation parameters in some aromatic polyesters (Mikhailov and Eidelnant, 1960a). T_{max} is the temperature at which the maximum value of tan δ (tan δ_{max}) occurs at 20 kc/s

	α Process		β Process	
Compound	10^2 tan δ_{max}	T_{max} (°C)	10^2 tan δ_{max}	T_{max} (°C)
Polydian sebacate	3·1	47	3·0	−100
Polydian isophthalate	4·0	210	1·85	−9
Polydian terephthalate	7·5	260	1·30	−13
Polydian carbonate	2·65	174	0·6	−40
PET	7·3	106	—	−13

Schnell (1956) gave dielectric data for the series of polycarbonates listed in Table 13.3. This table also includes the T_g values listed by Schnell. Taking polydian carbonate as the reference compound, it is seen that replacement of a methyl group by an ethyl group hardly affects T_g. Replacement of a methyl group by a hydrogen unit lowers T_g slightly. The 1,1-butane carbonate has still lower T_g, but the 1,1-isobutane carbonate gives a T_g near that for polydian carbonate. The 1,1-cyclohexane carbonate has $T_g \simeq 171°$C, the rise being due to the stiffening action of the bulky cyclohexane group.

Reding and others (1961) studied the mechanical properties of the four polycarbonates listed in Table 13.3. The tan δ peaks at 1 c/s were not actually observed by these authors for the α relaxation, but the onset of the process was seen as a rapidly rising loss in a narrow temperature region. It was found that the onset of the α process in polydian carbonate and poly(bis-*ortho*-cresol A carbonate) occurred near 150°C, indicating that the methyl substitution in the rings has little effect on the chain mobility. However, poly-(*ortho*-tetrachlorobisphenol A carbonate) gave the onset of the α process near 240°C, the increase over that observed in polydian carbonate being due to the very heavy substitution in the rings of the repeat unit. Poly(bisphenol of acetophenone carbonate) gave the onset of the α process near 190°C, the benzene ring having a stiffening action on the chain motions. This result compares favourably with $T_g = 171°$C for poly-4,4′-dioxydiphenyl-1,1-cyclohexane carbonate, obtained by Schnell (1956) (see Table 13.3).

Thus variation in the structure of the repeat unit in the various dian polyesters and related polycarbonates can change the location of dielectric and mechanical α relaxations by a large amount. Increase in the aromatic content of the chain leads to high temperatures for the α process, the extreme being polydian terephthalate. Also, heavy atom substitution in the rings or cyclic groups off the chain can increase T_{max} by a large amount.

Polydian carbonate

This polymer has been extensively studied both by dielectric and by mechanical methods. It is therefore appropriate to discuss these results in some detail.

Polydian carbonate may be obtained in amorphous or partially crystalline states. O'Reilly, Karasz and Bair (1964) used a calorimetric technique to show that solution-grown samples were about 20 % crystalline. If a crystalline sample was heated above the melting temperature (227°C) and cooled to room temperature, the resultant sample was amorphous. They concluded that moulded polydian carbonate was amorphous. If an amorphous sample is exposed to the vapours of certain liquids, crystallization occurs (Nielsen, 1962). The reported dielectric and mechanical relaxation studies of polydian carbonate do not indicate the degree of crystallinity of the specimens.

Figure 13.12 gives the frequency–temperature location for the dielectric and mechanical α relaxations in polydian carbonate. The dielectric data cover a wide frequency range, but the mechanical studies are all near 1 c/s. Extrapolation of the dielectric data to 1 c/s gives a temperature which is essentially the same as that obtained by mechanical relaxation experiments.

It was noted earlier that an increase of degree of crystallinity for polydian carbonate has three effects, (a) the magnitude of the α relaxation is reduced, (b) the process shifts to higher temperatures, and (c) broadens. This behaviour was observed electrically by Mikhailov and Eidelnant (1960a) and mechanically by Nielsen (1962), for polydian carbonate. This is analogous to the behaviour observed for the polymethylene terephthalates.

Saito (1964) and Ishida and Matsuoka (1965) observed the dielectric α relaxation, and showed that the frequency–temperature location obeyed the WLF equation. The loss data of Ishida and Matsuoka are shown in Figure 13.13. Saito found that the relaxation frequency (f_{max}) decreased with increasing molecular weight for molecular weights near 3×10^4. The WLF parameters C_1 and C_2 were independent of molecular weight, and T_g was found to increase with increasing molecular weight. Saito made use of the relationship between (f_{max}) and molecular weight to evaluate the molecular weights of irradiated polydian carbonate.

The α relaxation of polydian carbonate may be modified by thermal and mechanical treatments. Krum and Müller (1959) have given dielectric

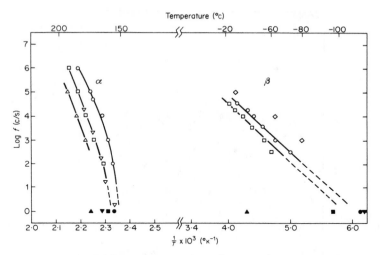

Figure 13.12. Log f against $(1/T°\text{K})$ for polydian carbonate. In the following key, (f) refers to frequency of maximum loss in constant temperature experiments and (T) refers to temperature of maximum loss in constant frequency experiments.

Dielectric

○ $\tan \delta_\varepsilon (T)$ amorphous (Mikhailov and Eidelnant, 1960a)

□ $\tan \delta_\varepsilon (T)$ partially crystalline (Mikhailov and Eidelnant, 1960a)

△ $\tan \delta_\varepsilon (T)$ annealed (Krum and Müller, 1959)

▽ $\varepsilon'' (f) \overline{M} = 3.5 \times 10^4$ (Saito, 1964)

◇ $\tan \delta_\varepsilon (T)$ annealed (Müller and Huff, 1959)

Mechanical

● $\tan \delta_G (T)$ (Reding and others, 1961)

■ $G'' (T)$ (Illers and Breuer, 1961)

▼ $\tan \delta_G (T)$ amorphous (Nielsen, 1962)

▲ $\tan \delta_G (T)$ crystalline (Nielsen, 1962)

data for annealed and stretched samples of polydian carbonate. A sample which had been repeatedly stretched and annealed gave an α relaxation which was broader and smaller and occurred at higher temperatures than that observed for the annealed sample. Presumably, repeated stretching and annealing does induce crystallinity, but the actual degree of crystallinity achieved was not known.

To summarize, the α relaxation in polydian carbonate shows certain similarities with the α relaxation observed in PET. At a given frequency,

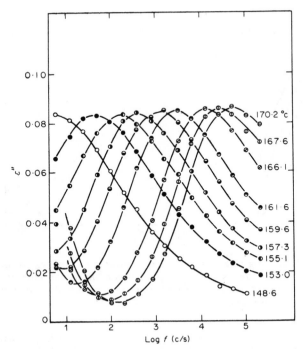

Figure 13.13. ε'' against $\log f$ at different temperatures in the α relaxation region of amorphous polydian carbonate (Ishida and Matsuoka, 1965).

the temperature of the α relaxation peak for polydian carbonate is higher than that obtained for PET, presumably due to the increased aromatic content of the chain. Increased crystallinity affects the α relaxation in polydian carbonate in a similar manner to that observed for PET, and the crystallites considerably restrict the motions of the chain.

13.2b The β Relaxation

Effect of Crystallinity

Mikhailov and Eidelnant (1960a) showed that the dielectric β relaxation was affected by the degree of crystallinity in polydian isophthalate, polydian terephthalate and polydian carbonate. Increase of crystallinity decreased the magnitude (tan δ_{max}) of the β process, and increased the temperature of maximum loss at a given frequency. A magnitude decrease was observed in PET (described above) but in this case the location of the β process was independent of crystallinity. In the mechanical case, Illers and Breuer (1961) studied the naphthalene polycarbonate described in Table 13.3. Their data are shown in Figure 13.11. The β peak was reduced in size on annealing and (T_{max}) for the tan δ peak was shifted to slightly *lower* values. Nielsen (1962), however, found that the mechanical β peak in polydian carbonate shifted from -110 to $+30°C$ on going from an amorphous to a crystalline sample. The reason for the difference in the results for these two polycarbonates is not clear.

The mechanical β relaxation appears to have structure in polydian carbonate (Illers and Breuer, 1961; Nielsen, 1962) and in the polycarbonate of 1,5-naphthalenedi-(β-oxyethyl ether) (Illers and Breuer, 1961). Structure was also observed mechanically for the β relaxation in the polymethylene terephthalates (see Section 13.1 above). In the latter case Illers and Breuer (1961) suggested that the methylene sequence and the conformational structures of the COO residue were responsible for the features within the mechanical β region. This interpretation does not apply to the polycarbonates.

Effect of Chemical Structure

Table 13.4 gives the values of tan δ_{max} and T_{max} for the β relaxation at 20 kc/s for the polymers studied by Mikhailov and Eidelnant (1960a). Taking PET as the reference compound, polydian sebacate gives T_{max} considerably lower than that observed for PET. A similar tendency was obtained for the α process in these polymers and, as described above in the discussion of the α relaxation, is due to the flexibility imposed on the carbonate group by the presence of the methylene sequence. Kawaguchi (1958) found that the mechanical β peak (tan δ) moved to lower temperatures on going from poly-*trans*-quinitol adipate to poly-*trans*-quinitol sebacate. This also indicates the role of the methylene sequence on the location of the β relaxation. These results are similar to those obtained by Farrow and others (1960) for the polymethylene terephthalates. The locations of the β process in polydian isophthalate and polydian terephthalate are similar to that observed for PET. The β process in polydian carbonate occurs at a lower temperature than that observed in PET, which would indicate greater freedom for the rotation of the carbonate group. This will be discussed further below in the section devoted to polycarbonate.

The mechanical β processes in polydian carbonate, poly(bis-*ortho*-cresol A carbonate) and poly(bisphenol of acetophenone carbonate) all occur near $-110°C$ at 1 c/s (Reding and others, 1961). However, poly-(*ortho*-tetrachlorobisphenol A carbonate) gives the β process near $+75°C$. Reding and others (1961) suggest that the β process is associated with the motion of the carbonate group. Substitution of a phenyl group for a methyl group in the dian residue, or *ortho*-methyl substitution, has little effect on the motion of the carbonate group. However, heavy *ortho* substitution as in the case of the chloro carbonate interferes strongly with the motions of the carbonate group, and the T_{max} value for the β process is increased by a large amount.

Polydian Carbonate

Figure 13.14 shows the plot ε'' against $\log f$ for amorphous polydian carbonate (Ishida and Matsuoka, 1965) in the β region. As the temperature is raised, the area below the peak increases and the curves become narrower. There is no real evidence of structure within the dielectric β process, but the mechanical β process shown in Figure 13.15 is quite non-symmetric and clearly has structure. A similar result was obtained in the case of PET and the higher polymethylene terephthalates. Ishida and

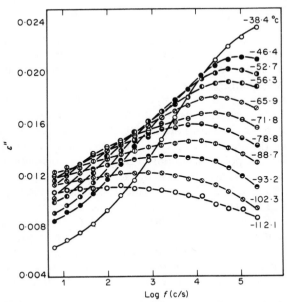

Figure 13.14. ε'' against $\log f$ at different temperatures in the β relaxation region of amorphous polydian carbonate (Ishida and Matsuoka, 1965).

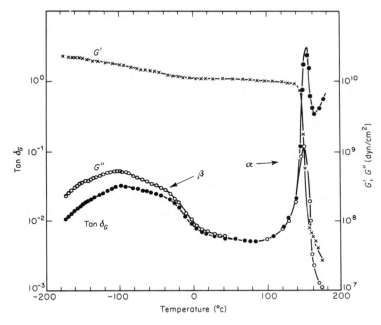

Figure 13.15. G', G'' and tan δ_G against temperature at 1 c/s for polydian carbonate (Illers and Breuer, 1961).

Matsuoka (1965) found that the free-volume fraction, $f(T_g)$ at the glass-transition temperature, derived from the WLF plot for the α relaxation, was 0·029. This is larger than the so called 'universal' value 0·025. They suggested that the large value of $f(T_g)$ reflected the open packing of chains in the glassy state of polydian carbonate. Such an open packing gives considerable freedom to the carbonate groups in the glassy state, and this allows the β process to occur. The poor packing characteristic of polydian carbonate may be the reason for the lower $(T_{max})_\beta$ value for this polymer compared with that observed for PET (see Table 13.4). Ishida and Matsuoka suggest that at low temperatures in the glass, the chains are constrained, which leads to a broad loss curve of small magnitude. As the temperature is increased, the constraints are relieved, and the β process narrows and increases in magnitude, a hypothesis which is obviously in agreement with the data of Figure 13.14.

Müller and Huff (1959) have shown that the magnitude of the dielectric β process can be changed by mechanical and thermal treatments. Samples were annealed and stretched at different temperatures. These samples gave smaller β relaxations than that observed in the untreated polymer, but the frequency–temperature location of the loss peak was unchanged by the treatment.

The β process in polydian carbonate is affected also by changes in the degree of crystallinity. Mikhailov and Eidelnant (1960a) found that the dielectric β process decreased in magnitude and also moved to higher temperatures at a given frequency on going from amorphous to partially crystalline polydian carbonate. In the mechanical case, Nielsen (1962) found that the β process at 1 c/s moved from -110 to $+30°$c on going from amorphous to partially crystalline polymer. The magnitude decrease observed by Mikhailov and Eidelnant is similar to that observed for PET, described above. However, the shift to higher temperatures with increase in crystallinity observed by these authors, in the electrical case, and also by Nielsen in the mechanical case is not observed in PET. The reason for this shift is not clear since, if local motions of the carbonate group give rise to the β process, these motions would not be expected to change significantly for only a moderate increase in crystallinity.

13.2c The 'Intermediate Temperature' Relaxation in Polydian Carbonate

A small relaxation region may be found both electrically and mechanically in polydian carbonate, and is situated between the α and β relaxation processes in this polymer. We refrain from terming this process the 'β' process. It is not observed in a well-annealed specimen, but is observed in oriented specimens and may in freshly moulded material be a function of time and temperature: it is not therefore characteristic of polycarbonate in its equilibrium state. Figure 13.16 shows the mechanical results of Illers and Breuer (1963) for untreated, annealed and oriented specimens. The untreated sample contains frozen stresses, and these give rise to the relaxation. Annealing at 160°c decreases the magnitude and annealing at 180°c results in the disappearance of the relaxation. Müller and Huff (1957) and Krum and Müller (1959) observed the dielectric process in untreated, annealed and stretched samples of polydian carbonate. Krum and Müller also found that an annealed sample did not exhibit this process. A similar process was observed for oriented PET (see above) which disappeared on heating a sample above 60°c. According to Illers and Breuer (1963) it is due to the freezing in of non-equilibrium stresses during the moulding or drawing of a specimen. These stresses relax out only

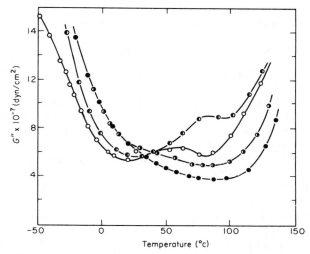

Figure 13.16. G'' against temperature for different samples of polydian carbonate. ◑ untreated; ◐ annealed at 160°C; ● annealed at 180°C; ○ drawn by 110% (Illers and Breuer, 1963).

slowly at temperatures below T_g. The intermediate temperature relaxation is associated with these frozen stresses, but the actual mechanism is not established.

13.3 LINEAR POLYESTERS

Table 13.5 summarizes the linear polyesters which have been studied by dielectric and dynamic mechanical techniques.

Yager and Baker (1942) investigated the dielectric properties of eleven linear polyesters in the range 10^3 to 10^8 c/s, at 25°C. It was found that polyethylene succinate gave a relaxation region centred at 10^6 c/s. Some evidence was seen for relaxation above 10^8 c/s in the other polymers. Evaluation of ε_R showed that as the number of ester dipoles per cc increased, so ε_R increased in a continuous fashion.

Polyethylene sebacate was investigated by these authors over the temperature range $+26.3$°C to -98.5°C. They found that the plots of ε' and ε'' as a function of frequency, broadened rapidly with decreasing temperature, and the intensity of the relaxation $(\varepsilon_R - \varepsilon_U)$ went from 2.2 at 26.3°C to 0.03 at -98.5°C. The frequency–temperature location of the loss peak varied uniformly from 10^5 c/s, -27°C to 7.5×10^7 c/s, $+17$°C,

Table 13.5

Compound	Type of measurement	Ref.
[—OC—(CH$_2$)$_8$COOCH$_2$CH$_2$—O—]$_n$ polyethylene sebacate	D	Yager and Baker (1942)
[—OC(CH$_2$)$_2$COO(CH$_2$)$_2$O—]$_n$ polyethylene succinate	D	Yager and Baker (1942) Pelmore and Simons (1940)
[—OC—(CH$_2$)$_8$COO(CH$_2$)$_{10}$—O—]$_n$ polydecamethylene sebacate	D	Yager and Baker (1942)
[—OC . COO(CH$_2$)$_{10}$—O—]$_n$ polydecamethylene oxalate	D	Yager and Baker (1942)
[—OC(CH$_2$)$_8$COO(CH$_2$)$_2$—O—]$_n$ polyethylene sebacate	D	Yager and Baker (1942)
[—OC(CH$_2$)$_7$—COO(CH$_2$)$_2$O—]$_n$ polyethylene azelate	D	Yager and Baker (1942)
[—OC(CH$_2$)$_4$COO(CH$_2$)$_2$—O—]$_n$ polyethylene adipate	D	Yager and Baker (1942) Pelmore and Simons (1940)
[—OC(CH$_2$)$_9$—O—]$_n$ poly-ω-hydroxydecanoic acid	D	Yager and Baker (1942)
[—OC(CH$_2$)$_4$COO(CH$_2$)$_4$O—]$_n$ polytetramethylene adipate	D	Würstlin (1948)

[—OC(CH$_2$)$_4$COOCH(CH$_2$)$_2$—O—]$_n$
        |
     CH$_3$
poly-(1-methyl)trimethylene adipate
	D	Würstlin (1948)

with an apparent activation energy of 12 kcal/mole. Yager and Baker attributed the relaxation to the 'oscillation' of small units of the chain. In the light of modern evidence, it seems more probable that the relaxation is due to micro-Brownian motions (glass–rubber relaxation) in the amorphous phase of the partially crystalline polymer. The rapid broadening of the relaxation curves with decreasing temperature is probably due to the interaction of the crystalline phase with the amorphous phase.

Pelmore and Simons (1940) studied the dielectric properties of several linear polyesters. They considered that if the polyester chains were in the planar zig-zag form, then, for a general structure [—O(CH$_2$)$_x$—OOC(CH$_2$)$_y$CO—]$_n$, there were four distinct possibilities. (a) if x and y are even numbers, the bond dipoles along the chain would be aligned antiparallel, thus the chain has zero moment. (b) If x is even and y odd, or x odd, y even, then the dipoles oppose again and the chain has zero moment. (c) If x and y are both odd, the bond dipoles all point along the same direction, giving maximum moment for the chain. Polytrimethylene malonate corresponds in their scheme to maximum moment, and the other polymers have zero moment. They found experimentally that tan δ_{max} in a temperature plot was near 4×10^{-2} for polyethylene succinate and polyethylene adipate, but was near $2 \cdot 5 \times 10^{-1}$ for polytrimethylene succinate, polyethylene malonate and for polytrimethylene malonate. Due to the large inconsistency between their predicted behaviour, and the experimental results, they concluded that their hypothesis was incorrect, and the relaxation actually observed was due (a) to end-groups or unknown sources or (b) the chains were not rigid and planar. We know in the light of modern evidence that the relaxation in these polymers does not correspond to the rotation of whole rigid chains, which their hypothesis requires. The relaxation is due to motion in the disordered phase of the polymer and involves the cooperative motion of smaller units in the chain.

Würstlin (1948) studied the dielectric properties of the two isomeric linear polyesters shown in the table. The polyester formed from 1,4-butanediol and adipic acid is highly crystalline. The polyester formed from 1,3-butanediol and adipic acid is amorphous due to the presence of the randomly attached (in terms of stereoregularity) methyl side-group. Würstlin showed that the 1,3-butanediol polymer had a large dispersion and absorption at 0°C (1 Mc/s) corresponding to the backbone motions of the chain. The 1,4-butanediol polymer crystallized on cooling to about 55°C, but a small dispersion and absorption region was found at the same location as that observed for the amorphous 1,3-butanediol polymer. It was concluded that the relaxation observed in the highly crystalline polymer is essentially the same process as that observed in the amorphous polymer, but the intensity of the relaxation is considerably reduced since the amorphous content of the crystalline polymer is small.

Mikhailov and Eidelnant (1960b) studied the dielectric properties of a number of mixed polyesters. These were made from a 1:1 mixture of two dibasic acids and a diol. The acids used were terephthalic, sebacic and adipic acids. The diols were ethylene glycol and 'dian'.* Five samples were studied, and one region of 'dipole elastic' (i.e. micro-Brownian motions) was observed in all samples. Two regions of 'dipole radical' losses were observed in the samples and were attributed to the motions of each acid residue in the polymer.

* See Section 13.2 for the structure of 'dian'.

14

Oxide Polymers

14.1   POLYOXYMETHYLENE

$$[-\overset{\overset{\displaystyle H}{|}}{\underset{\underset{\displaystyle H}{|}}{C}}-O-]_n$$

(1)

Polyoxymethylene (POM) crystallizes with a hexagonal unit cell in which the chains have a helical conformation with nine monomer units in the repeat distance of 17·3 Å (Hengstenburg, 1927; Sauter, 1933). Single crystals of POM, precipitated from solution, were shown by Geil, Symons and Scott (1959) to consist of hexagonal lamellae approximately 100 Å thick. Larger hexagonal structures, known as hedrites, can be grown by the slow crystallization of a molten film (Geil, 1960b). If the crystallization is more rapid, or the molten film is thicker than 30 microns, then spherulites are obtained instead of hedrites. Thus, POM crystallized in bulk usually has a spherulitic structure. Geil (1960b) has shown that both hedrites and spherulites include lamellae approximately 100 Å thick in which the molecular chains are folded. POM melts at about 183°C.

Dynamic mechanical studies of POM include those of Thurn (1960), Read and Williams (1961a), McCrum (1961), Takayanagi (1963), Heijboer (1963), Eby (1963), Arisawa and others (1963), Wetton and Allen (1966) and Bohn (1965). Figure 14.1 illustrates some results of McCrum for two specimens having respective densities of 1·444 g/cm^3 (open circles) and 1·404 g/cm^3 (closed circles). These density values cover the full range which may be obtained by slow cooling or quenching from the melt. The crystal (x-ray) density is 1·506 g/cm^3. Three loss peaks are clearly observed in Figure 14.1. These are labelled α, β and γ respectively. The frequency–temperature plot (Figure 14.2) summarizes the mechanical results of the various workers, the α, β and γ loss regions each being evident on this diagram.

The dielectric properties of POM have been investigated by Thurn (1960), Ishida (1960b), Mikhailov and Eidelnant (1960c), Read and Williams

540

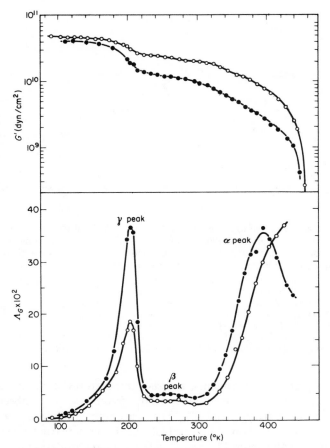

Figure 14.1. Temperature dependence of the shear modulus and logarithmic decrement for polyoxymethylene (McCrum, 1961).

(1961a), Ishida and others (1961), Williams (1963b) and Arisawa and others (1963). These studies have been largely concerned with the γ process which gives rise to the most dominant and well-defined dielectric loss peak. However, the α and β relaxations have also been detected dielectrically (Ishida and others, 1961; Arisawa and others, 1963). The frequency–temperature locations of the dielectric peaks are summarized in Figure 14.3. In the γ region at low temperatures, the curves marked f_{max} and T_{max} correspond, respectively, to the frequencies of maximum ε'' from

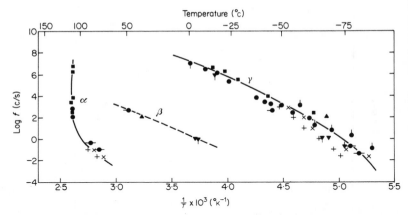

Figure 14.2. Plots of log f versus $1/T$ for the mechanical loss maxima of POM. In the following key (f) refers to the frequency of maximum loss from constant temperature experiments and (T) denotes the temperature of maximum loss in experiments at constant frequency.

■ E'' (T) (Thurn, 1960)
● tan δ_E (T) (Thurn, 1960)
▲ E'' (T) (Takayanagi, 1963)
▼ Λ (T) (McCrum, 1961)
▼ tan δ_G (T) (Heijboer, 1963)
–● tan δ_E (T) (Heijboer, 1963)
–●– G'' (T) (Read and Williams, 1961a)

▼ tan δ_G (T) (Read and Williams, 1961a)
● tan δ_E (T) (Eby, 1963)
+ tan δ_G (T) (Wetton and Allen, 1966)
× G'' (T) (Wetton and Allen, 1966)
● G'' (f) (Wetton and Allen, 1966)

constant temperature experiments, and the temperatures of maximum ε'' from measurements performed at constant frequency. The observed divergence between the f_{max} and T_{max} curves is commented on in Section 14.1c below. A comparison of Figure 14.2, which was constructed largely from T_{max} data, with Figure 14.3 reveals a fairly good correlation between the locations of the mechanical and dielectric loss peaks. For the purposes of discussion, it is convenient to consider separately the results for the α, β and γ relaxations.

14.1a   The α Relaxation

The α peak in POM is most marked in the mechanical results at low frequencies, as for example in Figure 14.1. From an analysis of areas beneath G'' versus $1/T$ plots (according to Equation 4.91) Read and Williams (1961b) have estimated that the activation energy lies in the range 65 to 92 kcal/mole, when determined at frequencies between about 0·08 and 0·42 c/s. The mechanical results of Thurn (1960) are also consistent with a very high activation energy since, according to his data

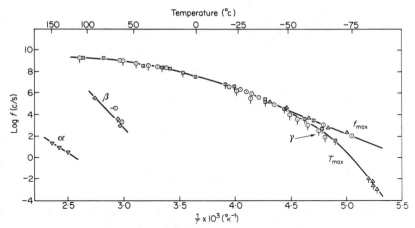

Figure 14.3. Plots of log f versus $1/T$ for the dielectric loss maxima of POM. In the following key (f) refers to the frequency of maximum loss from constant temperature experiments and (T) denotes the temperature of maximum loss in experiments at constant frequency.

⊙ ε'' (f) (Read and Williams, 1961a)
⊙ ε'' (T) (Read and Williams, 1961a)
△ ε'' (f) (Ishida, 1960b)
▫ tan δ_ε (f) (Thurn, 1960)
▫ tan δ_ε (T) (Thurn, 1960)

△ ε'' (T) (Williams, 1963b)
▽ ε'' (f) (Arisawa and others, 1963)
⊖ ε'' (f) (Ishida and others, 1961)
◇ (Quoted by Arisawa and others, 1963)

(see Figure 14.2), the α (E'') peak becomes essentially independent of temperature at frequencies above about 10^2 c/s.

Figure 14.4 shows plots of ε'' against frequency for POM at temperatures between 110 and 140°C. These results, which were obtained by Arisawa and others (1963), illustrate the dielectric α peak at frequencies below 10 c/s. It is probable, however, that the ε'' values in this plot are partly determined by a superposed conductivity process, as evidenced by the absence of a low-frequency maximum at 110°C and the fact that the maximum ε'' values are greater than typical ε' values ($\simeq 4$) in this region. Arisawa and others (1963) have quoted an activation energy of about 20 kcal/mole for this dielectric α process. This value is close to the value of 24 kcal/mole obtained by McCrum (1961) using the elastic after-effect method. The cause of the large discrepancy between the values obtained by Read and Williams (1961b) on the one hand and Arisawa and others (1963) and McCrum (1961) on the other has not been resolved.

At low frequencies the mechanical α peak in POM lies some 50°C to 100°C below the melting point, the exact temperature depending largely

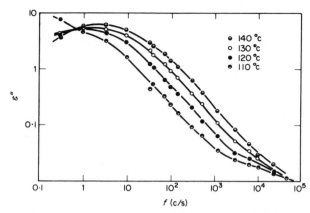

Figure 14.4. Plots of ε'' against frequency at various temperatures in the α relaxation region (Arisawa and others, 1963).

on whether G'' or tan δ is plotted. By way of comparison it may be recalled that a similar temperature interval separates the α peak from the melting temperature in polyethylene (Section 10.1). It now seems fairly generally accepted that the α mechanism in POM, as in the case of polyethylene, is in some way associated with the crystalline phase. Evidence in favour of this hypothesis comes from the observation that the α peak shifts to higher temperatures following annealing, since annealing is known to increase the crystallite size, or fold height. The latter observation is seen in Figure 14.1 and has also been reported by Read and Williams (1961a).

Further evidence in favour of a proposed crystalline mechanism is obtained from the work of Takayanagi (1963). Figure 14.5 shows that, whereas the α peak was hardly detected in the E'' versus temperature plot for a melt-crystallized film, it is more pronounced for a sintered film prepared from the powdered polymer, and most prominent for a single-crystal laminate. From these results, and also from observed effects produced by drawing POM at temperatures in the α region, Takayanagi concluded that the α mechanism was associated with the crystalline phase rnd proposed that it involved translational motions along the chain axis.

14.1b   The β Relaxation

The mechanical β peak is the smallest of the three peaks in POM (Figure 14.1). A very weak dielectric peak observed by Ishida and others (1961) at about $10^{3\cdot3}$ c/s (62°C) and $10^{4\cdot6}$ c/s (68·5°C) would seem to correspond

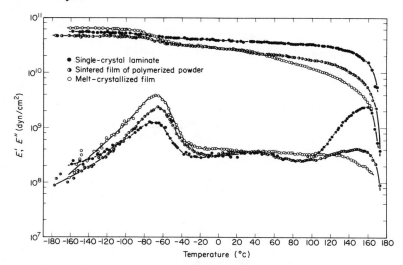

Figure 14.5. Temperature dependence of E' and E'' for different samples of POM at 138 c/s (Takayanagi, 1963).

to the β relaxation (cf. Figures 14.2 and 14.3), though this peak is labelled α by the authors. Another dielectric observation of the β relaxation has been reported by Arisawa and others (1963).

A detailed study by McCrum (1961) has shown that the magnitude of the mechanical β peak depends both on the amount of absorbed water and upon thermal history. Figure 14.6 shows the variation of G' and Λ for a dried specimen and for a specimen containing water. The wet and dry specimens were prepared by maintaining the samples at 23°C, at 100% and 0% relative humidity, respectively, until equilibrium was achieved. The equilibrium water uptake of the wet specimen was about 1 weight per cent. It is seen that the β peak is considerably smaller for the dried specimen than for the specimen containing water.

The dependence of the β peak on thermal history is illustrated in Figure 14.7. A specimen was heated at 430°K for 30 min and then cooled to room temperature. The variation of the logarithmic decrement was then immediately observed, first cooling from room temperature (296°K) to 90°K and then heating from room temperature to 440°K. The results obtained are shown by the open circles in Figure 14.7. After storage for several weeks at room temperature the temperature dependence of Λ_G was again determined, the results of these measurements being represented

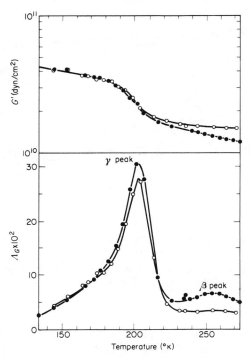

Figure 14.6. Temperature dependence of the shear modulus and logarithmic decrement for a wet (●) and dry (○) specimen of POM at 1 c/s (McCrum, 1961).

by the closed circles in Figure 14.7. It is seen that storage at room temperature depresses the magnitude of the β peak. The magnitude of the β peak of a freshly cooled specimen was found not to change if the sample was stored in liquid nitrogen.

The effect of thermal history on the β peak is not due to the uptake or loss of water. This was checked by observing that the effect was independent of whether the sample took up or lost water during the heating. A very small density increase occurs upon storage at room temperature after cooling from a high temperature. This is possibly due to an internal reordering, and leads to the conclusion that the β relaxation is associated with the disordered regions of the polymer. Takayanagi (1963) has ascribed the β relaxation to main–chain micro-Brownian motions in the disordered interlamellar regions of POM, an assignment which implies that the β process may be related to the 'glass transition'.

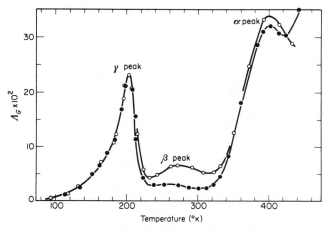

Figure 14.7. Illustration of the effect of thermal history on the β peak of POM. Logarithmic decrement plotted against temperature for (○) freshly cooled specimen; (●) aged specimen (McCrum, 1961).

14.1c The γ Relaxation

Plots of ε'' against frequency are shown in Figure 14.8 at several temperatures in the γ region. The most striking feature of this diagram is the rapid broadening of the γ peak as the temperature is lowered. This phenomenon is further illustrated in Figure 14.9 by the rapid decrease in the Fuoss–Kirkwood distribution parameter with decreasing temperature (below about $-20°$C). The marked asymmetry of the mechanical γ peak (Figure 14.1) could well be indicative of a similar broadening of the mechanical distribution. Figure 14.10 shows that the dielectric relaxation magnitude $\varepsilon_R - \varepsilon_U$ is also decreasing with decreasing temperature.

The observed divergence between the f_{max} and T_{max} curves (Figure 14.3) for the γ relaxation has been ascribed largely to the rapid broadening of the dielectric relaxation spectrum with decreasing temperature (Read and Williams, 1961a). As discussed in Chapter 4 the determination of average relaxation times $[\tau_{av} = (2\pi f_{max})^{-1}]$, and hence the activation energy, can only be performed from the f_{max} curve. From this curve Read and Williams (1961a) obtained an activation energy of 19·3 kcal/mole independent of temperature in the range -20 to $-80°$C. This value compares well with the value of 20 kcal/mole reported by Ishida (1960b) for the dielectric γ process. For the mechanical γ relaxation Read and Williams (1961b) estimated an activation energy of about 20 kcal/mole from an analysis of

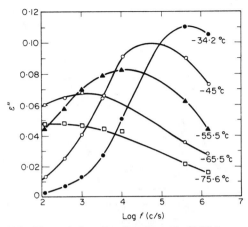

Figure 14.8. Plots of ε'' against frequency for POM at several temperatures in the γ relaxation region (Read and Williams, 1961a).

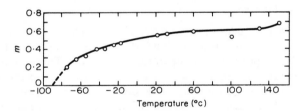

Figure 14.9. Temperature dependence of the Fuoss–Kirkwood distribution parameter for the dielectric γ relaxation of POM (Read and Williams, 1961a).

areas under G'' versus $1/T$ curves (Equation 4.91). From the slope and curvature of the T_{max} curve in Figure 14.3 it is clear that an apparent activation energy derived (erroneously) from this curve will be larger than the correct value at low temperatures and will appear to vary with temperature. A similar result would be (erroneously) obtained from the curved mechanical plot in Figure 14.2 since this contains largely T_{max} data. Furthermore, analyses of the curved $\log f$ versus $1/T_{max}$ plots in terms of WLF-type equations (Thurn, 1960; Saitô and others, 1963) would seem to be invalid.

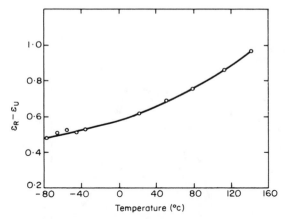

Figure 14.10. Temperature dependence of $\varepsilon_R - \varepsilon_U$ for the dielectric γ relaxation in POM (Read and Williams, 1961a).

McCrum's data shown in Figure 14.1, and illustrated in more detail in Table 14.1, shows that the intensity of the γ peak decreases with increasing density. A similar result has been obtained by Read and Williams (1961a), who found that the modulus increment $G_U - G_R$ decreased after annealing. These results suggest that the γ mechanism is associated with the disordered regions of the polymer. Takayanagi's data (Figure 14.5) are also consistent with this conclusion since the intensity of the γ relaxation is lower for the single-crystal laminate than for the melt-crystallized film.

The effects produced by swelling POM with dioxane liquid are illlustrated in Figure 14.11. Here it is observed that the dielectric γ peak shifts to lower temperatures with increasing dioxane concentration up to 6.1%.

Table 14.1. The height of the γ peak (Λ_{max}) in polyoxymethylene as a function of density (McCrum, 1961)

Density (g/cc) (23°C)	$\Lambda_{max} \times 10^2$
1·404	36·5
1·414	29·0
1.420	28·0
1.430	23·2
1·444	18·5

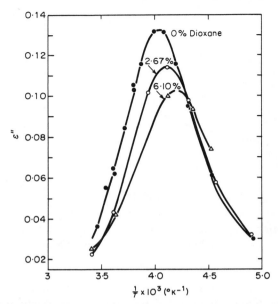

Figure 14.11. Plots of ε'' against temperature at 1.42 Mc/s for POM containing different amounts of absorbed dioxane (Read and Williams, 1961a).

A similar (plasticizing) effect has been observed on the mechanical γ peak (Read and Williams, 1961a). Since there is evidence to suggest that liquids are absorbed only into the disordered regions of POM (Alsup, Punderson and Leverett, 1959), these results are also suggestive of a process occurring in the disordered regions.

Arisawa and others (1963) have reported some ultrasonic attenuation results for solution-grown crystals of POM and have compared their results with dielectric data of Yamafuji and coworkers on similar specimens. Like Takayanagi (1963), these workers also observed the γ peak for the solution-grown crystals. Their results also indicated that the temperature of the γ peak was very slightly lower for the solution-grown crystals than for the POM crystallized in bulk.

14.1d   The Question of the Glass Transition

Having established that both the β and γ relaxations result from motions in the disordered regions of the polymer, and noting that these motions must occur within the main chain (side-branches being absent in POM), the question arises as to which, if either, of these relaxations can be associated with a glass transition. Read and Williams (1961a) initially proposed that the γ relaxation may be related to a glass transition on the grounds that it has a relatively large intensity. This hypothesis would seem to be supported by the fact that the γ relaxation is plasticized by dioxane, and that the thermal expansion coefficient changes abruptly at temperatures a few degrees below that of the γ peak (Leksina and Novikova, 1959). On the other hand, Takayanagi (1963) considers that the γ relaxation results from the local mode mechanism (Section 5.5) and occurs largely at defect regions within the crystal lamellae. This view certainly appears consistent with the observation of the γ peak for the solution-grown crystals. As noted above (Section 14.1b), Takayanagi's assignment of the β process to the amorphous interlamellar regions implies that the β mechanism may be related to T_g. In fact Bohn (1965) has recently reported that the intensity of the β peak is enhanced when the crystallinity is lowered by copolymerizing POM with polyethylene oxide. Bohn also concluded from this study that the β process was related to the glass transition.

From the above discussion it may be concluded that somewhat diverse views have been reported concerning the question of a glass transition in POM. A similar problem exists with other highly crystalline polymers, notably with linear polyethylene (Chapter 10). As discussed in connection with polyethylene, the question of the glass transition is partly a matter of definition and the glass-transition concept is only strictly valid for polymers in the amorphous state or having a fairly low degree of crystallinity. For polymers having a very high degree of crystallinity, the concept of a glass transition seems of doubtful significance, and attempts to assign a T_g must inevitably lead to some confusion.

14.2 POLYETHYLENE OXIDE

$$[-\overset{\overset{\displaystyle H}{|}}{\underset{\underset{\displaystyle H}{|}}{C}}-\overset{\overset{\displaystyle H}{|}}{\underset{\underset{\displaystyle H}{|}}{C}}-O-]_n$$

(2)

Polyethylene oxide crystallizes in a form in which the chain has a helical conformation. There are seven ethylene oxide units and two turns in a fibre period of 19·3 Å. This structure corresponds to the succession of nearly *trans, trans, gauche* conformations for the sequence C—O—C—C—O (see Tadokoro and others, 1963). This helix is more open than that of

POM and this may partly explain the low density of PEO compared with POM.

Single crystals of PEO may be obtained from dilute solution, hedrites are obtained from the crystallization of thin films, and spherulites are obtained by crystallization from the bulk (see Geil, 1963). A feature of PEO is that because of its low spherulite nucleation rate, extremely large spherulites (up to 1 cm diameter) may be grown from the melt (Price, 1961). PEO can be obtained commercially in the molecular-weight range 2×10^2 to 2×10^6. These materials have narrow molecular-weight distributions up to molecular weights near 10^4, but at higher molecular weights narrow fractions may only be obtained by careful fractionation procedures (Allen, 1965).

Melt-crystallized PEO is highly crystalline. Table 14.2 gives the x-ray estimate of the degree of crystallinity for samples of different molecular weight (Turner–Jones, 1963). These samples were obtained by cooling at 6°C per hour from the melt at 100°C. It is seen that the degree of crystallinity is greatest at molecular weights of 4×10^3 and 3×10^4.

Table 14.2. Characteristics of polyethylene oxides used by Connor, Read and Williams (1964)

Molecular weight	X-ray crystallinity (%)
$2 \cdot 8 \times 10^6$	61
$8 \cdot 4 \times 10^5$	70
$2 \cdot 8 \times 10^5$	71
30,000	76
4000 ± 50	80
1000 ± 50	56

Dynamic mechanical studies of PEO include those of McCrum (1961), Read (1962b), Wetton (1962) and Connor, Read and Williams (1964). Figure 14.12 illustrates some of the results of Read (1962) for samples of different molecular weight. The G'' plots indicate one main relaxation region at each molecular weight, where the magnitude and temperature locations for the process are a complex function of molecular weight. However, for the sample of highest molecular weight ($2 \cdot 8 \times 10^6$), the Λ_G plots indicate that a second loss peak occurs at higher temperatures ($0 \rightarrow -10$°C) than that of the main peak. This high-temperature process is labelled α in Figure 14.12 and the dominant peak is labelled β. Both the Λ_G and G'' plots also give some indication of a very low and broad loss peak

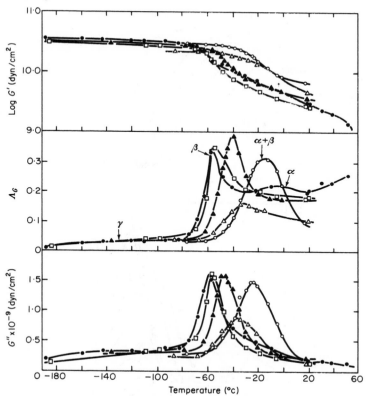

Figure 14.12. G', G'' and the logarithmic decrement Λ_G against temperature for PEO ($f \simeq 0.5$ c/s) (Read, 1962b). Point symbols refer to molecular weights as follows.

● 2.8×10^6 □ 8.4×10^5 ▲ 2.8×10^5 ○ 3×10^4 △ 4×10^3

centred at about $-130°$c. This region is tentatively labelled γ in Figure 14.12. The samples of lowest molecular weight (3×10^4 and 4×10^3) exhibit only a single broad peak in Figure 14.12. For reasons given below (Section 14.2b) this peak is labelled $\alpha + \beta$.

The dielectric properties of melt-crystallized PEO have been described by Ishida, Matsuo, Togami, Yamafuji and Takayanagi (1962), Connor, Read and Williams (1964) and by Hikichi and Furuichi (1965). Figure 14.13 shows ε' and ε'' against $\log f$ for a 2.8×10^6 molecular-weight sample (Connor, Read and Williams, 1964). One broad relaxation is observed, which narrows and increases in magnitude ($\int \varepsilon'' \mathrm{d} \log f$) as the temperature

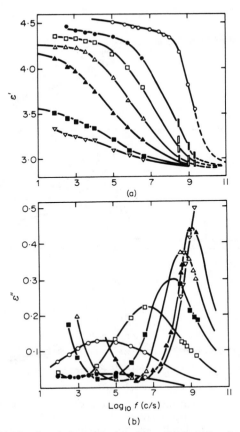

Figure 14.13. ε' and ε'' against $\log f$ for a high molecular weight PEO ($M_w = 2\cdot 8 \times 10^6$) (Connor, Read and Williams, 1964). Point symbols refer to temperatures of measurements as follows.

(a)	○	20°C	●	−15°C	□	−30°C	△	−40°C
	▲	−50°C	■	−60°C	▽	−75°C		
(b)	▽	35°C	▲	20°C	△	1˘C	■	−16°C
	□	−35°C	○	−50°C	●	−75°C		

is raised. As illustrated below, this relaxation correlates with the mechanical β relaxation of Figure 14.12. Ishida, Matsuo and Takayanagi (1965) have studied melt-crystallized and single-crystal mat samples of a high molecular weight PEO. Figure 14.14 shows their results at 12.8 kc/s. The single-crystal mat gives a low-temperature (γ) process at about

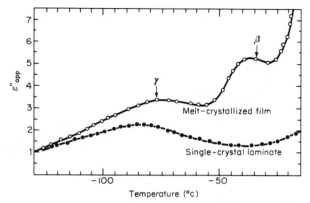

Figure 14.14. ε''_{app} against temperature at 12·8 kc/s for two samples of PEO. ○ and ● refer to melt-crystallized and single-crystal mat samples, respectively (Ishida, Matsuo and Takayanagi, 1965).

−80°C which is also observed in the melt-crystallized sample. The latter sample gives a further (β) loss process near −40°C, and this process is absent in the single-crystal mat.

In order to present the mechanical and dielectric results in a parallel manner, we first attempt to correlate the relaxations in the plot of $\log f$ against $(1/T_{max})$. Figure 14.15 shows this plot for various molecular weights. The dielectric data of Connor, Read and Williams (1964) covers an extensive frequency range, and relates to the higher-temperature process observed by Ishida and others (1965). Inspection of the dielectric and mechanical data of Connor, Read and Williams (1964) shown in Figure 14.15 indicates that the 'principal' (β) mechanical and dielectric loss processes are a function of molecular weight, but at a given high molecular weight ($\geqslant 2 \times 10^5$) the mechanical points lie on a smooth extrapolation of the dielectric curve. The low molecular weight results are not sufficiently extensive to enable such a correlation to be made. Therefore, the following discussion will deal first with the data obtained for high molecular weight PEO ($M_w > 10^5$), and the lower molecular weight data will be considered separately.

14.2a   High Molecular Weight PEO ($M_w > 10^5$)

The α Relaxation

From Figure 14.12 the α process is seen as a small high-temperature shoulder to the principal loss process for G'' plots, and is observed as a

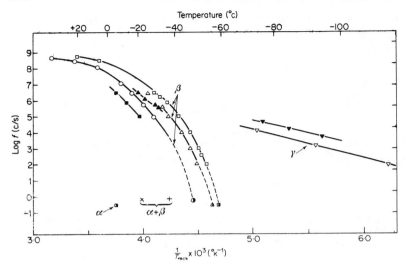

Figure 14.15. Log f against $1/T_{max}$ for polyethylene oxides.

Dielectric

☐ ε'' melt-crystallized, molecular weight 2.8×10^6 (Connor and others, 1964)

△ ε'' melt-crystallized, molecular weight 8.4×10^5 (Connor and others, 1964)

○ ε'' melt-crystallized, molecular weight 2.8×10^5 (Connor and others, 1964)

Mechanical

◨ G'' molecular weight 2.8×10^6 (Read, 1962b; Connor and others, 1964)

▲ G'' molecular weight 8.4×10^5 (Read, 1962b; Connor and others, 1964)

◑ G'' molecular weight 2.8×10^5 (Read, 1962b; Connor and others, 1964)

■ ε'' melt-crystallized, molecular weight 3×10^4 (Connor and others, 1964)

▲ ε'' melt-crystallized, molecular weight 4×10^3 (Connor and others, 1964)

▽ ε'' melt-crystallized (Ishida and others, 1965)

▼ ε'' single-crystal mat (Ishida and others, 1965)

× G'' molecular weight 3×10^4 (Read, 1962b; Connor and others, 1964)

+ G'' molecular weight 4×10^3 (Read, 1962b; Connor and others, 1964)

The data of Connor and others (1964) refer to the β relaxation process observed at the different molecular weights. The data of Ishida and others (1965) for the β relaxation process in a melt-crystallized sample superpose the data of Connor and others (1964) for the 8.4×10^5 molecular-weight sample and therefore are not indicated on the figure. $\triangledown$ and ▼ refer to the γ relaxation process.

region of high Λ in the temperature region $0°c$ to $-10°c$ at high molecular weights. Read (1962b) showed that annealing of the 2×10^6 and 8×10^5 molecular-weight samples resulted in a shift to higher temperatures and an increase in the magnitude of this loss region for both G'' and Λ_G representations of the data. This may be taken as evidence for the suggestion that this process is associated with the crystalline phase. It was suggested (Connor and others, 1964) that a very highly annealed high molecular weight PEO would give the α process in the same location, with roughly the same magnitude, as the principal loss region observed in a PEO of molecular weight 4×10^3. Since the low molecular weight polymer has a high degree of crystallinity, the conclusion might be drawn that the relaxation process associated with the crystalline phase occurs both in the low and high molecular weight polymers.

The α process is not observed in the dielectric studies. This may be a result of the large rising loss due to conductivity effects as is seen in Figure 14.14 above $-25°c$.

For POM, the α relaxation was clearly observed in the mechanical Λ_G plots (at 1 c/s) near $120°c$ (see Figure 14.1). The α process in high molecular weight PEO occurs in the $0°$ to $-10°c$ region (for Λ_G). Hence for both POM and PEO the α relaxation occurs about $60 \rightarrow 70°c$ below the melting temperature. However, owing to the poor resolution of the α peak in PEO, the correlation between the α processes in these two polymers must be tentative.

The β Relaxation

The β relaxation in PEO was observed by Read (1962b), Ishida and others (1962) and by Connor and others (1964) in melt-crystallized PEO. The most striking result is that of Ishida and others (1965) for single-crystal mats of PEO. Figure 14.14 shows that the dielectric β process is removed on going from a melt-crystallized to a single-crystal mat sample. This clearly indicates that the β process arises from the disordered regions of melt-crystallized high molecular weight PEO. Ishida and others (1965) suggest that this process is due to the micro-Brownian motions of the chain in the amorphous regions of PEO. A similar conclusion was reached by Read (1962b) and by Connor and others (1964) from the following experimental results. The mechanical β process moved to higher temperatures and decreased in magnitude on annealing the polymer. This indicates that the crystallites exert an effect on the β process, and the process occurs in the disordered regions of the polymer. In the dielectric case, the β process was reduced in magnitude but not in position on annealing high molecular weight PEO.

Absorption of dioxane into melt-crystallized PEO resulted in a shift of the mechanical and dielectric β process to lower temperatures, accompanied in the mechanical case by an increase in the magnitude of the loss process. Similarly, a small amount of absorbed water shifted the mechanical loss peak to lower temperatures and increased the magnitude of the process. These effects are consistent with a plasticization of the disordered regions of PEO. Therefore, the early mechanical and dielectric studies indicated that the β process was due to micro-Brownian motions of the main chains in the disordered phase of the polymer, and these motions were influenced by the presence of crystallites. Read (1962b) quotes a dilatometric T_g value for quenched PEO having molecular weight $M_w = 2 \times 10^6$ as $-67°C$. This is about $10°C$ lower than the temperature of $(G''_{max})_\beta$ at 0.3 c/s given by Read (1962b) for the same polymer. This result suggests that the β process in PEO is related to the glass transition.

The feature of the β process in high molecular weight PEO is that its frequency–temperature location is a function of molecular weight even in the molecular-weight range 10^5 to 10^6, as is seen from Figure 14.15. As the molecular weight is increased, so the β process moves to lower temperatures. Also, from Table 14.2 it is found that the degree of crystallinity also falls as molecular weight is raised, which may indicate that the location of the β process changes with *degree of crystallinity*, and molecular weight is merely a means of achieving such a change. This suggestion is supported by the mechanical studies of Read (1962b) and Connor and others (1964) where it was shown that the temperature of G''_{max} was increased by as much as $7°C$ by annealing a specimen for one week under nitrogen. Thus, as the crystallinity is raised (by lowering molecular weight or by annealing) so the temperature of maximum loss also increases. This implies that the crystallites exert a constraint on the disordered regions, as is the case for the partially crystalline aromatic polyesters (see Chapter 13). Read (1962b) suggested that the disordered regions of high molecular weight PEO were under a constraint from the crystallites, but distinguished between orientational and compressional strains. It was suggested that the expansion of crystallites on annealing introduced additional compressional strains in the disordered regions, which reduce the free volume and hence reduce the mobility of the chains in the disordered phase.

With regard to the detailed shape and frequency–temperature behaviour of the dielectric β process, the only comprehensive study has been made by Connor, Read and Williams (1964). Figure 14.13 shows the dielectric data as a function of frequency at given temperatures for a melt-crystallized high molecular weight PEO. As the temperature is reduced so the loss curves broaden and the magnitude $\int \varepsilon'' \, d \log f$ decreases. This behaviour is very similar to that observed for the γ relaxation in POM described above (see Figure 14.8), and might indicate a correlation between the mechanisms for these processes. Both arise from motions in the disordered phase, but the explanation of the broadening is not clear. It is possible in the case of PEO that the unusual broadening is associated with a resolution of a broad process into two component processes, β and γ, at very low temperatures. The data of Ishida and others (1965) shows β and γ relaxation regions where the γ region moves rapidly into the β region as the temperature is raised. Figure 14.15 shows that the γ region moves into the β region near $-35°$C for the polymer studied by these authors. There is no experimental evidence for the two relaxation regions in Figure 14.13, but the broadening could be a result of the separation of two processes. In polytetramethylene oxide two processes were qualitatively resolved from data similar in form to those shown in Figure 14.13 (Wetton and Williams, 1965). This will be further discussed below. Due to the rapid broadening of the β process as temperature is reduced, a discrepancy will exist between plots of $\log f$ against $(1/T_{\max})$ and $\log f_{\max}$ against $(1/T)$ as was discussed above for the γ relaxation of POM. Figure 14.16 shows the dielectric plot of $\log f_{\max}$ against $1/T$ for the three high molecular weight PEO samples studied by Connor, Read and Williams (1964). If Figures 14.15 and 14.16 are compared, it will be found that the plot $\log f$ against $1/T_{\max}$ lies to the left of the plot $\log f_{\max}$ against $1/T$ at each molecular weight. A similar result occurs for the γ relaxation in POM (see Section 14.1c above).

It is noted that the dielectric plot in Figure 14.16 curves as the temperature is lowered. This is reminiscent of the behaviour of amorphous polymers in the glass-transition region and may be taken as further evidence that the β relaxation is due to micro-Brownian motions in the non-crystalline regions of the polymer.

To conclude, there is a fairly good correlation between the dielectric and mechanical β relaxations in high molecular weight PEO. The relaxation is almost certainly due to the micro-Brownian motions of the chain (glass–rubber relaxation) in the non-crystalline regions of the polymer.

The γ Relaxation

Figure 14.14 illustrates the dielectric γ relaxation in high molecular weight PEO. It is present in both the melt-crystallized and single crystal mat samples, being slightly larger for the former sample (after the β process is subtracted from the total loss curve). Extrapolation of the

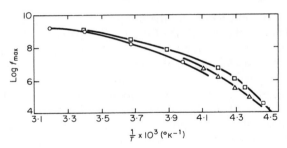

Figure 14.16. Log f_{max} against $1/T$ for the dielectric β relaxation in polyethylene oxides (Connor and others, 1964). Point symbols refer to molecular weights as follows.

□ $2\cdot8 \times 10^6$ △ $8\cdot4 \times 10^5$ ○ $2\cdot8 \times 10^5$

location of the γ process to 1 c/s in Figure 14.15 gives a temperature around $-130°$C. Inspection of the mechanical data of Figure 14.12 for high molecular weight PEO gives some indication of a low broad peak in Λ_G in this temperature region, and the G'' results also indicate a small broad process which peaks near $-140°$C for the $2\cdot8 \times 10^6$ molecular weight sample. The correlation between this poorly defined mechanical process and the well-defined dielectric γ process must be regarded as tentative.

Ishida and others (1965) suggest that the dielectric γ process is due to the local twisting motion of main chains in both the non-crystalline regions and the defective regions within the crystal phase. This is consistent with the observation that the γ process is present in the melt-crystallized and single-crystal mat samples.

14.2b   Low Molecular Weight PEO ($M_w < 10^5$)

Figure 14.12 shows the mechanical data for samples having molecular weights of 3×10^4 and 4×10^3. One loss process is observed in both polymers both for the G'' and tan δ presentations of the results. The peak occurs at higher temperatures for the 3×10^4 sample than the 4×10^3 sample. It is not clearly established if this loss process is to be correlated with the dominant (β) loss process which occurs in the high molecular weight samples near $-60°$C, or if it should be regarded as a separate process which might be correlated with the α process discussed above for the high molecular weight polymers. The dielectric data of Connor, Read and Williams (1964) for the low molecular weight samples was only obtained over a narrow frequency range, due to a large conductivity process in these polymers. Figure 14.15 shows that the dielectric loss process occurs at higher temperatures for the 3×10^4 sample than the 4×10^3 sample, in accord with the mechanical data; however, the dielectric plots do not extrapolate to the location of the mechanical points near 1 c/s. Therefore, it is questionable if the dielectric and mechanical processes have the same origin in low molecular weight PEO.

Read (1962b) found that the G'' loss peak in a sample of molecular weight 3×10^4 moved by 15°C to higher temperatures on annealing. Since annealing increases the crystallite size, it was suggested that the relaxation might be due to overlapping mechanisms involving motions in both the disordered and crystalline phases of the polymer, but is concerned mainly with the crystalline phase (Connor, Read and Williams, 1964).

14.3 POLYTRIMETHYLENE OXIDE (3) AND POLYTETRAMETHYLENE OXIDE (4)

$$[-\overset{\overset{\displaystyle H}{|}}{\underset{\underset{\displaystyle H}{|}}{C}}-\overset{\overset{\displaystyle H}{|}}{\underset{\underset{\displaystyle H}{|}}{C}}-\overset{\overset{\displaystyle H}{|}}{\underset{\underset{\displaystyle H}{|}}{C}}-O-]_n$$

(3)

$$[-\overset{\overset{\displaystyle H}{|}}{\underset{\underset{\displaystyle H}{|}}{C}}-\overset{\overset{\displaystyle H}{|}}{\underset{\underset{\displaystyle H}{|}}{C}}-\overset{\overset{\displaystyle H}{|}}{\underset{\underset{\displaystyle H}{|}}{C}}-\overset{\overset{\displaystyle H}{|}}{\underset{\underset{\displaystyle H}{|}}{C}}-O-]_n$$

(4)

Both polytrimethylene oxide (PTriMO) and polytetramethylene oxide (PTetraMO) are crystallizable polymers. PTriMO melts a little below 30°C and PTetraMO melts at 35°C (Tadokoro and others, 1963). The chain conformation in the crystal has been reported to be planar zig-zag for both polymers (Bunn and Holmes 1958; Tadokoro and others, 1963).

The mechanical behaviour of PTriMO has been studied by Willbourn (1958), and the mechanical behaviour of PTetraMO has been studied by Willbourn (1958) and Wetton (1962). The dielectric behaviour of PTetraMO has been studied by Wetton and Williams (1965).

Figure 14.17 shows the mechanical $\tan \delta_E$ against temperature at about 100 c/s for both polymers (Willbourn, 1958). Two loss regions are observed in both polymers which we label as α, β in descending order of temperature. Figure 14.18 shows the plot of ε'' against $\log f$ at different temperatures for PTetraMO (Wetton and Williams, 1965). As the

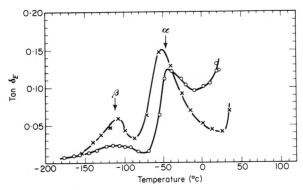

Figure 14.17. Tan δ_E against temperature for PTriMO (○) and PTetraMO (×) ($f \simeq 10^2$ c/s) (Willbourn, 1958).

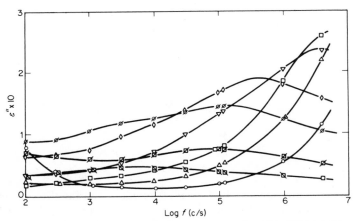

Figure 14.18. ε'' against log f for partially crystalline PTetraMO, at different temperatures (33 % crystalline sample) (Wetton and Williams, 1965).

| ⊠ −99.5°C | ⊠ −81.4°C | ⊘ −67.2°C | ◇ −58.6°C |
| ▽ −47.9°C | □ −37.1°C | △ −27.2°C | ○ −9.8°C |

temperature is reduced, the loss curves broaden considerably. This behaviour is consistent with two overlapping loss regions which separate as the temperature is decreased. At −67.2°C the loss curve appears to show the two component relaxations. Wetton and Williams (1965) resolved the data of Figure 14.18 into α and β processes in a semi-empirical manner.

Figure 14.19 shows the frequency–temperature locations of the dielectric and mechanical α and β relaxations. The dielectric and mechanical α relaxations occur in the same frequency–temperature locations, and the same is true of the β regions. Therefore, it seems reasonable to conclude that the mechanisms responsible for the mechanical α and β processes also give rise to the α and β dielectric processes. We shall now discuss the α and β processes in greater detail.

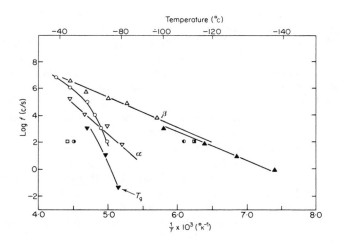

Figure 14.19. Log f against $1/T$ for PTriMO and PTetraMO. In the following key (f) refers to the frequency of maximum loss at a given temperature and (T) refers to the temperature of maximum loss at a given frequency.

Dielectric

○ ε'' (T) combined (α, β) process (Wetton and Williams, 1965)

▽ ε'' (f) resolved α process (Wetton and Williams, 1965)

△ ε'' (f) resolved β process (Wetton and Williams, 1965)
All data refer to PTetraMO

Mechanical

▼ G'' (T) α process PTetraMO (Wetton, 1962)

▲ G'' (T) β process PTetraMO (Wetton, 1962)

◑ tan δ_E α process PTetraMO (Willbourn, 1958)

◐ tan δ_E β process PTetraMO (Willbourn, 1958)

□ tan $\delta_E(T)$ α relaxation, PTriMO (Willbourn, 1958)

◪ tan δ_E (T) β relaxation, PTriMO (Willbourn, 1958)

14.3a   The α Relaxation

The mechanical α peak shown in Figure 14.17 is far larger than the β peak. The glass transition of PTetraMO occurs at $-84°$C, and inspection of Figure 14.19 shows that the α relaxation occurs in this temperature region at very low frequencies. It is therefore considered that the mechanical α relaxation is due to the micro-Brownian motions of the chain in the amorphous phase of the polymer.

The dielectric α relaxation in PTetraMO overlapped with the β relaxation for the larger part of the temperature range studied by Wetton and Williams (1965). However, it was found that the loss curves had the same shape and location for an amorphous and a 33% crystalline sample though the loss curves in the amorphous sample were 30% larger than curves obtained for the 33% crystalline sample. This behaviour is consistent with both the α and β mechanisms occurring in the amorphous phase of the polymer. In view of the correlation in location of the dielectric and mechanical α relaxations shown in Figure 14.19, the dielectric α process is associated with the micro-Brownian motions of the chain in the non-crystalline regions of the polymer.

Wetton and Williams (1965) evaluated the effective mean square dipole moment of the PTetraMO chain using static dielectric constant data at 20°C for an amorphous specimen. The theory of Read (1965) (see Equation 3.114), which relates the mean square moment of PTetraMO to the conformation of the chain, was applied to the experimental data. The results indicated that the *trans* conformation is more favoured than the *gauche*, the energy difference being from 0·8 to 1·0 kcal/mole.

14.3b   The β Relaxation

The mechanical β process occurs in the same temperature region for PTriMO and PTetraMO, but a decrease in magnitude occurs on going from PTetraMO to PTriMO (see Figure 14.17). Willbourn (1958) suggested that if a polymer contained the $(CH_2)_x$ sequence, a process corresponding to the motions of this sequence would be observed near $-110°$C ($f \simeq 10^2$ c/s). This suggestion was substantiated by Willbourn for a number of polymers containing this sequence. Since PTriMO and PTetraMO give the β process at $-110°$C, this process may be associated with the motion of the $(CH_2)_x$ sequence. The actual mechanism is not established however, and it is not clear if the mechanical β relaxation occurs in the crystalline or non-crystalline phase of the polymer.

As indicated above, the dielectric β relaxation appears to occur in the non-crystalline phase of PTetraMO. This was concluded from the magnitude increase in the overall (α, β) relaxation on going from an amorphous to a crystalline sample (Wetton and Williams, 1965). It was considered that the dielectric β process could not be due to the independent motions of $(CH_2)_n$ units, since in order to be dielectrically active, motion of the ether group must occur. These authors suggested that the β process might be due to the local motions of the chains in accord with the theory of Yamafuji and Ishida (1962) (Section 5.5b). The magnitude $(\varepsilon_R - \varepsilon_U)$ for the β process was calculated according to this theory and was in reasonable agreement with the experimental value.

The frequency–temperature locations of the dielectric and mechanical β relaxations are in good agreement, and suggest that the mechanisms of relaxation are similar. The plots are approximately linear, with an apparent activation energy of 9·7 kcal/mole. The γ relaxation in PEO of high molecular weight (see Figure 14.15) has an apparent activation energy of 9 kcal/mole. Comparison of Figures 14.15 and 14.19 shows that the γ relaxation of PEO occurs near the β relaxation in PTetraMO, and suggests that the two processes have similar mechanisms. Ishida and others (1965) suggested that the γ relaxation in PEO was due to the local motions of the chain in *both* the crystalline and non-crystalline regions of the polymer. The data for PTetraMO indicate a similar motion, but in this case the motions are probably restricted to the non-crystalline phase.

14.4 POLYACETALDEHYDE

$$[-\overset{\displaystyle H}{\underset{\displaystyle CH_3}{C}}-O-]_n$$

(5)

Polyacetaldehyde (PAc) may be obtained in the partially crystalline or amorphous state, depending upon the method used for polymerization (Weissermel and Schmieder, 1962). The polymer is unstable and decomposes at a rapid rate if heated above 30°C. The mechanical relaxation of amorphous (atactic) PAc has been studied by Read (1962b, 1964b) and the dielectric relaxation of a similar sample has been studied by Williams (1963a).

Figure 14.20 shows the mechanical data of Read (1962b). One relaxation region is observed, the peak in Λ being observed near $-20°C$ (0·58 c/s). The dilatometric glass-transition temperature occurs at $-30·4°C$ (Williams, 1963a), and it is clear that the mechanical relaxation is associated with the glass transition, and is due to the micro-Brownian motions

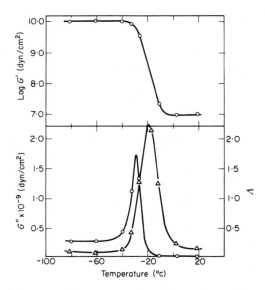

Figure 14.20. G', $\Lambda(\triangle)$ and $G''(\bigcirc)$ as a function of temperature for polyacetaldehyde (Read, 1962b).

of the chain. Figure 14.21 shows the dielectric data of Williams (1963a), and again one relaxation region is observed. A feature of the results is the large losses observed in this polymer, and this will be discussed below.

Figure 14.22 shows the frequency–temperature location of the mechanical and dielectric relaxations. The mechanical data are presented in terms of E'', $\tan \delta_E$ and D'' maxima. The $\tan \delta_E$ data agree in position with the dielectric loss data. The curvature in the dielectric plot corresponds to WLF behaviour, and is further evidence that the relaxation process is due to the micro-Brownian motions of the chain. This relaxation process will be considered under the heading 'the α relaxation'.

14.4a   The α Relaxation

Figure 14.23 shows the master plot for E' and E'' as a function of reduced frequency at a reference temperature of $0°\mathrm{c}$ (Read, 1964b). The process clearly has the characteristics of a glass–rubber relaxation. Figure 14.24 shows the dielectric master plots for

$$J_{\omega} = \left(\frac{\varepsilon'_{\omega} - \varepsilon_{\mathrm{U}}}{\varepsilon_{\mathrm{R}} - \varepsilon_{\mathrm{U}}}\right) \text{ and } \frac{H_{\omega}}{H_{\omega_{\max}}} = \varepsilon''_{\omega}/\varepsilon''_{\max}$$

against $\log \omega/\omega_{\text{max}}$ (Williams, 1963a). It was found that the Kirkwood–Fuoss molecular theory of relaxation (Kirkwood and Fuoss, 1941; Section 5.3b) fitted the data of Figure 14.24. Also the empirical relations of Cole and Cole (1941) and Fuoss and Kirkwood (1941) (see Chapter 4) fitted the data of Figure 14.24 for suitably chosen distribution parameters. However, the Fröhlich barrier theory (1949) (Section 5.4a) did not give a satisfactory fit to the data.

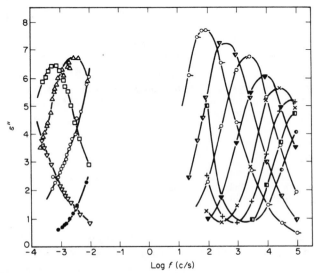

Figure 14.21. ε'' against $\log f$ for polyacetaldehyde (Williams, 1963a).
◐ 34·8°c ◨ 30·5°c + 25·0°c × 18·5°c ▼ 9·7°c ◑ 3·25°c
▼ −3·5°c ◯− −9·0°c ● −19·2°c ◯ −21·8°c △ −24·5°c □ −26·4°c
▽ −28·7°c

Williams (1963a) analysed the static dielectric constant data for PAc above $-10°$c in terms of the mean square dipole moment $\overline{P^2}$ of the chain. Using the theory of Read (see Williams, 1963a) which related $\overline{P^2}$ to the structure of the chain, it was shown that the experimental dielectric constant data above $-10°$c was consistent with the chain having *trans* conformations preferred over *gauche* conformations (Figure 2.3b) by an amount $u = 1·0 \pm 0·1$ kcal/mole.

The static dielectric constant increased with decreasing temperature down to $-15°$c. Below $-15°$c the dielectric constant decreased with decreasing temperature. This behaviour is consistent with a 'dielectric

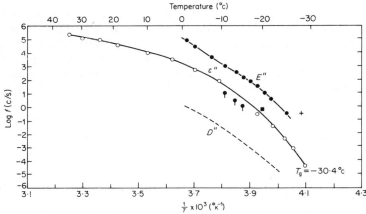

Figure 14.22. Log f against $1/T°$K for polyacetaldehyde. In the following key, (f) represents the frequency of maximum loss and (T) represents the temperature of maximum loss.

Dielectric

○ ε'' (f) data (Williams, 1963a)

Mechanical

● E'' (f) + G'' (T) (Read, 1962b)
– – – –Young's compliance data   ■ tan δ (T) (Read, 1962b)
● tan δ_E (f) (Read, 1964b)

transition' as discussed by Saito (1964) and in Chapter 13 above for PET. Williams (1963a) used the Gibbs and DiMarzio theory (1958) (Section 2.3a) for second-order transitions in linear polymers, and calculated the second-order transition temperature T_2 using the u value given above together with thermal expansion data for PAc. T_2 was calculated to be $-33°$C, and this is in the temperature region where ε_R is rapidly decreasing with decreasing temperature for PAc.

The static dielectric constant ε_R for PAc is very large, being near 30 at $-10°$C. This is a consequence of the large dipole concentration along the chain, and the preferred *trans* conformation of the chain.

14.5 POLYPROPYLENE OXIDE

$$[-\underset{\underset{CH_3}{|}}{\overset{\overset{H}{|}}{C}}-\underset{\underset{H}{|}}{\overset{\overset{H}{|}}{C}}-O-]_n$$

(6)

Stereoregular polypropylene oxide is thought to crystallize with an

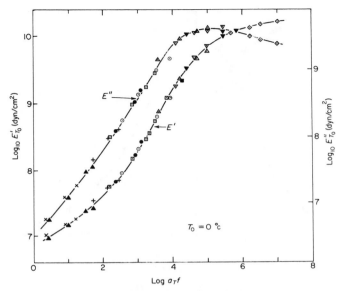

Figure 14.23. Master plots of E' and E'' as a function of reduced frequency for polyacetaldehyde (Read, 1964b).

Reduction temperature $T_0 = 0°$c
◇ $-25.0°$c ▼ $-21.5°$c ▽ $-20.0°$c △ $-18.7°$c ■ $-16.6°$c
⊙ $-15.0°$c ▣ $-13.1°$c ● $-10.8°$c + $-7.2°$c ▲ $-3.0°$c
× $0.0°$c

orthorhombic unit cell having two planar zig-zag chain units which lie parallel to the c axis (Stanley and Litt, 1960). Bulk high molecular weight polypropylene oxide (PPO) is normally partially crystalline, but careful fractionation procedures have been used to separate samples into their amorphous and crystalline components (Allen, Booth and Jones, 1964). The melting point of PPO occurs at 72–74°c (Price, Osgan, Hughes and Shambelan, 1956; Price and Osgan, 1956) and the dilatometric glass-transition temperature has been given as $-78°$c (St. Pierre and Price, 1956), $-73°$c (Work, McCammon and Saba, 1963) and $-75°$c (Allen, 1963).

The mechanical relaxation of PPO has been studied by Read (1962b), Wetton (1962), Saba, Sauer and Woodward (1963) and Crissman, Sauer and Woodward (1964). The dielectric relaxation of PPO was studied by Work, McCammon and Saba (1963) and by Williams (1965).

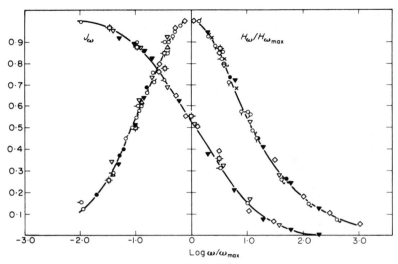

Figure 14.24.

$$J_\omega = \left(\frac{\varepsilon'_\omega - \varepsilon_U}{\varepsilon_R - \varepsilon_U}\right) \quad \text{and} \quad \frac{H_\omega}{H_{\omega_{max}}} = \frac{\varepsilon''}{\varepsilon''_{max}}$$

against $\log \omega/\omega_{max}$ for polyacetaldehyde (Williams, 1963a).

Sample 1

⊞ 18·5°c   ⦵ 9·7°c   ▽ 3·2°c   ▼ −3·5°c   ◇ −9·0°c   ◔ −24·5°c   ◯ −26·4°c

Sample 2

△ 18·7°c   ◯ 11·5°c   ● 4·5°c   ◷ −3·5°c   × −26·2°c

Figure 14.25 shows the results of Saba, Sauer and Woodward (1963) for PPO. The loss peak at 211°K is clearly associated with the glass–rubber relaxation of the polymer and the rising loss above 300°K is associated with the onset of the melting of the polymer. Therefore the 211°K peak is labelled α and the small peak at 110°K is labelled β. Crissman, Sauer and Woodward (1964) have detected two further low-temperature processes which we shall label γ and δ, respectively, in decreasing order of temperature.

Figure 14.26 shows the dielectric results of Williams (1965) for an amorphous PPO sample. One loss region is observed and corresponds to the α relaxation observed mechanically in Figure 14.25.

Figure 14.27 shows the collected frequency–temperature locations of the dielectric and mechanical relaxation processes. The dielectric and mechanical α and β relaxations agree in their locations, indicating a

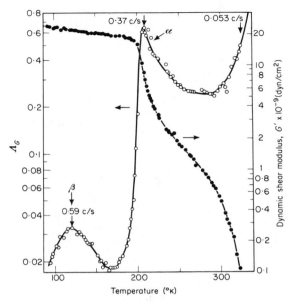

Figure 14.25. G' and Λ_G as a function of temperature for PPO (Saba, Sauer and Woodward, 1963).

correlation of these processes between the two techniques. The α relaxation occurs in the region of 10^{-3} c/s at the dilatometric glass-transition temperature.

14.5a   The α Relaxation

Read (1962b) found that the mechanical Λ peak was more pronounced in an amorphous than in a 39% crystalline specimen, but the temperature location of Λ_{max} was unaffected by this change in crystallinity. A similar result was obtained by Wetton (1962) for samples of 21 and 54% crystallinity. This behaviour is consistent with the α relaxation occurring in the amorphous phase of the polymer. The large modulus drop associated with the α process (Figure 14.25) and the proximity of the dilatometric glass-transition temperature ($\simeq -75°$C) confirms that the α process is related to the glass transition. Saba, Sauer and Woodward (1963) noted that the α relaxation in polypropylene occurs at $0°$C (Λ_{max}, 0.025 c/s) which is considerably higher than that obtained for PPO at a comparable frequency. This would imply that the introduction of a —O— unit

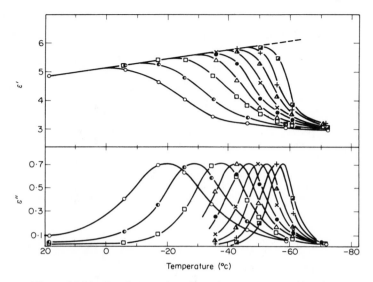

Figure 14.26. ε' and ε'' as a function of temperature for PPO (Williams, 1965).

◯ $3\cdot3 \times 10^6$ c/s	◖ $6\cdot3 \times 10^5$ c/s	☐ $1\cdot0 \times 10^5$ c/s
△ $3\cdot0 \times 10^4$ c/s	● $1\cdot0 \times 10^4$ c/s	× $3\cdot0 \times 10^3$ c/s
▲ $1\cdot0 \times 10^3$ c/s	+ $3\cdot0 \times 10^2$ c/s	◪ $1\cdot0 \times 10^2$ c/s

into the chain leads to an overall increase in the mobility of the chain. It seems probable that this effect is determined largely by the increased separation of the —CH_3 substituents along the chain.

Work and others (1963) and Williams (1965) studied the dielectric α relaxation of PPO over a range of frequency and temperature. Williams (1965) also studied the α relaxation as a function of hydrostatic pressure up to 3000 atm. It was found that the shape of the α relaxation (in frequency plots) was unaffected by variations in temperature or pressure. This shows that the broad distribution associated with the micro-Brownian motions of the chain arises due to the cooperative motions of the units along the chain, where the mobility coefficient of each unit has the same dependence on temperature and pressure.

Williams (1965) calculated the effective mean square moment per repeat unit $\overline{P^2}/n$ from the static dielectric constant data included in Figure 14.26. $\overline{P^2}/n$ was related to the conformation parameters of the chain using the relation (Equation 3.112) derived initially by Marchal and Benoit (1955). This relation assumes that the potential contours for rotation

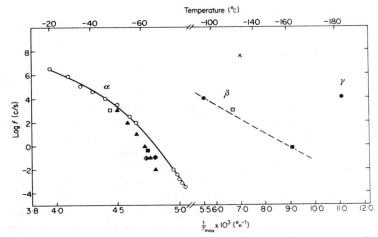

Figure 14.27. Log f against $(1/T_{max}$°K) for PPO. Here, T_{max} corresponds to the temperature of maximum loss in a constant frequency experiment.

Dielectric

○ ε'' data (Williams, 1965)   □ tan δ data (Work, McCammon and Saba, 1963)

Mechanical

■ tan δ_G data (Saba, Sauer and Woodward, 1963)   ▲ G'' data (Wetton, 1962)

● tan δ_E data (Crissman, Sauer and Woodward, 1964)   ◆ G'' data (Read, 1962b)

◈ tan δ_G data (Read, 1962b)

Point marked × corresponds to the temperature of minimum nuclear spin lattice relaxation time (T_1) (Connor, Blears and Allen, 1965).

around C—C and C—O bonds are the same. In this way, it was found that the experimental data were consistent with the chain having the *trans* conformation preferred over the *gauche* conformation (Figure 2.3b) by an energy $u \simeq 600$ cal/mole.

14.5b The β Relaxation

In Figure 14.25 the mechanical β process is seen as a minor feature compared with the α process. Work, McCammon and Saba (1963) found that the dielectric β process was also considerably smaller than the dielectric α process. Inspection of Figure 14.27 reveals a correlation in the locations of the β processes detected by the two techniques. The mechanism for this process is not well established. Saba, Sauer and Woodward (1963) suggest that it might be associated with the cooperative motion of possibly three or four units of the chain in the amorphous regions of the polymer. A similar process occurs in polypropylene near $-63°$C (0.32 c/s) and these authors suggest that the introduction of —O— into the chain lowers the temperature of the process.

14.5c The γ Relaxation

Crissman, Sauer and Woodward (1964) found a very small mechanical process at $-183°$C (tan δ_E, 10^4 c/s). They suggest that this process is associated with the reorientation of methyl groups in the polymer. This suggestion is based on their unpublished n.m.r. data for PPO, and is possibly substantiated by recent spin echo n.m.r. measurements of Connor, Blears and Allen (1965). The latter authors observed two n.m.r. relaxation regions in high molecular weight PPO. The higher-temperature process is to be correlated with the α relaxation described above. The lower-temperature process occurs in the region -110 to $-130°$C (3×10^7 c/s), the actual location depending upon the degree of crystallinity of the polymer. It was shown by deuteration of the (CH_3) group on the chain that the low-temperature n.m.r. process is associated with the rotation of the methyl groups in PPO. Inspection of Figure 14.27 shows the β relaxation is unlikely to be associated with the lower-temperature n.m.r. process, whereas the γ relaxation region may possibly be associated with this process. This interpretation must, however, be regarded as tentative.

14.5d The δ Relaxation

Crissman, Sauer and Woodward (1964) have reported mechanical data for PPO down to $6°$K. There is evidence of a very small loss peak near $20°$K (10^4 c/s).

REFERENCES

Adam, G. and F. H. Müller (1962). *Z. Elektrochem.*, **66**, 844.
Alexandrov, A. P. and J. S. Lazurkin (1940). *Acta. Phys. Chem. USSR*, **12**, 647.
Alfrey, T. (1944). *J. Chem. Phys.*, **12**, 374.
Alfrey, T. (1948). In *Mechanical Behavior of High Polymers*, Interscience, New York.
Alfrey, Jnr., T., N. Wiederhorn, R. Stein and A. Tobolsky (1949). *J. Colloid. Sci.*, **4**, 211.
Allen, G. (1963). *Soc. Chem. Ind. Monograph*, **17**, 167.
Allen, G., U. Bianchi and C. Price (1963). *Trans. Faraday Soc.*, **59**, 2493.
Allen, G. (1965). *Private Commun.*
Allen, G., C. Booth and M. N. Jones (1964). *Polymer (London)*, **5**, 195, 257.
Alsup, R. G., J. O. Punderson and G. F. Leverett (1959). *J. Appl. Polymer Sci.*, **1**, 185.
Andrade, E. N. da C. and B. Chalmers (1932). *Proc. Roy. Soc.* A, **138**, 348.
Angelo, R. J., R. M. Ikeda and M. L. Wallach (1965). *Polymer (London)*, **6**, 141.
Anthony, R. L., R. H. Gaston and E. Guth (1942). *J. Phys. Chem.*, **46**, 826.
Arisawa, K., K. Tsuge and Y. Wada (1963). *Rept. Progr. Polymer Phys.*, *(Japan)*, **VI**, 151.
Asahina, M. and S. Enomoto (1962). *J. Polymer Sci.*, **59**, 101.
Baccaredda, M. and E. Butta (1958). *Chim. Ind.*, *(Milan)*, **40**, 6.
Baccaredda, M. and E. Butta (1958). *J. Polymer Sci.*, **31**, 189.
Baccaredda, M. and E. Butta (1960). *J. Polymer Sci.*, **44**, 421.
Baccaredda, M., E. Butta and V. Frosini (1963). *Makromol. Chem.*, **61**, 14.
Baccaredda, M., E. Butta and V. Frosini (1965). *J. Polymer Sci.*, Pt. B, **3**, 185.
Baer, M. (1964). *J. Polymer Sci.*, **42**, 417.
Baker, E. B. (1949). *Rev. Sci. Instr.*, **20**, 716.
Baker, E. B., R. P. Auty and G. J. Ritenour (1953). *J. Chem. Phys.*, **21**, 159.
Baker, W. O. and W. A. Yager (1942). *J. Am. Chem. Soc.*, **64**, 2171.
Barb, W. G. (1959). *J. Polymer Sci.*, **37**, 515.
Bares, J. and J. Janacek (1965). *Collection Czech. Chem. Commun.*, **30**, 1604.
Bares, J., J. Janacek and P. Cefelin (1965). *Collection Czech. Chem. Commun.*, **30**, 2582.
Barriault, R. J. and L. F. Gronholz (1955). *J. Polymer Sci.*, **18**, 393.
Barton, D. H. R. and R. C. Cookson (1956). *Quart. Rev.*, **10**, No. 1, 44.
Baumann, V., H. Schreiber and K. Tessmar (1959). *Makromol. Chemie.*, **36**, 81.
Baur, M. and W. H. Stockmayer (1959). *Meeting Am. Chem. Soc.*, Boston, USA, April.
Bawn, C. E. H., A. Ledwith and P. Matthies (1959). *J. Polymer Sci.*, **34**, 93.
Beaman, R. G. (1953). *J. Polymer Sci.*, **9**, 472.
Berestneva, G. L., L. L. Burshtein, P. V. Kozlov, G. P. Mikhailov and K. E. Nordbeck (1960). *High Mol. Wt. Compds.*, **2**, 1739.
Becker, G. W. (1955). *Kolloid Z.*, **140**, 1.
Becker, G. W. and H. Oberst (1957). *Kolloid Z.*, **152**, 1.
Beckett, C. W., K. S. Pitzer and R. Spitzer (1947). *J. Am. Chem. Soc.*, **69**, 2488.
Beevers, R. B. (1962). *Trans. Faraday Soc.*, **58**, 1465.
Beevers, R. B. and E. F. T. White (1960). *Trans. Faraday Soc.*, **56**, 1529.
Benoit, H. (1947). *J. Chim. Phys.*, **44**, 18.

Bevington, J. C., H. W. Melville and R. P. Taylor (1954). *J. Polymer Sci.*, 12, 449.
Beyer, R. T. (1957). *J. Acoust. Soc. (America)*, 29, 243.
Biggs, B. S. (1953). *N.B.S. Circular*, 525, 137.
Billmeyer, F. W. (1962). *Textbook of Polymer Science*, Wiley, New York.
Bloomfield G. F. (1947). *J. Chem. Soc.*, 1947, 1547.
Bloomfield, G. F. (1949). *Soc. Chem. Ind.*, 68, 66.
Bohn, L. (1965). *Kolloid Z.*, 201, 20.
Boltzmann, L. (1876). *Pogg. Ann. Physik*, 7, 108.
Booij, H. C. (1965). *Personal Commun.*
Bordoni, P. G. (1954). *J. Acoust. Soc. Am.*, 26, 495.
Borisova, T. I., L. L. Burshtein and G. P. Mikhailov (1962). *Vysokomolekul. Soyedin.*, 4, 1479.
Borisova, T. I. and G. P. Mikhailov (1959). *Vysokomolekul. Soyedin.*, 1, 563.
Bos, F. F., Thesis, Leiden, 1958.
Böttcher, C. J. F. (1952). *Theory of Electric Polarization*, Elsevier, Amsterdam.
Bovey, F. A. (1958). *The Effects of Ionizing Radiation on Natural and Synthetic High Polymers*, Interscience, New York.
Bovey, F. A. and G. V. D. Tiers (1960). *J. Polymer Sci.*, 44, 173.
Bovey, F. A., E. W. Anderson and D. C. Douglass (1963). *J. Chem. Phys.*, 39, 1199.
Boyd, R. H. (1959). *J. Chem. Phys.*, 30, 1276.
Boyer, R. F. and R. S. Spencer (1944). *J. Appl. Phys.*, 15, 398.
Boyer, R. F. (1952). 'Changements de Phases' Soc. de Chimie Physik, Paris 1952.
Boyer, R. F. (1963). *Rubber Rev.*, 34, 1303.
Brandt, W. (1957). *J. Chem. Phys.*, 26, 262.
Brinkman, H. C. (1956). *Physica, 's Grav.*, 22, 29, 149.
Broadhurst, M. G. (1960). *J. Chem. Phys.*, 33, 221.
Broens, O. and F. H. Müller (1955). *Kolloid Z.*, 140, 121; 141, 20.
Brouckere, L. de and M. Mandel (1958). *Advan. Chem. Phys.*, 1, 77.
Brouckere, L. and G. Offergeld (1958). *J. Polymer Sci.*, 30, 105.
Bryant, W. M. D. and R. C. Voter (1953). *J. Am. Chem. Soc.*, 75, 6113.
Buchdahl, R. and L. E. Nielsen (1955). *J. Polymer Sci.*, 15, 1.
Buchdahl, R. (1958). In F. R. Eirich (Ed.), *Rheology*, Vol. 2, Academic Press, New York.
Buckley, F. and A. A. Maryott (1958). Tables of Dielectric Dispersion Data for Pure Liquids and Dilute Solutions, *N.B.S. Circular* 589, November 1st (1958).
Buckely, G. D. and N. H. Ray (1952). *J. Chem. Soc.*, Pt. 3, 3701.
Bueche, F. (1952). *J. Chem. Phys.*, 20, 1959.
Bueche, F. (1954). *J. Chem. Phys.*, 22, 603.
Bueche, F. (1955). *J. Appl. Phys.*, 26, 738.
Bueche, F. (1956). *J. Polymer Sci.*, 22, 113.
Bueche, F. (1956). *J. Chem. Phys.*, 24, 418.
Bueche, F. (1959). *J. Chem. Phys.*, 30, 748.
Bueche, F. (1961). *J. Polymer Sci.*, 54, 597.
Bunn, C. W. (1939). *Trans. Faraday Soc.*, 35, 482.
Bunn, C. W. and T. C. Alcock (1945). *Trans. Faraday Soc.*, 41, 317.
Bunn, C. W. (1946). *Chemical Crystallography*, Clarendon Press, Oxford.
Bunn, C. W. and E. V. Garner (1947). *Proc. Roy. Soc.*, A189, 39.
Bunn, C. W. (1948). *Nature*, 161, 929.

Bunn, C. W. and E. R. Howells (1954). *Nature*, **174**, 549.
Bunn, C. W. (1957). In A. Renfrew and P. Morgan (Eds.), *Polythene*, Iliffe, London.
Bunn, C. W. (1960). In A. Renfrew and P. Morgan (Eds.), *Polythene*, Chapt. 5, Iliffe, London.
Bunn, C. W., A. J. Cobbald and R. P. Palmer (1958). *J. Polymer Sci.*, **28**, 365.
Bunn, C. W. and D. R. Holmes (1958). *Discussions Faraday Soc.*, **25**, 95.
Bunn, C. W. (1959). *Soc. Chem. Ind. Monograph* No. **5**, 3.
Campbell, T. W. and A. C. Haven (1959). *J. Appl. Polymer Sci.*, **1**, 73.
Carothers, W. H. (1929). *J. Am. Chem. Soc.*, **51**, 2548.
Carter, W. C., M. Magat, W. G. Schmeider and C. P. Smyth (1946). *Trans. Faraday Soc.*, **42A**, 213.
Catsiff, E. and A. V. Tobolsky (1954). *J. Appl. Phys.*, **25**, 1092.
Catsiff, E. and A. V. Tobolsky (1955). *J. Colloid Sci.*, **10**, 375.
Catsiff, E., J. Offenbach and A. V. Tobolsky (1956). *J. Colloid Sci.*, **11**, 48.
Cerf, R. (1955). *Compt. Rend. Acad. Sci.*, **241**, 496.
Cerf, R. (1959). *Advan. Polymer Sci.*, **1**, 382.
Chasset, R., P. Thirion and V. B. Nqyuen (1959). *P. Rev. Gen. Caoutechouc.*, **36**, 857.
Child, W. C. and J. D. Ferry (1957a). *J. Colloid Sci.*, **12**, 327.
Child, W. C. and J. D. Ferry (1957b). *J. Colloid Sci.*, **12**, 389.
Christopher, W. F. and D. W. Fox (1962). *Polycarbonates*, Reinhold, New York.
Chujo, R., S. Satoh, T. Ozeki and E. Nagai (1962). *Rept. Prog. Polymer Phys.*, Japan, **5**, 248.
Ciferri, A. (1961). *J. Polymer Sci.*, **54**, 149.
Clark, E. S. and L. T. Muus (1957). *Meeting Am. Chem. Soc.*, New York, September.
Clark, E. S. and L. T. Muus (1958). *Meeting Am. Cryst. Assoc.*, Milwaukee, Wisconsin, June.
Clark, E. S. (1959). *Symp. on Helices in Macromolecular Systems, Polytech. Inst.*, Brooklyn, May.
Clark, K. J., A. Turner Jones and D. H. Sandiford (1962). *Chem. Ind.*, 2010.
Claver, G. C., R. Buchdahl and R. L. Miller (1956). *J. Polymer Sci.*, **20**, 202.
Cohen, M. H. and D. Turnbull (1959). *J. Chem. Phys.*, **31**, 1164.
Cole, E. A. and D. R. Holmes (1960). *J. Polymer Sci.*, **46**, 245.
Cole, R. H. and K. S. Cole (1941). *J. Chem. Phys.*, **9**, 341.
Cole, R. H. and K. S. Cole (1942). *J. Chem. Phys.*, **10**, 98.
Cole, R. H. and P. M. Gross (1949). *Rev. Sci. Instr.*, **20**, 252.
Cole, R. H. (1955). *J. Chem. Phys.*, **23**, 493.
Coleman, B. D. (1958). *J. Polymer Sci.*, **31**, 155.
Connor, T. M., B. E. Read and G. Williams (1964). *J. Appl. Chem.*, **14**, 74.
Connor, T. M., D. J. Blears and G. Allen (1965). *Trans. Faraday Soc.*, **61**, 1097.
Cooper, W., F. R. Johnston and G. Vaughan (1963). *J. Polymer Sci.*, **1**, 1509.
Cox, W. P., R. Isaksen and E. H. Merz (1960). *J. Polymer Sci.*, **44**, 149.
Crissman, J. M., J. A. Sauer and A. E. Woodward (1964). *J. Polymer Sci.*, **2A**, 5075.
Curie, J. (1888, 1889). *Ann. Chem. Phys.*, [**6**], **17**, 385; **18**, 203.
Curtis, A. J. (1960). *Progr. Dielectrics*, **2**, 31.
Curtis, A. J. (1961). *J. Res. Nat. Bur. Std.*, **65A**, 185.
Curtis, A. J. (1962). *Soc. Plastics Engrs. Trans.*, **18**, 82.
Daal, W. and F. H. Müller (1961). *Z. Elektrochem.*, **65**, 652.

Dakin, T. W. and C. N. Works (1947). *J. Appl. Phys.*, **18**, 789.
Dannhauser, W., W. C. Child and J. D. Ferry (1958). *J. Colloid Sci.*, **13**, 103.
Dannis, M. L. (1959). *J. Appl. Polymer Sci.*, **1**, 121.
Danusso, F., G. Moraglio and G. Talamini (1956). *J. Polymer Sci.*, **21**, 139.
Daubeny, R. de P., C. W. Bunn and C. J. Brown (1954). *Proc. Roy. Soc.*, **A 226**, 531.
Dauscher, R., E. W. Fischer and H. A. Stuart (1960). *Z. Naturforsch.*, **15a**, 116.
Davidson, D. W., R. P. Auty and R. H. Cole (1951). *Rev. Sci. Instr.*, **22**, 678.
Davidson, D. W. and R. H. Cole (1951). *J. Chem. Phys.*, **19**, 1484.
Davidson, D. W. and R. H. Cole (1950). *J. Chem. Phys.*, **18**, 1417.
Davies, J. M., R. F. Miller and W. F. Busse (1941). *J. Am. Chem. Soc.*, **63**, 361.
Davies, R. O. and J. Lamb (1957). *Quart. Rev.*, **11**, 134.
Debye, P. (1929). *Polar Molecules, Chem. Catalog*, New York.
Debye, P. (1945). *Polar Molecules*, Dover Publications, New York.
Debye, P. and F. Bueche (1951). *J. Chem. Phys.*, **19**, 589.
Deeley, C. W., A. E. Woodward and J. A. Sauer (1957). *J. Appl. Phys.*, **28**, 1124.
Desanges, H., R. Chasset and P. Thirion (1957). *Rev. Gen. Caoutechouc.*, **34**, 893.
Desanges, H., R. Chasset and P. Thirion (1958). *Rubber Chem. Technol.*, **31**, 631.
Deutsch, K., E. A. W. Hoff and W. Reddish (1954). *J. Polymer Sci.*, **13**, 565.
De Vos, F. C. (1958). *Thesis*, Leiden.
Di Marzio, E. A. and J. H. Gibbs (1959). *J. Polymer Sci.*, **40**, 121.
Dogadkin, B. A. and A. I. Lukomskaya (1953). *Kolloid Z.*, **15**, 3183.
Dole, M. and others (1952). *J. Chem. Phys.*, **20**, 781.
Doolittle, A. K. (1951). *J. Appl. Phys.*, **22**, 1471.
Doolittle, A. K. (1952). *J. Appl. Phys.*, **23**, 236.
Dulmage, W. J. and L. E. Contois (1958). *J. Polymer Sci.*, **28**, 275.
Dunn, C. M. R., W. M. Mills and S. Turner (1964). *Brit. Plastics*, July, 2.
Dyson, A. (1951). *J. Polymer Sci.*, **7**, 133.
Eastham, A. M. (1960). *Advan. Polymer Sci.*, **2**, 18.
Eby, R. K. (1963). *J. Chem. Phys.*, **37**, 2785.
Elliott, D. R. and S. A. Lippmann (1945). *J. Appl. Phys.*, **16**, 50.
Eppe, R., E. W. Fischer and H. A. Stuart (1959). *J. Polymer Sci.*, **34**, 721.
Erlikh, Y. M. and P. N. Shcherbak (1955). *Zh. Tekhn. Fiz.*, **25**, 1575.
Eyring, H. (1932). *Phys. Rev.*, **39**, 746.
Farmer, E. H. and F. W. Shipley (1947). *J. Chem. Soc.*, Pt. 2, 1519.
Farrow, G., J. McIntosh and I. M. Ward (1960). *Makromol. Chem.*, **38**, 147.
Farrow, G., J. McIntosh and I. M. Ward (1959). *Symp. Macromolecules*, Wiesbaden 1959, Verlag Chemie, Weinheim. Section I.
Fattakov, K. Z. (1952). *Zh. Tekhn. Fiz.*, **22**, 313.
Faucher, J. A. (1959). *Trans. Soc. Rheology*, **3**, 81.
Ferry, J. D. and G. S. Parks (1936). *J. Chem. Phys.*, **4**, 70.
Ferry, J. D. and E. R. Fitzgerald (1953). *Proc. 2nd. Intern. Congr. Rheol.*, 140.
Ferry, J. D., M. L. Williams and E. R. Fitzgerald (1955). *J. Phys. Chem.*, **59**, 403.
Ferry, J. D. (1956). In H. A. Stuart (Ed.), *Die Physik der Hochpolyneren*, Springer-Verlag, Berlin, Vol. **4**, p. 96.
Ferry, J. D. and others (1957). *J. Colloid Sci.*, **12**, 53.
Ferry, J. D. and S. Strella (1958). *J. Colloid Sci.*, **13**, 459.
Ferry, J. D. (1961). *Viscoelastic Properties of Polymers*, Wiley, New York.
Ferry, J. D., R. G. Mancke, E. Maekawa, Y. Oyanagi and R. A. Dickie (1964). *J. Phys. Chem.*, **68**, 3414.

Findley, W. M. (1962). *Trans. J. Plastics Inst.*, **30**, 138.
Fischer, E. W. (1957). *Z. Naturforsch.*, **12a**, 753.
Fischer, E. W. and G. F. Schmidt (1962). *Angew. Chem.*, **74**, 551.
Fischer, E. W. (1966). *Private Commun.*
Fitzgerald, E. R. and R. F. Miller (1953). *J. Colloid Sci.*, **8**, 148.
Fitzgerald, E. R. and J. D. Ferry (1953). *J. Colloid Sci.*, **8**, 1.
Fitzgerald, E. R., L. D. Grandine and J. D. Ferry (1953). *J. Appl. Phys.*, **24**, 650, 911.
Fitzhugh, A. F. and R. N. Crozier (1952). *J. Polymer Sci.*, **8**, 225.
Foerster, F. (1937). *Z. Metallk.*, **29**, 109.
Foerster, F. (1955). *Z. Metallk.*, **46**, 297.
Flocke, H. A. (1962). *Kolloid Z.*, **180**, 118.
Flocke, H. A. (1963). *Kolloid Z.*, **188**, 114.
Flory, P. J. (1944). *Chem. Rev.*, **35**, 51.
Flory, P. J. and F. S. Leutner (1948). *J. Polymer Sci.*, **3**, 880.
Flory, P. J. (1953). *Principles of Polymer Chemistry*, Cornell University Press, New York.
Flory, P. J., C. A. J. Hoeve and A. Ciferri (1959). *J. Polymer Sci.*, **34**, 337.
Flory, P. J., A. Ciferri and C. A. J. Hoeve (1960). *J. Polymer Sci.*, **45**, 235.
Flory, P. J. (1961). *Trans. Faraday Soc.*, **57**, 829.
Flory, P. J., V. Crescenzi and J. E. Mark (1964). *J. Am. Chem. Soc.*, **86**, 146.
Flory, P. J. and J. E. Mark (1964). *Makromol. Chem.*, **75**, 11.
Flory, P. J. and R. L. Jernigan (1965). *J. Chem. Phys.*, **42**, 3509.
Fordham, J. W. L., G. H. McCain and L. E. Alexander (1959). *J. Polymer Sci.*, **39**, 335.
Fordham, J. W. L., P. H. Burleigh and C. L. Sturm (1959). *J. Polymer Sci.*, **41**, 73.
Foster, F. C. and J. R. Binder (1957). *Advan. Chem. Ser.*, No. **19**, 26.
Fowler, R. and E. A. Guggenheim (1956). *Statistical Thermodynamics*, Cambridge University Press.
Fox, T. G. and P. J. Flory (1950). *J. Appl. Phys.*, **21**, 581.
Fox, T. G. and P. J. Flory (1954). *J. Polymer Sci.*, **14**, 315.
Fox, T. G., S. Gratch and S. Loshaek (1956). In F. Eirich (Ed.), *Rheology*, Chapt. 12, Academic Press, New York.
Fox, T. G. (1956). *Bull. Am. Phys. Soc.*, **1**, No. 3, 123.
Fox, T. G. and others (1958). *J. Am. Chem. Soc.*, **80**, 1768.
Fox, T. G. and H. W. Schnecko (1962). *Polymer* **3**, 575.
Frank, F. C., A. Keller and A. O'Connor (1959). *Phil. Mag.*, **4**, 200.
Frenkel, J. (1946). *Kinetic Theory of Liquids*, Oxford University Press, Oxford.
Fröhlich, H. (1942). *Proc. Phys. Soc.*, **54**, 422.
Fröhlich, H. (1949). *Theory of Dielectrics*, Oxford University Press, Oxford.
Fröhlich, H. (1958). *Theory of Dielectrics*, 2nd ed., Oxford University Press, Oxford.
Fujino, K., H. Kawai, T. Horino and K. Miyamoto (1956). *Textile Res. J.*, **26**, 852.
Fujino, K., T. Horino, K. Miyamoto and H. Kawai (1962). *Rept. Progr. Polymer Phys.*, (*Japan*), **5**, 111.
Fujino, K., K. Senshu and H. Kawai (1962). *Rept. Progr. Polymer Phys. (Japan)*, **5**, 107.
Fujino, K., K. Senshu, T. Horino and H. Kawai (1962). *Rept. Progr. Polymer Phys. (Japan)*, **5**, 115.

Fujita, H. and A. Kishimoto (1958). *J. Colloid Sci.*, **13**, 418.
Fukada, E. (1954). *J. Phys. Soc. (Japan)*, **9**, 786.
Fukada, E. and T. Takamatsu (1962). *Rept. Progr. Polymer Phys. (Japan)*, **5**, 149.
Fuller, C. S. (1940). *Chem. Rev.*, **26**, 143.
Funt, B. L. (1952). *Can. J. Chem.*, **30**, 84.
Funt, B. L. and T. H. Sutherland (1952). *Can. J. Chem.*, **30**, 940.
Fuoss, R. M. (1941a). *J. Am. Chem. Soc.*, **63**, 369.
Fuoss, R. M. (1941b). *J. Am. Chem. Soc.*, **63**, 378.
Fuoss, R. M. and J. G. Kirkwood (1941). *J. Am. Chem. Soc.*, **63**, 385.
Furukawa, G. T., R. E. McCoskey and G. J. King (1952). *J. Res. Nat. Bur. Std.*, **49**, 273.
Furukawa, J. and T. Saegusa (1963). *Polymerisation of Aldehydes and Oxides*, *Polymer Reviews*, Vol. 3, Interscience, New York.
Gall, W. G. and N. G. McCrum (1961). *J. Polymer Sci.*, **50**, 489.
Gee, G. (1946). *Trans. Faraday Soc.*, **42**, 585.
Gee, G., W. C. E. Higginson and G. T. Merrall (1959). *J. Chem. Soc.*, Pt. 2, 1345.
Geil, P. H., N. K. J. Symons and R. G. Scott (1959). *J. Appl. Phys.*, **30**, 1516.
Geil, P. H. (1960a). *J. Polymer Sci.*, **44**, 449.
Geil, P. H. (1960b). *J. Polymer Sci.*, **47**, 65.
Geil, P. H. (1961). *Private Commun.*
Geil, P. H. (1963). *Polymer Single Crystals*, Interscience, New York.
General Radio Expts. (1962). **36**, August–September, P. 3.
George, M. R. H., R. J. Grisenthwaite and R. F. Hunter (1958). *Chem. Ind. (London)*, 1114.
Germar, H. (1963). *Kolloid Z.*, **193**, 25.
Gerngross, O., K. Herrmann and W. Abitz (1930). *Z: Phys. Chem.*, **B10**, 371.
Gevers, M. (1946). *Philips Res. Repts.*, **1**, 298, 447.
Gibbs, J. (1956). *J. Chem. Phys.*, **25**, 185.
Gibbs, J. and E. DiMarzio (1958). *J. Chem. Phys.*, **28**, 373, 807.
Gibbs, J. H. and E. A. DiMarzio (1959). *J. Polymer Sci.*, **40**, 121.
Gibbs, J. H. (1960). In J. D. Mackenzie (Ed.), *Modern Aspects of The Vitreous State*, Butterworths, Washington, Chap. 7.
Glarum, S. H. (1960). *J. Chem. Phys.*, **33**, 639, 1371.
Glasstone, S., K. J. Laidler and H. Eyring (1941). *The Theory of Rate Processes*, McGraw-Hill, New York.
Gohn, G. R. and J. D. Cummings (1960). *Am. Soc. Testing Mater., Bull.* July, 64.
Gordon, M. and J. S. Taylor (1952). *J. Appl. Chem.*, **2**, 493.
Gotlib, Yu. Ya. and M. V. Volkenstein (1953). *Zh. Tekhn. Fiz.*, **23**, 1936.
Gotlib, Yu. Ya. and K. M. Salikhov (1962). *Vysokomolekul. Soedin.*, **4**, 1163; (Translated in *Polymer Sci. USSR*, **4**, 348, (1963)).
Grime, D. and I. M. Ward (1958). *Trans. Faraday Soc.*, **54**, 959.
Gross, B. (1948). *J. Appl. Phys.*, **19**, 257.
Gross, B. (1953). *Kolloid Z.*, **131**, 168 : **134**, 65.
Guth, E. and H. Mark (1934). *Monatsh. Chem.*, **65**, 93.
Haas, H. C., E. S. Emerson and N. W. Schuler (1956). *J. Polymer Sci.*, **22**, 291.
Hammerle, W. G. and J. G. Kirkwood (1955). *J. Chem. Phys.*, **23**, 1743.
Hamon, B. V. (1952). *Proc. Inst. Elec. Engrs.*, **99**, Pt. IV, *Monograph* 27.
Hartley, F. D., F. W. Lord and L. B. Morgan (1955). *Internl. Symp. Macromolecular Chemistry*, LaRicerca Scientifica **25**, 577.

Hartmann, A. (1956). *Kolloid Z.*, **148**, 30.
Hartmann, A. (1957). *Kolloid Z.*, **153**, 157.
Hartshorn, L. and W. H. Ward (1936). *J. Inst. Elec. Engrs.*, **79**, 597.
Hartshorn, L., J. V. L. Parry and E. Rushton (1953). *J. Inst. Elec. Engrs.*, **100**, Pt. IIA, No. 3.
Hearmon, R. F. S. (1956). *Phil. Mag. Suppl.*, **5**, 323.
Heijboer, J. (1952). *Chem. Weekblad.*, **48**, 264.
Heijboer, J., P. Dekking and A. J. Staverman (1954). In V. G. W. Harrison (Ed.), *Proc. 2nd Internl. Congr. Rheology*, Academic Press, New York. p. 123.
Heijboer, J. (1956). *Kolloid Z.*, **134**, 149.
Heijboer, J. (1956). *Kolloid Z.*, **148**, 36.
Heijboer, J. (1960a). *Makromol. Chem.*, **35A**, 86.
Heijboer, J. (1960b). *Kolloid Z.*, **171**, 7.
Heijboer, J. and F. Schwarzl (1962). In K. Wolf (Ed.), *Struktur und Verhalten der Kunststoffe*, Springer Verlag, p. 383.
Heijboer, J. (1963). *Private Commun.*
Heijboer, J. (1965). In *Physics of Non Crystalline Solids*, North Holland, Amsterdam, p. 231.
Heijboer, J. (1966). *Private Commun.*
Heinze, H. D., K. Schmieder, G. Schnell and K. A. Wolf (1962). *Rubber Chem. Technol.*, **35**, No. 1.
Hellwege, K. H., R. Kaiser and K. Kuphal (1958). *Kolloid Z.*, **157**, 27.
Hellwege, K. H. and G. Langebin (1960). *Kolloid Z.*, **172**, 44.
Hendus, H., G. Schnell, H. Thurn and K. Wolf (1959). *Ergeb. Exakt. Naturw.*, **31**, 5, 220.
Hendus, H., K. Schmieder, G. Schnell and K. Wolf (1960). *Festschr. Carl Wurster*, *B.A.S.F.* Ludwigshafen, p. 293.
Hengstenburg, J. (1927). *Ann. Phys.*, **84**, 245.
Heydemann, P. (1963). *Kolloid Z.*, **193**, 12.
Higasi, K. (1961). In *Dielectric Relaxation and Molecular Structure*, Monograph Ser. Res. Inst. Appl. Elec. No. 9, Hokkaido Univ. Sapporo, Japan. Kasai, Tokyo, Japan.
Hikichi, K. and J. Furuichi (1961). *Rept. Progr. Polymer Phys.*, (*Japan*), **4**, 69.
Hikichi, K. and J. Furuichi (1965). *J. Polymer Sci.*, Pt. A, **3**, 3003.
Hildebrand, J. H. and R. L. Scott (1950). *The Solubility of Non-electrolytes*, Reinhold New York, p. 123, 424.
Hill, F. N., F. E. Bailey and J. T. Fitzpatrick (1958). *Ind. Eng. Chem.*, **50**, 5.
Hill, F. N., F. E. Bailey, G. M. Powell and K. L. Smith (1958). *Ind. Eng. Chem.*, **50**, 8.
Hill, N. E. (1961). *Proc. Phys. Soc.*, **78**, 311.
Hill, N. E. (1963). *Proc. Phys. Soc.*, **82**, 723.
Hoeve, C. A. J. (1960). *J. Chem. Phys.*, **32**, 888.
Hoff, E. A. W., D. W. Robinson and A. H. Willbourn (1955). *J. Polymer Sci.*, **18**, 161.
Hoff, E. A. W., P. L. Clegg and K. Sherrard-Smith (1958). *Brit. Plastics*, **31**, 384.
Hoffman, J. D. (1952). *J. Chem. Phys.*, **20**, 541.
Hoffman, J. D. (1954). *J. Chem. Phys.*, **22**, 156.
Hoffman, J. D. and H. G. Pfeiffer (1954). *J. Chem. Phys.*, **22**, 132.
Hoffman, J. D. (1955). *J. Chem. Phys.*, **23**, 1331.

Hoffman, J. D. and B. M. Axilrod (1955). *J. Res. Nat. Bur. Stds.*, **54**, 357. RP2598.
Hoffman, J. D. and J. J. Weeks (1958). *J. Res. Nat. Bur. Stds.*, **60**, 465.
Hoffman, J. D. (1959). *8th Ampère Colloq.*, **12**, 36.
Holmes, D. R., C. W. Bunn and D. J. Smith (1955). *J. Polymer Sci.*, **17**, 159.
Holzmüller, W. (1940). *Kunststoffe*, **30**, 177.
Holzmüller, W. (1941). *Phys. Z.*, **42**, 281.
Hopkinson, J. (1877). *Phil. Trans. Roy. Soc.*, (London), **167**, 599.
Hosemann, R. (1950). *Z. Phys.*, **128**, 1 and 465.
Hosemann, R., R. Bonart and G. Schoknecht (1956). *Z. Phys.*, **146**, 588.
Hosemann, R. and R. Bonart (1957). *Kolloid Z.*, **152**, 53.
Hopkins, T. L. and W. O. Baker (1960). In F. R. Eirich (Ed.), *Rheology*, Academic Press, New York, Vol. 3, p. 365.
Horner, F., T. A. Taylor, R. Dunsmuir, J. Lamb and W. Jackson (1946). *J. Inst. Elec. Engrs.*, *(London)*, **93**, 53.
Howard, W. H. and M. L. Williams (1963). *Textile Res. J.*, **33**, 689.
Huff, K. and F. H. Müller (1957). *Kolloid Z.*, **153**, 5.
Hughes, L. J. and G. L. Brown (1961). *J. Appl. Polymer Sci.*, **5**, 580.
Ilavski, M. and J. Janacek (1965). *Collection Czech. Chem. Commun.*, **30**, 833.
Illers, K. H. and E. Jenckel (1958). *Rheol. Acta*, **1**, 322.
Illers, K. H. and E. Jenckel (1958a). *Kolloid Z.*, **160**, 97.
Illers, K. H. and E. Jenckel (1959b). I.U.P.A.C. Kurzmitteilungen, Sec. 1: Physik Makromolekular Stoffe, Verlag Chemie, GmbH, Weinheim. (Quoted by Curtis (1961)).
Illers, K. H. and E. Jenckel (1959). *J. Polymer Sci.*, **41**, 528.
Illers, K. H. and E. Jenckel (1959a). *Kolloid Z.*, **165**, 84.
Illers, K. H. (1960). *Makromol. Chem.*, **38**, 168.
Illers, K. H. and H. Jacobs (1960). *Makromol. Chem.*, **39**, 234.
Illers, K. H. (1961). *Z. Elektrochem.*, **65**, 679.
Illers, K. H. and H. Breuer (1961). *Kolloid Z.*, **176**, 110.
Illers, K. H. and H. Breuer (1963). *J. Colloid. Sci.*, **18**, 1.
Illers, K. H. (1963). *Kolloid Z.*, **190**, 16.
Illers, K. H. (1964). *Rheol. Acta*, **3**, 194.
Inoue, Y. and Y. Kobatake (1958). *Kolloid Z.*, **159**, 18.
Ishida, Y. (1960a). *Kolloid Z.*, **168**, 29.
Ishida, Y., M. Yamamoto and M. Takayanagi (1960). *Kolloid Z.*, **168**, 124.
Ishida, Y., Y. Takada and M. Takayanagi (1960). *Kolloid Z.*, **168**, 121.
Ishida, Y. (1960b). *Kolloid. Z.*, **171**, 149.
Ishida, Y. and K. Yamafuji (1961). *Kolloid Z.*, **177**, 97.
Ishida, Y., and others (1961). *Kolloid Z.*, **174**, 162.
Ishida, Y. (1961). *Kolloid. Z.*, **174**, 124.
Ishida, Y., O. Amano and M. Takayanagi (1961). *Kolloid Z.*, **176**, 62.
Ishida, Y., M. Matsuo, S. Togami, K. Yamatuji and M. Takayanagi (1962). *Kolloid Z.*, **183**, 74.
Ishida, Y., M. Matsuo and K. Yamafuji (1962). *Kolloid Z.*, **180**, 108.
Ishida, Y., K. Yamafuji, H. Ito and M. Takayanagi (1962). *Kolloid Z.*, **184**, 97.
Ishida, Y., M. Watanabe and K. Yamafuji (1964). *Kolloid Z.*, **200**, 48.
Ishida, Y. and K. Yamafuji (1964). *Kolloid Z.*, **200**, 50.
Ishida, Y. and S. Matsuoka (1965). *Am. Chem. Soc.*, *Polymer Preprints*, **6**, No. 2, 795.

Ishida, Y., M. Matsuo and M. Takayanagi (1965). *J. Polymer Sci.*, Pt. B, **3**, 321.
Iwasaki, K. (1962). *J. Polymer Sci.*, **56**, 27.
Iwayanagi, S. and T. Hideshima (1953a). *J. Phys. Soc. (Japan)*, **8**, 365. (1953b)
J. Phys. Soc. (Japan), **8**, 368.
Iwayanagi, S. (1955a). *J. Sci. Res. Inst. (Tokyo)*, **49**, 13.
Iwayanagi, S. (1955b). *J. Sci. Res. Inst. (Tokyo)*, **49**, 23.
Iwayanagi, S. (1962). *Rep. Progr. Polymer Phys. (Japan)*, **5**, 131.
Iwayanagi, S. (1962). *Progr. Polymer Phys. (Japan)*, V, 135.
Jackson, J. B., P. J. Flory, R. Chaing and M. J. Richardson (1963). *Polymer*, **4**, 237.
Jackson, S. B. and J. C. McMillan (1963). *Soc. Plastics Engrs. J.*, **19**, 203.
Jackson, W. and J. S. A. Forsyth (1945). *J. Inst. Elec. Eng.*, **92**, III, 23.
Jacobs, H. and E. Jenckel (1961a). *Makromel. Chem.*, **43**, 132.
Jacobs, H. and E. Jenckel (1961b). *Makromel. Chem.*, **47**, 72.
Jaeger, J. C. (1964). In *Elasticity, Fracture and Flow*, Methuen, London. Reprinted 2nd ed.
James, H. M. and E. Guth (1943). *J. Chem. Phys.*, **11**, 455.
James, H. M. and E. Guth (1947). *J. Chem. Phys.*, **15**, 669.
Janacek, J. and J. Kolarik (1965). *Collection Czech. Chem. Commun.*, **30**, 1597.
Jenckel, E. (1942). *Kolloid Z.*, **100**, 163.
Jenckel, E. and K. H. Illers (1954). *Z. Naturforsch.*, **9a**, 440.
Jenckel, E. (1954). *Kolloid Z.*, **136**, 142.
Johnson, U. (1961). *J. Polymer Sci.*, **54**, 56.
Jones, G. O. (1956). In *Glass*, Methuen, London.
Kabin, S. P. and G. P. Mikhailov (1956). *Zh. Tekhn. Fiz.*, **26**, 511.
Kabin, S. P. (1956). *Zh. Tekhn. Fiz.*, **26**, 2628.
Kabin, S. P. (1956). *Sov. Phys. Tech. Phys.*, **1**, 2542.
Kabin, S. P. (1960). *Vysokomolekul. Soedin.*, **2**, 1324.
Kabin, S. P., S. G. Malkevich, G. P. Mikhailov, B. I. Sazhin, A. L. Smolyanskii and L. V. Chereshkevich (1961). *Vysokomolekul. Soedin.*, **3**, 618.
Kahlbaum, G. W. A. (1880). *Ber.*, **13**, 2348.
Kajiyama, T., S. Togami, Y. Ishida and M. Takayanagi (1965). *J. Polymer Sci.*, Part B, **3**, 103.
Kakudo, M. and N. Kasai (1955). "Polyvinylalcohol," (I. Sakurada, ed.) p. 223, *Soc. Polymer Sci. (Japan)*.
Kanig, G. (1963). *Kolloid. Z.*, **190**, 1.
Kargin, V. A. and G. L. Slonimskii (1948). *Ber. Acad. Wiss. UdSSR*, **62**, 239.
Kargin, V. A. and G. L. Slonimskii (1949). *Zh. Fiz. Khim.*, **23**, 526.
Kargin, V. A., A. I. Kitaigorodskii and G. L. Slonomskii (1957). *Kolloid Zh.*, **19**, 131.
Kargin, V. A. and others (1960). High Mol. Wt. *Compds.*, **2**, 1280.
Karpovich, J. (1954). *J. Chem. Phys.*, **22**, 1767.
Kästner, S. (1961a). *Kolloid Z.*, **178**, 24.
Kästner, S. (1961b). *Kolloid. Z.*, **178**, 119.
Kästner, S. (1962b). *Kolloid Z.*, **184**, 109.
Kästner, S. (1962). *Kolloid. Z.*, **185**, 126.
Kästner, S. (1963). *Kolloid. Z.*, **187**, 27.
Kästner, S., E. Schlosser and G. Pohl (1963). *Kolloid Z.*, **192**, 21.
Kaufman, H. S. (1953). *J. Am. Chem. Soc.*, **75**, 1447.
Kauzmann, W. and H. Eyring (1940). *J. Am. Chem. Soc.*, **62**, 3113.

Kauzmann, W. (1948). *Chem. Rev.*, **43**, 219.
Kawaguchi, I. (1958). *J. Polymer Sci.*, **32**, 417.
Kawaguchi, T. (1959). *J. Appl. Polymer Sci.*, **2**, 56.
Kawai, T. (1961). *J. Phys. Soc. (Japan)*, **16**, 1220.
Kê, T. S. (1947). *Phys. Rev.*, **71**, 533.
Keith, H. D. and F. J. Padden (1959). *J. Polymer Sci.*, **39**, 101, 123.
Keith, H. D. and F. J. Padden (1961). *J. Polymer Sci.*, **51**, 54.
Keith, H. D. and F. J. Padden (1963). *J. Appl. Phys.*, **34**, 2409.
Keith, H. D. (1963). In D. Fox (Ed.), *Physics and Chemistry of the Organic Solid State*, Interscience, New York.
Keith, H. D. and F. J. Padden (1964a). *J. Appl. Phys.*, **35**, 1270.
Keith, H. D. and F. J. Padden (1964b). *J. Appl. Physics*, **35**, 1286.
Keller, A., G. R. Lester and L. B. Morgan (1954). *Phil. Trans. Roy. Soc.*, **A247**, 1.
Keller, A. (1955). *J. Polymer Sci.*, **17**, 291.
Keller, A. (1955). *J. Polymer Sci.*, **17**, 351.
Keller, A. (1957). *Phil. Mag.*, **2**, 1171.
Keller, A. (1958). In R. H. Doremus and others (Eds.), *Growth and Perfection of Crystals*, Wiley, New York.
Keller, A. (1959). *J. Polymer Sci.*, **39**, 151.
Keller, A. (1959). *J. Polymer Sci.*, **36**, 361.
Keller, A. and S. Sawada (1964). *Makromol. Chem.*, **74**, 190.
Kelly, A. (1966). In *Strong Solids*, Clarendon Press, Oxford.
Kern, R. J., J. J. Hawkins and J. D. Calfee (1963). *Makromol. Chem.*, **66**, 126.
Ketley, A. D. (1962). *J. Polymer Sci.*, **62**, 581.
Kiessling, D. (1961). *Kolloid Z.*, **176**, 119.
Kilian, H. G., H. Halboth and E. Jenckel (1960). *Kolloid Z.*, **172**, 166.
Kinoshita, Y. (1959). *Makromol. Chem.*, **33**, 1.
Kirkwood, J. G. (1939). *J. Chem. Phys.*, **7**, 911.
Kirkwood, J. G. and R. M. Fuoss (1941). *J. Chem. Phys.*, **9**, 329.
Kirkwood, J. G. (1946). *J. Chem. Phys.*, **14**, 51.
Kirkwood, J. G. and J. Riseman (1948). *J. Chem. Phys.*, **16**, 565.
Kirkwood, J. G. (1949). *Rec. Trav. Chim.*, **68**, 649.
Kirkwood, J. G. (1944). *J. Polymer Sci.*, **12**, 1.
Kishi, N. and N. Uchida (1963). *Rept. Progr. Polymer Phys. (Japan)*, **VI**, 233.
Kline, D. E. (1956). *J. Polymer Sci.*, **22**, 449.
Kline, D. E., J. A. Sauer and A. E. Woodward (1956). *J. Polymer Sci.*, **22**, 455.
Klyne, W. (1954). In *Progress in Stereochemistry*, Butterworths, London.
Kohlrausch, F. (1955). *Prakt. Phys. (Stuttgart)*, **I**, **129**.
Kolarik, J. and J. Janacek (1965). *Collection Czech. Chem. Commun.*, **30**, 2388.
Kolb, H. J. and E. F. Izard (1949). *J. Appl. Phys.*, **20**, 564.
Kolsky, H. (1953). In *Stress Waves in Solids*, Clarendon Press, Oxford.
Kono, R. (1960). *J. Phys. Soc. (Japan.)*, **15**, 718.
Kono, R. (1961). *J. Phys. Soc. (Japan)*, **16**, 1580.
Köster, W. (1948). *Z. Metallk.*, **39**, 1.
Koppelmann, J. (1955). *Kolloid Z.*, **144**, 12.
Koppelmann, J. (1958). *Rheol. Acta.*, **1**, 20.
Koppelmann, J. and J. Gielessen (1961a). *Kolloid Z.*, **175**, 97.
Koppelmann, J. and J. Gielessen (1961b). *Z. Elektrochem.*, **65**, 689.
Koppelmann, J. (1963). *Kolloid Z.*, **189**, 1.

Koppelmann, J. (1965). In *Physics of Non Crystalline Solids*, North Holland, Amsterdam, p. 255.
Kovacs, A. J. (1958). *J. Polymer Sci.*, **30**, 131.
Kovacs, A. J., R. A. Stratton and J. D. Ferry (1963). *J. Phys. Chem.*, **67**, 152.
Kovacs, A. J. (1964). *Advan. Polymer Sci.*, **3**, 394.
Kramer, H. and K. E. Helf (1962). *Kolloid Z.*, **180**, 114.
Kramers, H. A. (1927). *Atti Congr. Fis. Bom.*, p. 545.
Kramers, H. A. (1940). *Phys.*, **7**, 284.
Krebs, K. and J. Lamb (1958). *Proc. Roy. Soc.*, **A244**, 558.
Krimm, S. (1960). *Advan. Polymer Sci.*, **2**, 51.
Kronig, R. de L. (1926). *J. Opt. Soc. Am.*, **12**, 547.
Krum, F. and F. H. Müller (1959). *Kolloid Z.*, **164**, 8.
Kuhn, W. (1934). *Kolloid Z.*, **68**, 2.
Kuhn, W. (1936). *Kolloid Z.*, **76**, 258.
Kuhn, W. (1939). *Kolloid Z.*, **87**, 3.
Kuhn, W. and H. Kuhn (1945). *Helv. Chim. Acta*, **28**, 1533.
Kuhn, W. and H. Kuhn (1946). *Helv. Chim. Acta*, **29**, 71, 609, 830.
Kuhn, W. (1948). *Helv. Chim. Acta*, **31**, 1092, 1259.
Kuhn, W. (1950). *Helv. Chim. Acta*, **33**, 2057.
Kurata, M. and Stockmayer, W. H. (1963). *Advan. Polymer Sci.*, **3**, 196.
Kurath, S. F., E. Passaglia and R. Pariser (1957). *J. Appl. Phys.*, **28**, 499.
Kurath, S. F., T. P. Yin, J. W. Berge and J. D. Ferry (1959). *J. Colloid Sci.*, **14**, 147.
Kurosaki, S. and T. Furumaya (1960). *J. Polymer Sci.*, **43**, 137.
Laible, R. C. and H. M. Morgan (1961). *J. Polymer Sci.*, **54**, 53.
Lamb, J. and J. Sherwood (1955). *Trans. Faraday Soc.*, **51**, 1674.
Lamb, J. and A. J. Matheson (1964). *Proc. Roy. Soc.*, **A281**, 207.
Lamont, H. (1942). In *Waveguides*, Methuen Monograph, 1942.
Lauritzen, J. I. (1958). *J. Chem. Phys.*, **28**, 118.
Lawson, K. D., J. A. Sauer and A. E. Woodward (1963). *J. Appl. Phys.*, **34**, 2492.
Leaderman, H. (1943). In *Elastic and Creep Properties of Filamentous Materials and other High Polymers*, Textile Foundation, Washington, D.C.
Leaderman, H. (1958). In F. R. Eirich (Ed.), *Rheology*, Academic Press, New York, Vol. 2.
Leksina, I. E. and S. I. Novikova (1959). *Sov. Phys. Solid State*, **1**, 453.
Lennard-Jones, J. E. and A. F. Devonshire (1939). *Proc. Roy. Soc. (London)* **A169**, 317; **A170**, 464.
Lethersich, W. (1947). *J. Sci. Inst.*, **24**, 66.
Lethersich, W. (1950). *Brit. J. Appl. Phys.*, **1**, 294.
Liang, C. Y. and S. Krimm (1956). *J. Chem. Phys.*, **25**, 563.
London, F. (1937). *Trans. Faraday Soc.*, **33**, 8.
Loshaek, S. (1955). *J. Polymer Sci.*, **15**, 391.
Lukomskaya, A. I. and B. A. Dogadkin (1960). *Kolloid Z.*, **22**, 576.
Luther, H. and G. Weisel (1957). *Kolloid Z.*, **154**, 15.
Lynch, A. C. (1957). *Proc. Inst. Elec. Engrs. (London)*, **104**, Pt. B., 363.
Lyons, W. J. (1959). *J. Appl. Phys.*, **30**, 796.
McCall, D. W. and E. W. Anderson (1960). *J. Chem. Phys.*, **32**, 237.
McCall, D. W. (1964). Paper given at a Conference held at Mortonhampstead, Devon, England.
McCall, D. W. (1965). *Private Commun.*

McCrum, N. G. (1958). *J. Polymer Sci.*, 27, 555.

McCrum, N. G. (1959a). *Am. Soc. Testing Mater. Bull.*, No. 242.

McCrum, N. G. (1959b). *J. Polymer Sci.*, 34, 355.

McCrum, N. G. (1959c). *Makromol. Chem.*, 34, 50.

McCrum, N. G. (1961). *J. Polymer Sci.*, 54, 561.

McCrum, N. G. (1962). *J. Polymer Sci.*, 60, 53.

McCrum, N. G. (1964). *Polymer Letters*, 2, 495.

McCrum, N. G. and E. L. Morris (1964). *Proc. Roy. Soc.*, A, 281, 258.

McCrum, N. G. and E. L. Morris (1965). *J. Appl. Phys. (Japan)*, 4, 542.

McIntire, A. D. (1961). *J. Appl. Polymer Sci.*, 5, 195.

McKinney, J. E., S. Edelman and R. S. Marvin (1956). *J. Appl. Phys.*, 27, 425.

McKinney, J. E. and C. S. Bowyer (1960). *J. Acoust. Soc. Am.*, 32, 56.

McKinney, J. E. and H. V. Belcher (1963). *J. Res. Nat. Bur. Stds.*, 67A, 43.

McLoughlin, J. R. and A. V. Tobolsky (1952). *J. Colloid Sci.*, 7, 555.

McSkimin, H. J. (1951). *J. Acoust. Soc. Am.*, 23, 429.

Maeda, Y. (1957). *Chem. High Polymers (Japan)*, 14, 442.

Manaresi, P. and V. Giannella (1960). *J. Appl. Polymer Sci.*, 4, 251.

Mandelkern, L., G. M. Martin and F. A. Quinn (1957). *J. Res. Nat. Bur. Stds.*, 58, 137.

Mann, J. and L. Roldan-Gonzalez (1962). *Polymer*, 3, 549.

Manning, M. F. and M. E. Bell (1940). *Rev. Mod. Phys.*, 12, 215.

Marchal, J. and H. Benoit (1955). *J. Chim. Phys.*, 52, 518.

Mark, H. (1940). In *High Polymers*, Interscience, New York, Vol. II.

Mark, J. E. and P. J. Flory (1965). *J. Am. Chem. Soc.*, 87, 1415.

Marvel, C. S., J. H. Sample and M. F. Roy (1939). *J. Am. Chem. Soc.*, 61, 3241.

Marvel, C. S. (1943). In R. E. Burk and O. Grummit (Eds.), *The Chemistry of Large Molecules*, Interscience, New York.

Marvin, R. S., R. Aldrich and H. S. Sack (1954). *J. Res. Nat. Bur. Stds.*, 25, 1213.

Marvin, R. S. and J. E. McKinney (1964). Volume Relaxations in Amorphous Polymers. In *Physical Acoustics*, Academic Press, New York, Vol. 2B, Chap. 9.

Maryott, A. A. and F. Buckley (1953). In *Table of Dielectric Constants and Electric Dipole Moments of Substances in the Gaseous State*, Nat. Bur. Stds. Circ., 537, June.

Mason, P. (1964). *Polymer*, 5, 625.

Mason, W. P. and H. J. McSkimin (1952). *Bell Syst. Tech. J.*, 21, 112.

Mason, W. P. (1958). In *Physical Acoustics and the Properties of Solids*, Van Nostrand, Princeton.

Matsuoka, S. (1962). *J. Polymer Sci.*, 57, 569.

Matthews, J. L., H. S. Peiser and R. B. Richards (1949). *Acta Cryst.*, 2, 85.

Maxwell, B. (1956). *J. Polymer Sci.*, 20, 551.

Maxwell, B. (1956). *Am. Soc. Testing Mater. Bull.*, No. 215, July.

Mayo, F. R. and C. Walling (1950). *Chem. Revs.*, 46, 191.

Mead, D. J. and R. M. Fuoss (1941). *J. Am. Chem. Soc.*, 63, 2832.

Mead, D. J. and R. M. Fuoss (1942). *J. Am. Chem. Soc.*, 64, 2389.

Meredith, R. and B. S. Hsu (1962). *J. Polymer Sci.*, 61, 253, 271.

Merz, E. H., L. E. Nielsen and R. Buchdahl (1951). *Ind. Eng. Chem.*, 43, 1396.

Meyer, K. H. and W. Lotmar (1936). *Helv. Chim. Acta*, 19, 68.

Mikhailov, G. P. (1951). *J. Tech. Phys. (USSR)*, 21, 1395.

Mikhailov, G. P. and T. I. Borisova (1953). *J. Tech. Phys. (USSR)*, **23**, 2159.
Mikhailov, G. P., A. M. Lobanov and B. I. Sazhin (1954), *J. Tech. Phys. (USSR)*, **24**, 1553.
Mikhailov, G. P., S. P. Kabin and B. I. Sazhin (1955). *J. Tech. Phys. (USSR)*, **25**, 590.
Mikhailov, G. P. (1955). *Usp. Khim.*, **24**, 875.
Mikhailov, G. P. and others (1956). *J. Tech. Phys.*, *(USSR)*, **26**, 1924; (*see* (1956) *Sov. Phys. Tech. Phys.*, **1**, 1857).
Mikhailov, G. P., T. I. Borisova and D. A. Dmitrochenko (1956). *J. Tech. Phys. (USSR)*, **26**, 1924; (*see* (1956) *Sov. Phys. Tech. Phys.*, **1**, 1857.
Mikhailov, G. P. and B. I. Sazhin (1957). *Sov. Phys. Tech. Phys.*, **1**, 1670.
Mikhailov, I. G. and V. A. Soloview (1957). *Akust. Zh.*, **3**, 65.
Mikhailov, G. P., S. P. Kabin and T. A. Krylova (1957). *Sov. Phys. Tech. Phys.*, **2**, 1899.
Mikhailov, G. P., S. P. Kabin and T. A. Krylova (1957). *J. Tech. Phys. (USSR)*, **27**, 2050.
Mikhailov, G. P. and T. I. Borisova (1958). *Sov. Phys. Tech. Phys.*, **3**, 120.
Mikhailov, G. P. and A. M. Lobanov (1958). *Sov. Phys. Tech. Phys.*, **3**, 249.
Mikhailov, G. P. (1958). *J. Polymer Sci.*, **30**, 605.
Mikhailov, G. P., B. I. Sazhin and V. S. Presnyakova (1958). *Kolloidn. Zh.*, **20**, 461.
Mikhailov, G. P. and B. I. Sazhin (1959). *Vysokomolekul. Soedin.*, **1**, 9, 29.
Mikhailov, G. P. and T. I. Borisova (1960). *Vysokomolekul. Soedin.*, **2**, 1772.
Mikhailov, G. P. (1960). *Makromol. Chem.*, **35**, 26.
Mikhailov, G. P. and M. P. Eidelnant (1960a). *Vysokomolekul. Soedin.*, **2**, 287.
Mikhailov, G. P. and M. P. Eidelnant (1960b). *Vysokomolekul. Soedin.*, **2**, 295.
Mikhailov, G. P. and M. P. Eidelnant (1960c). *Vysokomolekul. Soedin.*, **10**, 1552.
Mikhailov, G. P. and T. I. Borisova (1961). *Polymer Sci.*, *USSR*, **2**, 387.
Mikhailov, G. P. and T. I. Borisova (1962). *Vysokomolekul. Soedin.*, **4**, 1732.
Mikhailov, G. P., A. M. Lobanov and H. A. Shevelev (1961). *High Mol. Wt. Compds.*, **3**, 794.
Mikhailov, G. P. and L. V. Krasner (1962). *Vysokomolekul. Soedin.*, **4**, 1071.
Mikhailov, G. P. and L. V. Krasner (1963). *Vysokomolekul. Soedin.*, **5**, 1085.
Mikhailov, G. P. and T. I. Borisova (1964). *Polymer Sci.*, *USSR*, **6**, 1971, 1979.
Mikhailov, G. P. (1965). In *Physics of Non-Crystalline Solids*, North Holland, Amsterdam, p. 270.
Miller, A. A. (1963). *J. Polymer Sci.*, Part A, **1**, 1857, 1865.
Miller, R. G. J. and H. A. Willis (1956). *J. Polymer Sci.*, **19**, 485:
Miller, R. L. and L. E. Nielsen (1960). *J. Polymer Sci.*, **44**, 391.
Miyake, A. (1960). *Sci. Rept. Tokyo Rayon Co.*, **15**, 229.
Mizushima, S.-I. and H. Okazaki (1949). *J. Am. Chem. Soc.*, **71**, 3411.
Mizushima, S.-I. and T. Shimanouchi (1949). *J. Am. Chem. Sci.*, **71**, 1320.
Mizushima, S.-I. (1952). In *Internal Rotation*, The University of Notre Dame, Notre Dame, Ind.
Mizushima, S.-I. (1954). In *Structure of Molecules and Internal Rotation*, Academic Press, New York.
Mooney, M. (1959). *J. Polymer Sci.*, **34**, 599.
Moore, C. G. and W. F. Watson (1956). *J. Polymer Sci.*, **19**, 237.
Morrison, T. E., L. J. Zapas and T. W. DeWitt (1955). *Rev. Sci. Instr.*, **26**, 357.

Moynihan, R. E. (1959). *J. Am. Chem. Soc.*, **81**, 1045.
Müller, A. (1928). *Proc. Roy. Soc.*, **A120**, 437.
Müller, A. (1936). *Proc. Roy. Soc.*, **A154**, 624.
Müller, A. (1941). *Proc. Roy. Soc.*, **A178**, 227.
Müller, F. H. and E. Hellmuth (1955). *Kolloid. Z.*, **144**, 125.
Müller, F. H. and K. Huff (1959). *Kolloid Z.*, **164**, 34.
Mulvaney, J. E., C. G. Overberger and A. M. Schiller (1961). *Advan. Polymer Sci.*, **3**, 106.
Murahashi, S. and others (1962). *J. Polymer Sci.*, **62**, 877.
Muus, L. T., N. G. McCrum and F. C. McGrew (1959). *Soc. Plastics Engrs., J.*, **15**, May.
Nagai, E. (1955). In I. Sakurada (Ed.), *Polyvinyl Alcohol*, 252, Soc. Polymer Sci. (*Japan*).
Nagai, A. and M. Takayanagi (1964). *Rept. Progr. Polymer Phys.* (*Japan*), **7**, 249.
Nagamatsu, K., T. Takemura, T. Yoshitomi and T. Takemoto (1958). *J. Polymer Sci.*, **33**, 515.
Nagamatsu, K., T. Yoshitomi and T. Takemoto (1958). *J. Colloid Sci.*, **13**, 257.
Nagata, N., K. Hikichi, M. Kaneri and J. Furuichi (1963). *Rept. Progr. Polymer Phys.* (*Japan*), **6**, 235.
Nakada, O. (1955). *J. Phys. Soc.* (*Japan*), **10**, 804.
Nakada, O. (1960). *J. Polymer Sci.*, **43**, 149.
Nakajima, T. and S. Saito (1958). *J. Polymer Sci.*, **31**, 423.
Nakayasu, H., H. Markovitz and D. J. Plazek (1961). *Trans. Soc. Rheol.*, **5**, 261.
Natta, G. and P. Corradini (1955). *Makromol. Chem.*, **16**, 77.
Natta, G. and P. Corradini (1955). *Atti. Accad. Naz. Lincei*, Ser. VIII, IV°, Sez. II, Fasc. 5, 73.
Natta, G. and others (1955). *J. Am. Chem. Soc.*, **77**, 1708.
Natta, G. and P. Corradini (1956). *J. Polymer Sci.*, **20**, 251.
Natta, G. (1956). *Angew Chem.*, **68**, 393.
Natta, G., F. Danusso and G. Moraglio (1958). *Makromol. Chem.*, **28**, 166.
Natta, G., F. Danusso and G. Moraglio (1957). *J. Polymer Sci.*, **25**, 119.
Natta, G., M. Baccaredda and E. Butta (1959). *Chim. Ind.*, **41**, 737.
Natta, G. and F. Danusso (1959). *J. Polymer Sci.*, **34**, 3.
Natta, G. (1959). *J. Polymer Sci.*, **34**, 21.
Natta, G. (1960). *Makromol. Chem.*, **35**, 94.
Natta, G. and G. Crespi (1960). *Rubber Age*, **87**, 459.
Newman, S. and W. P. Cox (1960). *J. Polymer Sci.*, **46**, 29.
Nielsen, L. E., R. Buchdahl and R. Lavreault (1950). *J. Appl. Phys.*, **21**, 607.
Nielsen, L. E. (1951). *Rev. Sci. Instr.*, **22**, 690.
Nielsen, L. E., G. Claver and R. Buchdahl (1951). *Ind. Eng. Chem.*, **43**, 341.
Nielsen, L. E. (1960). *J. Polymer Sci.*, **42**, 357.
Nielsen, L. E. (1962). *Mechanical Properties of Polymers*, Reinhold, New York.
Nielsen, L. E. and F. D. Stockton (1963). *J. Polymer Sci.*, **1**, 1995.
Nishioka, A. and M. Watanabe (1957). *J. Polymer Sci.*, **24**, 298.
Nolle, A. W. and S. C. Mowry (1948). *J. Acoust. Soc. Am.*, **20**, 432.
Nolle, A. W. (1950). *J. Polymer Sci.*, **5**, 1.
Norman, R. H. (1953). *Proc. Inst. Elec. Engrs.* (*London*), **100**, IIA, 341.
Nowick, A. S. (1953). In B. Chalmers (Ed.), *Progress in Metal Physics*, Interscience, New York, p. 1.

Oakes, W. G. and D. W. Robinson (1954). *J. Polymer Sci.*, **14**, 505.
Oberst, H. and L. Bohm (1961). *Rheol. Acta*, **1**, 608.
Odajima, A., J. Sohma and M. Koike (1957). *J. Phys. Soc. (Japan)*, **12**, 272.
Odajima, A., A. E. Woodward and J. A. Sauer (1961). *J. Polymer Sci.*, **55**, 181.
Okamoto, S. and K. Takeuchi (1959). *J. Phys. Soc. Japan*, **14**, 378.
Okuyama, M. and T. Yanagida (1963). *Rept. Progr. Polymer Phys. (Japan)*, **6**, 125.
Ohlberg, S. M. and S. S. Fenstermaker (1958). *J. Polymer Sci.*, **32**, 516.
Ohzawa, Y. and Y. Wada (1963). *Rept. Progr. Polymer Phys. (Japan)*, **6**, 147.
Ohzawa, Y. and Y. Wada (1964). *J. Appl. Phys. (Japan)*, **3**, 436.
Onogi, S., K. Sasaguri, T. Adachi and S. Ogihara (1962). *J. Polymer Sci.*, **58**, 1.
Onsager, L. (1936). *J. Am. Chem. Soc.*, **58**, 1486.
O'Reilly, J. M. (1962). *J. Polymer Sci.*, **57**, 429.
O'Reilly, J. M., F. E. Karasz and H. E. Bair (1964). *J. Polymer Sci., Pt. C*, **6**, 189.
Oskin, E. T. and B. Maxwell (1957). Princeton University *Rept.*, **44A**.
Oster, G. and J. G. Kirkwood (1943). *J. Chem. Phys.* **11**, 175.
Overberger, C., Frazier, C., J. Mandelman and H. Smith (1953). *J. Am. Chem. Soc.*, **75**, 3326.
Padden, F. J. and H. D. Keith (1959). *J. Appl. Phys.*, **30**, 1479.
Pake, G. E. (1956). In F. Seitz (Ed.), *Solid State Physics*, Academic Press, New York, Vol. 2, p. 1.
Palmer, R. P. and A. J. Cobbald (1964). *Makromol. Chem.*, **74**, 174.
Palmer, R. P. (1965). *Personal Commun.*
Panofsky, W. K. H. and M. Phillips (1955). In *Classical Electricity and Magnetism*, Addison-Wesley, Mass.
Pao, Y. H. (1957). *J. Appl. Phys.*, **28**, 591.
Pao, Y. H. (1962). *J. Polymer Sci.*, **61**, 413.
Parry, J. V. L. (1951). *Proc. Inst. Elec. Engrs.*, Pt. III, 303.
Passaglia, E. and G. M. Martin (1964). *J. Res. Nat. Bur. Stds.*, **68**, 519.
Passaglia, E., J. M. Crissman and R. R. Stromberg (1965). *Polymer Preprints*, *Am. Chem. Soc.* **6**, No. 2, 590.
Pauling, L. (1960). In *Nature of the Chemical Bond*, Cornell University Press, 3rd ed.
Payne, A. R. (1958). In P. Mason and N. Wookey (Eds.), *Rheology of Elastomers*, Pergamon, London, p. 86.
Payne, A. R. (1961). *Mater. Res. Std.*, **1**, 942.
Payne, A. R. (1960). *J. Appl. Polymer Sci.*, **24**, 127.
Payne, A. R. (1962a). *J. Appl. Polymer Sci.*, **6**, 57.
Payne, A. R. (1962b). *J. Appl. Polymer Sci.*, **6**, 368.
Pechold, W., S. Blasenbrey and S. Woerner (1963). *Kolloid Z.*, **189**, 14.
Peck, V. and W. Kaye (1954). *J. Appl. Phys.*, **25**, 1465.
Pelmore, O. R. and E. L. Simon (1940). *Proc. Roy. Soc. (London)*, **A175**, 468.
Perrin, F. (1934). *J. Phys. Radium*, **5**, 497.
Peterlin, A. and H. G. Olf (1964). *Polymer Letters*, **2**, 409.
Pierce, R. H. H., E. S. Clark, J. F. Whitney and W. M. D. Bryant (1956). Meeting of the *Am. Chem. Soc.*, Atlantic City, September.
Powers, R. W. and M. V. Doyle (1958). *Acta Met.*, **6**, 643.
Philippoff, W. (1953). *J. Appl. Phys.*, **24**, 685.
Phillips, W. D. (1955). *J. Chem. Phys.*, **23**, 1363.
Phillips, W. D. (1958). *Ann. N. Y. Acad. Sci.*, **70**, 817.

Pickett, G. (1945). *Am. Soc. Testing Mater. Proc.*, **45**, 846.
Polder, D. (1945). *Philips Res. Repts.*, **1**, 5.
Pople, J. A. (1951). *Proc. Roy. Soc. (London)*, **205A**, 163.
Powles, J. G. (1956). *J. Polymer Sci.*, **22**, 19.
Powles, J. G. and P. Mansfield (1962). *Polymer*, **3**, 336.
Powles, J. G., J. H. Strange and D. J. H. Sandiford (1963). *Polymer*, **4**, 401.
Powles, J. G., B. I. Hunt and D. J. H. Sandiford (1964). *Polymer*, **5**, 505.
Price, C. C., M. Osgan, R. E. Hughes and C. Shambelan (1956). *J. Am. Chem. Soc.*, **78**, 690.
Price, C. C. and M. Osgan (1956). *J. Am. Chem. Soc.*, **78**, 4787.
Price, F. P. (1959). *J. Polymer Sci.*, **37**, 71.
Price, F. P. (1959). *J. Polymer Sci.*, **39**, 139.
Price, F. P. (1962). *J. Polymer Sci.*, **57**, 395.
Ptitsyn, O. B. and Yu. A. Sharanov (1957). *Zh. Tekhn. Fiz.*, **27**, 2744, 2762.
Ptitsyn, O. B. (1959). *Sov. Phys. Solid State*, **1**, 843.
Quinn, F. A. and L. Mandelkern (1958). *J. Am. Chem. Soc.*, **80**, 3178.
Quistwater, J. M. R. and B. A. Dunell (1959). *J. Appl. Polymer Sci.*, **1**, 267.
Rasmussen, R. S. (1948). *J. Chem. Phys.*, **16**, 712.
Rayleigh, Lord (1877). In *Theory of Sound*, Macmillan, London, Chap. 7, reprinted 1929.
Read, B. E. and G. Williams (1961a). *Polymer*, **2**, 239.
Read, B. E. and G. Williams (1961b). *Trans. Faraday Soc.*, **57**, 1979.
Read, B. E. (1962a). *Polymer*, **3**, 143.
Read, B. E. (1962b), *Polymer*, **3**, 529.
Read, B. E. (1964a). *Polymer*, **5**, 1.
Read, B. E. (1964b), *J. Polymer Sci.*, Part C, Polymer Symposia No. 5, p. 87.
Read, B. E. (1965). *Trans. Faraday Soc.*, **61**, 2140.
Read, B. E. (1965a). Paper presented to the *Intern. Symp. Macromol. Chem.*, Prague. (To be published in *J. Polymer Sci.*, Part C.)
Reddish, W. (1950). *Trans. Faraday Soc.*, **46**, 459.
Reddish, W. (1958). *Soc. Chem. Ind. Symp.*, April 1958.
Reddish, W. and J. T. Barrie (1959). I.U.P.A.C. *Symp. über Macromol.*, Wiesbaden, Kurzmitteilung I.A.3.
Reddish, W. (1959). *Soc. Chem. Ind., London*, Monograph No. 5, p. 138, London.
Reddish, W. (1962). *Pure Appl. Chem.*, **5**, 723.
Reddish, W. (1965). *Am. Chem. Soc.*, Polymer Preprints, Sept. (1965) **6**, No. 2, 571.
Reding, F. P., J. A. Faucher and R. D. Whitman (1962). *J. Polymer Sci.*, **57**, 483.
Reding, F. P., J. A. Faucher and R. D. Whitman (1961). *J. Polymer Sci.*, **54**, 556.
Reding, F. P., E. R. Walter and F. J. Welch (1962). *J. Polymer Sci.*, **56**, 225.
Rempel, R. C. and others (1957). *J. Appl. Phys.*, **28**, 1082.
Reneker, D. H. and P. H. Geil (1960). *J. Appl. Phys.*, **31**, 1916.
Reuss, A. (1929). *Z. Angew., Math. Mech.*, **9**, 49.
Reynolds, S., V. G. Thomas, A. H. Sharbaugh and R. M. Fuoss (1951). *J. Am. Chem. Soc.*, **73**, 3714.
Reynolds, S. I. (1947). *Gen. Elec. Rev.*, **50**, 34.
Richardson, A. and A. Sacher (1953). *J. Polymer Sci.*, **10**, 353.
Riddle, E. H. (1954). In *Monomeric Acrylic Esters*, Reinhold, New York.
Rigby, H. A. and C. W. Bunn (1949). *Nature*, **164**, 583.
Roberts, S. and A. R. Von Hippel (1946). *J. Appl. Phys.*, **17**, 610.

Robinson, D. W. (1955). *J. Sci. Inst.*, **32**, 2.
Roedel, M. J. (1953). *J. Am. Chem. Soc.*, **75**, 6110.
Roeiling, J. A. (1965). *Polymer* **6**, 311.
Rogers, S. S. and L. Mandelkern (1957). *J. Phys. Chem.*, **61**, 985.
Rouse, P. E. (1953). *J. Chem. Phys.*, **21**, 1272.
Rushton, E. and E. J. Pratt (1940). *J. Sci. Inst.*, **17**, 297.
Rushton, E. (1954). *Brit. Elec. Ind. Res. Assoc. Rept.*, L/T 315.
Rushton, E. and G. Russell (1956). *Brit. Elec. Ind. Res. Assoc., Rept.*, L/T 355.
Rydel, E. (1965). *Thesis*, University of Strasbourg.
Saba, R. G., J. A. Sauer and A. E. Woodward (1963). *J. Polymer Sci.*, **1**, 1483.
Sabia, R. and F. R. Eirich (1963). *J. Polymer Sci.*, Part A, **1**, 2497, 2511.
Sack, R. A. (1952). *Australian J. Sci. Res.*, **5**, 135.
Saito, N., K. Okano, S. Iwayanagi and T. Hideshima (1963). In *Solid State Physics*, Academic Press Inc., New York, **14**, 343.
Saito, S. and T. Nakajima (1959). *J. Polymer Sci.*, **36**, 533.
Saito, S. and T. Nakajima (1959a). *J. Polymer Sci.*, **37**, 229.
Saito, S. and T. Nakajima (1959b). *J. Appl. Polymer Sci.*, **2**, 93.
Saito, S. and T. Nakajima (1959c). *Japan Soc. Testing Mater.*, **67**, 315.
Saito, S. and T. Nakajima (1959d). *J. Soc. Rheol. (Japan)*, **8**, 313.
Saito, S. (1962). *Rept. Progr. Polymer Phys.*, *(Japan)*, **5**, 205.
Saito, S. (1963). *Kolloid Z.*, **189**, 116.
Saito, S. (1964). *Res. Electrotech. Lab.*, *(Tokyo)*, No. 648.
St. Pierre, L. E. and C. C. Price (1956). *J. Am. Chem. Soc.*, **78**, 3432.
Sakurada, I., T. Itoho and N. Nukushina (1961). Paper presented at the 10th Annual Meeting of the *Soc. Polymer Sci.*, *(Japan)*. (See Shimanouchi *et al.* 1962).
Sandeman, I. and A. Keller (1956). *J. Polymer Sci.*, **19**, 401.
Sandiford, D. J. H. and A. H. Willbourn (1960). In A. Renfrew and P. Morgan (Eds.), *Polythene*, Iliffe, London, Chap. 8.
Sato, K. and others (1954). *J. Phys. Soc. (Japan)*, **9**, 413.
Sato, K., H. Nakane, T. Hideshima and S. Iwayanagi (1954). *J. Phys. Soc. (Japan)*, **9**, 413.
Satoh, S., R. Chûjô, T. Ozeki and E. Nagai (1962). *Rept. Polymer Phys.*, *(Japan)*, **5**, 251.
Satoh, S. (1964). *J. Polymer Sci.*, **2A**, 5221.
Satokawa, T. and S. Koizumi (1962). *J. Chem. Soc. Japan, Ind. Chem*, Sect., **65**, 1211.
Sauer, J. A. and D. E. Kline (1955). *J. Polymer Sci.*, **18**, 491.
Sauer, J. A. and D. E. Kline (1956). *Intern. Congr. Appl. Mech.*, Brussels, Belgium, **5**, 368.
Sauer, J. A., R. A. Wall, N. Fuschillo and A. E. Woodward (1958). *J. Appl. Phys.*, **29**, 1385.
Sauer, J. A., L. J. Merril and A. E. Woodward (1962). *J. Polymer Sci.*, **58**, 1.
Sauter, E. (1933). *Phys. Chem.*, **2**, 186, Part B.
Sazhin, B. I., V. A. Skurikhina and Y. I. Illin (1959). *Vysokomolekul Soedin.*, 1959, **1**, 1383.
Schallamach, A. (1946). *Trans. Faraday Soc.*, **42A**, 495.
Schallamach, A. (1951). *Trans. Inst. Rubber Ind.*, **27**, 40.
Schatzki, T. F. (1962). *J. Polymer Sci.*, **57**, 496.
Schatzki, T. F. (1965). Meeting of *Am. Chem. Soc.*, *Div. Polymer Chem.*, Atlantic City, September. Polymer Preprints, **6**, 646.

Scheiber, D. J. and D. J. Mead (1957). *J. Chem. Phys.*, **27**, 326.
Scheiber, D. J. (1957). *Dissert. Univ. Notre Dame,* Indiana.
Scheiber, D. J. (1961). *J. Res. Nat. Bur. Stds.*, (Washington), **65C**, 23.
Schmieder, K. and K. Wolf (1952). *Kolloid Z.*, **127**, 65.
Schmieder, K. and K. Wolf (1953). *Kolloid Z.*, **134**, 149.
Schmieder, K. and K. Wolf (1955). *Ric. Sci.*, (Suppl. A) **25**, 732.
Schmieder, W. G., W. C. Carter, M. Magat and C. R. Smyth (1945). *J. Am. Chem. Soc.*, **67**, 959.
Schmidt, P. G. and F. P. Gay (1962). *Angew. Chem.*, **74**, 638.
Schnell, H. (1956). *Angew. Chem.*, **68**, 633.
Schultz, A. K. (1956). *J. Chim. Phys.*, **53**, 933.
Schwarzl, F. and A. J. Staverman (1952). *J. Appl. Phys.*, **23**, 838.
Schwarzl, F. and A. J. S. Staverman (1956). In H. A. Stuart, (Ed.), *Die Physik der Hochpolymeren*, Springer-Verlag, Berlin, Band IV.
Schweitzer, C. E., R. N. MacDonald and J. O. Punderson (1959). *J. Appl. Polymer Sci.*, **1**, 158.
Scott, A. H., A. T. McPherson and H. L. Curtis (1933). *J. Nat. Bur. Stds.*, Washington, **11**, 373.
Scott, A. H. and others (1962). *J. Res. Nat. Bur. Stds.*, **66A**, 269.
Sherby, O. D. and J. E. Dorn (1958). *J. Mech. Phys. Solids*, **6**, 145.
Shetter, J. A. (1963). *J. Polymer Sci.*, Part B, **1**, 209.
Shimanouchi, T., M. Asahina and S. Enomoto (1962). *J. Polymer Sci.*, **59**, 93.
Sillars, R. W. (1939). *Proc. Roy. Soc.*, **A169**, 66.
Sinnott, K. M. (1958). (*See* Muus, McCrum and McGrew 1959.)
Sinnott, K. M. (1959). *J. Polymer Sci.*, **35**, 273.
Sinnott, K. M. (1960). *J. Polymer Sci.*, **42**, 3.
Sinnott, K. M. (1961). *Private Commun.*
Sinnott, K. M. (1962). *Soc. Plastics Engrs. Trans.*, **2**, 65.
Slichter, W. P. (1959). *J. Polymer Sci.*, **36**, 259.
Slichter, W. P. and E. R. Mandell (1959). *J. Appl. Phys.*, **30**, 1473.
Smets, G. and R. Hart (1960). *Advan. Polymer Sci.*, **2**, 173.
Smith, D. C. (1956). Paper presented at 129th Meeting *Am. Chem. Soc.*, Dallas, April 8–13.
Smith, J. A. S. (1955). *Discuss. Faraday Soc.*, **19**, 207.
Smith, J. W. (1955). In *Electric Dipole Moments*, Butterworths, London.
Smyth, C. P. (1955). In *Dielectric Behaviour and Structure*, McGraw-Hill, New York.
Snoek, J. L. (1941). *Phys.*, **8**, 711.
Sochava, I. V. and O. D. Trapeznikova (1958). Vestnik Leningrad Univ. 13, No. **16**, Ser. Fiz. Khim., No. 3, 65–72.
Sommer, W. (1959). *Kolloid Z.*, **167**, 97.
Sorensen, W. and T. W. Campbell (1961). In *Preparative Methods of Polymer Chemistry*, Interscience, New York.
Sperati, C. A., W. A. Franta and H. W. Starkweather (1953). *J. Am. Chem. Soc.*, **75**, 6127.
Sperati, C. A. and H. W. Starkweather (1961). *Advan. Polymer Sci.*, **2**, 465.
Stanley, E. and M. Litt (1960). *J. Polymer Sci.*, **43**, 453.
Starkweather, H. W. Jr., and others. (1956). *J. Polymer Sci.*, **21**, 189.
Starkweather, H. W. and R. E. Moynihan (1956). *J. Polymer Sci.*, **22**, 363.

Starkweather, H. W. (1959). *J. Appl. Polymer Sci.*, **2**, 129.
Starkweather, H. W. and R. E. Brooks (1959). *J. Appl. Polymer Sci.*, **1**, 236.
Starkweather, H. W. and R. H. Boyd (1960). *J. Phys. Chem.*, **64**, 410.
Staudinger, H. (1920). *Ber.*, **53**, 1973.
Staverman, A. J. (1953). *Kolloid Z.*, **134**, 189.
Staverman, A. J. and F. Schwarzl (1956). In H. A. Stuart (Ed.), *Die Physik der Hochpolymenen*, Springer-Verlag, Berlin. Vol. 4, Chap. 1.
Stein, R. S. (1964). In C. F. Baconke, (Ed.), *Newer Methods of Polymer Characterization*, Interscience, New York, Chap. IV.
Stockmayer, W. H. and M. Baur (1964). *J. Am. Chem. Soc.*, **86**, 3485.
Stratton, R. A. and J. D. Ferry (1963). *J. Phys. Chem.*, **67**, 2781.
Strella, S. and R. Zand (1957a). *J. Polymer Sci.*, **25**, 97.
Strella, S. and R. Zand (1957b). *J. Polymer Sci.*, **25**, 105.
Strella, S. and S. N. Chinai (1958). *J. Polymer Sci.*, **31**, 45.
Stroupe, J. D. and R. E. Hughes (1958). *J. Am. Chem. Soc.*, **80**, 2341.
Stuart, H. A. (1956). In *Die Physik der Hochpolymeren*, Springer-Verlag, Berlin.
Stuart, H. A. (1959). *Ann. N.Y. Acad. Sci.*, **83**, 1.
Studebaker, M. L. and L. G. Nabors (1959). *Rubber Chem. Technol.*, **32**, 941.
Sutherland, T. H. and B. L. Funt (1953). *J. Polymer Sci.*, **11**, 177.
Szwarc, M. (1960). *Advan. Polymer Sci.*, **2**, 275.
Szwarc, M. (1960). *Makromol. Chem.*, **35**, 132.
Tadokoro, H. and others (1963). *Rept. Progr. Polymer Phys. (Japan)*, **6**, 303.
Takayanagi, M. (1961). *High Polymers (Japan)*, **10**, 289.
Takahashi, Y. (1961a). *J. Appl. Polymer Sci.*, **5**, 468.
Takahashi, Y. (1961b). *J. Phys. Soc. Japan*, **16**, 1024.
Takayanagi, M. and others (1963). *Rept. Progr. Polymer Phys. (Japan)*, **6**, 121.
Takayanagi, M. (1963). *Mem. Fac. Eng. Kyushu Univ.*, **23**, No. 1, p. 1.
Takeda, M. and K. Iimura (1962). *J. Polymer Sci.*, **57**, 383.
Tammann, G. and W. Hesse (1926). *Z. Anorg. Allgem. Chem.*, **156**, 245.
Tanaka, K. (1962). *Rept. Progr. Polymer Phys. (Japan)*, **5**, 138.
Tanaka, H. and A. Matsumoto (1963). *Progr. Polymer Phys.*, *(Japan)*, **6**, 133.
Taylor, W. J. (1948). *J. Chem. Phys.*, **16**, 257.
Thirion, P. and R. Chasset (1951). *Trans. Inst. Rubber Ind.*, **27**, 364.
Thomas, Ann M. (1957). *Nature*, **179**, 862.
Thomas, P. E., J. F. Lontz, C. A. Sperati and J. L. McPherson (1956). *J. Soc. Plastics Engrs.*, **12**, No. 6, June.
Thompson, A. B. and D. W. Woods (1956). *Trans. Faraday Soc.*, **52**, 1383.
Thompson, A. M. (1956). *Proc. Inst. Elec. Engrs.*, (London), **103**, Pt. B. 704.
Thurn, H. (1955). *Z. Angew. Phys.*, **7**, 44.
Thurn, H. and K. Wolf (1956). *Kolloid Z.*, **148**, d6.
Thurn, H. and F. Würstlin (1958). *Kolloid Z.*, **156**, 21.
Thurn, H. (1960). *Festschr. Carl Wurster, BASF, Ludwigshafen*, 321.
Thurn, H. (1960). *Kolloid Z.*, **173**, 72.
Thurn, H. and K. Wolf (1962). In *Struktur und Verhalten der Kunststoffe*, Springer-Verlag, p. 399.
Tiers, G. V. D. and F. A. Bovey (1963). *J. Polymer Sci.*, Part A, **1**, 833.
Till, P. H. (1957). *J. Polymer Sci.*, **24**, 301.
Timoshenko, S. and J. N. Goodier (1951). In *Theory of Elasticity*, New York: Wiley, 2nd ed.

Tincher, W. C. (1962). *J. Polymer Sci.*, **62**, S148.
Tobolsky, A. V. and R. D. Andrews (1945). *J. Chem. Phys.*, **13**, 3.
Tobolsky, A. V. and J. R. McLoughlin (1955). *J. Phys. Chem.*, **59**, 989.
Tobolsky, A. V. (1959). *J. Polymer Sci.*, **35**, 555.
Tobolsky, A. V. (1960). In *Properties and Structure of Polymers*, Wiley, New York.
Tobolsky, A. V., D. W. Carlson and N. Indicator (1961). *J. Polymer Sci.*, **54**, 175.
Tobolsky, A. V. and J. J. Aklonis (1964). *J. Phys. Chem.*, **68**, 1970.
Tokita, N. (1956). *J. Polymer Sci.*, **20**, 515.
Treloar, L. R. G. (1949). In *The Physics of Rubber Elasticity*, Oxford University Press, Oxford.
Treloar, L. R. G. (1960a). *Polymer*, **1**, 95.
Treloar, L. R. G. (1960b). *Polymer*, **1**, 279.
Treloar, L. R. G. (1960c). *Polymer*, **1**, 290.
Treloar, L. R. G. (1958). In *The Physics of Rubber Elasticity*, Oxford University Press, Oxford, 2nd ed.
Trifan, D. S. and J. F. Terenzi (1958). *J. Polymer Sci.*, **28**, 443.
Tschoegl, N. W. (1963). *J. Chem. Phys.*, **39**, 149.
Tschoegl, N. W. and J. D. Ferry (1963). *Kolloid Z.*, **189**, 37.
Tsuge, K. and others (1962). *J. Appl. Phys. (Japan)*, **1**, 270.
Tuijnman, C. A. F. (1963). *Polymer*, **4** (a); 259 and (b) 315.
Tuijnman, C. A. F. (1965). *J. Polymer Sci.* (in press).
Turnbull, D. and M. H. Cohen (1961). *J. Chem. Phys.*, **34**, 120.
Turner, A. and F. E. Bailey (1963). *J. Polymer Sci.*, Part B, **1**, 601.
Turner, S. (1964a). *Brit. Plastics*, June, 322.
Turner, S. (1964b). *Brit. Plastics*, September–October, 2.
Turner-Jones, A. (1963). *Private Commun.* (see Connor and others, 1964).
Uchida, T., Y. Kurita and M. Kubo (1956). *J. Polymer Sci.*, **19**, 365.
Uematsu, Y. and I. Uematsu (1960). *Rept. Progr. Polymer Phys.*, (Japan), **3**, 80, 102.
Uematsu, Y. and I. Uematsu (1959). *Rept. Progr. Polymer Phys. (Japan)*, **2**, 27.
Van Beek, L. K. H. and J. J. Hermans (1957). *J. Polymer Sci.*, **23**, 211.
Van Schooten, J., H. van Hoorn and J. Boerma (1960). *Polymer*, **2**, 161.
Veselovskii, P. F. and A. I. Slusker (1955). *Zh. Tech.* **25**, 939; **25**, 1204.
Veselovskii, P. F. (1956). *Ber. Tomsk. Polytech. Inst.*, U.S.S.R., **91**, 399.
Vogelsong, D. C. and Pearce, E. M. (1960). *J. Polymer Sci.*, **45**, 546.
Voigt, W. (1910). In *Lehrbuch der Kristallphysik*, Teubner, Leipzig.
Volkenstein, M. V. and O. B. Ptitsyn (1955). *Zh. Tekhn. Fiz.*, **25**, 649.
Volkenstein, M. V. (1959a). *Sov. Phys. Doklady*, **4**, 351.
Volkenstein, M. V. (1959b). *Sov. Phys. Usp.*, **67**, (2), No. 1, 59.
Volkenstein, M. V. (1958). *J. Polymer Sci.*, **29**, 441.
Volkenstein, M. V. (1963). In *Configurational Statistics of Polymeric Chains*, Interscience, New York.
Von Hippel, A. R. (1954). In *Dielectric Materials and Applications*, Wiley, New York.
Von Schweidler, E. (1907). *Ann. Phys.*, **24**, 711.
Wachtman, J. and D. G. Lam (1959). *J. Am. Ceram. Soc.*, **42**, 254.
Wada, Y. and K. Yamamoto (1956). *J. Phys. Soc. (Japan)*, **11**, 887.
Wada, Y., H. Hirose, T. Asano and S. Fukutomi (1959). *J. Phys. Soc. (Japan)*, **11**, 887.

Wada, Y., H. Enjoji and H. Terada (1962). *Progr. Polymer Phys. (Japan)*, V, 131.
Wada, Y. and K. Tsuge (1962). *J. Appl. Phys. (Japan)*, 1, 64.
Wall, R. A., J. A. Sauer and A. E. Woodward (1959). *J. Polymer Sci.*, 35, 281.
Walter, E. R. and F. P. Reding (1956). *J. Polymer Sci.*, 21, 561.
Ward, I. M. (1960). *Trans. Faraday Soc.*, 56, 648.
Waring, J. R. S. (1951). *Trans. Inst. Rubber Ind.*, 27, 16.
Weissermel, K. and W. Schmieder (1962). *Makromol. Chem.*, 51, 39.
Wert, C. and C. Zener (1949). *Phys. Rev.*, 76, 1169.
Westphal, W. H. (1947). *Phys. Pract.* (Braunschweig), 292.
Westphal, W. H. (1954). In A. R. Von Hippel (Ed.), *Dielectric Materials and Applications*, Wiley, New York.
Wetton, R. E. (1962). *Thesis* (Manchester).
Wetton, R. E. (1964). *Private Commun.*
Wetton, R. and G. Williams (1965). *Trans. Faraday Soc.*, 61, 2132.
Wetton, R. E. and G. Allen (1966), *Polymer*, 7, 331.
Willbourn, A. H. (1958). *Trans. Faraday Soc.*, 54, 717.
Willbourn, A. H. (1959). *J. Polymer Sci.*, 34, 569.
Williams, G. (1959). *J. Phys. Chem.*, 63, 534.
Williams, G. (1962). *Trans. Faraday Soc.*, 58, 1041.
Williams, G. (1963b). *Polymer*, 4, 27.
Williams, G. (1963a). *Trans. Faraday Soc.*, 59, 1397.
Williams, G. (1964). *Trans. Faraday Soc.*, 60, 1548, 1556.
Williams, G. (1965). *Trans. Faraday Soc.*, 61, 1564.
Williams, M. L. and J. D. Ferry (1954). *J. Colloid Sci.*, 9, 479.
Williams, M. L. and J. D. Ferry (1955). *J. Colloid Sci.*, 10, 474.
Williams, M. L., R. F. Landel and J. D. Ferry (1955). *J. Am. Chem. Soc.*, 77, 3701.
Williams, M. L. (1962a). *J. Polymer Sci.*, 62, S7.
Wilson, C. W. and G. E. Pake (1953). *J. Polymer Sci.*, 10, 503.
Wilson, E. B. (1959). *Advan. Chem. Phys.*, II, 367.
Wolf, K. (1951). *Kunstoffe-Plastics*, 41, 89.
Wolf, K. (1956). In *Verband deutscher physikalischer Gesellschaften. Hauptvorträge der Physikertagung.* Mosbach, Physik Verlag, Münich, Band 4, 141–165.
Wolf, K. and K. Schmieder (1955). *Simp. Intern. Chim. Macromol., Suppl. Ric. Sci.*, 3.
Wood, L. A. (1958). *J. Polymer Sci.*, 28, 319.
Woodward, A. E., J. A. Sauer, C. W. Deeley and D. E. Kline (1957). *J. Colloid Sci.*, 12, 363.
Woodward, A. E. and J. A. Sauer (1958). *Advan. Polymer Sci.*, 1, 114.
Woodward, A. E., J. M. Crissman and J. A. Sauer (1960). *J. Polymer Sci.*, 44, 23.
Woodward, A. E., J. A. Sauer and R. A. Wall (1961). *J. Polymer Sci.*, 50, 117.
Work, R. N. (1956). *J. Appl. Phys.*, 27, 69.
Work, R. N., R. D. McCammon and R. G. Saba (1963). *Bull. Am. Phys. Soc.*, 8, 266.
Works, C. N., T. W. Dakin and F. W. Boggs (1944). *Trans. Am. Inst. Elec. Engrs.*, 63, 1092, 1952.
Wunderlich, B. (1962). *J. Chem. Phys.*, 37, 2429.
Würstlin, F. (1943). *Kolloid Z.*, 105, 9.
Würstlin, F. (1948). *Kolloid Z.*, 110, 71.

Würstlin, F. (1949). *Kolloid Z.*, **113**, 18.
Würstlin, F. (1950). *Z. Angew. Phys.*, **2**, 131.
Würstlin, F. (1951). *Kolloid Z.*, **120**, 84.
Würstlin, F. (1953). *Kolloid Z.*, **134**, 135.
Yager, W. A. and W. O. Baker (1942). *J. Am. Chem. Soc.*, **64**, 2164.
Yamamoto, K. and Y. Wada (1957). *J. Phys. Soc.*, (*Japan*), **12**, 374.
Yamamura, H. and N. Kuramoto (1959). *J. Appl. Polymer Sci.*, **2**, 71.
Yamafuji, K. (1960). *J. Phys. Soc.* (*Japan*), **15**, 2295.
Yamafuji, K. and Y. Ishida (1962). *Kolloid Z.*, **183**, 15.
Yamafuji, K. (1960). *J. Phys. Soc.* (*Japan*), **15**, 2295.
Yoshino, M. and M. Takayanagi (1959). *J. Soc. Testing Mat.* (*Japan*), **10**, 330.
Yoshitomi, T., K. Nagamatsu and K. Kosiyama (1958). *J. Polymer Sci.*, **27**, 35.
Young, J. M. and A. A. Petrauskas (1956). *J. Chem. Phys.*, **25**, 943.
Zener, C. (1947). *J. Appl. Phys.*, **18**, 1022.
Zener, C. (1941). *Phys. Rev.*, **60**, 906.
Zener, C. (1948). In *Elasticity and Anelasticity of Metals*, University of Chicago Press, Chicago.
Zener, C. (1950). In *Thermodynamics in Physical Metallurgy*, *Am. Soc. Mech. Engrs.*, Pages 16–27.
Ziabicki, A. (1959). *Kolloid Z.*, **167**, 132.
Ziegler, K., E. Kolzkamp, H. Breil and H. Martin (1955). *Chem.*, **67**, 541.
Zimm, B. H. (1956). *J. Chem. Phys.*, **24**, 269.
Zimm, B. H., G. M. Roe and L. E. Epstein (1956). *J. Chem. Phys.*, **24**, 279.
Zimm, B. H. (1960). In F. R. Eirich (Ed.), *Rheology*, Academic Press, New York, Vol. 3, p. 1.
Zutty, N. L. and C. J. Whitworth (1964). *J. Polymer Sci.*, Part B., **2**, 709.

Author Index

Subject Index

A CATALOG OF SELECTED

DOVER BOOKS
IN SCIENCE AND MATHEMATICS

A CATALOG OF SELECTED
DOVER BOOKS
IN SCIENCE AND MATHEMATICS

QUALITATIVE THEORY OF DIFFERENTIAL EQUATIONS, V.V. Nemytskii and V.V. Stepanov. Classic graduate-level text by two prominent Soviet mathematicians covers classical differential equations as well as topological dynamics and erqodic theory. Bibliographies. 523pp. 5⅜ × 8½. 65954-2 Pa. $10.95

MATRICES AND LINEAR ALGEBRA, Hans Schneider and George Phillip Barker. Basic textbook covers theory of matrices and its applications to systems of linear equations and related topics such as determinants, eigenvalues and differential equations. Numerous exercises. 432pp. 5⅜ × 8½. 66014-1 Pa. $8.95

QUANTUM THEORY, David Bohm. This advanced undergraduate-level text presents the quantum theory in terms of qualitative and imaginative concepts, followed by specific applications worked out in mathematical detail. Preface. Index. 655pp. 5⅜ × 8½. 65969-0 Pa. $10.95

ATOMIC PHYSICS (8th edition), Max Born. Nobel laureate's lucid treatment of kinetic theory of gases, elementary particles, nuclear atom, wave-corpuscles, atomic structure and spectral lines, much more. Over 40 appendices, bibliography. 495pp. 5⅜ × 8½. 65984-4 Pa. $11.95

ELECTRONIC STRUCTURE AND THE PROPERTIES OF SOLIDS: The Physics of the Chemical Bond, Walter A. Harrison. Innovative text offers basic understanding of the electronic structure of covalent and ionic solids, simple metals, transition metals and their compounds. Problems. 1980 edition. 582pp. 6⅛ × 9¼. 66021-4 Pa. $14.95

BOUNDARY VALUE PROBLEMS OF HEAT CONDUCTION, M. Necati Özisik. Systematic, comprehensive treatment of modern mathematical methods of solving problems in heat conduction and diffusion. Numerous examples and problems. Selected references. Appendices. 505pp. 5⅜ × 8½. 65990-9 Pa. $11.95

A SHORT HISTORY OF CHEMISTRY (3rd edition), J.R. Partington. Classic exposition explores origins of chemistry, alchemy, early medical chemistry, nature of atmosphere, theory of valency, laws and structure of atomic theory, much more. 428pp. 5⅜ × 8½. (Available in U.S. only) 65977-1 Pa. $10.95

A HISTORY OF ASTRONOMY, A. Pannekoek. Well-balanced, carefully reasoned study covers such topics as Ptolemaic theory, work of Copernicus, Kepler, Newton, Eddington's work on stars, much more. Illustrated. References. 521pp. 5⅜ × 8½. 65994-1 Pa. $11.95

PRINCIPLES OF METEOROLOGICAL ANALYSIS, Walter J. Saucier. Highly respected, abundantly illustrated classic reviews atmospheric variables, hydrostatics, static stability, various analyses (scalar, cross-section, isobaric, isentropic, more). For intermediate meteorology students. 454pp. 6⅛ × 9¼. 65979-8 Pa. $12.95

CATALOG OF DOVER BOOKS

CHALLENGING MATHEMATICAL PROBLEMS WITH ELEMENTARY SOLUTIONS, A.M. Yaglom and I.M. Yaglom. Over 170 challenging problems on probability theory, combinatorial analysis, points and lines, topology, convex polygons, many other topics. Solutions. Total of 445pp. 5⅜ × 8½. Two-vol. set.
Vol. I 65536-9 Pa. $5.95
Vol. II 65537-7 Pa. $5.95

FIFTY CHALLENGING PROBLEMS IN PROBABILITY WITH SOLUTIONS, Frederick Mosteller. Remarkable puzzlers, graded in difficulty, illustrate elementary and advanced aspects of probability. Detailed solutions. 88pp. 5⅜ × 8½.
65355-2 Pa. $3.95

EXPERIMENTS IN TOPOLOGY, Stephen Barr. Classic, lively explanation of one of the byways of mathematics. Klein bottles, Moebius strips, projective planes, map coloring, problem of the Koenigsberg bridges, much more, described with clarity and wit. 43 figures. 210pp. 5⅜ × 8½. 25933-1 Pa. $4.95

RELATIVITY IN ILLUSTRATIONS, Jacob T. Schwartz. Clear non-technical treatment makes relativity more accessible than ever before. Over 60 drawings illustrate concepts more clearly than text alone. Only high school geometry needed. Bibliography. 128pp. 6⅛ × 9¼. 25965-X Pa. $5.95

AN INTRODUCTION TO ORDINARY DIFFERENTIAL EQUATIONS, Earl A. Coddington. A thorough and systematic first course in elementary differential equations for undergraduates in mathematics and science, with many exercises and problems (with answers). Index. 304pp. 5⅜ × 8¼. 65942-9 Pa. $7.95

FOURIER SERIES AND ORTHOGONAL FUNCTIONS, Harry F. Davis. An incisive text combining theory and practical example to introduce Fourier series, orthogonal functions and applications of the Fourier method to boundary-value problems. 570 exercises. Answers and notes. 416pp. 5⅜ × 8½. 65973-9 Pa. $8.95

THE THOERY OF BRANCHING PROCESSES, Theodore E. Harris. First systematic, comprehensive treatment of branching (i.e. multiplicative) processes and their applications. Galton-Watson model, Markov branching processes, electron-photon cascade, many other topics. Rigorous proofs. Bibliography. 240pp. 5⅜ × 8½. 65952-6 Pa. $6.95

AN INTRODUCTION TO ALGEBRAIC STRUCTURES, Joseph Landin. Superb self-contained text covers "abstract algebra": sets and numbers, theory of groups, theory of rings, much more. Numerous well-chosen examples, exercises. 247pp. 5⅜ × 8½. 65940-2 Pa. $6.95

GAMES AND DECISIONS: Introduction and Critical Survey, R. Duncan Luce and Howard Raiffa. Superb non-technical introduction to game theory, primarily applied to social sciences. Utility theory, zero-sum games, n-person games, decision-making, much more. Bibliography. 509pp. 5⅜ × 8½. 65943-7 Pa. $10.95

Prices subject to change without notice.
Available at your book dealer or write for free Mathematics and Science Catalog to Dept. GI, Dover Publications, Inc., 31 East 2nd St., Mineola, N.Y. 11501. Dover publishes more than 175 books each year on science, elementary and advanced mathematics, biology, music, art, literary history, social sciences and other areas.